Mining Text Data

Charu C. Aggarwal • ChengXiang Zhai
Editors

Mining Text Data

 Springer

Editors
Charu C. Aggarwal
IBM T.J. Watson Research Center
Yorktown Heights, NY, USA
charu@us.ibm.com

ChengXiang Zhai
University of Illinois at Urbana-Champaign
Urbana, IL, USA
czhai@cs.uiuc.edu

ISBN 978-1-4614-3222-7   e-ISBN 978-1-4614-3223-4
DOI 10.1007/978-1-4614-3223-4
Springer New York Dordrecht Heidelberg London

Library of Congress Control Number: 2012930923

© Springer Science+Business Media, LLC 2012

All rights reserved. This work may not be translated or copied in whole or in part without the written permission of the publisher (Springer Science+Business Media, LLC, 233 Spring Street, New York, NY 10013, USA), except for brief excerpts in connection with reviews or scholarly analysis. Use in connection with any form of information storage and retrieval, electronic adaptation, computer software, or by similar or dissimilar methodology now known or hereafter developed is forbidden.

The use in this publication of trade names, trademarks, service marks, and similar terms, even if they are not identified as such, is not to be taken as an expression of opinion as to whether or not they are subject to proprietary rights.

Printed on acid-free paper

Springer is part of Springer Science+Business Media (www.springer.com)

Contents

1
An Introduction to Text Mining   1
Charu C. Aggarwal and ChengXiang Zhai
    1. Introduction   1
    2. Algorithms for Text Mining   4
    3. Future Directions   8
    References   10

2
Information Extraction from Text   11
Jing Jiang
    1. Introduction   11
    2. Named Entity Recognition   15
        2.1 Rule-based Approach   16
        2.2 Statistical Learning Approach   17
    3. Relation Extraction   22
        3.1 Feature-based Classification   23
        3.2 Kernel Methods   26
        3.3 Weakly Supervised Learning Methods   29
    4. Unsupervised Information Extraction   30
        4.1 Relation Discovery and Template Induction   31
        4.2 Open Information Extraction   32
    5. Evaluation   33
    6. Conclusions and Summary   34
    References   35

3
A Survey of Text Summarization Techniques   43
Ani Nenkova and Kathleen McKeown
    1. How do Extractive Summarizers Work?   44
    2. Topic Representation Approaches   46
        2.1 Topic Words   46
        2.2 Frequency-driven Approaches   48
        2.3 Latent Semantic Analysis   52
        2.4 Bayesian Topic Models   53
        2.5 Sentence Clustering and Domain-dependent Topics   55
    3. Influence of Context   56
        3.1 Web Summarization   57
        3.2 Summarization of Scientific Articles   58

	3.3	Query-focused Summarization	58
	3.4	Email Summarization	59
4.	Indicator Representations and Machine Learning for Summarization		60
	4.1	Graph Methods for Sentence Importance	60
	4.2	Machine Learning for Summarization	62
5.	Selecting Summary Sentences		64
	5.1	Greedy Approaches: Maximal Marginal Relevance	64
	5.2	Global Summary Selection	65
6.	Conclusion		66
	References		66

4
A Survey of Text Clustering Algorithms   77
Charu C. Aggarwal and ChengXiang Zhai

1.	Introduction		77
2.	Feature Selection and Transformation Methods for Text Clustering		81
	2.1	Feature Selection Methods	81
	2.2	LSI-based Methods	84
	2.3	Non-negative Matrix Factorization	86
3.	Distance-based Clustering Algorithms		89
	3.1	Agglomerative and Hierarchical Clustering Algorithms	90
	3.2	Distance-based Partitioning Algorithms	92
	3.3	A Hybrid Approach: The Scatter-Gather Method	94
4.	Word and Phrase-based Clustering		99
	4.1	Clustering with Frequent Word Patterns	100
	4.2	Leveraging Word Clusters for Document Clusters	102
	4.3	Co-clustering Words and Documents	103
	4.4	Clustering with Frequent Phrases	105
5.	Probabilistic Document Clustering and Topic Models		107
6.	Online Clustering with Text Streams		110
7.	Clustering Text in Networks		115
8.	Semi-Supervised Clustering		118
9.	Conclusions and Summary		120
	References		121

5
Dimensionality Reduction and Topic Modeling   129
Steven P. Crain, Ke Zhou, Shuang-Hong Yang and Hongyuan Zha

1.	Introduction		130
	1.1	The Relationship Between Clustering, Dimension Reduction and Topic Modeling	131
	1.2	Notation and Concepts	132
2.	Latent Semantic Indexing		133
	2.1	The Procedure of Latent Semantic Indexing	134
	2.2	Implementation Issues	135
	2.3	Analysis	137
3.	Topic Models and Dimension Reduction		139
	3.1	Probabilistic Latent Semantic Indexing	140
	3.2	Latent Dirichlet Allocation	142
4.	Interpretation and Evaluation		148

		4.1	Interpretation	148
		4.2	Evaluation	149
		4.3	Parameter Selection	150
		4.4	Dimension Reduction	150
	5.	Beyond Latent Dirichlet Allocation		151
		5.1	Scalability	151
		5.2	Dynamic Data	151
		5.3	Networked Data	152
		5.4	Adapting Topic Models to Applications	154
	6.	Conclusion		155
	References			156

6
A Survey of Text Classification Algorithms 163
Charu C. Aggarwal and ChengXiang Zhai

	1.	Introduction		163
	2.	Feature Selection for Text Classification		167
		2.1	Gini Index	168
		2.2	Information Gain	169
		2.3	Mutual Information	169
		2.4	χ^2-Statistic	170
		2.5	Feature Transformation Methods: Supervised LSI	171
		2.6	Supervised Clustering for Dimensionality Reduction	172
		2.7	Linear Discriminant Analysis	173
		2.8	Generalized Singular Value Decomposition	175
		2.9	Interaction of Feature Selection with Classification	175
	3.	Decision Tree Classifiers		176
	4.	Rule-based Classifiers		178
	5.	Probabilistic and Naive Bayes Classifiers		181
		5.1	Bernoulli Multivariate Model	183
		5.2	Multinomial Distribution	188
		5.3	Mixture Modeling for Text Classification	190
	6.	Linear Classifiers		193
		6.1	SVM Classifiers	194
		6.2	Regression-Based Classifiers	196
		6.3	Neural Network Classifiers	197
		6.4	Some Observations about Linear Classifiers	199
	7.	Proximity-based Classifiers		200
	8.	Classification of Linked and Web Data		203
	9.	Meta-Algorithms for Text Classification		209
		9.1	Classifier Ensemble Learning	209
		9.2	Data Centered Methods: Boosting and Bagging	210
		9.3	Optimizing Specific Measures of Accuracy	211
	10.	Conclusions and Summary		213
	References			213

7
Transfer Learning for Text Mining 223
Weike Pan, Erheng Zhong and Qiang Yang

	1.	Introduction		224
	2.	Transfer Learning in Text Classification		225
		2.1	Cross Domain Text Classification	225

	2.2	Instance-based Transfer	231
	2.3	Cross-Domain Ensemble Learning	232
	2.4	Feature-based Transfer Learning for Document Classification	235
3.	Heterogeneous Transfer Learning		239
	3.1	Heterogeneous Feature Space	241
	3.2	Heterogeneous Label Space	243
	3.3	Summary	244
4.	Discussion		245
5.	Conclusions		246
References			247

8
Probabilistic Models for Text Mining — 259
Yizhou Sun, Hongbo Deng and Jiawei Han

1.	Introduction		260
2.	Mixture Models		261
	2.1	General Mixture Model Framework	262
	2.2	Variations and Applications	263
	2.3	The Learning Algorithms	266
3.	Stochastic Processes in Bayesian Nonparametric Models		269
	3.1	Chinese Restaurant Process	269
	3.2	Dirichlet Process	270
	3.3	Pitman-Yor Process	274
	3.4	Others	275
4.	Graphical Models		275
	4.1	Bayesian Networks	276
	4.2	Hidden Markov Models	278
	4.3	Markov Random Fields	282
	4.4	Conditional Random Fields	285
	4.5	Other Models	286
5.	Probabilistic Models with Constraints		287
6.	Parallel Learning Algorithms		288
7.	Conclusions		289
References			290

9
Mining Text Streams — 297
Charu C. Aggarwal

1.	Introduction		297
2.	Clustering Text Streams		299
	2.1	Topic Detection and Tracking in Text Streams	307
3.	Classification of Text Streams		312
4.	Evolution Analysis in Text Streams		316
5.	Conclusions		317
References			318

10
Translingual Mining from Text Data — 323
Jian-Yun Nie, Jianfeng Gao and Guihong Cao

1.	Introduction	324
2.	Traditional Translingual Text Mining – Machine Translation	325

	2.1	SMT and Generative Translation Models	325
	2.2	Word-Based Models	327
	2.3	Phrase-Based Models	329
	2.4	Syntax-Based Models	333
3.	Automatic Mining of Parallel texts		336
	3.1	Using Web structure	337
	3.2	Matching parallel pages	339
4.	Using Translation Models in CLIR		341
5.	Collecting and Exploiting Comparable Texts		344
6.	Selecting Parallel Sentences, Phrases and Translation Words		347
7.	Mining Translingual Relations From Monolingual Texts		349
8.	Mining using hyperlinks		351
9.	Conclusions and Discussions		353
References			354

11
Text Mining in Multimedia 361
Zheng-Jun Zha, Meng Wang, Jialie Shen and Tat-Seng Chua

1.	Introduction		362
2.	Surrounding Text Mining		364
3.	Tag Mining		366
	3.1	Tag Ranking	366
	3.2	Tag Refinement	367
	3.3	Tag Information Enrichment	369
4.	Joint Text and Visual Content Mining		370
	4.1	Visual Re-ranking	371
5.	Cross Text and Visual Content Mining		374
6.	Summary and Open Issues		377
References			379

12
Text Analytics in Social Media 385
Xia Hu and Huan Liu

1.	Introduction		385
2.	Distinct Aspects of Text in Social Media		388
	2.1	A General Framework for Text Analytics	388
	2.2	Time Sensitivity	390
	2.3	Short Length	391
	2.4	Unstructured Phrases	392
	2.5	Abundant Information	393
3.	Applying Text Analytics to Social Media		393
	3.1	Event Detection	393
	3.2	Collaborative Question Answering	395
	3.3	Social Tagging	397
	3.4	Bridging the Semantic Gap	398
	3.5	Exploiting the Power of Abundant Information	399
	3.6	Related Efforts	401
4.	An Illustrative Example		402
	4.1	Seed Phrase Extraction	402
	4.2	Semantic Feature Generation	404
	4.3	Feature Space Construction	406
5.	Conclusion and Future Work		407
References			408

13
A Survey of Opinion Mining and Sentiment Analysis   415
Bing Liu and Lei Zhang

1.	The Problem of Opinion Mining	416
	1.1 Opinion Definition	416
	1.2 Aspect-Based Opinion Summary	420
2.	Document Sentiment Classification	422
	2.1 Classification based on Supervised Learning	422
	2.2 Classification based on Unsupervised Learning	424
3.	Sentence Subjectivity and Sentiment Classification	426
4.	Opinion Lexicon Expansion	429
	4.1 Dictionary based approach	429
	4.2 Corpus-based approach and sentiment consistency	430
5.	Aspect-Based Sentiment Analysis	432
	5.1 Aspect Sentiment Classification	433
	5.2 Basic Rules of Opinions	434
	5.3 Aspect Extraction	438
	5.4 Simultaneous Opinion Lexicon Expansion and Aspect Extraction	440
6.	Mining Comparative Opinions	441
7.	Some Other Problems	444
8.	Opinion Spam Detection	447
	8.1 Spam Detection Based on Supervised Learning	448
	8.2 Spam Detection Based on Abnormal Behaviors	449
	8.3 Group Spam Detection	450
9.	Utility of Reviews	451
10.	Conclusions	452
	References	453

14
Biomedical Text Mining: A Survey of Recent Progress   465
Matthew S. Simpson and Dina Demner-Fushman

1.	Introduction	466
2.	Resources for Biomedical Text Mining	467
	2.1 Corpora	467
	2.2 Annotation	469
	2.3 Knowledge Sources	470
	2.4 Supporting Tools	471
3.	Information Extraction	472
	3.1 Named Entity Recognition	473
	3.2 Relation Extraction	478
	3.3 Event Extraction	482
4.	Summarization	484
5.	Question Answering	488
	5.1 Medical Question Answering	489
	5.2 Biological Question Answering	491
6.	Literature-Based Discovery	492
7.	Conclusion	495
	References	496

Index   519

Preface

The importance of text mining applications has increased in recent years because of the large number of web-enabled applications which lead to the creation of such data. While classical applications have focussed on processing and mining raw text, the advent of web enabled applications requires novel methods for mining and processing, such as the use of linkage, multi-lingual information or the joint mining of text with other kinds of multimedia data such as images or videos. In many cases, this has also lead to the development of other related areas of research such as heterogeneous transfer learning.

An important characteristic of this area is that it has been explored by multiple communities such as data mining, machine learning and information retrieval. In many cases, these communities tend to have some overlap, but are largely disjoint and carry on their research independently. One of the goals of this book is to bring together researchers of different communities together in order to maximize the cross-disciplinary understanding of this area.

Another aspect of the text mining area is that there seems to be a distinct set of researchers working on newer aspects of text mining in the context of emerging platforms such as data streams and social networks. This book is also an attempt to discuss both the classical and modern aspects of text mining in a unified way. Chapters are devoted to many classical methods such as clustering, classification and topic modeling. In addition, we also study different aspects of text mining in the context of modern applications in social and information networks, and social media. Many new applications such as data streams have also been explored for the first time in this book.

Each chapter in the book is structured as a comprehensive survey which discusses the key models and algorithms for the particular area. In addition the future trends and research directions are presented in each chapter. It is hoped that this book will provide a comprehensive understanding of the area to students, professors and researchers.

Chapter 1

AN INTRODUCTION TO TEXT MINING

Charu C. Aggarwal
IBM T. J. Watson Research Center
Yorktown Heights, NY
charu@us.ibm.com

ChengXiang Zhai
University of Illinois at Urbana-Champaign
Urbana, IL
czhai@cs.uiuc.edu

Abstract
   The problem of text mining has gained increasing attention in recent years because of the large amounts of text data, which are created in a variety of social network, web, and other information-centric applications. Unstructured data is the easiest form of data which can be created in any application scenario. As a result, there has been a tremendous need to design methods and algorithms which can effectively process a wide variety of text applications. This book will provide an overview of the different methods and algorithms which are common in the text domain, with a particular focus on mining methods.

1. Introduction

Data mining is a field which has seen rapid advances in recent years [8] because of the immense advances in hardware and software technology which has lead to the availability of different kinds of data. This is particularly true for the case of text data, where the development of hardware and software platforms for the web and social networks has enabled the rapid creation of large repositories of different kinds of data. In particular, the web is a technological enabler which encourages the creation of a large amount of text content by different users in a form which is easy to store and process. The increasing amounts of text data available from different applications has created a need for advances in algorithmic design which can learn interesting patterns from the data in a dynamic and scalable way.

While structured data is generally managed with a database system, text data is typically managed via a search engine due to the lack of structures [5]. A search engine enables a user to find useful information from a collection conveniently with a keyword query, and how to improve the effectiveness and efficiency of a search engine has been a central research topic in the field of information retrieval [13, 3], where many related topics to search such as text clustering, text categorization, summarization, and recommender systems are also studied [12, 9, 7].

However, research in information retrieval has traditionally focused more on facilitating information access [13] rather than analyzing information to discover patterns, which is the primary goal of text mining. The goal of information access is to connect the right information with the right users at the right time with less emphasis on processing or transformation of text information. Text mining can be regarded as going beyond information access to further help users analyze and digest information and facilitate decision making.There are also many applications of text mining where the primary goal is to analyze and discover any interesting patttterns, including trends and outliers, in text data, and the notion of a query is not essential or even relevant.

Technically, mining techniques focus on the primary models, algorithms and applications about what one can learn from different kinds of text data. Some examples of such questions are as follows:

- What are the primary supervised and unsupervised models for learning from text data? How are traditional clustering and classification problems different for text data, as compared to the traditional database literature?

- What are the useful tools and techniques used for mining text data? Which are the useful mathematical techniques which one should know, and which are repeatedly used in the context of different kinds of text data?

- What are the key application domains in which such mining techniques are used, and how are they effectively applied?

A number of key characteristics distinguish text data from other forms of data such as relational or quantitative data. This naturally affects the mining techniques which can be used for such data. The most important characteristic of text data is that it is *sparse* and *high dimensional*. For example, a given corpus may be drawn from a lexicon of about 100,000 words, but a given text document may contain only a few hundred words. Thus, a corpus of text documents can be represented as a *sparse term-document matrix* of size $n \times d$, when n is the number of documents, and d is the size of the lexicon vocabulary. The (i,j)th entry of this matrix is the (normalized) frequency of the jth word in the lexicon in document i. The large size and the sparsity of the matrix has immediate implications for a number of data analytical techniques such as dimensionality reduction. In such cases, the methods for reduction should be specifically designed while taking this characteristic of text data into account. The variation in word frequencies and document lengths also lead to a number of issues involving document representation and normalization, which are critical for text mining.

Furthermore, text data can be analyzed at different levels of representation. For example, text data can easily be treated as a bag-of-words, or it can be treated as a string of words. However, in most applications, it would be desirable to represent text information *semantically* so that more meaningful analysis and mining can be done. For example, representing text data at the level of named entities such as people, organizations, and locations, and their relations may enable discovery of more interesting patterns than representing text as a bag of words. Unfortunately, the state of the art methods in natural language processing are still not robust enough to work well in unrestricted text domains to generate accurate semantic representation of text. Thus most text mining approaches currently still rely on the more shallow word-based representations, especially the bag-of-wrods approach, which, while losing the positioning information in the words, is generally much simpler to deal with from an algorithmic point of view than the string-based approach. In special domains (e.g., biomedical domain) and for special mining tasks (e.g., extraction of knowledge from the Web), natural language processing techniques, especially information extraction, are also playing an important role in obtaining a semantically more meaningful representation of text.

Recently, there has been rapid growth of text data in the context of different web-based applications such as social media, which often occur in the context of multimedia or other heterogeneous data domains. Therefore, a number of techniques have recently been designed for the *joint mining* of text data in the context of these different kinds of data domains. For example, the Web contains text and image data which are often intimately connected to each other and these links can be used to improve the learning process from one domain to another. Similarly, cross-lingual linkages between documents of different languages can also be used in order to transfer knowledge from one language domain to another. This is closely related to the problem of transfer learning [11].

The rest of this chapter is organized as follows. The next section will discuss the different kinds of algorithms and applications for text mining. We will also point out the specific chapters in which they are discussed in the book. Section 3 will discuss some interesting future research directions.

2. Algorithms for Text Mining

In this section, we will explore the key problems arising in the context of text mining. We will also present the organization of the different chapters of this book in the context of these different problems. We intentionally leave the definition of the concept "text mining" vague to broadly cover a large set of related topics and algorithms for text analysis, spanning many different communities, including natural language processing, information retrieval, data mining, machine learning, and many application domains such as the World Wide Web and Biomedical Science. We have also intentionally allowed (sometimes significant) overlaps between chapters to allow each chapter to be relatively self contained, thus useful as a standing-alone chapter for learning about a specific topic.

Information Extraction from Text Data: Information Extraction is one of the key problems of text mining, which serves as a starting point for many text mining algorithms. For example, extraction of entities and their relations from text can reveal more meaningful semantic information in text data than a simple bag-of-words representation, and is generally needed to support inferences about knowledge buried in text data. Chapter 2 provides an survey of key problems in Information Extraction and the major algorithms for extracting entities and relations from text data.

Text Summarization: Another common function needed in many text mining applications is to summarize the text documents in order to obtain a brief overview of a large text document or a set of documents on a topic. Summarization techniques generally fall into two categories. In extractive summarization, a summary consists of information units extracted from the original text; in contrast, in abstractive summarization, a summary may contain "synthesized" information units that may not necessarily occur in the text documents. Most existing summarization methods are extractive, and in Chapter 3, we give a brief survey of these commonly used summarization methods.

Unsupervised Learning Methods from Text Data: Unsupervised learning methods do not require any training data, thus can be applied to any text data without requiring any manual effort. The two main unsupervised learning methods commonly used in the context of text data are *clustering* and *topic modeling*. The problem of clustering is that of segmenting a corpus of documents into partitions, each corresponding to a topical cluster. The problems of clustering and topic modeling are closely related. In topic modeling we use a probabilistic model in order to determine a *soft* clustering, in which each document has a membership probability of the cluster, as opposed to a hard segmentation of the documents. Topic models can be considered as the process of clustering with a generative probabilistic model. Each *topic* can be considered a probability distribution over words, with the representative words having the highest probability. Each document can be expressed as a probabilistic combination of these different topics. Thus, a topic can be considered to be analogous to a cluster, and the membership of a document to a cluster is probabilistic in nature. This also leads to a more elegant cluster membership representation in cases in which the document is known to contain distinct topics. In the case of hard clustering, it is sometimes challenging to assign a document to a single cluster in such cases. Furthermore, topic modeling relates elegantly to the dimension reduction problem, where each topic provides a conceptual dimension, and the documents may be represented as a linear probabilistic combination of these different topics. Thus, topic-modeling provides an extremely general framework, which relates to both the clustering and dimension reduction problems. In chapter 4, we study the problem of clustering, while topic modeling is covered in two chapters (Chapters 5 and 8). In Chapter 5, we discuss topic modeling from the perspective of dimension reduction since the discovered topics can serve as a low-dimensional space representation of text data, where semantically related words can "match" each other, which is hard to achieve with bag-of-words representation. In chapter 8, topic modeling is discussed as a general probabilistic model for text mining.

LSI and Dimensionality Reduction for Text Mining: The problem of dimensionality reduction is widely studied in the database literature as a method for representing the underlying data in compressed format for indexing and retrieval [10]. A variation of dimensionality reduction which is commonly used for text data is known as *latent semantic indexing* [6]. One of the interesting characteristics of latent semantic indexing is that it brings our the key semantic aspects of the text data, which makes it more suitable for a variety of mining applications. For example, the noise effects of synonymy and polysemy are reduced because of the use of such dimensionality reduction techniques. Another family of dimension reduction techniques are probabilistic topic models, notably PLSA, LDA, and their variants; they perform dimension reduction in a probabilistic way with potentially more meaningful topic representations based on word distributions. In chapter 5, we will discuss a variety of LSI and dimensionality reduction techniques for text data, and their use in a variety of mining applications.

Supervised Learning Methods for Text Data: Supervised learning methods are general machine learning methods that can exploit training data (i.e., pairs of input data points and the corresponding desired output) to learn a classifier or regression function that can be used to compute predictions on unseen new data. Since a wide range of application problems can be cast as a classification problem (that can be solved using supervised learning), the problem of supervised learning is sometimes also referred to as classification. Most of the traditional methods for text mining in the machine learning literature have been extended to solve problems of text mining. These include methods such as rule-based classifier, decision trees, nearest neighbor classifiers, maximum-margin classifiers, and probabilistic classifiers. In Chapter 6, we will study machine learning methods for automated text categorization, a major application area of supervised learning in text mining. A more general discussion of supervised learning methods is given in Chapter 8. A special class of techniques in supervised learning to address the issue of lack of training data, called *transfer learning*, are covered in Chapter 7.

Transfer Learning with Text Data: The afore-mentioned example of cross-lingual mining provides a case where the attributes of the text collection may be heterogeneous. Clearly, the feature representations in the different languages are heterogeneous, and it can often provide useful to transfer knowledge from one domain to another, especially when their is paucity of data in one domain. For example, labeled English documents are copious and easy to find. On the other hand, it is much harder to obtain labeled Chinese documents. The problem of transfer learning attempts to *transfer* the learned knowledge from one domain to another. Some other scenarios in which this arises is the case where we have a mixture of text and multimedia data. This is often the case in many web-based and social media applications such as *Flickr*, *Youtube* or other multimedia sharing sites. In such cases, it may be desirable to transfer the learned knowledge from one domain to another with the use of cross-media transfer. Chapter 7 provides a detailed survey of such learning techniques.

Probabilistic Techniques for Text Mining: A variety of probabilistic methods, particularly unsupervised topic models such as PLSA and LDA and supervised learning methods such as conditional random fields are used frequently in the context of text mining algorithms. Since such methods are used frequently in a wide variety of contexts, it is useful to create an organized survey which describes the different tools and techniques that are used in this context. In Chapter 8, we introduce the basics of the common probabilistic models and methods which are often used in the context of text mining. The material in this chapter is also relevant to many of the clustering, dimensionality reduction, topic modeling and classification techniques discussed in Chapters 4, 5, 6 and 7.

Mining Text Streams: Many recent applications on the web create massive streams of text data. In particular web applications such as social networks which allow the simultaneous input of text from a wide variety of users can result in a continuous stream of large volumes of text data. Similarly, news streams such as *Reuters* or aggregators such as *Google news* create large volumes of streams which can be mined continuously. Such text data are more challenging to mine, because they need to be processed in the context of a one-pass constraint [1]. The one-pass constraint essentially means that it may sometimes be difficult to store the data offline for processing, and it is necessary to perform the mining tasks continuously, as the data comes in. This makes algorithmic design a much more challenging task. In chapter 9, we study the common techniques which are often used in the context of a variety of text mining tasks.

Cross-Lingual Mining of Text Data: With the proliferation of web-based and other information retrieval applications to other applications, it has become particularly useful to apply mining tasks in different languages, or use the knowledge or corpora in one language to another. For example, in cross-language mining, it may be desirable to cluster a group of documents in different languages, so that documents from different languages but similar semantic topics may be placed in the same cluster. Such cross-lingual applications are extremely rich, because they can often be used to leverage knowledge from one data domain into another. In chapter 10, we will study methods for cross-lingual mining of text data, covering techniques such as machine translation, cross-lingual information retrieval, and analysis of comparable and parallel corpora.

Text Mining in Multimedia Networks: Text often occurs in the context of many multimedia sharing sites such as *Flickr* or *Youtube*. A natural question arises as to whether we can enrich the underlying mining process by simultaneously using the data from other domains together with the text collection. This is also related to the problem of transfer learning, which was discussed earlier. In chapter 11, a detailed survey will be provided on mining other multimedia data together with text collections.

Text Mining in Social Media: One of the most common sources of text on the web is the presence of social media, which allows human actors to express themselves quickly and freely in the context of a wide range of subjects [2]. Social media is now exploited widely by commercial sites for influencing users and targeted marketing. The process of mining text in social media requires the special ability to mine dynamic data which often contains poor and non-standard vocabulary. Furthermore, the text may occur in the context of linked social networks. Such links can be used in order to improve the quality of the underlying mining process. For example, methods that use both link and content [4] are widely known to provide much more effective results which use only content or links. Chapter 12 provides a detailed survey of text mining methods in social media.

Opinion Mining from Text Data: A considerable amount of text on web sites occurs in the context of product reviews or opinions of different users. Mining such opinionated text data to reveal and summarize the opinions about a topic has widespread applications, such as in supporting consumers for optimizing decisions and business intelligence. spam opinions which are not useful and simply add noise to the mining process. Chapter 13 provides a detailed survey of models and methods for opinion mining and sentiment analysis.

Text Mining from Biomedical Data: Text mining techniques play an important role in both enabling biomedical researchers to effectively and efficiently access the knowledge buried in large amounts of literature and supplementing the mining of other biomedical data such as genome sequences, gene expression data, and protein structures to facilitate and speed up biomedical discovery. As a result, a great deal of research work has been done in adapting and extending standard text mining methods to the biomedical domain, such as recognition of various biomedical entities and their relations, text summarization, and question answering. Chapter 14 provides a detailed survey of the models and methods used for biomedical text mining.

3. Future Directions

The rapid growth of online textual data creates an urgent need for powerful text mining techniques. As an interdisciplinary field, text data mining spans multiple research communities, especially data mining, natural language processing, information retrieval, and machine learning with applications in many different areas, and has attracted much attention recently. Many models and algorithms have been developed for various text mining tasks have been developed as we discussed above and will be surveyed in the rest of this book.

Looking forward, we see the following general future directions that are promising:

- **Scalable and robust methods for natural language understanding:** Understanding text information is fundamental to text mining. While the current approaches mostly rely on bag of words representation, it is clearly desirable to go beyond such a simple representation. Information extraction techniques provide one step forward toward semantic representation, but the current information extraction methods mostly rely on supervised learning and generally only work well when sufficient training data are available, restricting its applications. It is thus important to develop effective and robust information extraction and other natural language processing methods that can scale to multiple domains.

- **Domain adaptation and transfer learning:** Many text mining tasks rely on supervised learning, whose effectiveness highly depends on the amount of training data available. Unfortunately, it is generally labor-intensive to create large amounts of training data. Domain adaptation and transfer learning methods can alleviate this problem by attempting to exploit training data that might be available in a related domain or for a related task. However, the current approaches still have many limitations and are generally inadequate when there is no or little training data in the target domain. Further development of more effective domain adaptation and transfer learning methods is necessary for more effective text mining.

- **Contextual analysis of text data:** Text data is generally associated with a lot of context information such as authors, sources, and time, or more complicated information networks associated with text data. In many applications, it is important to consider the context as well as user preferences in text mining. It is thus important to further extend existing text mining approaches to further incorporate context and information networks for more powerful text analysis.

- **Parallel text mining:** In many applications of text mining, the amount of text data is huge and is likely increasing over time, thus it is infeasible to store the data in one machine, making it necessary to develop parallel text mining algorithms that can run on a cluster of computers to perform text mining tasks in parallel. In particular, how to parallelize all kinds of text mining algorithms, including both unsupervised and supervised learning methods is a major future challenge. This direction is clearly related to cloud computing and data-intensive computing, which are growing fields themselves.

References

[1] C. Aggarwal. *Data Streams: Models and Algorithms*, Springer, 2007.

[2] C. Aggarwal. *Social Network Data Analytics*, Springer, 2011.

[3] R. A. Baeza-Yates, B. A. Ribeiro-Neto, *Modern Information Retrieval - the concepts and technology behind search, Second edition*, Pearson Education Ltd., Harlow, England, 2011.

[4] S. Chakrabarti, B. Dom, P. Indyk. Enhanced Hypertext Categorization using Hyperlinks, *ACM SIGMOD Conference*, 1998.

[5] W. B. Croft, D. Metzler, T. Strohma, *Search Engines - Information Retrieval in Practice*, Pearson Education, 2009.

[6] S. Deerwester, S. Dumais, T. Landauer, G. Furnas, R. Harshman. Indexing by Latent Semantic Analysis. *JASIS*, 41(6), pp. 391–407, 1990.

[7] D. A. Grossman, O. Frieder, *Information Retrieval: Algorithms and Heuristics (The Kluwer International Series on Information Retrieval)*, Springer-Verlag New York, Inc, 2004.

[8] J. Han, M. Kamber. *Data Mining: Concepts and Techniques*, 2nd Edition, Morgan Kaufmann, 2005.

[9] C. Manning, P. Raghavan, H. Schutze, *Introduction to Information Retrieval*, Cambridge University Press, 2008.

[10] I. T. Jolliffee. Principal Component Analysis. *Springer*, 2002.

[11] S. J. Pan, Q. Yang. A Survey on Transfer Learning, *IEEE Transactions on Knowledge and Data Engineering*, 22(10): pp 1345–1359, Oct. 2010.

[12] G. Salton. *An Introduction to Modern Information Retrieval*, Mc Graw Hill, 1983.

[13] K. Sparck Jones P. Willett (ed.). *Readings in Information Retrieval*, Morgan Kaufmann Publishers Inc, 1997.

Chapter 2

INFORMATION EXTRACTION FROM TEXT

Jing Jiang
Singapore Management University
jingjiang@smu.edu.sg

Abstract   Information extraction is the task of finding structured information from unstructured or semi-structured text. It is an important task in text mining and has been extensively studied in various research communities including natural language processing, information retrieval and Web mining. It has a wide range of applications in domains such as biomedical literature mining and business intelligence. Two fundamental tasks of information extraction are named entity recognition and relation extraction. The former refers to finding names of entities such as people, organizations and locations. The latter refers to finding the semantic relations such as `FounderOf` and `HeadquarteredIn` between entities. In this chapter we provide a survey of the major work on named entity recognition and relation extraction in the past few decades, with a focus on work from the natural language processing community.

Keywords: Information extraction, named entity recognition, relation extraction

1.   Introduction

Information extraction from text is an important task in text mining. The general goal of information extraction is to discover structured information from unstructured or semi-structured text. For example, given the following English sentence,

   In 1998, Larry Page and Sergey Brin founded Google Inc.

we can extract the following information,

   `FounderOf`(*Larry Page*, *Google Inc.*),
   `FounderOf`(*Sergey Brin*, *Google Inc.*),
   `FoundedIn`(*Google Inc.*, *1998*).

Such information can be directly presented to an end user, or more commonly, it can be used by other computer systems such as search engines and database management systems to provide better services to end users.

Information extraction has applications in a wide range of domains. The specific type and structure of the information to be extracted depend on the need of the particular application. We give some example applications of information extraction below:

- Biomedical researchers often need to sift through a large amount of scientific publications to look for discoveries related to particular genes, proteins or other biomedical entities. To assist this effort, simple search based on keyword matching may not suffice because biomedical entities often have synonyms and ambiguous names, making it hard to accurately retrieve relevant documents. A critical task in biomedical literature mining is therefore to automatically identify mentions of biomedical entities from text and to link them to their corresponding entries in existing knowledge bases such as the FlyBase.

- Financial professionals often need to seek specific pieces of information from news articles to help their day-to-day decision making. For example, a finance company may need to know all the company takeovers that take place during a certain time span and the details of each acquisition. Automatically finding such information from text requires standard information extraction technologies such as named entity recognition and relation extraction.

- Intelligence analysts review large amounts of text to search for information such as people involved in terrorism events, the weapons used and the targets of the attacks. While information retrieval technologies can be used to quickly locate documents that describe terrorism events, information extraction technologies are needed to further pinpoint the specific information units within these documents.

- With the fast growth of the Web, search engines have become an integral part of people's daily lives, and users' search behaviors are much better understood now. Search based on bag-of-word representation of documents can no longer provide satisfactory results. More advanced search problems such as entity search, structured search and question answering can provide users with better search experience. To facilitate these search capabilities, information ex-

Terrorism Template	
Slot	Fill Value
Incident: Date	07 Jan 90
Incident: Location	Chile: Molina
Incident: Type	robbery
Incident: Stage of execution	accomplished
Incident: Instrument type	gun
Human Target: Name	"Enrique Ormazabal Ormazabal"
Human Target: Description	"Businessman": "Enrique Ormazabal Ormazabal"
Human Target: Type	civilian: "Enrique Ormazabal Ormazabal"
Human Target: Number	1: "Enrique Ormazabal Ormazabal"
...	...
A Sample Document	
Santiago, 10 Jan 90 – Police are carrying out intensive operations in the town of Molina in the seventh region in search of a gang of alleged extremists who could be linked to a recently discovered arsenal. It has been reported that Carabineros in Molina raided the house of of 25-year-old worker Mario Munoz Pardo, where they found a fal rifle, ammunition clips for various weapons, detonators, and material for making explosives. It should be recalled that a group of armed individuals wearing ski masks robbed a businessman on a rural road near Molina on 7 January. The businessman, Enrique Ormazabal Ormazabal, tried to resist; The men shot him and left him seriously wounded. He was later hospitalized in Curico. Carabineros carried out several operations, including the raid on Munoz' home. The police are continuing to patrol the area in search of the alleged terrorist command.	

Figure 2.1. Part of the terrorism template used in MUC-4 and a sample document that contains a terrorism event.

traction is often needed as a preprocessing step to enrich document representation or to populate an underlying database.

While extraction of structured information from text dates back to the '70s (e.g. DeJong's FRUMP program [28]), it only started gaining much attention when DARPA initiated and funded the Message Understanding Conferences (MUC) in the '90s [33]. Since then, research efforts on this topic have not declined. Early MUCs defined information extraction as filling a predefined template that contains a set of predefined slots. For example, Figure 2.1 shows a subset of the slots in the terrorism template used in MUC-4 and a sample document from which template slot fill values were extracted. Some of the slot fill values such as *"Enrique Ormazabal Ormazabal"* and *"Businessman"* were extracted directly from the text while others such as *robbery, accomplished* and *gun* were selected from a predefined value set for the corresponding slot based on the document.

Template filling is a complex task and systems developed to fill one template cannot directly work for a different template. In MUC-6, a number of template-independent subtasks of information extraction were defined [33]. These include named entity recognition, coreference resolution and relation extraction. These tasks serve as building blocks to support full-fledged, domain-specific information extraction systems.

Early information extraction systems such as the ones that participated in the MUCs are often rule-based systems (e.g. [32, 42]). They use linguistic extraction patterns developed by humans to match text and locate information units. They can achieve good performance on the specific target domain, but it is labor intensive to design good extraction rules, and the developed rules are highly domain dependent. Realizing the limitations of these manually developed systems, researchers turned to statistical machine learning approaches. And with the decomposition of information extraction systems into components such as named entity recognition, many information extraction subtasks can be transformed into classification problems, which can be solved by standard supervised learning algorithms such as support vector machines and maximum entropy models. Because information extraction involves identifying segments of text that play different roles, sequence labeling methods such as hidden Markov models and conditional random fields have also been widely used.

Traditionally information extraction tasks assume that the structures to be extracted, e.g. the types of named entities, the types of relations, or the template slots, are well defined. In some scenarios, we do not know in advance the structures of the information we would like to extract and would like to mine such structures from large corpora. For example, from a set of earthquake news articles we may want to automatically discover that the date, time, epicenter, magnitude and casualty of an earthquake are the most important pieces of information reported in news articles. There have been some recent studies on this kind of unsupervised information extraction problems but overall work along this line remains limited.

Another new direction is open information extraction, where the system is expected to extract *all* useful entity relations from a large, diverse corpus such as the Web. The output of such systems includes not only the arguments involved in a relation but also a description of the relation extracted from the text. Recent advances in this direction include systems like TEXTRUNNER [6], WOE [66] and REVERB [29].

Information extraction from semi-structured Web pages has also been an important research topic in Web mining (e.g. [40, 45, 25]). A major difference of Web information extraction from information extraction

Information Extraction from Text   15

studied in natural language processing is that Web pages often contain structured or semi-structured text such as tables and lists, whose extraction relies more on HTML tags than linguistic features. Web information extraction systems are also called *wrappers* and learning such systems is called *wrapper induction*. In this survey we only cover information extraction from purely unstructured natural language text. Readers who are interested in wrapper induction may refer to [31, 20] for in-depth surveys.

In this chapter we focus on the two most fundamental tasks in information extraction, namely, named entity recognition and relation extraction. The state-of-the-art solutions to both tasks rely on statistical machine learning methods. We also discuss unsupervised information extraction, which has not attracted much attention traditionally. The rest of this chapter is organized as follows. Section 2 discusses current approaches to named entity recognition, including rule-based methods and statistical learning methods. Section 3 discusses relation extraction under both a fully supervised setting and a weakly supervised setting. We then discuss unsupervised relation discovery and open information extraction in Section 4. In Section 5 we discuss evaluation of information extraction systems. We finally conclude in Section 6.

2. Named Entity Recognition

A named entity is a sequence of words that designates some real-world entity, e.g. "California," "Steve Jobs" and "Apple Inc." The task of named entity recognition, often abbreviated as NER, is to identify named entities from free-form text and to classify them into a set of predefined types such as *person*, *organization* and *location*. Oftentimes this task cannot be simply accomplished by string matching against pre-compiled gazetteers because named entities of a given entity type usually do not form a closed set and therefore any gazetteer would be incomplete. Another reason is that the type of a named entity can be context-dependent. For example, "JFK" may refer to the person "John F. Kennedy," the location "JFK International Airport," or any other entity sharing the same abbreviation. To determine the entity type for "JFK" occurring in a particular document, its context has to be considered.

Named entity recognition is probably the most fundamental task in information extraction. Extraction of more complex structures such as relations and events depends on accurate named entity recognition as a preprocessing step. Named entity recognition also has many applications apart from being a building block for information extraction. In question answering, for example, candidate answer strings are often named entities that need to be extracted and classified first [44]. In entity-oriented search, identifying named entities in documents as well as in queries is the first step towards high relevance of search results [34, 21].

Although the study of named entity recognition dates back to the early '90s [56], the task was formally introduced in 1995 by the sixth Message Understanding Conference (MUC-6) as a subtask of information extraction [33]. Since then, NER has drawn much attention in the research community. There have been several evaluation programs on this task, including the Automatic Content Extraction (ACE) program [1], the shared task of the Conference on Natural Language Learning (CoNLL) in 2002 and 2003 [63], and the BioCreAtIvE (Critical Assessment of Information Extraction Systems in Biology) challenge evaluation [2].

The most commonly studied named entity types are *person, organization* and *location*, which were first defined by MUC-6. These types are general enough to be useful for many application domains. Extraction of expressions of dates, times, monetary values and percentages, which was also introduced by MUC-6, is often also studied under NER, although strictly speaking these expressions are not named entities. Besides these general entity types, other types of entities are usually defined for specific domains and applications. For example, the GENIA corpus uses a fine-grained ontology to classify biological entities [52]. In online search and advertising, extraction of product names is a useful task.

Early solutions to named entity recognition rely on manually crafted patterns [4]. Because it requires human expertise and is labor intensive to create such patterns, later systems try to automatically learn such patterns from labeled data [62, 16, 23]. More recent work on named entity recognition uses statistical machine learning methods. An early attempt is Nymble, a name finder based on hidden Markov models [10]. Other learning models such as maximum entropy models [22], maximum entropy Markov models [8, 27, 39, 30], support vector machines [35] and conditional random fields [59] have also been applied to named entity recognition.

2.1 Rule-based Approach

Rule-based methods for named entity recognition generally work as follows: A set of rules is either manually defined or automatically learned. Each token in the text is represented by a set of features. The text is then compared against the rules and a rule is fired if a match is found.

A rule consists of a pattern and an action. A pattern is usually a regular expression defined over features of tokens. When this pattern matches a sequence of tokens, the specified action is fired. An action can be labeling a sequence of tokens as an entity, inserting the start or end label of an entity, or identifying multiple entities simultaneously. For example, to label any sequence of tokens of the form "Mr. X" where X is a capitalized word as a person entity, the following rule can be defined:

(token = "Mr." orthography type = $FirstCap$) → person name.

The left hand side is a regular expression that matches any sequence of two tokens where the first token is "Mr." and the second token has the orthography type $FirstCap$. The right hand side indicates that the matched token sequence should be labeled as a person name.

This kind of rule-based methods has been widely used [4, 62, 16, 61, 23]. Commonly used features to represent tokens include the token itself, the part-of-speech tag of the token, the orthography type of the token (e.g. first letter capitalized, all letters capitalized, number, etc.), and whether the token is inside some predefined gazetteer.

It is possible for a sequence of tokens to match multiple rules. To handle such conflicts, a set of policies has to be defined to control how rules should be fired. One approach is to order the rules in advance so that they are sequentially checked and fired.

Manually creating the rules for named entity recognition requires human expertise and is labor intensive. To automatically learn the rules, different methods have been proposed. They can be roughly categorized into two groups: top-down (e.g. [61]) and bottom-up (e.g. [16, 23]). With either approach, a set of training documents with manually labeled named entities is required. In the top-down approach, general rules are first defined that can cover the extraction of many training instances. However, these rules tend to have low precision. The system then iteratively defines more specific rules by taking the intersections of the more general rules. In the bottom-up approach, specific rules are defined based on training instances that are not yet covered by the existing rule set. These specific rules are then generalized.

2.2 Statistical Learning Approach

More recent work on named entity recognition is usually based on statistical machine learning. Many statistical learning-based named entity recognition algorithms treat the task as a sequence labeling problem. Sequence labeling is a general machine learning problem and has been used to model many natural language processing tasks including part-of-speech tagging, chunking and named entity recognition. It can be formulated as follows. We are given a sequence of observations, denoted as $\boldsymbol{x} = (x_1, x_2, \ldots, x_n)$. Usually each observation is represented as a

Steve	Jobs	was	a	co-founder	of	Apple	Inc.
B-PER	I-PER	O	O	O	O	B-ORG	I-ORG

Figure 2.2. An example sentence with NER labels in the BIO notation. `PER` stands for person and `ORG` stands for organization.

feature vector. We would like to assign a label y_i to each observation x_i. While one may apply standard classification to predict the label y_i based solely on x_i, in sequence labeling, it is assumed that the label y_i depends not only on its corresponding observation x_i but also possibly on other observations and other labels in the sequence. Typically this dependency is limited to observations and labels within a close neighborhood of the current position i.

To map named entity recognition to a sequence labeling problem, we treat each word in a sentence as an observation. The class labels have to clearly indicate both the boundaries and the types of named entities within the sequence. Usually the BIO notation, initially introduced for text chunking [55], is used. With this notation, for each entity type `T`, two labels are created, namely, `B-T` and `I-T`. A token labeled with `B-T` is the beginning of a named entity of type `T` while a token labeled with `I-T` is inside (but not the beginning of) a named entity of type `T`. In addition, there is a label `O` for tokens outside of any named entity. Figure 2.2 shows an example sentence and its correct NER label sequence.

2.2.1 Hidden Markov Models.

In a probabilistic framework, the best label sequence $\boldsymbol{y} = (y_1, y_2, \ldots, y_n)$ for an observation sequence $\boldsymbol{x} = (x_1, x_2, \ldots, x_n)$ is the one that maximizes the conditional probability $p(\boldsymbol{y}|\boldsymbol{x})$, or equivalently, the one that maximizes the joint probability $p(\boldsymbol{x}, \boldsymbol{y})$. One way to model the joint probability is to assume a Markov process where the generation of a label or an observation is dependent only on one or a few previous labels and/or observations. If we treat $\boldsymbol{y}$ as hidden states, then we essentially have a hidden Markov model [54].

An example is the Nymble system developed by BBN, one of the earliest statistical learning-based NER systems [10]. Nymble assumes the following generative process:

(1) Each y_i is generated conditioning on the previous label y_{i-1} and the previous word x_{i-1}.

(2) If x_i is the first word of a named entity, it is generated conditioning on the current and the previous labels, i.e. y_i and y_{i-1}.

(3) If x_i is inside a named entity, it is generated conditioning on the previous observation x_{i-1}.

Information Extraction from Text

For subsequences of words outside of any named entity, Nymble treats them as a NOT-A-NAME class. Nymble also assumes that there is a magical +end+ word at the end of each named entity and models the probability of a word being the final word of a named entity. With the generative process described above, the probability $p(\boldsymbol{x}, \boldsymbol{y})$ can be expressed in terms of various conditional probabilities.

Initially x_i is simply the word at position i. Nymble further augments it into $x_i = \langle w, f \rangle_i$, where w is the word at position i and f is a word feature characterizing w. For example, the feature FourDigitNum indicates that the word is a number with four digits. The rationale behind introducing word features is that these features may carry strong correlations with entity types. For example, a four-digit number is likely to be a year.

The model parameters of Nymble are essentially the various multinomial distributions that govern the generation of x_i and y_i. Nymble uses supervised learning to learn these parameters. Given sentences labeled with named entities, Nymble performs maximum likelihood estimation to find the model parameters that maximize $p(\boldsymbol{X}, \boldsymbol{Y})$ where $\boldsymbol{X}$ denotes all the sentences in the training data and $\boldsymbol{Y}$ denotes their true label sequences. Parameter estimation essentially becomes counting. For example,

$$p(y_i = \mathtt{c}_1 | y_{i-1} = \mathtt{c}_2, x_{i-1} = \mathtt{w}) = \frac{c(\mathtt{c}_1, \mathtt{c}_2, \mathtt{w})}{c(\mathtt{c}_2, \mathtt{w})}, \quad (2.1)$$

where $\mathtt{c}_1$ and $\mathtt{c}_2$ are two class labels and $\mathtt{w}$ is a word. $p(y_i = \mathtt{c}_1 | y_{i-1} = \mathtt{c}_2, x_{i-1} = \mathtt{w})$ is the probability of observing the class label $\mathtt{c}_1$ given that the previous class label is $\mathtt{c}_2$ and the previous word is $\mathtt{w}$. $c(\mathtt{c}_1, \mathtt{c}_2, \mathtt{w})$ is the number of times we observe class label $\mathtt{c}_1$ when the previous class label is $\mathtt{c}_2$ and the previous word is $\mathtt{w}$, and $c(\mathtt{c}_2, \mathtt{w})$ is the number of times we observe the previous class label to be $\mathtt{c}_2$ and the previous word to be $\mathtt{w}$ regardless of the current class label.

During prediction, Nymble uses the learned model parameters to find the label sequence $\boldsymbol{y}$ that maximizes $p(\boldsymbol{x}, \boldsymbol{y})$ for a given $\boldsymbol{x}$. With the Markovian assumption, dynamic programming can be used to efficiently find the best label sequence.

2.2.2 Maximum Entropy Markov Models.

The hidden Markov models described above are generative models. In general, researchers have found that when training data is sufficient, compared with generative models that model $p(x|y)$, discriminative models that directly model $p(y|x)$ tend to give a lower prediction error rate and thus are preferable [65]. For named entity recognition, there has also been such a shift from generative models to discriminative models. A commonly used discriminative model for named entity recognition is the maximum entropy model [9] coupled with a Markovian assumption. Existing work using such a model includes [8, 27, 39, 30].

Specifically, with a Markovian assumption, the label y_i at position i is dependent on the observations within a neighborhood of position i as well as a number of previous labels:

$$p(\boldsymbol{y}|\boldsymbol{x}) = \prod_i p(y_i|\boldsymbol{y}_{i-k}^{i-1}, \boldsymbol{x}_{i-l}^{i+l}). \qquad (2.2)$$

In the equation above, $\boldsymbol{y}_{i-k}^{i-1}$ refers to $(y_{i-k}, y_{i-k+1}, \ldots, y_{i-1})$ and $\boldsymbol{x}_{i-l}^{i+l}$ refers to $(x_{i-l}, x_{i-l+1}, \ldots, x_{i+l})$. And with maximum entropy models, the functional form of $p(y_i|\boldsymbol{y}_{i-k}^{i-1}, \boldsymbol{x}_{i-l}^{i+l})$ follows an exponential model:

$$p(y_i|\boldsymbol{y}_{i-k}^{i-1}, \boldsymbol{x}_{i-l}^{i+l}) = \frac{\exp\left(\sum_j \lambda_j f_j(y_i, \boldsymbol{y}_{i-k}^{i-1}, \boldsymbol{x}_{i-l}^{i+l})\right)}{\sum_{y'} \exp\left(\sum_j \lambda_j f_j(y', \boldsymbol{y}_{i-k}^{i-1}, \boldsymbol{x}_{i-l}^{i+l})\right)}. \qquad (2.3)$$

In the equation above, $f_j(\cdot)$ is a feature function defined over the current label, the previous k labels as well as $2l+1$ observations surrounding the current observation, and λ_j is the weight for feature f_j. An example feature is below:

$$f(y_i, y_{i-1}, x_i) = \begin{cases} 1 & \text{if } y_{i-1} = \text{O and } y_i = \text{B-PER and } \text{word}(x_i) = \text{``Mr.''}, \\ 0 & \text{otherwise}. \end{cases}$$

The model described above can be seen as a variant of the maximum entropy Markov models (MEMMs), which were formally introduced by McCallum et al. for information extraction [48].

To train a maximum entropy Markov model, we look for the feature weights $\Lambda = \{\lambda_j\}$ that can maximize the conditional probability $p(\boldsymbol{Y}|\boldsymbol{X})$ where $\boldsymbol{X}$ denotes all the sentences in the training data and $\boldsymbol{Y}$ denotes their true label sequences. Just like for standard maximum entropy models, a number of optimization algorithms can be used to train maximum entropy Markov models, including Generalized Iterative Scaling (GIS), Improved Iterative Scaling (IIS) and limited memory quasi-Newton methods such as L-BFGS [15]. A comparative study of these optimization methods for maximum entropy models can be found in [46]. L-BFGS is a commonly used method currently.

2.2.3 Conditional Random Fields.

Conditional random fields (CRFs) are yet another popular discriminative model for sequence labeling. They were introduced by Lafferty et al. to also address information extraction problems [41]. The major difference between CRFs

Information Extraction from Text 21

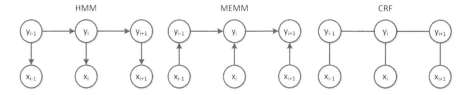

Figure 2.3.   Graphical representations of linear-chain HMM, MEMM and CRF.

and MEMMs is that in CRFs the label of the current observation can depend not only on previous labels but also on future labels. Also, CRFs are undirected graphical models while both HMMs and MEMMs are directed graphical models. Figure 2.3 graphically depicts the differences between linear-chain (i.e. first-order) HMM, MEMM and CRF. Ever since they were first introduced, CRFs have been widely used in natural language processing and some other research areas.

Usually linear-chain CRFs are used for sequence labeling problems in natural language processing, where the current label depends on the previous one and the next one labels as well as the observations. There have been many studies applying conditional random fields to named entity recognition (e.g. [49, 59]). Specifically, following the same notation used earlier, the functional form of $p(\boldsymbol{y}|\boldsymbol{x})$ is as follows:

$$p(\boldsymbol{y}|\boldsymbol{x}) = \frac{1}{Z(\boldsymbol{x})} \exp\left(\sum_i \sum_j \lambda_j f_j(y_i, y_{i-1}, \boldsymbol{x}, i)\right), \quad (2.4)$$

where $Z(\boldsymbol{x})$ is a normalization factor of all possible label sequences:

$$Z(\boldsymbol{x}) = \sum_{\boldsymbol{y'}} \exp\left(\sum_i \sum_j \lambda_j f_j(y'_i, y'_{i-1}, \boldsymbol{x}, i)\right). \quad (2.5)$$

To train CRFs, again maximum likelihood estimation is used to find the best model parameters that maximize $p(\boldsymbol{Y}|\boldsymbol{X})$. Similar to MEMMs, CRFs can be trained using L-BFGS. Because the normalization factor $Z(\boldsymbol{x})$ is a sum over all possible label sequences for $\boldsymbol{x}$, training CRFs is more expensive than training MEMMs.

In linear-chain CRFs we cannot define long-range features. General CRFs allow long-range features but are too expensive to perform exact inference. Sarawagi and Cohen proposed semi-Markov conditional random fields as a compromise [58]. In semi-Markov CRFs, labels are assigned to segments of the observation sequence $\boldsymbol{x}$ and features can measure properties of these segments. Exact learning and inference on semi-Markov CRFs is thus computationally feasible. Sarawagi and Cohen applied Semi-Markov CRFs to named entity recognition and achieved better performance than standard CRFs.

3. Relation Extraction

Another important task in information extraction is relation extraction. Relation extraction is the task of detecting and characterizing the semantic relations between entities in text. For example, from the following sentence fragment,

Facebook co-founder Mark Zuckerberg

we can extract the following relation,

`FounderOf`(*Mark Zuckerberg*, *Facebook*).

Much of the work on relation extraction is based on the task definition from the Automatic Content Extraction (ACE) program [1]. ACE focuses on binary relations, i.e. relations between two entities. The two entities involved are also referred to as *arguments*. A set of major relation types and their subtypes are defined by ACE. Examples of ACE major relation types include `physical` (e.g. an entity is physically near another entity), `personal/social` (e.g. a person is a family member of another person), and `employment/affiliation` (e.g. a person is employed by an organization). ACE makes a distinction between relation extraction and relation mention extraction. The former refers to identifying the semantic relation between a pair of entities based on *all* the evidence we can gather from the corpus, whereas the latter refers to identifying individual mentions of entity relations. Because corpus-level relation extraction to a large extent still relies on accurate mention-level relation extraction, in the rest of this chapter we do not make any distinction between these two problems unless necessary.

Various techniques have been proposed for relation extraction. The most common and straightforward approach is to treat the task as a classification problem: Given a pair of entities co-occurring in the same sentence, can we classify the relation between the two entities into one of the predefined relation types? Although it is also possible for relation mentions to cross sentence boundaries, such cases are less frequent and hard to detect. Existing work therefore mostly focuses on relation extraction within sentence boundaries.

There have been a number of studies following the classification approach [38, 71, 37, 18, 19]. Feature engineering is the most critical step of this approach. An extension of the feature-based classification approach is to define kernels rather than features and to apply kernel machines such as support vector machines to perform classification. Kernels defined over word sequences [14], dependency trees [26], dependency paths [13] and parse trees [67, 68] have been proposed.

Both feature-based and kernel-based classification methods require a large amount of training data. Another major line of work on relation extraction is weakly supervised relation extraction from large corpora that does not rely on the availability of manually labeled training data. One approach is the bootstrapping idea to start with a small set of seed examples and iteratively find new relation instances as well as new extraction patterns. Representative work includes the Snowball system [3]. Another approach is distant supervision that makes use of known relation instances from existing knowledge bases such as Freebase [50].

3.1 Feature-based Classification

A typical approach to relation extraction is to treat the task as a classification problem [38, 71, 37, 18, 19]. Specifically, any pair of entities co-occurring in the same sentence is considered a candidate relation instance. The goal is to assign a class label to this instance where the class label is either one of the predefined relation types or *nil* for unrelated entity pairs. Alternatively, a two-stage classification can be performed where at the first stage whether two entities are related is determined and at the second stage the relation type for each related entity pair is determined.

Classification approach assumes that a training corpus exists in which all relation mentions for each predefined relation type have been manually annotated. These relation mentions are used as positive training examples. Entity pairs co-occurring in the same sentence but not labeled are used as negative training examples. Each candidate relation instance is represented by a set of features that are carefully chosen. Standard learning algorithms such as support vector machines and logistic regression can then be used to train relation classifiers.

Feature engineering is a critical step for this classification approach. Researchers have examined a wide range of lexical, syntactic and semantic features. We summarize some of the most commonly used features as follows:

- **Entity features:** Oftentimes the two argument entities, including the entity words themselves and the entity types, are correlated with certain relation types. In the ACE data sets, for example, entity words such as *father*, *mother*, *brother* and *sister* and the `person` entity type are all strong indicators of the `family` relation subtype.

- **Lexical contextual features:** Intuitively the contexts surrounding the two argument entities are important. The simplest way to incorporate evidence from contexts is to use lexical features. For example, if the word *founded* occurs between the two arguments, they are more likely to have the `FounderOf` relation.

- **Syntactic contextual features:** Syntactic relations between the two arguments or between an argument and another word can often be useful. For example, if the first argument is the subject of the verb *founded* and the second argument is the object of the verb *founded*, then one can almost immediately tell that the `FounderOf` relation exists between the two arguments. Syntactic features can be derived from parse trees of the sentence containing the relation instance.

- **Background knowledge:** Chan and Roth studied the use of background knowledge for relation extraction [18]. An example is to make use of Wikipedia. If two arguments co-occur in the same Wikipedia article, the content of the article can be used to check whether the two entities are related. Another example is word clusters. For example, if we can group all names of companies such as *IBM* and *Apple* into the same word cluster, we achieve a level of abstraction higher than words and lower than the general entity type `organization`. This level of abstraction may help extraction of certain relation types such as `Acquire` between two companies.

Jiang and Zhai proposed a framework to organize the features used for relation extraction such that a systematic exploration of the feature space can be conducted [37]. Specifically, a relation instance is represented as a labeled, directed graph $G = (V, E, A, B)$, where V is the set of nodes in the graph, E is the set of directed edges in the graph, and A and B are functions that assign labels to the nodes.

First, for each node $v \in V$, $A(v) = \{a_1, a_2, \ldots, a_{|A(v)|}\}$ is a set of attributes associated with node v, where $a_i \in \Sigma$, and Σ is an alphabet that contains all possible attribute values. For example, if node v represents a token, then $A(v)$ can include the token itself, its morphological base form, its part-of-speech tag, etc. If v also happens to be the head word of arg_1 or arg_2, then $A(v)$ can also include the entity type. Next, function $B : V \to \{0, 1, 2, 3\}$ is introduced to distinguish argument nodes from non-argument nodes. For each node $v \in V$, $B(v)$ indicates how node v is related to arg_1 and arg_2. 0 indicates that v does not cover any argument, 1 or 2 indicates that v covers arg_1 or arg_2, respectively, and 3 indicates that v covers both arguments. In a constituency parse tree, a

Information Extraction from Text   25

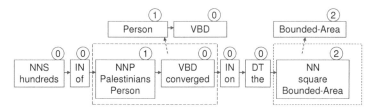

Figure 2.4. An example sequence representation. The subgraph on the left represents a bigram feature. The subgraph on the right represents a unigram feature that states the entity type of arg_2.

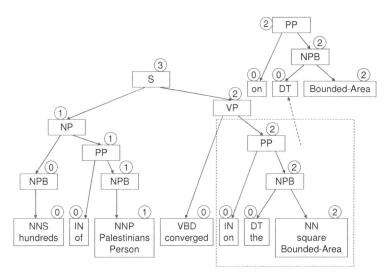

Figure 2.5. An example constituency parse tree representation. The subgraph represents a subtree feature (grammar production feature).

node v may represent a phrase and it can possibly cover both arguments. Figures 2.4, 2.5 and 2.6 show three relation instance graphs based on the token sequence, the constituency parse tree and the dependency parse tree, respectively.

Given the above definition of relation instance graphs, a feature of a relation instance captures part of the attributive and/or structural properties of the relation instance graph. Therefore, it is natural to define a feature as a subgraph of the relation instance graph. Formally, given a graph $G = (V, E, A, B)$, which represents a single relation instance, a feature that exists in this relation instance is a subgraph $G' = (V', E', A', B')$ that satisfies the following conditions: $V' \subseteq V$, $E' \subseteq E$, and $\forall v \in V', A'(v) \subseteq A(v), B'(v) = B(v)$.

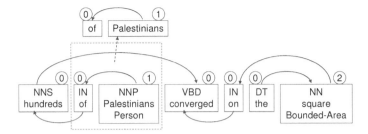

Figure 2.6. An example dependency parse tree representation. The subgraph represents a dependency relation feature between arg_1 *Palestinians* and *of*.

It can be shown that many features that have been explored in previous work on relation extraction can be transformed into this graphic representation. Figures 2.4, 2.5 and 2.6 show some examples.

This framework allows a systematic exploration of the feature space for relation extraction. To explore the feature space, Jiang and Zhai considered three levels of small unit features in increasing order of their complexity: unigram features, bigram features and trigram features. They found that a combination of features at different levels of complexity and from different sentence representations, coupled with task-oriented feature pruning, gave the best performance.

3.2 Kernel Methods

An important line of work for relation extraction is kernel-based classification. In machine learning, a kernel or kernel function defines the inner product of two observed instances represented in some underlying vector space. It can also be seen as a similarity measure for the observations. The major advantage of using kernels is that observed instances do not need to be explicitly mapped to the underlying vector space in order for their inner products defined by the kernel to be computed. We will use the convolution tree kernel to illustrate this idea below.

There are generally three types of kernels for relation extraction: sequence-based kernels, tree-based kernels and composite kernels.

3.2.1 Sequence-based Kernels. Bunescu and Mooney defined a simple kernel based on the shortest dependency paths between two arguments [13]. Two dependency paths are similar if they have the same length and they share many common nodes. Here a node can be represented by the word itself, its part-of-speech tag, or its entity type. Thus the two dependency paths "protestors → seized ← stations" and "troops → raided ← churches" have a non-zero similarity value because they can both be represented as "Person → VBD ← Facility," although they do not share any common word. A limitation of this kernel is that any two dependency paths with different lengths have a zero similarity.

In [14], Bunescu and Mooney introduced a subsequence kernel where the similarity between two sequences is defined over their similar subsequences. Specifically, each node in a sequence is represented by a feature vector and the similarity between two nodes is the inner product of their feature vectors. The similarity between two subsequences of the same length is defined as the product of the similarities of each pair of their nodes in the same position. The similarity of two sequences is then defined as a weighted sum of the similarities of all the subsequences of the same length from the two sequences. The weights are introduced to penalize long common subsequences. Bunescu and Mooney tested their subsequence kernel for protein-protein interaction detection.

3.2.2 Tree-based Kernels. Tree-based kernels use the same idea of using common substructures to measure similarities. Zelenko et al. defined a kernel on the constituency parse trees of relation instances [67]. The main motivation is that if two parse trees share many common subtree structures then the two relation instances are similar to each other. Culotta and Sorensen extended the idea to dependency parse trees [26]. Zhang et al. [68] further applied the convolution tree kernel initially proposed by Collins and Duffy [24] to relation extraction. This convolution tree kernel-based method was later further improved by Qian et al. [53] and achieved a state-of-the-art performance of around 77% of F-1 measure on the benchmark ACE 2004 data set.

We now briefly discuss the convolution tree kernels. As we explained earlier, a kernel function corresponds to an underlying vector space in which the observed instances can be represented. For convolution tree kernels, each dimension of this underlying vector space corresponds to a subtree. To map a constituency parse tree to a vector in this vector space, we simply enumerate all the subtrees contained in the parse tree. If a subtree i occurs k times in the parse tree, the value for the dimension corresponding to i is set to k. Only subtrees containing complete grammar production rules are considered. Figure 2.7 shows an example parse tree and all the subtrees under the NP "the company."

Formally, given two constituency parse trees T_1 and T_2, the convolution tree kernel K is defined as follows:

$$K(T_1, T_2) = \sum_{n_1 \in \mathcal{N}_1} \sum_{n_2 \in \mathcal{N}_2} \sum_i I_i(n_1) I_i(n_2). \qquad (2.6)$$

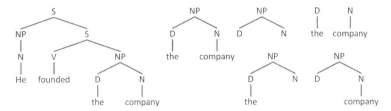

Figure 2.7. Left: The constituency parse tree of a simple sentence. Right: All the subtrees of the NP "the company" considered in convolution tree kernels.

Here $\mathcal{N}_1$ and $\mathcal{N}_2$ are the sets of all nodes in T_1 and T_2 respectively. i denotes a subtree in the feature space. $I_i(n)$ is 1 if subtree i is seen rooted at node n and 0 otherwise.

It is not efficient to directly compute K as defined in Equation 2.6. Instead, we can define $C(n_1, n_2) = \sum_i I_i(n_1) I_i(n_2)$. $C(n_1, n_2)$ can then be computed in polynomial time based on the following recursive property:

- If the grammar productions at n_1 and n_2 are different, then the value of $C(n_1, n_2)$ is 0.

- If the grammar productions at n_1 and n_2 are the same and n_1 and n_2 are pre-terminals, then $C(n_1, n_2)$ is 1. Here pre-terminals are nodes directly above words in a parse tree, e.g. the N, V and D in Figure 2.7.

- If the grammar productions at n_1 and n_2 are the same and n_1 and n_2 are not pre-terminals,

$$C(n_1, n_2) = \prod_{j=1}^{nc(n_1)} (1 + C(ch(n_1, j), ch(n_2, j))), \quad (2.7)$$

where $nc(n)$ is the number of child-nodes of n, and $ch(n, j)$ is the j-th child-node of n. Note that here $nc(n_1) = nc(n_2)$.

With this recursive property, convolution tree kernels can be efficiently computed in $O(|\mathcal{N}_1||\mathcal{N}_2|)$ time.

3.2.3 Composite Kernels. It is possible to combine different kernels into a composite kernel. This is when we find it hard to include all the useful features into a single kernel. Zhao and Grishman defined several syntactic kernels such as argument kernel and dependency path kernel before combing them into a composite kernel [70]. Zhang et al. combined an entity kernel with the convolution tree kernel to form a composite kernel [69].

3.3 Weakly Supervised Learning Methods

Both feature-based and kernel-based classification methods for relation extraction rely on a large amount of training data, which is expensive to obtain. A solution to this problem is weakly supervised learning methods that work with much less training data. The most notable weakly supervised method for relation extraction is bootstrapping, which starts from a small set of seed relation instances and iteratively learns more relation instances and extraction patterns. It has been widely explored [12, 3]. More recently, another learning paradigm called distant supervision has been proposed to make use of a large number of known relation instances in existing large knowledge bases to create training data [50]. For both bootstrapping and distant supervision, noisy training data is automatically generated. To achieve good performance, careful feature selection and pattern filtering need to be carried out.

3.3.1 Bootstrapping. A representative work on bootstrapping for relation extraction is the Snowball system developed by Agichtein and Gravano [3], which improved over an earlier system called DIPRE developed by Brin [12]. The idea behind Snowball is simple. We start with a set of seed entity pairs that are related through the target relation. For example, if the target relation is HeadquarteredIn, we may use seed pairs such as ⟨*Microsoft, Redmond*⟩, ⟨*Google, Mountain View*⟩ and ⟨*Facebook, Palo Alto*⟩. Given a large corpus, we then look for co-occurrences of these entity pairs within close proximity. The assumption is that if two entities related through the target relation co-occur closely, the context in which they co-occur is likely to be a pattern for the target relation. For example, we may find sentence fragments such as "Google's headquarters in Mountain View" and "Redmond-based Microsoft" and extract patterns like "ORG*'s headquarters in* LOC" and "LOC-*based* ORG." With these patterns, we can search the corpus and find more ⟨ORG, LOC⟩ entity pairs that have the HeadquarteredIn relation. We add these entity pairs to the set of seed relation instances and repeat the process. More patterns and entity pairs are added to the results until a certain condition is satisfied.

An important step in bootstrapping methods is to evaluate the quality of extraction patterns so as not to include many noisy patterns during the extraction process. For example, from the seed entity pair ⟨*Google, Mountain View*⟩ we may also find "Google, Mountain View" in the corpus. However, the pattern "ORG, LOC" is not a reliable one and thus should not be used. Heuristic methods have been proposed to judge the quality of an extraction pattern. Usually two factors are considered, coverage and precision. Coverage is related to the percentage of true relation instances that can be discovered by the pattern. Precision is related to the percentage of correct relation instances among all the relation instances discovered by the pattern.

3.3.2 Distant Supervision. In bootstrapping only a small set of seed entity pairs is used. With the growth of the social Web, much human knowledge has been contributed by a large crowd of users and stored in knowledge bases. A well-known example is Wikipedia. Another example is Freebase, a knowledge base that stores structured human knowledge such as entity relations [11]. With such freely available knowledge, it becomes possible to use a large set of entity pairs known to have a target relation to generate training data. Mintz et al. proposed distant supervision for relation extraction based on this idea [50]. They assume that if two entities participate in a relation, any sentence that contain these two entities express that relation. Because this assumption does not always hold, Mintz et al. use features extracted from different sentences containing the entity pair to create a richer feature vector that is supposed to be more reliable. They define lexical, syntactic and named entity tag features. They use standard multi-class logistic regression as the classification algorithm. Their experiments show that this method can reach almost 70% of precision based on human judgment. Nguyen and Moschitti further used knowledge from both YAGO and Wikipedia documents for distant supervision and achieved around 74% F-1 measure [51].

4. Unsupervised Information Extraction

In Section 2 and Section 3, we discussed named entity recognition and relation extraction where the entity types and relation types are well defined in advance based on the application. A large amount of labeled training data is also required in order to learn a good named entity recognizer or relation extractor. However, both defining the structures for the information to be extracted and annotating documents according to the defined structures require human expertise and are time consuming. To alleviate this problem, recently there has been an increasing amount of interest in unsupervised information extraction from large corpora.

In this section we review some recent studies along this line. We first discuss relation discovery and template induction where the goal is to discover salient relation types or templates for a given domain. The key idea is to cluster entities or entity pairs based on their lexico-syntactic contextual features. We then discuss open information extraction where the goal is to extract *any* type of relation from a large, diverse corpus such as the Web.

4.1 Relation Discovery and Template Induction

In Section 3 we discussed relation extraction when the types of relations to be extracted are known in advance. There are also cases where we do not have any specific relation types in mind but would like to discover salient relation types from a given corpus. For example, given a set of articles reporting hurricane events, it would be useful if we could automatically discover that one of the most important relations for this domain is the `hit` relation between a hurricane and the place being hit.

Shinyama and Sekine first proposed to study this problem, which they referred to as Unrestricted Relation Discovery [60]. They started by collecting a large number of news articles from different news sources on the Web. They then used simple clustering based on lexical similarity to find articles talking about the same event. In this way they could enrich the feature representation of an entity using its multiple occurrences in different articles. Next they performed syntactic parsing and extracted named entities from these articles. Each named entity could then be represented by a set of syntactic patterns as its features. For example, a pattern may indicate that the entity is the subject of the verb *hit*. Finally, they clustered pairs of entities co-occurring in the same article using their feature representations. The end results were tables in which rows corresponded to different articles and columns corresponded to different roles in a relation. They were able to achieve around 75% of accuracy for the discovered tables.

Rosenfeld and Feldman formulated unsupervised relation discovery in a more general way [57]. It is assumed that the input of the problem consists of entity pairs together with their contexts. An unsupervised relation discovery algorithm clusters these entity pairs into disjoint groups where each group represents a single semantic relation. There is also a garbage cluster to capture unrelated entity pairs or unimportant relations. The contexts for each entity pair consist of the contexts of each entity and the contexts of the two entities' co-occurrences. An entity pair can be represented by a set of features derived from the contexts. Rosenfeld and Feldman considered only surface pattern features. For example, "arg_1, based in arg_2" is a pattern to capture a co-occurrence context between the two entities. For clustering, Rosenfeld and Feldman considered hierarchical agglomerative clustering and K-means clustering. Their method was able to discover relations such as `CityOfState` and `EmployedIn`.

While relation discovery considers binary relations only, a more complex task is to automatically induce an information extraction template, which may contain multiple slots playing different semantic roles. The most straightforward solution is to identify candidates of role fillers first and then cluster these candidates into clusters. However, this simplified clustering approach does not consider an important observation, which is that a single document tends to cover different slots. To remedy this problem, Marx et al. proposed a cross-component clustering algorithm for unsupervised information extraction [47]. The algorithm assigns a candidate from a document to a cluster based on the candidate's feature similarity with candidates from *other documents* only. In other words, the algorithm prefers to separate candidates from the same document into different clusters. Leung et al. proposed a generative model to capture the same intuition [43]. Specifically, they assume a prior distribution over the cluster labels of candidates in the same document where the prior prefers a diversified label assignment. Their experiments show that clustering results are better with this prior than without using the prior.

The aforementioned two studies assume a single template and do not automatically label the discovered slots. Chambers and Jurafsky presented a complete method that is able to discover multiple templates from a corpus and give meaningful labels to discovered slots [17]. Specifically, their method performs two steps of clustering where the first clustering step groups lexical patterns that are likely to describe the same type of events and the second clustering step groups candidate role fillers into slots for each type of events. A slot can be labeled using the syntactic patterns of the corresponding slot fillers. For example, one of the slots discovered by their method for the bombing template is automatically labeled as "Person/Organization who raids, questions, discovers, investigates, diffuses, arrests." A human can probably infer from the description that this refers to the `police` slot.

4.2 Open Information Extraction

Relation discovery and template induction usually work on a corpus from a single domain, e.g. articles describing terrorism events, because the goal is to discover the most salient relations from such a domain-specific corpus. In some cases, however, our goal is to find all the potentially useful facts from a large and diverse corpus such as the Web. This is the focus of open information extraction, first introduced by Banko et al. [6].

Open information extraction does not assume any specific target relation type. It makes a single pass over the corpus and tries to extract as many relations as possible. Because no relation type is specified in advance, part of the extraction results is a phrase that describes the relation extracted. In other words, open information extraction generates $\langle \texttt{arg}_1, \texttt{rel}, \texttt{arg}_2 \rangle$ tuples.

In [7], Banko and Etzioni introduced an unlexicalized CRF-based method for open information extraction. The method is based on the observation that although different relation types have very different semantic meanings, there exists a small set of syntactic patterns that cover the majority of semantic relation mentions. It is therefore possible to train a relation extraction model that extracts arbitrary relations. The key is not to include lexical features in the model.

Later work on open information extraction introduced more heuristics to improve the quality of the extracted relations. In [29], for example, Fader et al. proposed the following two heuristics: (1) A multi-word relation phrase must begin with a verb, end with a preposition, and be a contiguous sequence of words in the sentence. (2) A binary relation phrase ought to appear with at least a minimal number of distinct argument pairs in a large corpus. It is found that the two heuristics can effectively lead to better extraction results.

5. Evaluation

To evaluate information extraction systems, manually annotated documents have to be created. For domain-specific information extraction systems, the annotated documents have to come from the target domain. For example, to evaluate gene and protein name extraction, biomedical documents such as PubMed abstracts are used. But if the purpose is to evaluate general information extraction techniques, standard benchmark data sets can be used. Commonly used evaluation data sets for named entity recognition include the ones from MUC [33], CoNLL-2003 [63] and ACE [1]. For relation extraction, ACE data sets are usually used.

The typical evaluation metrics for information extraction are precision, recall and F-1 scores. Precision measures the percentage of correct instances among the identified positive instances. Recall measures the percentage of correct instances that can be identified among all the positive instances. F-1 is the geometric mean of precision and recall.

For named entity recognition, strictly speaking a correctly identified named entity must satisfy two criteria, namely, correct entity boundary and correct entity type. Most evaluation is based on the exact match of entity boundaries. However, it is worth nothing that in some cases credit should also be given to partial matches, e.g. when the goal is only to tell whether an entity is mentioned in a document or a sentence [64].

For relation extraction, as we have mentioned, there are two levels of extraction, corpus-level and mention-level. While evaluation at mention level requires annotated relation mention instances, evaluation at corpus level requires only truly related entity pairs, which may be easier to obtain or annotate than relation mentions.

Currently, the state-of-the-art named entity recognition methods can achieve around 90% of F-1 scores when trained and tested on the same domain [63]. It is generally observed that person entities are easier to extract, followed by locations and then organizations. It is important to note that when there is domain change, named entity recognition performance can drop substantially. There have been several studies addressing the domain adaptation problem for named entity recognition (e.g. [36, 5]).

For relation extraction, the state-of-the-art performance is lower than that of named entity recognition. On the ACE 2004 benchmark data set, for example, the best F-1 score is around 77% for the seven major relation types [53].

6. Conclusions and Summary

Information extraction is an important text mining problem and has been extensively studied in areas such as natural language processing, information retrieval and Web mining. In this chapter we reviewed some representative work on information extraction, in particular work on named entity recognition and relation extraction. Named entity recognition aims at finding names of entities such as people, organizations and locations. State-of-the-art solutions to named entity recognition rely on statistical sequence labeling algorithms such as maximum entropy Markov models and conditional random fields. Relation extraction is the task of finding the semantic relations between entities from text. Current state-of-the-art methods use carefully designed features or kernels and standard classification to solve this problem.

Although supervised learning has been the dominating approach to information extraction, weakly supervised methods have also drawn much attention. Bootstrapping is a major technique for semi-supervised relation extraction. More recently, with large amounts of knowledge made available in online knowledge bases, distant supervision provides a new paradigm of learning without training data.

Unsupervised information extraction aims to automatically induce the structure of the information to be extracted such as the relation types and the templates. Clustering is the main technique used for unsupervised information extraction.

With the fast growth of textual data on the Web, it is expected that future work on information extraction will need to deal with even more diverse and noisy text. Weakly supervised and unsupervised methods will play a larger role in information extraction. The various user-generated content on the Web such as Wikipedia articles will also become important resources to provide some kind of supervision.

References

[1] Automatic content extraction (ACE) evaluation. http://www.itl.nist.gov/iad/mig/tests/ace/.

[2] BioCreAtIvE. http://www.biocreative.org/.

[3] Eugene Agichtein and Luis Gravano. Snowball: Extracting relations from large plain-text collections. In *Proceedings of the 5th ACM Conference on Digital Libraries*, pages 85–94, 2000.

[4] Douglas E. Appelt, Jerry R. Hobbs, John Bear, David Israel, and Mabry Tyson. FASTUS: A finite-state processor for information extraction from real-world text. In *Proceedings of the 13th International Joint Conference on Artificial Intelligence*, 1993.

[5] Andrew Arnold, Ramesh Nallapati, and William W. Cohen. Exploiting feature hierarchy for transfer learning in named entity recognition. In *Proceedings of the 46th Annual Meeting of the Association for Computational Linguistics*, pages 245–253, 2008.

[6] Michele Banko, Michael J. Cafarella, Stephen Soderland, Matthew Broadhead, and Oren Etzioni. Open information extraction from the Web. In *Proceedings of the 20th International Joint Conference on Artificial Intelligence*, pages 2670–2676, 2007.

[7] Michele Banko and Oren Etzioni. The tradeoffs between open and traditional relation extraction. In *Proceedings of the 46th Annual Meeting of the Association for Computational Linguistics*, pages 28–36, 2008.

[8] Oliver Bender, Franz Josef Och, and Hermann Ney. Maximum entropy models for named entity recognition. In *Proceedings of the 7th Conference on Natural Language Learning*, 2003.

[9] Adam L. Bergert, Vincent J. Della Pietra, and Stephen A. Della Pietra. A maximum entropy approach to natural language processing. *Computational Linguistics*, 22(1):39–71, March 1996.

[10] Daniel M. Bikel, Scott Miller, Richard Schwartz, and Ralph Weischedel. Nymble: a high-performance learning name-finder. In

Proceedings of the 5th Conference on Applied Natural Language Processing, pages 194–201, 1997.

[11] Kurt Bollacker, Colin Evans, Praveen Paritosh, Tim Sturge, and Jamie Taylor. Freebase: a collaboratively created graph database for structuring human knowledge. In *Proceedings of the 2008 ACM SIGMOD International Conference on Management of Data*, pages 1247–1250, 2008.

[12] Sergey Brin. Extracting patterns and relations from the World Wide Web. In *Proceedings of the 1998 International Workshop on the Web and Databases*, 1998.

[13] Razvan Bunescu and Raymond Mooney. A shortest path dependency kernel for relation extraction. In *Proceedings of the Human Language Technology Conference and the Conference on Empirical Methods in Natural Language Processing*, pages 724–731, 2005.

[14] Razvan Bunescu and Raymond Mooney. Subsequence kernels for relation extraction. In *Advances in Neural Information Processing Systems 18*, pages 171–178. 2006.

[15] Richard H. Byrd, Jorge Nocedal, and Robert B. Schnabel. Representations of quasi-newton matrices and their use in limited memory methods. *Journal of Mathematical Programming*, 63(2):129–156, January 1994.

[16] Mary Elaine Califf and Raymond J. Mooney. Relational learning of pattern-match rules for information extraction. In *Proceedings of the 16th National Conference on Artificial Intelligence and the 11th Innovative Applications of Artificial Intelligence Conference*, pages 328–334, 1999.

[17] Nathanael Chambers and Dan Jurafsky. Template-based information extraction without the templates. In *Proceedings of the 49th Annual Meeting of the Association for Computational Linguistics: Human Language Technologies*, pages 976–986, 2011.

[18] Yee Seng Chan and Dan Roth. Exploiting background knowledge for relation extraction. In *Proceedings of the 23rd International Conference on Computational Linguistics*, pages 152–160, 2010.

[19] Yee Seng Chan and Dan Roth. Exploiting syntactico-semantic structures for relation extraction. In *Proceedings of the 49th Annual Meeting of the Association for Computational Linguistics*, pages 551–560, 2011.

[20] Chia-Hui Chang, Mohammed Kayed, Moheb Ramzy Girgis, and Khaled F. Shaalan. A survey of Web information extraction sys-

tems. *IEEE Transactions on Knowledge and Data Engineering*, 18(10):1411–1428, October 2006.

[21] Tao Cheng, Xifeng Yan, and Kevin Chen-Chuan Chang. Supporting entity search: a large-scale prototype search engine. In *Proceedings of the 2007 ACM SIGMOD International Conference on Management of Data*, pages 1144–1146, 2007.

[22] Hai Leong Chieu and Hwee Tou Ng. Named entity recognition with a maximum entropy approach. In *Proceedings of the Seventh Conference on Natural Language Learning*, pages 160–163, 2003.

[23] Fabio Ciravegna. Adaptive information extraction from text by rule induction and generalisation. In *Proceedings of the 17th International Joint Conference on Artificial Intelligence - Volume 2*, pages 1251–1256, 2001.

[24] Michael Collins and Nigel Duffy. Convolution kernels for natural language. In *Advances in Neural Information Processing Systems 13*. 2001.

[25] Valter Crescenzi, Giansalvatore Mecca, and Paolo Merialdo. RoadRunner: Towards automatic data extraction from large Web sites. In *Proceedings of the 27th International Conference on Very Large Data Bases*, pages 109–118, 2001.

[26] Aron Culotta and Jeffrey Sorensen. Dependency tree kernels for relation extraction. In *Proceedings of the 42nd Annual Meeting of the Association for Computational Linguistics*, pages 423–429, 2004.

[27] James R. Curran and Stephen Clark. Language independent NER using a maximum entropy tagger. In *Proceedings of the 7th Conference on Natural Language Learning*, 2003.

[28] Gerald DeJong. Prediction and substantiation: A new approach to natural language processing. *Cognitive Science*, 3:251–173, 1979.

[29] Anthony Fader, Stephen Soderland, and Oren Etzioni. Identifying relations for open information extraction. In *Proceedings of the 2011 Conference on Empirical Methods in Natural Language Processing*, pages 1535–1545, 2011.

[30] Jenny Finkel, Shipra Dingare, Christopher D. Manning, Malvina Nissim, Beatrice Alex, and Claire Grover. Exploring the boundaries: gene and protein identification in biomedical text. *BMC Bioinformatics*, 6(Suppl 1)(S5), 2005.

[31] Sergio Flesca, Giuseppe Manco, Elio Masciari, Eugenio Rende, and Andrea Tagarelli. Web wrapper induction: a brief survey. *AI Communications*, 17(2):57–61, April 2004.

[32] Ralph Grishman, John Sterling, and Catherine Macleod. New York University: Description of the PROTEUS system as used for MUC-3. In *Proceedings of the 3rd Message Understadning Conference*, pages 183–190, 1991.

[33] Ralph Grishman and Beth Sundheim. Message understanding conference-6: A brief history. In *Proceedings of the 16th International Conference on Computational Linguistics*, pages 466–471, 1996.

[34] Guoping Hu, Jingjing Liu, Hang Li, Yunbo Cao, Jian-Yun Nie, and Jianfeng Gao. A supervised learning approach to entity search. In *Proceedings of the 3rd Asia Information Retrieval Symposium*, pages 54–66, 2006.

[35] Hideki Isozaki and Hideto Kazawa. Efficient support vector classifiers for named entity recognition. In *Proceedings of the 19th International Conference on Computational Linguistics*, 2002.

[36] Jing Jiang and ChengXiang Zhai. Exploiting domain structure for named entity recognition. In *Proceedings of the Human Language Technology Conference of the North American Chapter of the Association for Computational Linguistics*, pages 74–81, 2006.

[37] Jing Jiang and ChengXiang Zhai. A systematic exploration of the feature space for relation extraction. In *Proceedings of the Human Language Technology Conference of the North American Chapter of the Association for Computational Linguistics*, pages 113–120, 2007.

[38] Nanda Kambhatla. Combining lexical, syntactic, and semantic features with maximum entropy models for extracting relations. In *The Companion Volume to the Proceedings of 42st Annual Meeting of the Association for Computational Linguistics*, pages 178–181, 2004.

[39] Dan Klein, Joseph Smarr, Huy Nguyen, and Christopher D. Manning. Named entity recognition with character-level models. In *Proceedings of the 7th Conference on Natural Language Learning*, 2003.

[40] Nicholas Kushmerick, Daniel S. Weld, and Robert Doorenbos. Wrapper induction for information extraction. In *Proceedings of the 15th International Joint Conference on Artificial Intelligence*, 1997.

[41] John D. Lafferty, Andrew McCallum, and Fernando C. N. Pereira. Conditional random fields: Probabilistic models for segmenting and

labeling sequence data. In *Proceedings of the 18th International Conference on Machine Learning*, pages 282–289, 2001.

[42] Wendy Lehnert, Claire Cardie, Divid Fisher, Ellen Riloff, and Robert Williams. University of Massachusetts: Description of the CIRCUS system as used for MUC-3. In *Proceedings of the 3rd Message Understadning Conference*, pages 223–233, 1991.

[43] Cane Wing-ki Leung, Jing Jiang, Kian Ming A. Chai, Hai Leong Chieu, and Loo-Nin Teow. Unsupervised information extraction with distributional prior knowledge. In *Proceedings of the 2011 Conference on Empirical Methods in Natural Language Processing*, pages 814–824, 2011.

[44] Xin Li and Dan Roth. Learning question classifiers. In *Proceedings of the 19th International Conference on Computational Linguistics*, pages 1–7, 2002.

[45] Liu Ling, Calton Pu, and Wei Han. XWRAP: An XML-enabled wrapper construction system for Web information sources. In *Proceedings of the 16th International Conference on Data Engineering*, pages 611–621, 2000.

[46] Robert Malouf. A comparison of algorithms for maximum entropy parameter estimation. In *Proceedings of the 6th Conference on Natural Language Learning*, 2002.

[47] Zvika Marx, Ido Dagan, and Eli Shamir. Cross-component clustering for template learning. In *Proceedings of the 2002 ICML Workshop on Text Learning*, 2002.

[48] Andrew McCallum, Dayne Freitag, and Fernando C. N. Pereira. Maximum entropy Markov models for information extraction and segmentation. In *Proceedings of the 17th International Conference on Machine Learning*, pages 591–598, 2000.

[49] Andrew McCallum and Wei Li. Early results for named entity recognition with conditional random fields, feature induction and web-enhanced lexicons. In *Proceedings of the 7th Conference on Natural Language Learning*, 2003.

[50] Mike Mintz, Steven Bills, Rion Snow, and Daniel Jurafsky. Distant supervision for relation extraction without labeled data. In *Proceedings of the Joint Conference of the 47th Annual Meeting of the Association for Computational Linguistics and the 4th International Joint Conference on Natural Language Processing of the AFNLP*, pages 1003–1011, 2009.

[51] Truc Vien T. Nguyen and Alessandro Moschitti. End-to-end relation extraction using distant supervision from external semantic

repositories. In *Proceedings of the 49th Annual Meeting of the Association for Computational Linguistics*, pages 277–282, 2011.

[52] Tomoko Ohta, Yuka Tateisi, and Jin-Dong Kim. The GENIA corpus: an annotated research abstract corpus in molecular biology domain. In *Proceedings of the 2nd International Conference on Human Language Technology Research*, pages 82–86, 2002.

[53] Longhua Qian, Guodong Zhou, Fang Kong, Qiaoming Zhu, and Peide Qian. Exploiting constituent dependencies for tree kernel-based semantic relation extraction. In *Proceedings of the 22nd International Conference on Computational Linguistics*, pages 697–704, 2008.

[54] Lawrence R. Rabiner. A tutorial on hidden Markov models and selected applications in speech recognition. 77, 77(2):257–286, 1989.

[55] Lance A. Ramshaw and Mitch P. Marcus. Text chunking using transformation-based learning. In *Proceedings of the 3rd Workshop on Very Large Corpora*, pages 82–94, 1995.

[56] Lisa F. Rau. Extracting company names from text. In *Proceedings of the 7th IEEE Conference on Artificial Intelligence Applications*, pages 29–32, 1991.

[57] Benjamin Rosenfeld and Ronen Feldman. Clustering for unsupervised relation identification. In *Proceedings of the 16th ACM conference on Conference on Information and Knowledge Management*, pages 411–418, 2007.

[58] Sunita Sarawagi and William W. Cohen. Semi-markov conditional random fields for information extraction. In *Advances in Neural Information Processing Systems 17*, pages 1185–1192. 2005.

[59] Burr Settles. Biomedical named entity recognition using conditional random fields and rich feature sets. In *Proceedings of the International Joint Workshop on Natural Language Processing in Biomedicine and Its Applications*, pages 104–107, 2004.

[60] Yusuke Shinyama and Satoshi Sekine. Preemptive information extraction using unrestricted relation discovery. In *Proceedings of the Human Language Technology Conference of the North American Chapter of the Association for Computational Linguistics*, pages 304–311, 2006.

[61] Stephen Soderland. Learning information extraction rules for semistructured and free text. *Machine Learning*, 34(1-3):233–272, February 1999.

[62] Stephen Soderland, David Fisher, Jonathan Aseltine, and Wendy Lehnert. CRYSTAL inducing a conceptual dictionary. In *Proceed-*

ings of the 14th International Joint Conference on Artificial Intelligence, pages 1314–1319, 1995.

[63] Erik F. Tjong Kim Sang and Fien De Meulder. Introduction to the CoNLL-2003 shared task: Language-independent named entity recognition. In *Proceedings of the 7th Conference on Natural Language Learning*, pages 142–147, 2003.

[64] Richard Tzong-Han Tsai, Shih-Hung Wu, Wen-Chi Chou, Yu-Chun Lin, Ding He, Jieh Hsiang, Ting-Yi Sung, and Wen-Lian Hsu. Various criteria in the evaluation of biomedical named entity recognition. *BMC Bioinformatics*, 7(92), 2006.

[65] Vladimir Vapnik. *Statistical Learning Theory*. John Wiley & Sons, 2008.

[66] Fei Wu and Daniel S. Weld. Open information extraction using Wikipedia. In *Proceedings of the 48th Annual Meeting of the Association for Computational Linguistics*, pages 118–127, 2010.

[67] Dmitry Zelenko, Chinatsu Aone, and Anthony Richardella. Kernel methods for relation extraction. *Journal of Machine Learning Research*, 3:1083–1106, February 2003.

[68] Min Zhang, Jie Zhang, and Jian Su. Exploring syntactic features for relation extraction using a convolution tree kernel. In *Proceedings of the Human Language Technology Conference of the North American Chapter of the Association for Computational Linguistics*, pages 288–295, 2006.

[69] Min Zhang, Jie Zhang, Jian Su, and GuoDong Zhou. A composite kernel to extract relations between entities with both flat and structured features. In *Proceedings of the 21st International Conference on Computational Linguistics and the 44th Annual Meeting of the Association for Computational Linguistics*, pages 825–832, 2006.

[70] Shubin Zhao and Ralph Grishman. Extracting relations with integrated information using kernel methods. In *Proceedings of the 43rd Annual Meeting of the Association for Computational Linguistics*, pages 419–426, 2005.

[71] GuoDong Zhou, Jian Su, Jie Zhang, and Min Zhang. Exploring various knowledge in relation extraction. In *Proceedings of the 43rd Annual Meeting of the Association for Computational Linguistics*, pages 427–434, 2005.

Chapter 3

A SURVEY OF TEXT SUMMARIZATION TECHNIQUES

Ani Nenkova
University of Pennsylvania
nenkova@seas.upenn.edu

Kathleen McKeown
Columbia University
kathy@cs.columbia.edu

Abstract Numerous approaches for identifying important content for automatic text summarization have been developed to date. Topic representation approaches first derive an intermediate representation of the text that captures the topics discussed in the input. Based on these representations of topics, sentences in the input document are scored for importance. In contrast, in indicator representation approaches, the text is represented by a diverse set of possible indicators of importance which do not aim at discovering topicality. These indicators are combined, very often using machine learning techniques, to score the importance of each sentence. Finally, a summary is produced by selecting sentences in a greedy approach, choosing the sentences that will go in the summary one by one, or globally optimizing the selection, choosing the best set of sentences to form a summary. In this chapter we give a broad overview of existing approaches based on these distinctions, with particular attention on how representation, sentence scoring or summary selection strategies alter the overall performance of the summarizer. We also point out some of the peculiarities of the task of summarization which have posed challenges to machine learning approaches for the problem, and some of the suggested solutions[1].

---

[1]Portions of this chapter have already appeared in our more detailed overview of summarization research [67]. The larger manuscript includes sections on generation techniques

Keywords: Extractive text summarization, topic representation, machine learning for summarization

1. How do Extractive Summarizers Work?

Summarization systems need to produce a concise and fluent summary conveying the key information in the input. In this chapter we constrain our discussion to extractive summarization systems for short, paragraph-length summaries and explain how these systems perform summarization. These summarizers identify the most important sentences in the input, which can be either a single document or a cluster of related documents, and string them together to form a summary. The decision about what content is important is driven primarily by the input to the summarizer.

The choice to focus on extractive techniques leaves out the large body of text-to-text generation approaches developed for abstractive summarization, but allows us to focus on some of the most dominant approaches which are easily adapted to take users' information need into account and work for both single- and multi-document inputs. Moreover, by examining the stages in the operation of extractive summarizers we are able to point out commonalities and differences in summarization approaches which relate to critical components of a system and could explain the advantages of certain techniques over others.

In order to better understand the operation of summarization systems and to emphasize the design choices system developers need to make, we distinguish three relatively independent tasks performed by virtually all summarizers: creating an intermediate representation of the input which captures only the key aspects of the text, scoring sentences based on that representation and selecting a summary consisting of several sentences.

Intermediate representation Even the simplest systems derive some intermediate representation of the text they have to summarize and identify important content based on this representation. *Topic representation* approaches convert the text to an intermediate representation interpreted as the topic(s) discussed in the text. Some of the most popular summarization methods rely on topic representations and this class of approaches exhibits an impressive variation in sophistication and representation power. They include frequency, TF.IDF and topic word approaches in which the topic representation consists of a simple table

---

summarization, evaluation issues and genre specific summarization which we do not address in this chapter. http://dx.doi.org/10.1561/1500000015

of words and their corresponding weights, with more highly weighted words being more indicative of the topic; lexical chain approaches in which a thesaurus such as WordNet is used to find topics or concepts of semantically related words and then give weight to the concepts; latent semantic analysis in which patterns of word co-occurrence are identified and roughly construed as topics, as well as weights for each pattern; full blown Bayesian topic models in which the input is represented as a mixture of topics and each topic is given as a table of word probabilities (weights) for that topic. *Indicator representation* approaches represent each sentence in the input as a list of indicators of importance such as sentence length, location in the document, presence of certain phrases, etc. In graph models, such as LexRank, the entire document is represented as a network of inter-related sentences.

Score sentences Once an intermediate representation has been derived, each sentence is assigned a score which indicates its importance. For topic representation approaches, the score is commonly related to how well a sentence expresses some of the most important topics in the document or to what extent it combines information about different topics. For the majority of indicator representation methods, the weight of each sentence is determined by combining the evidence from the different indicators, most commonly by using machine learning techniques to discover indicator weights. In LexRank, the weight of each sentence is derived by applying stochastic techniques to the graph representation of the text.

Select summary sentences Finally, the summarizer has to select the best combination of important sentences to form a paragraph length summary. In the *best n* approaches, the top n most important sentences which combined have the desired summary length are selected to form the summary. In *maximal marginal relevance* approaches, sentences are selected in an iterative greedy procedure. At each step of the procedure the sentence importance score is recomputed as a linear combination between the original importance weight of the sentence and its similarity with already chosen sentences. Sentences that are similar to already chosen sentences are dispreferred. In *global selection* approaches, the optimal collection of sentences is selected subject to constraints that try to maximize overall importance, minimize redundancy, and, for some approaches, maximize coherence.

There are very few inherent dependencies between the three processing steps described above and a summarizer can incorporate any combination of specific choices on how to perform the steps. Changes in the way a specific step is performed can markedly change the performance of the summarizer, and we will discuss some of the known differences as we introduce the traditional methods.

In ranking the importance of sentences for summaries, other factors also come into play. If we have information about the context in which the summary is generated, this can help in determining importance. Context can take the form of information about user needs, often presented through a query. Context can include the environment in which an input document is situated, such as the links which point to a web page. Another factor which affects sentence ranking is the genre of a document. Whether the input document is a news article, an email thread, a web page or a journal article influences the strategies used to select sentences.

We begin with a discussion of topic representation approaches in Section 2. In these approaches the independence between the methods for deriving the intermediate representation and those for scoring sentences is most clear and we emphasize the range of choices for each as we discuss individual approaches. In Section 3 we discuss approaches that focus attention on the contextual information necessary for determining sentence importance rather than the topic representation itself. We follow with a presentation of indicator representation approaches in Section 4. We then discuss approaches to selecting the sentences of a summary in Section 5 before concluding.

2. Topic Representation Approaches

Topic representation approaches vary tremendously in sophistication and encompass a family of methods for summarization. Here we present some of the most widely applied topic representation approaches, as well as those that have been gaining popularity because of their recent successes.

2.1 Topic Words

In remarkably early work on text summarization [53], Luhn proposed the use of frequency thresholds to identify descriptive words in a document to be summarized, a simple representation of the document's topic. The descriptive words in his approach exclude the most frequent words in the document, which are likely to be determiners, prepositions, or domain specific words, as well as those occurring only a few times A modern statistical version of Luhn's idea applies the log-likelihood ratio test [22] for identification of words that are highly descriptive of the input. Such words have been traditionally called "topic signatures" in the summarization literature [46]. The use of topic signatures words as representation of the input has led to high performance in selecting important content for multi-document summarization of news [15, 38].

Topic signatures are words that occur often in the input but are rare in other texts, so their computation requires counts from a large collection of documents in addition to the input for summarization. One of the key strengths of the log-likelihood ratio test approach is that it provides a way of setting a threshold to divide all words in the input into either descriptive or not. The decision is made based on a test for statistical significance, to large extent removing the need for the arbitrary thresholds in the original approach.

Information about the frequency of occurrence of words in a large background corpus is necessary to compute the statistic on the basis of which topic signature words are determined. The likelihood of the input I and the background corpus is computed under two assumptions: (H1) that the probability of a word in the input is the same as in the background B or (H2) that the word has a different, higher probability, in the input than in the background.

H1: $P(w|I) = P(w|B) = p$ (w is not descriptive)

H2: $P(w|I) = p_I$ and $P(w|B) = p_B$ and $p_I > p_B$ (w is descriptive)

The likelihood of a text with respect to a given word of interest, w, is computed via the binomial distribution formula. The input and the background corpus are treated as a sequence of words w_i: $w_1 w_2 \ldots w_N$. The occurrence of each word is a Bernoulli trial with probability p of success, which occurs when $w_i = w$. The overall probability of observing the word w appearing k times in the N trials is given by the binomial distribution

$$b(k, N, p) = \binom{N}{k} p^k (1-p)^{N-k} \tag{3.1}$$

For H1, the probability p is computed from the input and the background collection taken together. For H2, p_1 is computed from the input, p_2 from the background, and the likelihood of the entire data is equal to the product of the binomial for the input and that for the background. More specifically, the likelihood ratio is defined as

$$\lambda = \frac{b(k, N, p)}{b(k_I, N_I, p_I) \cdot b(k_B, N_B, p_B)} \tag{3.2}$$

where the counts with subscript I are computed only from the input to the summarizer and those with index B are computed over the background corpus.

The statistic equal to $-2 \log \lambda$ has a known statistical distribution (χ^2), which can be used to determine which words are topic signatures.

Topic signature words are those that have a likelihood statistic greater than what one would expect by chance. The probability of obtaining a given value of the statistic purely by chance can be looked up in a χ^2 distribution table; for instance a value of 10.83 can be obtained by chance with probability of 0.001.

The importance of a sentence is computed as the number of topic signatures it contains or as the proportion of topic signatures in the sentence. Both of these sentence scoring functions are based on the same topic representation, the scores they assign to sentences may be rather different. The first approach is likely to score longer sentences higher, simply because they contain more words. The second approach favors density of topic words.

2.2 Frequency-driven Approaches

There are two potential modifications that naturally come to mind when considering the topic words approach. The weights of words in topic representations need not be binary (either 1 or 0) as in the topic word approaches. In principle it would even be beneficial to be able to compare the continuous weights of words and determine which ones are more related to the topic. The approaches we present in this section— word probability and TF.IDF—indeed assign non-binary weights related on the number of occurrences of a word or concept. Research has already shown that the binary weights give more stable indicators of sentence importance than word probability and TF.IDF [34]. Nonetheless we overview these approaches because of their conceptual simplicity and reasonable performance. We also describe the lexical chains approach to determining sentence importance. In contrast to most other approaches, it makes use of WordNet, a lexical database which records semantic relations between words. Based on the information derived from WordNet, lexical chain approaches are able to track the prominence, indicated by frequency, of different topics discussed in the input.

Word probability is the simplest form of using frequency in the input as an indicator of importance[2]. The probability of a word w, $p(w)$ is computed from the input, which can be a cluster of related documents or a single document. It is calculated as the number of occurrences of a word, $c(w)$ divided by the number of all words in the input, N:

---

[2]Raw frequency would be even simpler, but this measure is too strongly influenced by document length. A word appearing twice in a 10 word document may be important, but not necessarily so in a 1000 word document. Computing word probability makes an adjustment for document length.

$$p(w) = \frac{c(w)}{N} \tag{3.3}$$

SUMBASIC is one system developed to operationalize the idea of using frequency for sentence selection. It relies only on word probability to calculate importance [94]. For each sentence S_j in the input it assigns a weight equal to the average probability $p(w_i)$ of the content words in the sentence[3], estimated from the input for summarization:

$$Weight(S_j) = \frac{\sum_{w_i \in S_j} p(w_i)}{|\{w_i | w_i \in S_j\}|} \tag{3.4}$$

Then, in a greedy fashion, SUMBASIC picks the best scoring sentence that contains the word that currently has the highest probability. This selection strategy assumes that at each point when a sentence is selected, a single word—that with highest probability—represents the most important topic in the document and the goal is to select the best sentence that covers this word. After the best sentence is selected, the probability of each word that appears in the chosen sentence is adjusted. It is set to a smaller value, equal to the square of the probability of the word at the beginning of the current selection step, to reflect the fact that the probability of a word occurring twice in a summary is lower than the probability of the word occurring only once. This selection loop is repeated until the desired summary length is achieved.

With continuous weights, there are even greater number of possibilities for defining the sentence scoring function compared to the topic words method: the weights can be summed, multiplied, averaged, etc. In each case the scoring is derived by the same representation but the resulting summarizer performance can vary considerably depending on the choice [68]. The sentence selection strategy of SUMBASIC is a variation of the maximal marginal relevance strategy, but an approach that optimizes the occurrence of important words globally over the entire summary instead of greedy selection perform better [89]. Word probabilities can serve as the basis for increasingly complex views of summarization [50].

TF*IDF weighting (Term Frequency*Inverse Document Frequency)

The word probability approach relies on a stop word list to eliminate too common words from consideration. Deciding which words to include in a stop list, however, is not a trivial task and assigning TF*IDF weights to words [79, 87] provides a better alternative. This weighting

---

[3] Sentences that have fewer than 15 content words are assigned weight zero and a stop word list is used to eliminate very common words from consideration.

exploits counts from a background corpus, which is a large collection of documents, normally from the same genre as the document that is to be summarized; the background corpus serves as indication of how often a word may be expected to appear in an arbitrary text.

The only additional information besides the term frequency $c(w)$ that we need in order to compute the weight of a word w which appears $c(w)$ times in the input for summarization is the number of documents, $d(w)$, in a background corpus of D documents that contain the word. This allows us to compute the inverse document frequency:

$$TF*IDF_w = c(w).\log\frac{D}{d(w)} \qquad (3.5)$$

In many cases $c(w)$ is divided by the maximum number of occurrences of any word in the document, which normalizes for document length. Descriptive topic words are those that appear often in a document, but are not very common in other documents. Words that appear in most documents will have an IDF close to zero. The TF*IDF weights of words are good indicators of importance, and they are easy and fast to compute. These properties explain why TF*IDF is incorporated in one form or another in most current systems [25, 26, 28–30, 40].

Centroid summarization [73], which has become a popular baseline system, is also built on TF.IDF topic representation. In this approach, an empirically determined threshold is set, and all words with TF.IDF below that threshold are considered to have a weight of zero. In this way the centroid approach is similar to the topic word approach because words with low weight are treated as noise and completely ignored when computing sentence importance. It also resembles the word probability approach because it keeps differential weights (TF.IDF) for all word above the threshold. The sentence scoring function in the centroid method is the sum of weights of the words in it.

Lexical chains [3, 86, 31] and some related approaches represent topics that are discussed throughout a text by exploiting relations between words. They capture semantic similarity between nouns to determine the importance of sentences. The lexical chains approach captures the intuition that topics are expressed using not a single word but instead different related words. For example, the occurrence of the words "car", "wheel", "seat", "passenger" indicates a clear topic, even if each of the words is not by itself very frequent. The approach heavily relies on WordNet [63], a manually compiled thesaurus which lists the different senses of each word, as well as word relationships such as synonymy, antonymy, part-whole and general-specific.

A Survey of Text Summarization Techniques   51

A large part of Barzilay and Elhadad's original work on applying lexical chains for summarization [3] is on new methods for constructing good lexical chains, with emphasis on word sense disambiguation of words with multiple meanings (i.e. the word "bank" can mean a financial institution or the land near a river or lake). They develop an algorithm that improves on previous work by waiting to disambiguate polysemous words until all possible chains for a text have been constructed; word senses are disambiguated by selecting the interpretations with the most connections in the text. Later research further improved both the runtime of the algorithms for building of lexical chains, and the accuracy of word sense disambiguation [86, 31].

Barzilay and Elhadad claim that the most prevalent discourse topic will play an important role in the summary and argue that lexical chains provide a better indication of discourse topic than does word frequency simply because different words may refer to the same topic. They define the strength of a lexical chain by its length, which is equal to the number of words found to be members of the same chain, and its homogeneity, where homogeneity captures the number of distinct lexical items in the chain divided by its length. They build the summary by extracting one sentence for each highly scored chain, choosing the first sentence in the document containing a representative word for the chain.

This strategy for summary selection—one sentence per important topic—is easy to implement but possibly too restrictive. The question that stands out, and which Barzilay and Elhadad raise but do not address, is that maybe for some topics more than one sentence should be included in the summary. Other sentence scoring techniques for lexical chain summarization have not been explored, i.e. sentences that include several of the highly scoring chains may be even more informative about the connection between the discussed topics.

In later work, researchers chose to avoid the problem of word sense disambiguation altogether but still used WordNet to track the frequency of all members of a concept set [82, 102]. Even without sense disambiguation, these approaches were able to derive concepts like {war, campaign, warfare, effort, cause, operation, conflict}, {concern, carrier, worry, fear, scare} or {home, base, source, support, backing}. Each of the individual words in the concept could appear only once or twice in the input, but the concept itself appeared in the document frequently.

The heavy reliance on WordNet is clearly a bottleneck for the approaches above, because success is constrained by the coverage of WordNet. Because of this, robust methods such as latent semantic analysis that do not use a specific static hand-crafted resource have much appeal.

2.3 Latent Semantic Analysis

Latent semantic analysis (LSA) [19] is a robust unsupervised technique for deriving an implicit representation of text semantics based on observed co-occurrence of words. Gong and Liu [33] proposed the use of LSA for single and multi-document generic summarization of news, as a way of identifying important topics in documents without the use of lexical resources such as WordNet.

Building the topic representation starts by filling in a n by m matrix A: each row corresponds to a word from the input (n words) and each column corresponds to a sentence in the input (m sentences). Entry a_{ij} of the matrix corresponds to the weight of word i in sentence j. If the sentence does not contain the word, the weight is zero, otherwise the weight is equal to the TF*IDF weight of the word. Standard techniques for singular value decomposition (SVD) from linear algebra are applied to the matrix A, to represent it as the product of three matrices: $A = U\Sigma V^T$. Every matrix has a representation of this kind and many standard libraries provide a built-in implementation of the decomposition.

Matrix U is a n by m matrix of real numbers. Each column can be interpreted as a topic, i.e. a specific combination of words from the input with the weight of each word in the topic given by the real number. Matrix Σ is diagonal m by m matrix. The single entry in row i of the matrix corresponds to the weight of the "topic", which is the ith column of U. Topics with low weight can be ignored, by deleting the last k rows of U, the last k rows and columns of Σ and the last k rows of V^T. This procedure is called dimensionality reduction. It corresponds to the thresholds employed in the centroid and topic words approaches, and topics with low weight are treated as noise. Matrix V^T is a new representation of the sentences, one sentence per row, each of which is expressed not in terms of words that occur in the sentence but rather in terms of the topics given in U. The matrix $D = \Sigma V^T$ combines the topic weights and the sentence representation to indicate to what extent the sentence conveys the topic, with d_{ij} indicating the weight for topic i in sentence j.

The original proposal of Gong and Liu was to select one sentence for each of the most important topics. They perform dimensionality reduction, retaining only as many topics as the number of sentences they want to include in the summary. The sentence with the highest weight for each of the retained topics is selected to form the summary. This strategy suffers from the same drawback as the lexical chains approach because more than one sentence may be required to convey all information pertinent to that topic. Later researchers have proposed alternative procedures which have led to improved performance of the summarizer in content selection. One improvement is to use the weight of each topic in order to determine the relative proportion of the summary that should cover the topic, thus allowing for a variable number of sentences per topic. Another improvement was to notice that often sentences that discuss several of the important topics are good candidates for summaries [88]. To identify such sentences, the weight of sentence s_i is set to equal

$$Weight(s_i) = \sqrt{\sum_{j=1}^{m} d_{i,j}^2} \qquad (3.6)$$

Further variations of the LSA approach have also been explored [72, 35]. The systems that rely on LSA best exemplify the significance of the procedure for sentence scoring. In the many variants of the algorithm, the topic representation remains the same while the way sentences are scored and chosen varies, directly influencing the performance of the summarizer when selecting important content.

2.4 Bayesian Topic Models

Bayesian models are the most sophisticated approach for topic representation proposed for summarization which has been steadily gaining popularity [18, 36, 97, 11].

The original Bayesian model for multi-document summarization [18, 36], derives several distinct probabilistic distributions of words that appear in the input. One distribution is for general English (G), one for the entire cluster to be summarized (C) and one for each individual document i in that cluster (D_i). Each of G, C and D consist of tables of words and their probabilities, or weights, much like the word probability approach, but the weights are very different in G, C and D: a word with high probability in general English is likely to have (almost) zero weight in the cluster table C. The tables (probability distributions) are derived as a part of a hierarchical topic model [8]. It is an unsupervised model and the only data it requires are several multi-document clusters; the general English weights reflect occurrence of words across most of the input clusters.

The topic model representations are quite appealing because they capture information that is lost in most of the other approaches. They, for example, have an explicit representation of the individual documents that make up the cluster that is to be summarized, while it is customary in other approaches to treat the input to a multi-document summarizer as one long text, without distinguishing document boundaries. The detailed representation would likely enable the development of better summarizers which conveys the similarities and differences among the different documents that make up the input for multi-document summarization [55, 24, 54]. It is also flexible in the manner in which it derives the general English weights of words, without the need for a pre-determined stop word list, or IDF values from a background corpus.

In addition to the improved representation, the topic models highlight the use of a different sentence scoring procedure: Kullback-Lieber (KL) divergence. The KL divergence between two probability distributions captures the mismatch in probabilities assigned to the same events by the two distributions. In summarization, the events are the occurrence of words. The probability of words in the summary can be computed directly, as the number of times the word occurs divided by the total number of words.

In general the KL divergence of probability distribution Q with respect to distribution P over words w is defined as

$$KL(P||Q) = \sum_{w} P(w) \log \frac{P(w)}{Q(w)} \tag{3.7}$$

$P(w)$ and $Q(w)$ are the probabilities of w in P and Q respectively.

Sentences are scored and selected in a greedy iterative procedure [36]. In each iteration the best sentence i to be selected in the summary is determined as the one for which the KL divergence between C, the probabilities of words in the cluster to be summarized, and the summary so far, including i, is smallest.

KL divergence is appealing as a way of scoring and selecting sentence in summarization because it truly captures an intuitive notion that good summaries are similar to the input. Thinking about a good summary in this way is not new in summarization [21, 74] but KL provides a way of measuring how the importance of words, given by their probabilities, changes in the summary compared to the input. A good summary would reflect the importance of words according to the input, so the divergence between the two will be low. This intuition has been studied extensively in work on automatic evaluation of content selection in summarization, where another indicator of divergence—Jensen Shannon divergence—has proven superior to KL [45, 52].

Given all this, information theoretic measures for scoring sentences are likely to gain popularity even outside the domain on Bayesian topic model representations. All that is necessary in order to apply a divergence to score the summary is a table with word probabilities. The word probability approaches in the spirit of SUMBASIC [68] can directly apply divergence measures to score sentences rather than sum, multiply or average the probabilities of words; other methods that assign weights to words can normalize the weights to get a probability distribution of words. In the next section we will also discuss an approach for summarizing academic articles which uses KL divergence to score sentences.

2.5 Sentence Clustering and Domain-dependent Topics

In multi-document summarization of news, the input by definition consists of several articles, possibly from different sources, on the same topic. Across the different articles there will be sentences that contain similar information. Information that occurs in many of the input documents is likely important and worth selecting in a summary. Of course, verbatim repetition on the sentence level is not that common across sources. Rather, similar sentences can be clustered together [59, 39, 85]. In summarization, cosine similarity is standardly used to measure the similarity between the vector representations of sentences [78].

In this approach, clusters of similar sentences are treated as proxies for topics; clusters with many sentences represent important topic themes in the input. Selecting one representative sentence from each main cluster is one way to produce an extractive summary, while minimizing possible redundancy in the summary.

The sentence clustering approach to multi-document summarization exploits repetition at the sentence level. The more sentences there are in a cluster, the more important the information in the cluster is considered. Below is an example of a sentence cluster from different documents in the input to a multi-document summarizer. All four sentences share common content that should be conveyed in the summary.

S1 PAL was devastated by a pilots' strike in June and by the region's currency crisis.

S2 In June, PAL was embroiled in a crippling three-week pilots' strike.

S3 Tan wants to retain the 200 pilots because they stood by him when the majority of PAL's pilots staged a devastating strike in June.

S4 In June, PAL was embroiled in a crippling three-week pilots' strike.

Constraining each sentence to belong to only one cluster is a distinct disadvantage of the sentence clustering approach, and graph methods for summarization which we discuss in the next section, have proven to exploit the same ideas in a more flexible way.

For domain-specific summarization, however, clustering of sentences from many samples from the domain can give a good indication about the topics that are usually discussed in the domain, and the type of information that a summary would need to convey. In this case, Hidden Markov Models (HMM) that capture "story flow"—what topics are discussed in what order in the domain— can be trained [5, 28]. These models capitalize on the fact that within a specific domain, information in different texts is presented following a common presentation flow. For example, news articles about earthquakes often first talk about where the earthquake happened, what its magnitude was, then mention human casualties or damage, and finally discuss rescue efforts. Such "story flow" can be learned from multiple articles from the same domain. States in the HMM correspond to topics in the domain, which are discovered via iterative clustering of similar sentences from many articles from the domain of interest. Each state (topic) is characterized by a probability distribution which indicates how likely a given word is to appear in a sentence that discusses the topic. Transitions between states in the model correspond to topic transitions in typical texts. These HMM models do not require any labelled data for training and allow for both content selection and ordering in summarization. The sentences that have highest probability of conveying important topics are selected in the summary.

Even simpler approach to discovering the topics in a specific domain can be applied when there are available samples from the domain that are more structured and contain human-written headings. For example, there are plenty of Wikipedia articles about actors and diseases. Clustering similar section headings, where similarity is defined by cosine similarity for example, will identify the topics discussed in each type of article [80]. The clusters with most headings represent the most common topics, and the most common string in the cluster is used to label it. This procedure discovers for example that when talking about actors, writers most often include information about their biography, early life, career and personal life. Then to summarize web pages returned by a search for a specific actor, the system can create a Wikipedia-like web page on the fly, selecting sentences from the returned results that convey these topics.

3. Influence of Context

In many cases, the summarizer has available additional materials that can help determine the most important topics in the document to be summarized. For example in web page summarization, the augmented input consists of other web pages that have links to the pages that we want to summarize. In blog summarization, the discussion following the blog post is easily available and highly indicative of what parts of the blog post are interesting and important. In summarization of scholarly papers, later papers that cite the paper to be summarized and the citation sentences in particular, provide a rich context that indicate what sentences in the original paper are important. User interests are often taken into account in query-focused summarization, where the query provides additional context. All of these approaches relying on augmented input have been exploited for summarization.

3.1 Web Summarization

One type of web page context to consider is the text in pages that link to the one that has to be summarized, in particular the text surrounded by the hyperlink tag pointing to the page. This text often provides a descriptive summary of a web page (e.g., "Access to papers published within the last year by members of the NLP group"). Proponents of using context to provide summary sentences argue that a web site includes multimedia, may cover diverse topics, and it may be hard for a summarizer to distinguish good summary content from bad [20]. The earliest work on this approach was carried out to provide snippets for each result from a search engine [2]. To determine a summary, their system issued a search for a URL, selected all sentences containing a link to that URL and the best sentence was identified using heuristics. Later work has extended this approach through an algorithm that allows selection of a sentence that covers as many aspects of the web page as possible and that is on the same topic [20] . For coverage, Delort et al. used word overlap, normalized by sentence length, to determine which sentences are entirely covered by others and thus can be removed from consideration for the summary. To ensure topicality, Delort's system selects a sentence that is a reference to the page (e.g., "CNN is a news site") as opposed to content (e.g., "The top story for today..."). He computes topicality by measuring overlap between each context sentence and the text within the web page, normalizing by the number of words in the web page. When the web page does not have many words, instead he clusters all sentences in the context and chooses the sentence that is most similar to all others using cosine distance.

In summarization of blog posts, important sentences are identified based on word frequency [41]. The critical difference from other approaches is that here frequency is computed over the comments on the post rather then the original blog entry. The extracted sentences are those that elicited discussion.

3.2 Summarization of Scientific Articles

Impact summarization [60] is defined as the task of extracting sentences from a paper that represent the most influential content of that paper. Language models provide a natural way for solving the task. For each paper to be summarized, impact summarization methods find other papers in a large collection that cite that paper and extract the areas in which the references occur. A language model is built using the collection of all reference areas to a paper, giving the probability of each word to occur in a reference area. This language model gives a way of scoring the importance of sentences in the original article: important sentences are those that convey information similar to that which later papers discussed when referring to the original paper. The measure of similarity between a sentence and the language model is measured by KL divergence. In order to account for the importance of each sentence within the summarized article alone, the approach uses word probabilities estimated from the article. The final score of a sentence is a linear combination of impact importance coming from KL divergence and intrinsic importance coming from the word probabilities in the input article.

3.3 Query-focused Summarization

In query-focused summarization, the importance of each sentence will be determined by a combination of two factors: how relevant is that sentence to the user question and how important is the sentence in the context of the input in which it appears. There are two classes of approaches to this problem. The first adapts techniques for generic summarization of news. For example, an approach using topic signature words [15] is extended for query-focused summarization by assuming that the words that should appear in a summary have the following probability: a word has probability zero of appearing in a summary for a user defined topic if it neither appears in the user query nor is a topic signature word for the input; the probability of the word to appear in the summary is 0.5 if it either appears in the user query or is a topic signature, but not both; and the probability of a word to appear in a summary is 1 if it is both in the user query and in the list of topic signature words for the input. These probabilities are arbitrarily chosen, but in fact work well when used to assign weights to sentences equal to the average probability of words in the sentence. Graph-based approaches [71] have also been adapted for query-focused summarization with minor modifications.

Other approaches have been developed that use new methods for identifying relevant and salient sentences. These approaches have usually been developed for specific types of queries. For example, many people have worked on generation of biographical summaries, where the query is the name of the person for whom a biography should be generated. Most people use some balance of top-down driven approaches that search for patterns of information that might be found in a biography, often using machine learning to identify the patterns, combined with bottom-up approaches that sift through all available material to find sentences that are biographical in nature [7, 98, 81, 105, 6]. The most recent of these approaches uses language modeling of biographical texts found on Wikipedia and non-biographical texts in a news corpus to identify biographical sentences in input documents.

Producing snippets for search engines is a particularly useful query focused application [92, 95].

3.4 Email Summarization

Summarization must be sensitive to the unique characteristics of email, a distinct linguistic genre that exhibits characteristics of both written text and spoken conversation. A thread or a mailbox contains one or more conversations between two or more participants over time. As in summarization of spoken dialog, therefore, summarization needs to take the interactive nature of dialog into account; a response is often only meaningful in relation to the utterance it addresses. Unlike spoken dialog, however, the summarizer need not concern itself with speech recognition errors, the impact of pronunciation, or the availability of speech features such as prosody. Furthermore, responses and reactions are not immediate and due to the asynchronous nature of email, they may explicitly mark the previous email passages to which they are relevant.

In early research on summarization of email threads, [66] used an extractive summarizer to generate a summary for the first two levels of the discussion thread tree, producing relatively short "overview summaries." They extracted a sentence for each of the two levels, using overlap with preceding context. Later work on summarization of email threads [75] zeroed in on the dialogic nature of email. Their summarizer used machine learning and relied on email specific features in addition to traditional features, including features related to the thread and features related to email structure such as the number of responders to a message, similarity of a sentence with the subject, etc. Email conversations are a natural means of getting answers to one's questions and the asynchronous nature of email makes it possible for one to pursue several questions in parallel. As a consequence, question-answer exchanges figure as one of the dominant uses of email conversations. These observations led to research on identification of question and answer pairs in email [84, 64] and the integration of such pairs in extractive summaries of email [58].

Email summarizers have also been developed for a full mailbox or archive instead of just a thread. [69] present a system that can be used for browsing an email mailbox and that builds upon multi-document summarization techniques. They first cluster all email in topically related threads. Both an overview and a full-length summary are then generated for each cluster. A more recent approach to summarization of email within a folder uses a novel graph-based analysis of quotations within email [10]. Using this analysis, Carenini et al.'s system computes a graph representing how each individual email directly mentions other emails, at the granularity of fragments and sentences.

4. Indicator Representations and Machine Learning for Summarization

Indicator representation approaches do not attempt to interpret or represent the topics discussed in the input. Instead they come up with a representation of the text that can be used to directly rank sentences by importance. Graph methods are unique because in their most popular formulations they base summarization on a single indicator of importance, derived from the centrality of sentences in a graph representation of the input. In contrast other approaches employ a variety of indicators and combine them either heuristically or using machine learning to decide which sentences are worthy to be included in the summary.

4.1 Graph Methods for Sentence Importance

In the graph models inspired by the PageRank algorithm [25, 61], the input is represented as a highly connected graph. Vertices represent sentences and edges between sentences are assigned weights equal to the similarity between the two sentences. The method most often used to compute similarity is cosine similarity with TF*IDF weights for words. Sometimes, instead of assigning weights to edges, the connections between vertices can be determined in a binary fashion: the vertices are connected only if the similarity between the two sentences exceeds a predefined threshold. Sentences that are related to many other sentences are likely to be central and would have high weight for selection in the summary.

When the weights of the edges are normalized to form a probability distribution so that the weight of all outgoing edges from a given vertex sum up to one, the graph becomes a Markov chain and the edge weights correspond to the probability of transitioning from one state to another. Standard algorithms for stochastic processes can be used to compute the probability of being in each vertex of the graph at time t while making consecutive transitions from one vertex to next. As more and more transitions are made, the probability of each vertex converges, giving the stationary distribution of the chain. The stationary distribution gives the probability of (being at) a given vertex and can be computed using iterative approximation. Vertices with higher probabilities correspond to more important sentences that should be included in the summary.

Graph-based approaches have been shown to work well for both single-document and multi-document summarization [25, 61]. Since the approach does not require language-specific linguistic processing beyond identifying sentence and word boundaries, it can also be applied to other languages, for example, Brazilian Portuguese [62]. At the same time, incorporating syntactic and semantic role information in the building of the text graph leads to superior results over plain TF*IDF cosine similarity [13].

Using different weighting schemes for links between sentences that belong to the same article and sentences from different articles can help separate the notions of topicality within a document and recurrent topics across documents. This distinction can be easily integrated in the graph-based models for summarization [96].

Graph representations for summarization had been explored even before the PageRank models became popular. For example, the purpose of an older graph-based system for multi-document summarization [55] is to identify salient regions of each story related to a topic given by a user, and compare the stories by summarizing similarities and differences. The vertices in the graph are *words*, *phrases* and *named entities* rather than sentences and their initial weight is assigned using TF*IDF. Edges between vertices are defined using synonym and hypernym links in WordNet, as well as coreference links. Spreading activation is used to assign weights to non-query terms as a function of the weight of their neighbors in the graph and the type of relation connecting the nodes.

In order to avoid problems with coherence that may arise with the selection of single sentences, the authors of another approach [78] argue that a summarizer should select full paragraphs to provide adequate context. Their algorithm constructs a text graph for a document using cosine similarity between each pair of paragraphs in the document. The shape of the text graph determines which paragraphs to extract. In their experiments, they show that two strategies, selecting paragraphs that are well connected to other paragraphs or first paragraphs of topical text segments within the graph, both produce good summaries.

A combination of the subsentential granularity of analysis where nodes are words and phrases rather than sentences and edges are syntactic dependencies has also been explored [44]. Using machine learning techniques, the authors attempt to learn what portions of the input graph would be included in a summary. In their experiments on single document summarization of news articles, properties of the graph such as incoming and outgoing links, connectivity and PageRank weights are identified as the best class of features that can be used for content selection. This work provides an excellent example of how machine learning can be used to combine a range of indicators of importance rather than committing to a single one.

4.2 Machine Learning for Summarization

Edmundson's early work [23] set the direction for later investigation of applying machine learning techniques for summarization [43]. He proposed that rather than relying on a single representation of topics in the input, many different indicators of importance can be combined. Then a corpus of inputs and summaries written by people can be used to determine the weight of each indicator.

In supervised methods for summarization, the task of selecting important sentences is represented as a binary classification problem, partitioning all sentences in the input into summary and non-summary sentences. A corpus with human annotations of sentences that should be included in the summary is used to train a statistical classifier for the distinction, with each sentences represented as a list of potential indicators of importance. The likelihood of a sentence to belong to the summary class, or the confidence of the classifier that the sentence should be in the summary, is the score of the sentence. The chosen classifier plays the role of a sentence scoring function, taking as an input the intermediate representation of the sentence and outputting the score of the sentence. The most highly scoring sentences are selected to form the summary, possibly after skipping some because of high similarity to already chosen sentences.

Machine learning approaches to summarization offer great freedom because the number of indicators of importance is practically endless [40, 70, 104, 44, 27, 37, 99, 51]. Any of the topic representation approaches discussed above can serve as the basis of indicators. Some common features include the position of the sentence in the document (first sentences of news are almost always informative), position in the paragraph (first and last sentences are often important), sentence length, similarity of the sentence with the document title or headings, weights of the words in a sentence determined by any topic representation approach, presence of named entities or cue phrases from a predetermined list, etc.

It is hardly an exaggeration to say that every existing machine learning method has been applied for summarization. One important difference is whether the classifier assumes that the decision about inclusion in the summary is independently done for each sentence. This assumption is apparently not realistic, and methods that explicitly encode dependencies between sentences such as Hidden Markov Models and Conditional Random Fields outperform other learning methods [14, 30, 83].

A problem inherent in the supervised learning paradigm is the necessity of labeled data on which classifiers can be trained. Asking annotators to select summary-worthy sentences is a reasonable solution [93] but it is time consuming and even more importantly, annotator agreement is low and different people tend to choose different sentences when asked to construct an extractive summary of a text [76]. Partly motivated by this issue and partly because of their interest in ultimately developing abstractive methods for summarization many researchers have instead worked with abstracts written by people (often professional writers). Researchers concentrated their efforts on developing methods for automatic alignment of the human abstracts and the input [56, 42, 104, 4, 17] in order to provide labeled data of summary and non-summary sentences for machine learning. Some researchers have also proposed ways to leverage the information from manual evaluation of content selection in summarization in which multiple sentences can be marked as expressing the same fact that should be in the summary [16, 27]. Alternatively, one could compute similarity between sentences in human abstracts and those in the input in order to find very similar sentences, not necessarily doing full alignment [12].

Another option for training a classifier is to employ a semi-supervised approach. In this paradigm, a small number of examples of summary and non-summary sentences are annotated by people. Then two classifiers are trained on that data, using different sets of features which are independent given the class [100] or two different classification methods [99]. After that one of the classifiers is run on unannotated data, and its most confident predictions are added to the annotated examples to train the other classifier, repeating the process until some predefined halting condition is met.

Several modifications to standard machine learning approaches are appropriate for summarization. In effect formulating summarization as a binary classification problem, which scores individual sentences, is not equivalent to finding the best summary, which consists of several sentences. This is exactly the issue of selecting a summary that we discuss in the next section. In training a supervised model, the parameters may be optimized to lead to a summary that has the best score against a human model [1, 49].

For generic multi-document summarization of news, supervised methods have not been shown to outperform competitive unsupervised methods based on a single feature such as the presence of topic words and graph methods. Machine learning approaches have proved to be much more successful in single document or domain or genre specific summarization, where classifiers can be trained to identify specific types of information such as sentences describing literature background in scientific article summarization [90], utterances expressing agreement or disagreement in meetings [30], biographical information [105, 6, 80], etc.

5. Selecting Summary Sentences

Most summarization approaches choose content sentence by sentence: they first include the most informative sentence, and then if space constraints permit, the next most informative sentence is included in the summary and so on. Some process of checking for similarity between the chosen sentences is also usually employed in order to avoid the inclusion of repetitive sentences.

5.1 Greedy Approaches: Maximal Marginal Relevance

indexGreedy Approach to Summarization One of the early summarization approaches for both generic and query focused summarization that has been widely adopted is *Maximal Marginal Relevance* (MMR) [9]. In this approach, summaries are created using greedy, sentence-by-sentence selection. At each selection step, the greedy algorithm is constrained to select the sentence that is maximally relevant to the user query (or has highest importance score when a query is not available) and minimally redundant with sentences already included in the summary. MMR measures relevance and novelty separately and then uses a linear combination of the two to produce a single score for the importance of a sentence in a given stage of the selection process. To quantify both properties of a sentence, Carbonell and Goldstein use cosine similarity. For relevance, similarity is measured to the query, while for novelty, similarity is measured against sentences selected so far. The MMR approach was originally proposed for query-focused summarization in the context of information retrieval, but could easily be adapted for generic summarization, for example by using the entire input as a user [33]. In fact any of the previously discussed approaches for sentence scoring can be used to calculate the importance of a sentence. Many have adopted this seminal approach, mostly in its generic version, sometimes using different measures of novelty to select new sentences [91, 101, 65].

This greedy approach of sequential sentence selection might not be that effective for optimal content selection of the entire summary. One typical problematic scenario for greedy sentence selection (discussed in [57]) is when a very long and highly relevant sentence happens to be evaluated as the most informative early on. Such a sentence may contain several pieces of relevant information, alongside some not so relevant facts which could be considered noise. Including such a sentence in the summary will help maximize content relevance at the time of selection, but at the cost of limiting the amount of space in the summary remaining for other sentences. In such cases it is often more desirable to include several shorter sentences, which are individually less informative than the long one, but which taken together do not express any unnecessary information.

5.2 Global Summary Selection

Global optimization algorithms can be used to solve the new formulation of the summarization task, in which the best overall summary is selected. Given some constraints imposed on the summary, such as maximizing informativeness, minimizing repetition, and conforming to required summary length, the task would be to select the best summary. Finding an exact solution to this problem is NP-hard [26], but approximate solutions can be found using a dynamic programming algorithm [57, 103, 102]. Exact solutions can be found quickly via search techniques when the sentence scoring function is local, computable only from the given sentence [1].

Even in global optimization methods, informativeness is still defined and measured using features well-explored in the sentence selection literature. These include word frequency and position in the document [103], TF*IDF [26], similarity with the input [57], and concept frequency [102, 32]. Global optimization approaches to content selection have been shown to outperform greedy selection algorithms in several evaluations using news data as input, and have proved to be especially effective for extractive summarization of meetings [77, 32].

In a detailed study of global inference algorithms [57], it has been demonstrated that it is possible to find an exact solution for the optimization problem for content selection using Integer Linear Programming. The performance of the approximate algorithm based on dynamic programming was lower, but comparable to that of the exact solutions. In terms of running time, the greedy algorithm is very efficient, almost constant in the size of the input. The approximate algorithm scales linearly with the size of the input and is thus indeed practical to use. The running time for the exact algorithm grows steeply with the size of the input and is unlikely to be useful in practice [57]. However, when a monotone submodular function is used to evaluate the informativeness of the summary, optimal or near optimal solution can be found quickly [48, 47].

6. Conclusion

In this chapter we have attempted to give a comprehensive overview of the most prominent recent methods for automatic text summarization. We have outlined the connection to early approaches and have contrasted approaches in terms of how they represent the input, score sentences and select the summary. We have highlighted the success of KL divergence as a method for scoring sentences which directly incorporates an intuition about the characteristics of a good summary, as well as the growing interest in the development of methods that globally optimize the selection of the summary. We have shown how summarization strategies must be adapted to different genres, such as web pages and journal articles, taking into account contextual information that guides sentence selection. These three recent developments in summarization complement traditional topics in the field that concern intermediate representations and the application of appropriate machine learning methods for summarization.

References

[1] A. Aker, T. Cohn, and R. Gaizauskas. Multi-document summarization using a* search and discriminative training. In *Proceedings of the 2010 Conference on Empirical Methods in Natural Language Processing*, EMNLP'10, pages 482–491, 2010.

[2] E. Amitay and C. Paris. Automatically summarizing web sites - is there a way around it? In *Proceedings of the ACM Conference on Information and Knowledge Management*, pages 173–179, 2000.

[3] R. Barzilay and M. Elhadad. Text summarizations with lexical chains. In Inderjeet Mani and Mark Maybury, editors, *Advances in Automatic Text Summarization*, pages 111–121. MIT Press, 1999.

[4] R. Barzilay and N. Elhadad. Sentence alignment for monolingual comparable corpora. In *Proceedings of the Conference on Empirical Methods in Natural Language Processing*, pages 25–32, 2003.

[5] R. Barzilay and L. Lee. Catching the drift: Probabilistic content models, with applications to generation and summarization. In *Human Language Technology Conference of the North American Chapter of the Association for Computational Linguistics*, pages 113–120, 2004.

[6] F. Biadsy, J. Hirschberg, and E. Filatova. An unsupervised approach to biography production using wikipedia. In *Proceedings of the Annual Meeting of the Association for Computational Linguistics*, pages 807–815, 2008.

[7] S. Blair-Goldensohn, K. McKeown, and A. Schlaikjer. Defscriber: a hybrid system for definitional qa. In *Proceedings of the Annual International ACM SIGIR Conference on Research and Development in Information Retrieval*, pages 462–462, 2003.

[8] D. Blei, T. Griffiths, M. Jordan, and J. Tenenbaum. Hierarchical topic models and the nested chinese restaurant process. In *Advances in Neural Information Processing Systems*, page 2003, 2004.

[9] J. Carbonell and J. Goldstein. The use of mmr, diversity-based rerunning for reordering documents and producing summaries. In *Proceedings of the Annual International ACM SIGIR Conference on Research and Development in Information Retrieval*, pages 335–336, 1998.

[10] G. Carenini, R. Ng, and X. Zhou. Summarizing email conversations with clue words. In *Proceedings of the international conference on World Wide Web*, pages 91–100, 2007.

[11] A. Celikyilmaz and D. Hakkani-Tur. A hybrid hierarchical model for multi-document summarization. In *Proceedings of the 48th Annual Meeting of the Association for Computational Linguistics*, pages 815–824, 2010.

[12] Y. Chali, S. Hasan, and S. Joty. Do automatic annotation techniques have any impact on supervised complex question answering? In *Proceedings of the Joint Conference of the Annual Meeting of the ACL and the International Joint Conference on Natural Language Processing of the AFNLP*, pages 329–332, 2009.

[13] Y. Chali and S. Joty. Improving the performance of the random walk model for answering complex questions. In *Proceedings of the*

Annual Meeting of the Association for Computational Linguistics, Short Papers, pages 9–12, 2008.

[14] J. Conroy and D. O'Leary. Text summarization via hidden markov models. In *Proceedings of the Annual International ACM SIGIR Conference on Research and Development in Information Retrieval*, pages 406–407, 2001.

[15] J. Conroy, J. Schlesinger, and D. O'Leary. Topic-focused multi-document summarization using an approximate oracle score. In *Proceedings of the International Conference on Computational Linguistics and the annual meeting of the Association for Computational Linguistics*, pages 152–159, 2006.

[16] T. Copeck and S. Szpakowicz. Leveraging pyramids. In *Proceedings of the Document Understanding Conference*, 2005.

[17] H. Daumé III and D. Marcu. A phrase-based HMM approach to document/abstract alignment. In *Proceedings of the Conference on Empirical Methods in Natural Language Processing*, pages 119–126, 2004.

[18] H. Daumé III and D. Marcu. Bayesian query-focused summarization. In *Proceedings of the International Conference on Computational Linguistics and the annual meeting of the Association for Computational Linguistics*, pages 305–312, 2006.

[19] S. Deerwester, S. Dumais, G. Furnas, T. Landauer, and R. Harshman. Indexing by latent semantic analysis. *Journal of the American Society for Information Science*, pages 391–407, 1990.

[20] J.-Y. Delort, B. Bouchon-Meunier, and M. Rifqi. Enhanced web document summarization using hyperlinks. In *Proceedings of the ACM conference on Hypertext and hypermedia*, pages 208–215, 2003.

[21] R. Donaway, K. Drummey, and L. Mather. A comparison of rankings produced by summarization evaluation measures. In *Proceedings of the 2000 NAACL-ANLPWorkshop on Automatic summarization - Volume 4*, pages 69–78, 2000.

[22] T. Dunning. Accurate methods for the statistics of surprise and coincidence. *Computational Linguistics*, 19(1):61–74, 1994.

[23] H. Edmundson. New methods in automatic extracting. *Journal of the ACM*, 16(2):264–285, 1969.

[24] N. Elhadad, M.-Y. Kan, J. Klavans, and K. McKeown. Customization in a unified framework for summarizing medical literature. *Journal of Artificial Intelligence in Medicine*, 33:179–198, 2005.

[25] G. Erkan and D. Radev. Lexrank: Graph-based centrality as salience in text summarization. *Journal of Artificial Intelligence Research*, 2004.

[26] E. Filatova and V. Hatzivassiloglou. A formal model for information selection in multi-sentence text extraction. In *Proceedings of the International Conference on Computational Linguistic*, pages 397–403, 2004.

[27] M. Fuentes, E. Alfonseca, and H. Rodríguez. Support vector machines for query-focused summarization trained and evaluated on pyramid data. In *Proceedings of the Annual Meeting of the Association for Computational Linguistics, Companion Volume: Proceedings of the Demo and Poster Sessions*, pages 57–60, 2007.

[28] P. Fung and G. Ngai. One story, one flow: Hidden markov story models for multilingual multidocument summarization. *ACM Transactions on Speech and Language Processing*, 3(2):1–16, 2006.

[29] S. Furui, M. Hirohata, Y. Shinnaka, and K. Iwano. Sentence extraction-based automatic speech summarization and evaluation techniques. In *Proceedings of the Symposium on Large-scale Knowledge Resources*, pages 33–38, 2005.

[30] M. Galley. A skip-chain conditional random field for ranking meeting utterances by importance. In *Proceedings of the Conference on Empirical Methods in Natural Language Processing*, pages 364–372, 2006.

[31] M. Galley and K. McKeown. Improving word sense disambiguation in lexical chaining. In *Proceedings of the international joint conference on Artificial intelligence*, pages 1486–1488, 2003.

[32] D. Gillick, K. Riedhammer, B. Favre, and D. Hakkani-Tur. A global optimization framework for meeting summarization. In *Proceedings of the IEEE International Conference on Acoustics, Speech and Signal Processing*, pages 4769–4772, 2009.

[33] Y. Gong and X. Liu. Generic text summarization using relevance measure and latent semantic analysis. In *Proceedings of the Annual International ACM SIGIR Conference on Research and Development in Information Retrieval*, pages 19–25, 2001.

[34] S. Gupta, A. Nenkova, and D. Jurafsky. Measuring importance and query relevance in topic-focused multi-document summarization. In *Proceedings of the Annual Meeting of the Association for Computational Linguistics, Demo and Poster Sessions*, pages 193–196, 2007.

[35] B. Hachey, G. Murray, and D. Reitter. Dimensionality reduction aids term co-occurrence based multi-document summarization. In *SumQA '06: Proceedings of the Workshop on Task-Focused Summarization and Question Answering*, pages 1–7, 2006.

[36] A. Haghighi and L. Vanderwende. Exploring content models for multi-document summarization. In *Proceedings of Human Language Technologies: The 2009 Annual Conference of the North American Chapter of the Association for Computational Linguistics*, pages 362–370, 2009.

[37] D. Hakkani-Tur and G. Tur. Statistical sentence extraction for information distillation. In *Proceedings of the IEEE International Conference on Acoustics, Speech and Signal Processing*, volume 4, pages IV-1 –IV-4, 2007.

[38] S. Harabagiu and F. Lacatusu. Topic themes for multi-document summarization. In *Proceedings of the 28th annual international ACM SIGIR conference on Research and development in information retrieval*, SIGIR'05, pages 202–209, 2005.

[39] V. Hatzivassiloglou, J. Klavans, M. Holcombe, R. Barzilay, M. Kan, and K. McKeown. Simfinder: A flexible clustering tool for summarization. In *Proceedings of the NAACL Workshop on Automatic Summarization*, pages 41–49, 2001.

[40] E. Hovy and C.-Y. Lin. Automated text summarization in summarist. In *Advances in Automatic Text Summarization*, pages 82–94, 1999.

[41] M. Hu, A. Sun, and E.-P. Lim. Comments-oriented blog summarization by sentence extraction. In *Proceedings of the ACM Conference on Information and Knowledge Management*, pages 901–904, 2007.

[42] H. Jing. Using hidden markov modeling to decompose human-written summaries. *Computational linguistics*, 28(4):527–543, 2002.

[43] J. Kupiec, J. Pedersen, and F. Chen. A trainable document summarizer. In *Proceedings of the Annual International ACM SIGIR Conference on Research and Development in Information Retrieval*, pages 68–73, 1995.

[44] J. Leskovec, N. Milic-frayling, and M. Grobelnik. Impact of linguistic analysis on the semantic graph coverage and learning of document extracts. In *Proceedings of the national conference on Artificial intelligence*, pages 1069–1074, 2005.

[45] C.-Y. Lin, G. Cao, J. Gao, and J.-Y. Nie. An information-theoretic approach to automatic evaluation of summaries. In *Proceedings of the main conference on Human Language Technology Conference of the North American Chapter of the Association of Computational Linguistics (HLT-NAACL'06)*, pages 463–470, 2006.

[46] C.-Y. Lin and E. Hovy. The automated acquisition of topic signatures for text summarization. In *Proceedings of the International Conference on Computational Linguistic*, pages 495–501, 2000.

[47] H. Lin and J. Bilmes. Multi-document summarization via budgeted maximization of submodular functions. In *North American chapter of the Association for Computational Linguistics/Human Language Technology Conference (NAACL/HLT-2010)*, 2010.

[48] H. Lin, J. Bilmes, and S. Xie. Graph-based submodular selection for extractive summarization. In *Proc. IEEE Automatic Speech Recognition and Understanding (ASRU)*, 2009.

[49] S.-H. Lin, Y.-M. Chang, J.-W. Liu, and B. Chen. Leveraging evaluation metric-related training criteria for speech summarization. In *Proceedings of the IEEE International Conference on Acoustics, Speech, and Signal Processing, ICASSP 2010*, pages 5314–5317, 2010.

[50] S.-H. Lin and B. Chen. A risk minimization framework for extractive speech summarization. In *Proceedings of the 48th Annual Meeting of the Association for Computational Linguistics*, pages 79–87, 2010.

[51] A. Louis, A. Joshi, and A. Nenkova. Discourse indicators for content selection in summarization. In *Proceedings of the Annual Meeting of the Special Interest Group on Discourse and Dialogue*, pages 147–156, 2010.

[52] A. Louis and A. Nenkova. Automatically evaluating content selection in summarization without human models. In *Proceedings of the 2009 Conference on Empirical Methods in Natural Language Processing (EMNLP)*, pages 306–314, 2009.

[53] H. P. Luhn. The automatic creation of literature abstracts. *IBM Journal of Research and Development*, 2(2):159–165, 1958.

[54] M. Mana-López, M. De Buenaga, and J. Gómez-Hidalgo. Multidocument summarization: An added value to clustering in interactive retrieval. *ACM Transactions on Informations Systems*, 22(2):215–241, 2004.

[55] I. Mani and E. Bloedorn. Summarizing similarities and differences among related documents. *Information Retrieval*, 1(1-2):35–67, April 1999.

[56] D. Marcu. The automatic construction of large-scale corpora for summarization research. In *Proceedings of the Annual International ACM SIGIR Conference on Research and Development in Information Retrieval*, pages 137–144, 1999.

[57] R. McDonald. A study of global inference algorithms in multi-document summarization. In *Proceedings of the European Conference on IR Research*, pages 557–564, 2007.

[58] K. McKeown, L. Shrestha, and O. Rambow. Using question-answer pairs in extractive summarization of email conversations. In *Proceedings of the International Conference on Computational Linguistics and Intelligent Text Processing*, pages 542–550, 2007.

[59] K. McKeown, J. Klavans, V. Hatzivassiloglou, R. Barzilay, and E. Eskin. Towards multidocument summarization by reformulation: progress and prospects. In *Proceedings of the national conference on Artificial intelligence*, pages 453–460, 1999.

[60] Q. Mei and C. Zhai. Generating impact-based summaries for scientific literature. In *Proceedings of the Annual Meeting of the Association for Computational Linguistics*, pages 816–824, 2008.

[61] R. Mihalcea and P. Tarau. Textrank: Bringing order into texts. In *Proceedings of the Conference on Empirical Methods in Natural Language Processing*, pages 404–411, 2004.

[62] R. Mihalcea and P. Tarau. An algorithm for language independent single and multiple document summarization. In *Proceedings of the International Joint Conference on Natural Language Processing*, pages 19–24, 2005.

[63] G.A. Miller, R. Beckwith, C. Fellbaum, D. Gross, and K. J. Miller. Introduction to wordnet: An on-line lexical database. *International Journal of Lexicography (special issue)*, 3(4):235–312, 1990.

[64] H. Murakoshi, A. Shimazu, and K. Ochimizu. Construction of deliberation structure in email conversation. In *Proceedings of the Conference of the Pacific Association for Computational Linguistics*, pages 570–577, 2004.

[65] G. Murray, S. Renals, and J. Carletta. Extractive summarization of meeting recordings. In *Proc. 9th European Conference on Speech Communication and Technology*, pages 593–596, 2005.

[66] A. Nenkova and A. Bagga. Facilitating email thread access by extractive summary generation. In *Proceedings of the Recent Advances in Natural Language Processing Conference*, 2003.

[67] A. Nenkova and K. McKeown. Automatic Summarization. In *Foundations and Trends in Information Retrieval* 5(2-3), pages 103-233, 2011.

[68] A. Nenkova, L. Vanderwende, and K. McKeown. A compositional context sensitive multi-document summarizer: exploring the factors that influence summarization. In *Proceedings of the Annual International ACM SIGIR Conference on Research and Development in Information Retrieval*, pages 573-580, 2006.

[69] P. Newman and J. Blitzer. Summarizing archived discussions: a beginning. In *Proceedings of the international conference on Intelligent user interfaces*, pages 273-276, 2003.

[70] M. Osborne. Using maximum entropy for sentence extraction. In *Proceedings of the ACL Workshop on Automatic Summarization*, pages 1-8, 2002.

[71] J. Otterbacher, G. Erkan, and D. Radev. Biased lexrank: Passage retrieval using random walks with question-based priors. *Information Processing and Management*, 45:42-54, January 2009.

[72] M. Ozsoy, I. Cicekli, and F. Alpaslan. Text summarization of turkish texts using latent semantic analysis. In *Proceedings of the 23rd International Conference on Computational Linguistics (COLING 2010)*, pages 869-876, August 2010.

[73] D. Radev, H. Jing, M. Sty, and D. Tam. Centroid-based summarization of multiple documents. *Information Processing and Management*, 40:919-938, 2004.

[74] D. Radev, S. Teufel, H. Saggion, W. Lam, J. Blitzer, H. Qi, A. Çelebi, D. Liu, and E. Drabek. Evaluation challenges in large-scale document summarization. In *Proceedings of the 41st Annual Meeting on Association for Computational Linguistics (ACL'03)*, pages 375-382, 2003.

[75] O. Rambow, L. Shrestha, J. Chen, and C. Lauridsen. Summarizing email threads. In *Human Language Technology Conference of the North American Chapter of the Association for Computational Linguistics*, 2004.

[76] G. Rath, A. Resnick, and R. Savage. The formation of abstracts by the selection of sentences: Part 1: sentence selection by man and machines. *American Documentation*, 2(12):139-208, 1961.

[77] K. Riedhammer, D. Gillick, B. Favre, and D. Hakkani-Tur. Packing the meeting summarization knapsack. In *Proceedings of the Annual Conference of the International Speech Communication Association*, pages 2434–2437, 2008.

[78] G. Salton, A. Singhal, M. Mitra, and C. Buckley. Automatic text structuring and summarization. *Information Processing and Management*, 33(2):193–208, 1997.

[79] G. Salton and C. Buckley. Term-weighting approaches in automatic text retrieval. *Information Processing and Management*, 24:513–523, 1988.

[80] C. Sauper and R. Barzilay. Automatically generating wikipedia articles: A structure-aware approach. In *Proceedings of the Joint Conference of the 47th Annual Meeting of the ACL and the 4th International Joint Conference on Natural Language Processing of the AFNLP*, pages 208–216, 2009.

[81] B. Schiffman, I. Mani, and K. Concepcion. Producing biographical summaries: Combining linguistic knowledge with corpus statistics. In *Proceedings of the Annual Meeting of the Association for Computational Linguistics*, pages 458–465, 2001.

[82] B. Schiffman, A. Nenkova, and K. McKeown. Experiments in multidocument summarization. In *Proceedings of the international conference on Human Language Technology Research*, pages 52–58, 2002.

[83] D. Shen, J.-T. Sun, H. Li, Q. Yang, and Z. Chen. Document summarization using conditional random fields. In *Proceedings of the 20th international joint conference on Artifical intelligence*, pages 2862–2867, 2007.

[84] L. Shrestha and K. McKeown. Detection of question-answer pairs in email conversations. In *Proceedings of the International Conference on Computational Linguistic*, 2004.

[85] A. Siddharthan, A. Nenkova, and K. McKeown. Syntactic simplification for improving content selection in multi-document summarization. In *Proceedings of the International Conference on Computational Linguistic*, pages 896–902, 2004.

[86] H. Silber and K. McCoy. Efficiently computed lexical chains as an intermediate representation for automatic text summarization. *Computational Linguistics*, 28(4):487–496, 2002.

[87] K. Sparck Jones. A statistical interpretation of term specificity and its application in retrieval. *Journal of Documentation*, 28:11–21, 1972.

[88] J. Steinberger, M. Poesio, M. A. Kabadjov, and K. Jeek. Two uses of anaphora resolution in summarization. *Information Processing and Management*, 43(6):1663–1680, 2007.

[89] W. Yih, J. Goodman, L. Vanderwende, and H. Suzuki. Multi-document summarization by maximizing informative content-words. In *Proceedings of the international joint conference on Artificial intelligence*, pages 1776–1782, 2007.

[90] S. Teufel and M. Moens. Summarizing scientific articles: experiments with relevance and rhetorical status. *Computational Linguisics.*, 28(4):409–445, 2002.

[91] D. Radev, T. Allison, S. Blair-goldensohn, J. Blitzer, A. Celebi, S. Dimitrov, E. Drabek, A. Hakim, W. Lam, D. Liu, J. Otterbacher, H. Qi, H. Saggion, S. Teufel, A. Winkel, and Z. Zhang. Mead - a platform for multidocument multilingual text summarization. In *Proceedings of the International Conference on Language Resources and Evaluation*, 2004.

[92] A. Turpin, Y. Tsegay, D. Hawking, and H. Williams. Fast generation of result snippets in web search. In *Proceedings of the Annual International ACM SIGIR Conference on Research and Development in Information Retrieval*, pages 127–134, 2007.

[93] J. Ulrich, G. Murray, and G. Carenini. A publicly available annotated corpus for supervised email summarization. In *Proceedings of the AAAI EMAIL Workshop*, pages 77–87, 2008.

[94] L. Vanderwende, H. Suzuki, C. Brockett, and A. Nenkova. Beyond sumbasic: Task-focused summarization with sentence simplification and lexical expansion. *Information Processing and Managment*, 43:1606–1618, 2007.

[95] R. Varadarajan and V. Hristidis. A system for query-specific document summarization. In *Proceedings of the ACM Conference on Information and Knowledge Management*, 2006.

[96] X. Wan and J. Yang. Improved affinity graph based multi-document summarization. In *Human Language Technology Conference of the North American Chapter of the Association for Computational Linguistics, Companion Volume: Short Papers*, pages 181–184, 2006.

[97] D. Wang, S. Zhu, T. Li, and Y. Gong. Multi-document summarization using sentence-based topic models. In *Proceedings of the ACL-IJCNLP 2009 Conference Short Papers*, pages 297–300, 2009.

[98] R. Weischedel, J. Xu, and A. Licuanan. A hybrid approach to answering biographical questions. In Mark Maybury, editor, *New Directions In Question Answering*, pages 59–70, 2004.

[99] K. Wong, M. Wu, and W. Li. Extractive summarization using supervised and semi-supervised learning. In *Proceedings of the 22nd International Conference on Computational Linguistics (Coling 2008)*, pages 985–992, 2008.

[100] S. Xie, H. Lin, and Y. Liu. Semi-supervised extractive speech summarization via co-training algorithm. In *INTERSPEECH, the 11th Annual Conference of the International Speech Communication Association*, pages 2522–2525, 2010.

[101] S. Xie and Y. Liu. Using corpus and knowledge-based similarity measure in maximum marginal relevance for meeting summarization. In *Proceedings of the IEEE International Conference on Acoustics, Speech and Signal Processing*, pages 4985–4988, 2008.

[102] S. Ye, T.-S. Chua, M.-Y. Kan, and L. Qiu. Document concept lattice for text understanding and summarization. *Information Processing and Management*, 43(6):1643 – 1662, 2007.

[103] W. Yih, J. Goodman, L. Vanderwende, and H. Suzuki. Multi-document summarization by maximizing informative content-words. In *Proceedings of the international joint conference on Artificial intelligence*, pages 1776–1782, 2007.

[104] L. Zhou and E. Hovy. A web-trained extraction summarization system. In *Proceedings of the Conference of the North American Chapter of the Association for Computational Linguistics on Human Language Technology*, pages 205–211, 2003.

[105] L. Zhou, M. Ticrea, and E. Hovy. Multi-document biography summarization. In *Proceedings of the Conference on Empirical Methods in Natural Language Processing*, pages 434–441, 2004.

Chapter 4

A SURVEY OF TEXT CLUSTERING ALGORITHMS

Charu C. Aggarwal
IBM T. J. Watson Research Center
Yorktown Heights, NY
charu@us.ibm.com

ChengXiang Zhai
University of Illinois at Urbana-Champaign
Urbana, IL
czhai@cs.uiuc.edu

Abstract Clustering is a widely studied data mining problem in the text domains. The problem finds numerous applications in customer segmentation, classification, collaborative filtering, visualization, document organization, and indexing. In this chapter, we will provide a detailed survey of the problem of text clustering. We will study the key challenges of the clustering problem, as it applies to the text domain. We will discuss the key methods used for text clustering, and their relative advantages. We will also discuss a number of recent advances in the area in the context of social network and linked data.

Keywords: Text Clustering

1. Introduction

The problem of clustering has been studied widely in the database and statistics literature in the context of a wide variety of data mining tasks [50, 54]. The clustering problem is defined to be that of finding groups of similar objects in the data. The similarity between the objects is measured with the use of a similarity function. The problem of clustering can be very useful in the text domain, where the objects to be clusters can be of different granularities such as documents, paragraphs, sentences or terms. Clustering is especially useful for organizing documents to improve retrieval and support browsing [11, 26].

The study of the clustering problem precedes its applicability to the text domain. Traditional methods for clustering have generally focussed on the case of quantitative data [44, 71, 50, 54, 108], in which the attributes of the data are numeric. The problem has also been studied for the case of categorical data [10, 41, 43], in which the attributes may take on nominal values. A broad overview of clustering (as it relates to generic numerical and categorical data) may be found in [50, 54]. A number of implementations of common text clustering algorithms, as applied to text data, may be found in several toolkits such as *Lemur* [114] and *BOW* toolkit in [64]. The problem of clustering finds applicability for a number of tasks:

- **Document Organization and Browsing:** The hierarchical organization of documents into coherent categories can be very useful for systematic browsing of the document collection. A classical example of this is the *Scatter/Gather* method [25], which provides a systematic browsing technique with the use of clustered organization of the document collection.

- **Corpus Summarization:** Clustering techniques provide a coherent summary of the collection in the form of *cluster-digests* [83] or *word-clusters* [17, 18], which can be used in order to provide summary insights into the overall content of the underlying corpus. Variants of such methods, especially sentence clustering, can also be used for document summarization, a topic, discussed in detail in Chapter 3. The problem of clustering is also closely related to that of dimensionality reduction and topic modeling. Such dimensionality reduction methods are all different ways of summarizing a corpus of documents, and are covered in Chapter 5.

- **Document Classification:** While clustering is inherently an unsupervised learning method, it can be leveraged in order to improve the quality of the results in its supervised variant. In particular, word-clusters [17, 18] and co-training methods [72] can be used in order to improve the classification accuracy of supervised applications with the use of clustering techniques.

We note that many classes of algorithms such as the k-means algorithm, or hierarchical algorithms are general-purpose methods, which can be extended to any kind of data, including text data. A text document can be represented either in the form of binary data, when we use the presence or absence of a word in the document in order to create a binary vector. In such cases, it is possible to directly use a variety of categorical data clustering algorithms [10, 41, 43] on the binary representation. A more enhanced representation would include refined weighting methods based on the frequencies of the individual words in the document as well as frequencies of words in an entire collection (e.g., TF-IDF weighting [82]). Quantitative data clustering algorithms [44, 71, 108] can be used in conjunction with these frequencies in order to determine the most relevant groups of objects in the data.

However, such naive techniques do not typically work well for clustering text data. This is because text data has a number of unique properties which necessitate the design of specialized algorithms for the task. The distinguishing characteristics of the text representation are as follows:

- The dimensionality of the text representation is very large, but the underlying data is sparse. In other words, the lexicon from which the documents are drawn may be of the order of 10^5, but a given document may contain only a few hundred words. This problem is even more serious when the documents to be clustered are very short (e.g., when clustering sentences or tweets).

- While the lexicon of a given corpus of documents may be large, the words are typically correlated with one another. This means that the number of concepts (or principal components) in the data is much smaller than the feature space. This necessitates the careful design of algorithms which can account for word correlations in the clustering process.

- The number of words (or non-zero entries) in the different documents may vary widely. Therefore, it is important to normalize the document representations appropriately during the clustering task.

The sparse and high dimensional representation of the different documents necessitate the design of text-specific algorithms for document representation and processing, a topic heavily studied in the information retrieval literature where many techniques have been proposed to optimize document representation for improving the accuracy of matching a document with a query [82, 13]. Most of these techniques can also be used to improve document representation for clustering.

In order to enable an effective clustering process, the word frequencies need to be normalized in terms of their relative frequency of presence in the document and over the entire collection. In general, a common representation used for text processing is the *vector-space based* TF-IDF representation [81]. In the TF-IDF representation, the term frequency for each word is normalized by the *inverse document frequency*, or IDF. The inverse document frequency normalization reduces the weight of terms which occur more frequently in the collection. This reduces the importance of common terms in the collection, ensuring that the matching of documents be more influenced by that of more discriminative words which have relatively low frequencies in the collection. In addition, a sub-linear transformation function is often applied to the term-frequencies in order to avoid the undesirable dominating effect of any single term that might be very frequent in a document. The work on document-normalization is itself a vast area of research, and a variety of other techniques which discuss different normalization methods may be found in [86, 82].

Text clustering algorithms are divided into a wide variety of different types such as agglomerative clustering algorithms, partitioning algorithms, and standard parametric modeling based methods such as the EM-algorithm. Furthermore, text representations may also be treated as strings (rather than bags of words). These different representations necessitate the design of different classes of clustering algorithms. Different clustering algorithms have different tradeoffs in terms of effectiveness and efficiency. An experimental comparison of different clustering algorithms may be found in [90, 111]. In this chapter we will discuss a wide variety of algorithms which are commonly used for text clustering. We will also discuss text clustering algorithms for related scenarios such as dynamic data, network-based text data and semi-supervised scenarios.

This chapter is organized as follows. In section 2, we will present feature selection and transformation methods for text clustering. Section 3 describes a number of common algorithms which are used for distance-based clustering of text documents. Section 4 contains the description of methods for clustering with the use of word patterns and phrases. Methods for clustering text streams are described in section 5. Section 6 describes methods for probabilistic clustering of text data. Section 7 contains a description of methods for clustering text which naturally occurs in the context of social or web-based networks. Section 8 discusses methods for semi-supervised clustering. Section 9 presents the conclusions and summary.

2. Feature Selection and Transformation Methods for Text Clustering

The quality of any data mining method such as classification and clustering is highly dependent on the noisiness of the features that are used for the clustering process. For example, commonly used words such as *"the"*, may not be very useful in improving the clustering quality. Therefore, it is critical to select the features effectively, so that the noisy words in the corpus are removed before the clustering. In addition to feature *selection*, a number of feature *transformation* methods such as Latent Semantic Indexing (LSI), Probabilistic Latent Semantic Analysis (PLSA), and Non-negative Matrix Factorization (NMF) are available to improve the quality of the document representation and make it more amenable to clustering. In these techniques (often called dimension reduction), the correlations among the words in the lexicon are leveraged in order to create features, which correspond to the concepts or principal components in the data. In this section, we will discuss both classes of methods. A more in-depth discussion of dimension reduction can be found in Chapter 5.

2.1 Feature Selection Methods

Feature selection is more common and easy to apply in the problem of text categorization [99] in which supervision is available for the feature selection process. However, a number of simple unsupervised methods can also be used for feature selection in text clustering. Some examples of such methods are discussed below.

2.1.1 Document Frequency-based Selection. The simplest possible method for feature selection in document clustering is that of the use of *document frequency* to filter out irrelevant features. While the use of inverse document frequencies reduces the importance of such words, this may not alone be sufficient to reduce the noise effects of very frequent words. In other words, words which are too frequent in the corpus can be removed because they are typically common words such as "a", "an", "the", or "of" which are not discriminative from a clustering perspective. Such words are also referred to as *stop words*. A variety of methods are commonly available in the literature [76] for stop-word removal. Typically commonly available stop word lists of about 300 to 400 words are used for the retrieval process. In addition, words which occur extremely infrequently can also be removed from the collection. This is because such words do not add anything to the similarity computations which are used in most clustering methods. In some cases, such words may be misspellings or typographical errors in documents. Noisy text collections which are derived from the web, blogs or social networks are more likely to contain such terms. We note that some lines of research define document frequency based selection purely on the basis of very infrequent terms, because these terms contribute the least to the similarity calculations. However, it should be emphasized that very frequent words should also be removed, especially if they are not discriminative between clusters. Note that the TF-IDF weighting method can also naturally filter out very common words in a "soft" way. Clearly, the standard set of stop words provide a valid set of words to prune. Nevertheless, we would like a way of quantifying the importance of a term directly to the clustering process, which is essential for more aggressive pruning. We will discuss a number of such methods below.

2.1.2 Term Strength. A much more aggressive technique for stop-word removal is proposed in [94]. The core idea of this approach is to extend techniques which are used in supervised learning to the unsupervised case. The term strength is essentially used to measure how informative a word is for identifying two related documents. For example, for two related documents x and y, the term strength $s(t)$ of term t is defined in terms of the following probability:

$$s(t) = P(t \in y | t \in x) \tag{4.1}$$

Clearly, the main issue is how one might define the document x and y as related. One possibility is to use manual (or user) feedback to define when a pair of documents are related. This is essentially equivalent to utilizing supervision in the feature selection process, and may be practical in situations in which predefined categories of documents are available. On the other hand, it is not practical to manually create related pairs in large collections in a comprehensive way. It is therefore desirable to use an automated and purely unsupervised way to define the concept of when a pair of documents is related. It has been shown in [94] that it is possible to use automated similarity functions such as the cosine function [81] to define the relatedness of document pairs. A pair of documents are defined to be related if their cosine similarity is above a user-defined threshold. In such cases, the term strength $s(t)$ can be defined by randomly sampling a number of pairs of such related documents as follows:

$$s(t) = \frac{\text{Number of pairs in which } t \text{ occurs in both}}{\text{Number of pairs in which } t \text{ occurs in the first of the pair}} \tag{4.2}$$

Here, the first document of the pair may simply be picked randomly. In order to prune features, the term strength may be compared to the expected strength of a term which is randomly distributed in the training documents with the same frequency. If the term strength of t is not at least two standard deviations greater than that of the random word, then it is removed from the collection.

One advantage of this approach is that it requires no initial supervision or training data for the feature selection, which is a key requirement in the unsupervised scenario. Of course, the approach can also be used for feature selection in either supervised clustering [4] or categorization [100], when such training data is indeed available. One observation about this approach to feature selection is that it is particularly suited to similarity-based clustering because the discriminative nature of the underlying features is defined on the basis of similarities in the documents themselves.

2.1.3 Entropy-based Ranking. The entropy-based ranking approach was proposed in [27]. In this case, the quality of the term is measured by the entropy reduction when it is removed. Here the entropy $E(t)$ of the term t in a collection of n documents is defined as follows:

$$E(t) = -\sum_{i=1}^{n}\sum_{j=1}^{n}(S_{ij} \cdot \log(S_{ij}) + (1 - S_{ij}) \cdot \log(1 - S_{ij})) \quad (4.3)$$

Here $S_{ij} \in (0,1)$ is the similarity between the ith and jth document in the collection, after the term t is removed, and is defined as follows:

$$S_{ij} = 2^{-\frac{dist(i,j)}{\overline{dist}}} \quad (4.4)$$

Here $dist(i,j)$ is the distance between the terms i and j after the term t is removed, and $\overline{dist}$ is the average distance between the documents after the term t is removed. We note that the computation of $E(t)$ for each term t requires $O(n^2)$ operations. This is impractical for a very large corpus containing many terms. It has been shown in [27] how this method may be made much more efficient with the use of sampling methods.

2.1.4 Term Contribution. The concept of term contribution [62] is based on the fact that the results of text clustering are highly dependent on document similarity. Therefore, the contribution of a term can be viewed as its contribution to document similarity. For example, in the case of dot-product based similarity, the similarity between two documents is defined as the dot product of their normalized frequencies. Therefore, the contribution of a term of the similarity of two documents is the product of their normalized frequencies in the two documents. This needs to be summed over all pairs of documents in order to determine the term contribution. As in the previous case, this method requires $O(n^2)$ time for each term, and therefore sampling methods may be required to speed up the contribution. A major criticism of this method is that it tends to favor highly frequent words without regard to the specific discriminative power within a clustering process.

In most of these methods, the optimization of term selection is based on some pre-assumed similarity function (e.g., cosine). While this strategy makes these methods unsupervised, there is a concern that the term selection might be biased due to the potential bias of the assumed similarity function. That is, if a different similarity function is assumed, we may end up having different results for term selection. Thus the choice of an appropriate similarity function may be important for these methods.

2.2 LSI-based Methods

In feature selection, we attempt to explicitly select out features from the original data set. Feature transformation is a different method in which the new features are defined as a functional representation of the features in the original data set. The most common class of methods is that of dimensionality reduction [53] in which the documents are transformed to a new feature space of smaller dimensionality in which the features are typically a linear combination of the features in the original data. Methods such as Latent Semantic Indexing (LSI) [28] are based on this common principle. The overall effect is to remove a lot of dimensions in the data which are noisy for similarity based applications such as clustering. The removal of such dimensions also helps magnify the semantic effects in the underlying data.

Since LSI is closely related to problem of *Principal Component Analysis (PCA)* or *Singular Value Decomposition (SVD)*, we will first discuss this method, and its relationship to LSI. For a d-dimensional data set, PCA constructs the symmetric $d \times d$ covariance matrix C of the data, in which the (i,j)th entry is the covariance between dimensions i and j. This matrix is positive semi-definite, and can be diagonalized as follows:

$$C = P \cdot D \cdot P^T \qquad (4.5)$$

Here P is a matrix whose columns contain the orthonormal eigenvectors of C and D is a diagonal matrix containing the corresponding eigenvalues. We note that the eigenvectors represent a new orthonormal basis system along which the data can be represented. In this context, the eigenvalues correspond to the variance when the data is projected along this basis system. This basis system is also one in which the second order covariances of the data are removed, and most of variance in the data is captured by preserving the eigenvectors with the largest eigenvalues. Therefore, in order to reduce the dimensionality of the data, a common approach is to represent the data in this new basis system, which is further truncated by ignoring those eigenvectors for which the corresponding eigenvalues are small. This is because the variances along those dimensions are small, and the relative behavior of the data points is not significantly affected by removing them from consideration. In fact, it can be shown that the Euclidian distances between data points are not significantly affected by this transformation and corresponding truncation. The method of PCA is commonly used for similarity search in database retrieval applications.

LSI is quite similar to PCA, except that we use an approximation of the covariance matrix C which is quite appropriate for the sparse and high-dimensional nature of text data. Specifically, let A be the $n \times d$ term-document matrix in which the (i, j)th entry is the normalized frequency for term j in document i. Then, $A^T \cdot A$ is a $d \times d$ matrix which is close (scaled) approximation of the covariance matrix, in which the means have not been subtracted out. In other words, the value of $A^T \cdot A$ would be the same as a scaled version (by factor n) of the covariance matrix, if the data is mean-centered. While text-representations are not mean-centered, the sparsity of text ensures that the use of $A^T \cdot A$ is quite a good approximation of the (scaled) covariances. As in the case of numerical data, we use the eigenvectors of $A^T \cdot A$ with the largest variance in order to represent the text. In typical collections, only about 300 to 400 eigenvectors are required for the representation. One excellent characteristic of LSI [28] is that the truncation of the dimensions removes the noise effects of synonymy and polysemy, and the similarity computations are more closely affected by the semantic concepts in the data. This is particularly useful for a semantic application such as text clustering. However, if finer granularity clustering is needed, such low-dimensional space representation of text may not be sufficiently discriminative; in information retrieval, this problem is often solved by mixing the low-dimensional representation with the original high-dimensional word-based representation (see, e.g., [105]).

A similar technique to LSI, but based on probabilistic modeling is Probabilistic Latent Semantic Analysis (PLSA) [49]. The similarity and equivalence of PLSA and LSI are discussed in [49].

2.2.1 Concept Decomposition using Clustering. One interesting observation is that while feature transformation is often used as a pre-processing technique for clustering, the clustering itself can be used for a novel dimensionality reduction technique known as *concept decomposition* [2, 29]. This of course leads to the issue of circularity in the use of this technique for clustering, especially if clustering is required in order to perform the dimensionality reduction. Nevertheless, it is still possible to use this technique effectively for pre-processing with the use of two separate phases of clustering.

The technique of concept decomposition uses any standard clustering technique [2, 29] on the original representation of the documents. The frequent terms in the centroids of these clusters are used as *basis vectors* which are almost orthogonal to one another. The documents can then be represented in a much more concise way in terms of these basis vectors. We note that this condensed conceptual representation allows for enhanced clustering as well as classification. Therefore, a second phase of clustering can be applied on this reduced representation in order to cluster the documents much more effectively. Such a method has also been tested in [87] by using word-clusters in order to represent documents. We will describe this method in more detail later in this chapter.

2.3 Non-negative Matrix Factorization

The non-negative matrix factorization (NMF) technique is a latent-space method, and is particularly suitable to clustering [97]. As in the case of LSI, the NMF scheme represents the documents in a new axis-system which is based on an analysis of the term-document matrix. However, the NMF method has a number of critical differences from the LSI scheme from a conceptual point of view. In particular, the NMF scheme is a feature transformation method which is particularly suited to clustering. The main conceptual characteristics of the NMF scheme, which are very different from LSI are as follows:

- In LSI, the new basis system consists of a set of orthonormal vectors. This is not the case for NMF.

- In NMF, the vectors in the basis system directly correspond to cluster topics. Therefore, the cluster membership for a document may be determined by examining the largest component of the document along any of the vectors. The coordinate of any document along a vector is always non-negative. The expression of each document as an additive combination of the underlying semantics makes a lot of sense from an intuitive perspective. Therefore, the NMF transformation is particularly suited to clustering, and it also provides an intuitive understanding of the basis system in terms of the clusters.

Let A be the $n \times d$ term document matrix. Let us assume that we wish to create k clusters from the underlying document corpus. Then, the non-negative matrix factorization method attempts to determine the matrices U and V which minimize the following objective function:

$$J = (1/2) \cdot ||A - U \cdot V^T|| \tag{4.6}$$

Here $||\cdot||$ represents the sum of the squares of all the elements in the matrix, U is an $n \times k$ non-negative matrix, and V is a $m \times k$ non-negative matrix. We note that the columns of V provide the k basis vectors which correspond to the k different clusters.

What is the significance of the above optimization problem? Note that by minimizing J, we are attempting to factorize A approximately as:

$$A \approx U \cdot V^T \tag{4.7}$$

For each *row a* of A (document vector), we can rewrite the above equation as:

$$a \approx u \cdot V^T \tag{4.8}$$

Here u is the corresponding row of U. Therefore, the document vector a can be rewritten as an approximate linear (non-negative) combination of the basis vector which corresponds to the k columns of V^T. If the value of k is relatively small compared to the corpus, this can only be done if the column vectors of V^T discover the latent structure in the data. Furthermore, the non-negativity of the matrices U and V ensures that the documents are expressed as a non-negative combination of the key concepts (or clustered) regions in the term-based feature space.

Next, we will discuss how the optimization problem for J above is actually solved. The squared norm of any matrix Q can be expressed as the trace of the matrix $Q \cdot Q^T$. Therefore, we can express the objective function above as follows:

$$J = (1/2) \cdot tr((A - U \cdot V^T) \cdot (A - U \cdot V^T)^T)$$
$$= (1/2) \cdot tr(A \cdot A^T) - tr(A \cdot U \cdot V^T) + (1/2) \cdot tr(U \cdot V^T \cdot V \cdot U^T)$$

Thus, we have an optimization problem with respect to the matrices $U = [u_{ij}]$ and $V = [v_{ij}]$, the entries u_{ij} and v_{ij} of which are the variables with respect to which we need to optimize this problem. In addition, since the matrices are non-negative, we have the constraints that $u_{ij} \geq 0$ and $v_{ij} \geq 0$. This is a typical constrained non-linear optimization problem, and can be solved using the Lagrange method. Let $\alpha = [\alpha_{ij}]$ and $\beta = [\beta_{ij}]$ be matrices with the same dimensions as U and V respectively. The elements of the matrices α and β are the corresponding Lagrange multipliers for the non-negativity conditions on the different elements of U and V respectively. We note that $tr(\alpha \cdot U^T)$ is simply equal to $\sum_{i,j} \alpha_{ij} \cdot u_{ij}$ and $tr(\beta \cdot V^T)$ is simply equal to $\sum_{i,j} \beta_{ij} \cdot v_{ij}$. These correspond to the lagrange expressions for the non-negativity constraints. Then, we can express the Lagrangian optimization problem as follows:

$$L = J + tr(\alpha \cdot U^T) + tr(\beta \cdot V^T) \tag{4.9}$$

Then, we can express the partial derivative of L with respect to U and V as follows, and set them to 0:

$$\frac{\delta L}{\delta U} = -A \cdot V + U \cdot V^T \cdot V + \alpha = 0$$
$$\frac{\delta L}{\delta V} = -A^T \cdot U + V \cdot U^T \cdot U + \beta = 0$$

We can then multiply the (i,j)th entry of the above (two matrices of) conditions with u_{ij} and v_{ij} respectively. Using the Kuhn-Tucker conditions $\alpha_{ij} \cdot u_{ij} = 0$ and $\beta_{ij} \cdot v_{ij} = 0$, we get the following:

$$(A \cdot V)_{ij} \cdot u_{ij} - (U \cdot V^T \cdot V)_{ij} \cdot u_{ij} = 0$$
$$(A^T \cdot U)_{ij} \cdot v_{ij} - (V \cdot U^T \cdot U)_{ij} \cdot v_{ij} = 0$$

We note that these conditions are independent of α and β. This leads to the following iterative updating rules for u_{ij} and v_{ij}:

$$u_{ij} = \frac{(A \cdot V)_{ij} \cdot u_{ij}}{(U \cdot V^T \cdot V)_{ij}}$$
$$v_{ij} = \frac{(A^T \cdot U)_{ij} \cdot v_{ij}}{(V \cdot U^T \cdot U)_{ij}}$$

It has been shown in [58] that the objective function continuously improves under these update rules, and converges to an optimal solution.

One interesting observation about the matrix factorization technique is that it can also be used to determine word-clusters instead of document clusters. Just as the columns of V provide a basis which can be used to discover document clusters, we can use the columns of U to discover a basis which correspond to word clusters. As we will see later, document clusters and word clusters are closely related, and it is often useful to discover both simultaneously, as in frameworks such as *co-clustering* [30, 31, 75]. Matrix-factorization provides a natural way of achieving this goal. It has also been shown both theoretically and experimentally [33, 93] that the matrix-factorization technique is equivalent to another graph-structure based document clustering technique known as *spectral clustering*. An analogous technique called *concept factorization* was proposed in [98], which can also be applied to data points with negative values in them.

3. Distance-based Clustering Algorithms

Distance-based clustering algorithms are designed by using a similarity function to measure the closeness between the text objects. The most well known similarity function which is used commonly in the text domain is the cosine similarity function. Let $U = (f(u_1) \ldots f(u_k))$ and $V = (f(v_1) \ldots f(v_k))$ be the damped and normalized frequency term vector in two different documents U and V. The values $u_1 \ldots u_k$ and $v_1 \ldots v_k$ represent the (normalized) term frequencies, and the function $f(\cdot)$ represents the damping function. Typical damping functions for $f(\cdot)$ could represent either the square-root or the logarithm [25]. Then, the cosine similarity between the two documents is defined as follows:

$$cosine(U, V) = \frac{\sum_{i=1}^{k} f(u_i) \cdot f(v_i)}{\sqrt{\sum_{i=1}^{k} f(u_i)^2} \cdot \sqrt{\sum_{i=1}^{k} f(v_i)^2}} \qquad (4.10)$$

Computation of text similarity is a fundamental problem in information retrieval. Although most of the work in information retrieval has focused on how to assess the similarity of a keyword query and a text document, rather than the similarity between two documents, many weighting heuristics and similarity functions can also be applied to optimize the similarity function for clustering. Effective information retrieval models generally capture three heuristics, i.e., TF weighting, IDF weighting, and document length normalization [36]. One effective way to assign weights to terms when representing a document as a weighted term vector is the BM25 term weighting method [78], where the normalized TF not only addresses length normalization, but also has an upper bound which improves the robustness as it avoids overly rewarding the matching of any particular term. A document can also be represented with a probability distribution over words (i.e., unigram language models), and the similarity can then be measured based an information theoretic measure such as cross entropy or Kullback-Leibler divergencce [105]. For clustering, symmetric variants of such a similarity function may be more appropriate.

One challenge in clustering short segments of text (e.g., tweets or sentences) is that exact keyword matching may not work well. One general strategy for solving this problem is to expand text representation by exploiting related text documents, which is related to smoothing of a document language model in information retrieval [105]. A specific technique, which leverages a search engine to expand text representation, was proposed in [79]. A comparison of several simple measures for computing similarity of short text segments can be found in [66].

These similarity functions can be used in conjunction with a wide variety of traditional clustering algorithms [50, 54]. In the next subsections, we will discuss some of these techniques.

3.1 Agglomerative and Hierarchical Clustering Algorithms

Hierarchical clustering algorithms have been studied extensively in the clustering literature [50, 54] for records of different kinds including multidimensional numerical data, categorical data and text data. An overview of the traditional agglomerative and hierarchical clustering algorithms in the context of text data is provided in [69, 70, 92, 96]. An experimental comparison of different hierarchical clustering algorithms may be found in [110]. The method of agglomerative hierarchical clustering is particularly useful to support a variety of searching methods because it naturally creates a tree-like hierarchy which can be leveraged for the search process. In particular, the effectiveness of this method in improving the search efficiency over a sequential scan has been shown in [51, 77].

The general concept of agglomerative clustering is to successively merge documents into clusters based on their similarity with one another. Almost all the hierarchical clustering algorithms successively merge groups based on the best pairwise similarity between these groups of documents. The main differences between these classes of methods are in terms of how this pairwise similarity is computed between the different groups of documents. For example, the similarity between a pair of groups may be computed as the best-case similarity, average-case similarity, or worst-case similarity between documents which are drawn from these pairs of groups. Conceptually, the process of agglomerating documents into successively higher levels of clusters creates a cluster hierarchy (or dendogram) for which the leaf nodes correspond to individual documents, and the internal nodes correspond to the merged groups of clusters. When two groups are merged, a new node is created in this tree corresponding to this larger merged group. The two children of this node correspond to the two groups of documents which have been merged to it.

The different methods for merging groups of documents for the different agglomerative methods are as follows:

- **Single Linkage Clustering:** In single linkage clustering, the similarity between two groups of documents is the greatest similarity between any pair of documents from these two groups. In single link clustering we merge the two groups which are such that their closest pair of documents have the highest similarity compared to any other pair of groups. The main advantage of single linkage clustering is that it is extremely efficient to implement in practice. This is because we can first compute all similarity pairs and sort them in order of reducing similarity. These pairs are processed in this pre-defined order and the merge is performed successively if the pairs belong to different groups. It can be easily shown that this approach is equivalent to the single-linkage method. This is essentially equivalent to a spanning tree algorithm on the complete graph of pairwise-distances by processing the edges of the graph in a certain order. It has been shown in [92] how Prim's minimum spanning tree algorithm can be adapted to single-linkage clustering. Another method in [24] designs the single-linkage method in conjunction with the inverted index method in order to avoid computing zero similarities.

  The main drawback of this approach is that it can lead to the phenomenon of *chaining* in which a chain of similar documents lead to disparate documents being grouped into the same clusters. In other words, if A is similar to B and B is similar to C, it does not always imply that A is similar to C, because of lack of transitivity in similarity computations. Single linkage clustering encourages the grouping of documents through such transitivity chains. This can often lead to poor clusters, especially at the higher levels of the agglomeration. Effective methods for implementing single-linkage clustering for the case of document data may be found in [24, 92].

- **Group-Average Linkage Clustering:** In group-average linkage clustering, the similarity between two clusters is the *average* similarity between the pairs of documents in the two clusters. Clearly, the average linkage clustering process is somewhat slower than single-linkage clustering, because we need to determine the average similarity between a large number of pairs in order to determine group-wise similarity. On the other hand, it is much more robust in terms of clustering quality, because it does not exhibit the chaining behavior of single linkage clustering. It is possible to speed up the average linkage clustering algorithm by approximating the average linkage similarity between two clusters C_1 and C_2 by computing the similarity between the mean document of C_1

and the mean document of C_2. While this approach does not work equally well for all data domains, it works particularly well for the case of text data. In this case, the running time can be reduced to $O(n^2)$, where n is the total number of nodes. The method can be implemented quite efficiently in the case of document data, because the centroid of a cluster is simply the concatenation of the documents in that cluster.

- **Complete Linkage Clustering:** In this technique, the similarity between two clusters is the *worst-case* similarity between any pair of documents in the two clusters. Complete-linkage clustering can also avoid chaining because it avoids the placement of any pair of very disparate points in the same cluster. However, like group-average clustering, it is computationally more expensive than the single-linkage method. The complete linkage clustering method requires $O(n^2)$ space and $O(n^3)$ time. The space requirement can however be significantly lower in the case of the text data domain, because a large number of pairwise similarities are zero.

Hierarchical clustering algorithms have also been designed in the context of text data streams. A distributional modeling method for hierarchical clustering of streaming documents has been proposed in [80]. The main idea is to model the frequency of word-presence in documents with the use of a multi-poisson distribution. The parameters of this model are learned in order to assign documents to clusters. The method extends the COBWEB and CLASSIT algorithms [37, 40] to the case of text data. The work in [80] studies the different kinds of distributional assumptions of words in documents. We note that these distributional assumptions are required to adapt these algorithms to the case of text data. The approach essentially changes the distributional assumption so that the method can work effectively for text data.

3.2 Distance-based Partitioning Algorithms

Partitioning algorithms are widely used in the database literature in order to efficiently create clusters of objects. The two most widely used distance-based partitioning algorithms [50, 54] are as follows:

- **k-medoid clustering algorithms:** In k-medoid clustering algorithms, we use a set of points from the original data as the anchors (or medoids) around which the clusters are built. The key aim of the algorithm is to determine an optimal set of representative documents *from the original corpus* around which the clusters are built. Each document is assigned to its closest representative from the collection. This creates a running set of clusters from the corpus which are successively improved by a randomized process.

The algorithm works with an iterative approach in which the set of k representatives are successively improved with the use of randomized inter-changes. Specifically, we use the average similarity of each document in the corpus to its closest representative as the objective function which needs to be improved during this interchange process. In each iteration, we replace a randomly picked representative in the current set of medoids with a randomly picked representative from the collection, if it improves the clustering objective function. This approach is applied until convergence is achieved.

There are two main disadvantages of the use of k-medoids based clustering algorithms, one of which is specific to the case of text data. One general disadvantage of k-medoids clustering algorithms is that they require a large number of iterations in order to achieve convergence and are therefore quite slow. This is because each iteration requires the computation of an objective function whose time requirement is proportional to the size of the underlying corpus.

The second key disadvantage is that k-medoid algorithms do not work very well for sparse data such as text. This is because a large fraction of document pairs do not have many words in common, and the similarities between such document pairs are small (and noisy) values. Therefore, a single document medoid often does not contain all the concepts required in order to effectively build a cluster around it. This characteristic is specific to the case of the information retrieval domain, because of the sparse nature of the underlying text data.

- k-**means clustering algorithms:** The k-means clustering algorithm also uses a set of k representatives around which the clusters are built. However, these representatives are not necessarily obtained from the original data and are refined somewhat differently than a k-medoids approach. The simplest form of the k-means approach is to start off with a set of k seeds from the original corpus, and assign documents to these seeds on the basis of closest similarity. In the next iteration, the centroid of the assigned points to each seed is used to replace the seed in the last iteration. In other words, the new seed is defined, so that it is a better central point for this cluster. This approach is continued until convergence. One of the advantages of the k-means method over the k-medoids method is that it requires an extremely small number of iterations in order to converge. Observations from [25, 83] seem to suggest that for many large data sets, it is sufficient to use 5 or less iterations for an effective clustering. The main disadvantage of the k-means method is that it is still quite sensitive to the initial set of seeds picked during the clustering. Secondly, the centroid for a given cluster of documents may contain a large number of words. This will slow down the similarity calculations in the next iteration. A number of methods are used to reduce these effects, which will be discussed later on in this chapter.

The initial choice of seeds affects the quality of k-means clustering, especially in the case of document clustering. Therefore, a number of techniques are used in order to improve the quality of the initial seeds which are picked for the clustering process. For example, another lightweight clustering method such as an agglomerative clustering technique can be used in order to decide the initial set of seeds. This is at the core of the method discussed in [25] for effective document clustering. We will discuss this method in detail in the next subsection.

A second method for improving the initial set of seeds is to use some form of partial supervision in the process of initial seed creation. This form of partial supervision can also be helpful in creating clusters which are designed for particular application-specific criteria. An example of such an approach is discussed in [4] in which we pick the initial set of seeds as the centroids of the documents crawled from a particular category if the *Yahoo*! taxonomy. This also has the effect that the final set of clusters are grouped by the coherence of content within the different *Yahoo*! categories. The approach has been shown to be quite effective for use in a number of applications such as text categorization. Such semi-supervised techniques are particularly useful for information organization in cases where the starting set of categories is somewhat noisy, but contains enough information in order to create clusters which satisfy a pre-defined kind of organization.

3.3 A Hybrid Approach: The Scatter-Gather Method

While hierarchical clustering methods tend to be more robust because of their tendency to compare all pairs of documents, they are generally not very efficient, because of their tendency to require at least $O(n^2)$ time. On the other hand, k-means type algorithms are more efficient than hierarchical algorithms, but may sometimes not be very effective because of their tendency to rely on a small number of seeds.

The method in [25] uses both hierarchical and partitional clustering algorithms to good effect. Specifically, it uses a hierarchical clustering algorithm on a sample of the corpus in order to find a robust initial set of seeds. This robust set of seeds is used in conjunction with a standard k-means clustering algorithm in order to determine good clusters. The size of the sample in the initial phase is carefully tailored so as to provide the best possible effectiveness without this phase becoming a bottleneck in algorithm execution.

There are two possible methods for creating the initial set of seeds, which are referred to as *buckshot* and *fractionation* respectively. These are two alternative methods, and are described as follows:

- **Buckshot:** Let k be the number of clusters to be found and n be the number of documents in the corpus. Instead of picking the k seeds randomly from the collection, the buckshot scheme picks an overestimate $\sqrt{k \cdot n}$ of the seeds, and then agglomerates these to k seeds. Standard agglomerative hierarchical clustering algorithms (requiring quadratic time) are applied to this initial sample of $\sqrt{k \cdot n}$ seeds. Since we use quadratically scalable algorithms in this phase, this approach requires $O(k \cdot n)$ time. We note that this seed set is much more robust than one which simply samples for k seeds, because of the summarization of a large document sample into a robust set of k seeds.

- **Fractionation:** The fractionation algorithm initially breaks up the corpus into n/m buckets of size $m > k$ each. An agglomerative algorithm is applied to each of these buckets to reduce them by a factor of ν. Thus, at the end of the phase, we have a total of $\nu \cdot n$ agglomerated points. The process is repeated by treating each of these agglomerated points as an individual record. This is achieved by merging the different documents within an agglomerated cluster into a single document. The approach terminates when a total of k seeds remain. We note that the the agglomerative clustering of each group of m documents in the first iteration of the fractionation algorithm requires $O(m^2)$ time, which sums to $O(n \cdot m)$ over the n/m different groups. Since, the number of individuals reduces geometrically by a factor of ν in each iteration, the total running time over all iterations is $O(n \cdot m \cdot (1 + \mu + \nu^2 + \ldots))$. For constant $\nu < 1$, the running time over all iterations is still $O(n \cdot m)$. By picking $m = O(k)$, we can still ensure a running time of $O(n \cdot k)$ for the initialization procedure.

The *Buckshot* and *Fractionation* procedures require $O(k \cdot n)$ time which is also equivalent to running time of one iteration of the k means algorithm.

Each iteration of the K-means algorithm also requires $O(k \cdot n)$ time because we need to compute the similarity of the n documents to the k different seeds.

We further note that the fractionation procedure can be applied to a random grouping of the documents into n/m different buckets. Of course, one can also replace the random grouping approach with a more carefully designed procedure for more effective results. One such procedure is to sort the documents by the index of the jth most common word in the document. Here j is chosen to be a small number such as 3, which corresponds to medium frequency words in the data. The documents are then partitioned into groups based on this sort order by segmenting out continuous groups of m documents. This approach ensures that the groups created have at least a few common words in them and are therefore not completely random. This can sometimes provide a better quality of the centers which are determined by the fractionation algorithm.

Once the initial cluster centers have been determined with the use of the *Buckshot* or *Fractionation* algorithms we can apply standard k-means partitioning algorithms. Specifically, we each document is assigned to the nearest of the k cluster centers. The centroid of each such cluster is determined as the concatenation of the different documents in a cluster. These centroids replace the sets of seeds from the last iteration. This process can be repeated in an iterative approach in order to successively refine the centers for the clusters. Typically, only a smaller number of iterations are required, because the greatest improvements occur only in the first few iterations.

It is also possible to use a number of procedures to further improve the quality of the underlying clusters. These procedures are as follows:

- **Split Operation:** The process of splitting can be used in order to further refine the clusters into groups of better granularity. This can be achieved by applying the buckshot procedure on the individual documents in a cluster by using $k = 2$, and then re-clustering around these centers. This entire procedure requires $O(k \cdot n_i)$ time for a cluster containing n_i data points, and therefore splitting all the groups requires $O(k \cdot n)$ time. However, it is not necessary to split *all* the groups. Instead, only a subset of the groups can be split. Those are the groups which are not very coherent and contain documents of a disparate nature. In order to measure the coherence of a group, we compute the self-similarity of a cluster. This self-similarity provides us with an understanding of the underlying coherence. This quantity can be computed both in terms of the similarity of the documents in a cluster to its centroid or in terms of the similarity of the cluster documents to each other. The split criterion can then be applied selectively only to those clusters which have low self similarity. This helps in creating more coherent clusters.

- **Join Operation:** The join operation attempts to merge similar clusters into a single cluster. In order to perform the merge, we compute the *topical* words of each cluster by examining the most frequent words of the centroid. Two clusters are considered similar, if there is significant overlap between the topical words of the two clusters.

We note that the method is often referred to as the *Scatter-Gather* clustering method, but this is more because of how the clustering method has been presented in terms of its use for browsing large collections in the original paper [25]. The scatter-gather approach can be used for organized browsing of large document collections, because it creates a natural hierarchy of similar documents. In particular, a user may wish to browse the hierarchy of clusters in an interactive way in order to understand topics of different levels of granularity in the collection. One possibility is to perform a hierarchical clustering a-priori; however such an approach has the disadvantage that it is unable to merge and re-cluster related branches of the tree hierarchy on-the-fly when a user may need it. A method for constant-interaction time browsing with the use of the scatter-gather approach has been presented in [26]. This approach presents the keywords associated with the different keywords to a user. The user may pick one or more of these keywords, which also corresponds to one or more clusters. The documents in these clusters are merged and re-clustered to a finer-granularity on-the-fly. This finer granularity of clustering is presented to the user for further exploration. The set of documents which is picked by the user for exploration is referred to as the *focus set*. Next we will explain how this focus set is further explored and re-clustered on the fly in constant-time.

The key assumption in order to enable this approach is the *cluster refinement hypothesis*. This hypothesis states that documents which belong to the same cluster in a significantly finer granularity partitioning will also occur together in a partitioning with coarser granularity. The first step is to create a hierarchy of the documents in the clusters. A variety of agglomerative algorithms such as the buckshot method can be used for this purpose. We note that each (internal) node of this tree can be viewed as a meta-document corresponding to the concatenation of all the documents in the leaves of this subtree. The cluster-refinement hypothesis allows us to work with a smaller set of meta-documents rather than the entire set of documents in a particular subtree. The idea is to pick a constant M which represents the maximum number of meta-documents that we are willing to re-cluster with the use of the interactive approach. The tree nodes in the focus set are then expanded (with priority to the branches with largest degree), to a maximum of M nodes. These M nodes are then re-clustered on-the-fly with the scatter-gather approach. This requires constant time because of the use of a constant number M of meta-documents in the clustering process. Thus, by working with the meta-documents for M. we assume the cluster-refinement hypothesis of all nodes of the subtree at the lower level. Clearly, a larger value of M does not assume the cluster-refinement hypothesis quite as strongly, but also comes at a higher cost. The details of the algorithm are described in [26]. Some extensions of this approach are also presented in [85], in which it has been shown how this approach can be used to cluster arbitrary corpus subsets of the documents in constant time. Another recent online clustering algorithm called *LAIR2* [55] provides constant-interaction time for Scatter/Gather browsing. The parallelization of this algorithm is significantly faster than a corresponding version of the Buckshot algorithm. It has also been suggested that the *LAIR2* algorithm leads to better quality clusters in the data.

3.3.1 Projections for Efficient Document Clustering.

One of the challenges of the scatter-gather algorithm is that even though the algorithm is designed to balance the running times of the agglomerative and partitioning phases quite well, it sometimes suffer a slowdown in large document collections because of the massive number of distinct terms that a given cluster centroid may contain. Recall that a cluster centroid in the scatter-gather algorithm is defined as the concatenation of all the documents in that collection. When the number of documents in the cluster is large, this will also lead to a large number of distinct terms in the centroid. This will also lead to a slow down of a number of critical computations such as similarity calculations between documents and cluster centroids.

An interesting solution to this problem has been proposed in [83]. The idea is to use the concept of *projection* in order to reduce the dimensionality of the document representation. Such a reduction in dimensionality will lead to significant speedups, because the similarity computations will be made much more efficient. The work in [83] proposes three kinds of projections:

- **Global Projection:** In global projection, the dimensionality of the original data set is reduced in order to remove the least important (weighted) terms from the data. The weight of a term is defined as the aggregate of the (normalized and damped) frequencies of the terms in the documents.

- **Local Projection:** In local projection, the dimensionality of the documents in each cluster are reduced with a *locally specific approach* for that cluster. Thus, the terms in each cluster centroid are truncated separately. Specifically, the least weight terms in the different cluster centroids are removed. Thus, the terms removed from each document may be different, depending upon their local importance.

- **Latent Semantic Indexing:** In this case, the document-space is transformed with an LSI technique, and the clustering is applied to the transformed document space. We note that the LSI technique can also be applied either globally to the whole document collection, or locally to each cluster if desired.

It has been shown in [83] that the projection approaches provide competitive results in terms of effectiveness while retaining an extremely high level of efficiency with respect to all the competing approaches. In this sense, the clustering methods are different from similarity search because they show little degradation in quality, when projections are performed. One of the reasons for this is that clustering is a much less fine grained application as compared to similarity search, and therefore there is no perceptible difference in quality even when we work with a truncated feature space.

4. Word and Phrase-based Clustering

Since text documents are drawn from an inherently high-dimensional domain, it can be useful to view the problem in a dual way, in which important clusters of words may be found and utilized for finding clusters of documents. In a corpus containing d terms and n documents, one may view a term-document matrix as an $n \times d$ matrix, in which the (i,j)th entry is the frequency of the jth term in the ith document. We note that this matrix is extremely sparse since a given document contains an extremely small fraction of the universe of words. We note that the problem of clustering *rows* in this matrix is that of clustering documents, whereas that of clustering *columns* in this matrix is that of clustering words. In reality, the two problems are closely related, as good clusters of words may be leveraged in order to find good clusters of documents and vice-versa. For example, the work in [16] determines frequent itemsets of words in the document collection, and uses them to determine compact clusters of documents. This is somewhat analogous to the use of clusters of words [87] for determining clusters of documents. The most general technique for simultaneous word and document clustering is referred to as *co-clustering* [30, 31]. This approach simultaneous clusters the rows and columns of the term-document matrix, in order to create such clusters. This can also be considered to be equivalent to the problem of re-ordering the rows and columns of the term-document matrix so as to create dense rectangular blocks of non-zero entries in this matrix. In some cases, the ordering information among words may be used in order to determine good clusters. The work in [103] determines the frequent phrases in the collection and leverages them in order to determine document clusters.

It is important to understand that the problem of word clusters and document clusters are essentially dual problems which are closely related to one another. The former is related to dimensionality reduction, whereas the latter is related to traditional clustering. The boundary between the two problems is quite fluid, because good word clusters provide hints for finding good document clusters and vice-versa. For example, a more general probabilistic framework which determines word clusters and document clusters simultaneously is referred to as *topic modeling* [49]. Topic modeling is a more general framework than either clustering or dimensionality reduction. We will introduce the method of topic modeling in a later section of this chapter. A more detailed treatment is also provided in the next chapter in this book, which is on dimensionality reduction, and in Chapter 8 where a more general discussion of probabilistic models for text mining is given.

4.1 Clustering with Frequent Word Patterns

Frequent pattern mining [8] is a technique which has been widely used in the data mining literature in order to determine the most relevant patterns in transactional data. The clustering approach in [16] is designed on the basis of such frequent pattern mining algorithms. A frequent itemset in the context of text data is also referred to as a *frequent term set*, because we are dealing with documents rather than transactions. The main idea of the approach is to not cluster the high dimensional document data set, but consider the low dimensional frequent term sets as cluster candidates. This essentially means that a frequent terms set is a description of a cluster which corresponds to all the documents containing that frequent term set. Since a frequent term set can be considered a description of a cluster, a set of carefully chosen frequent terms sets can be considered a clustering. The appropriate choice of this set of frequent term sets is defined on the basis of the overlaps between the supporting documents of the different frequent term sets.

The notion of clustering defined in [16] does not necessarily use a strict partitioning in order to define the clusters of documents, but it allows a certain level of overlap. This is a natural property of many term- and phrase-based clustering algorithms because one does not directly control the assignment of documents to clusters during the algorithm execution. Allowing some level of overlap between clusters may sometimes be more appropriate, because it recognizes the fact that documents are complex objects and it is impossible to cleanly partition documents into specific clusters, especially when some of the clusters are partially related to one another. The clustering definition of [16] assumes that each document is covered by at least one frequent term set.

Let R be the set of chosen frequent term sets which define the clustering. Let f_i be the number of frequent term sets in R which are contained in the ith document. The value of f_i is at least one in order to ensure complete coverage, but we would otherwise like it to be as low as possible in order to minimize overlap. Therefore, we would like the average value of $(f_i - 1)$ for the documents in a given cluster to be as low as possible. We can compute the average value of $(f_i - 1)$ for the documents in the cluster and try to pick frequent term sets such that this value is as low as possible. However, such an approach would tend to favor frequent term sets containing very few terms. This is because if a term set contains m terms, then all subsets of it would also be covered by the document, as a result of which the standard overlap would be increased. The entropy overlap of a given term is essentially the sum of the values of $-(1/f_i) \cdot \log(1/f_i)$ over all documents in the cluster. This value is 0, when each document has $f_i = 1$, and increases monotonically with increasing f_i values.

It then remains to describe how the frequent term sets are selected from the collection. Two algorithms are described in [16], one of which corresponds to a flat clustering, and the other corresponds to a hierarchical clustering. We will first describe the method for flat clustering. Clearly, the search space of frequent terms is exponential, and therefore a reasonable solution is to utilize a greedy algorithm to select the frequent terms sets. In each iteration of the greedy algorithm, we pick the frequent term set with a cover having the minimum overlap with other cluster candidates. The documents covered by the selected frequent term are removed from the database, and the overlap in the next iteration is computed with respect to the remaining documents.

The hierarchical version of the algorithm is similar to the broad idea in flat clustering, with the main difference that each level of the clustering is applied to a set of term sets containing a fixed number k of terms. In other words, we are working only with frequent patterns of length k for the selection process. The resulting clusters are then further partitioned by applying the approach for $(k+1)$-term sets. For further partitioning a given cluster, we use only those $(k+1)$-term sets which contain the frequent k-term set defining that cluster. More details of the approach may be found in [16].

4.2 Leveraging Word Clusters for Document Clusters

A two phase clustering procedure is discussed in [87], which uses the following steps to perform document clustering:

- In the first phase, we determine word-clusters from the documents in such a way that most of mutual information between words and documents is preserved when we represent the documents in terms of word clusters rather than words.

- In the second phase, we use the condensed representation of the documents in terms of word-clusters in order to perform the final document clustering. Specifically, we replace the word occurrences in documents with word-cluster occurrences in order to perform the document clustering. One advantage of this two-phase procedure is the significant reduction in the noise in the representation.

Let $X = x_1 \ldots x_n$ be the random variables corresponding to the rows (documents), and let $Y = y_1 \ldots y_d$ be the random variables corresponding to the columns (words). We would like to partition X into k clusters, and Y into l clusters. Let the clusters be denoted by $\hat{X} = \hat{x_1} \ldots \hat{x_k}$ and $\hat{Y} = \hat{y_1} \ldots \hat{y_l}$. In other words, we wish to find the maps C_X and C_Y, which define the clustering:

$$C_X : x_1 \ldots x_n \Rightarrow \hat{x_1} \ldots \hat{x_k}$$
$$C_Y : y_1 \ldots y_d \Rightarrow \hat{y_1} \ldots \hat{y_l}$$

In the first phase of the procedure we cluster Y to $\hat{Y}$, so that most of the information in $I(X, Y)$ is preserved in $I(X, \hat{Y})$. In the second phase, we perform the clustering again from X to $\hat{X}$ using exactly the same procedure so that as much information as possible from $I(X, \hat{Y})$ is preserved in $I(\hat{X}, \hat{Y})$. Details of how each phase of the clustering is performed is provided in [87].

How to discover interesting word clusters (which can be leveraged for document clustering) has itself attracted attention in the natural language processing research community, with particular interests in discovering word clusters that can characterize word senses [34] or a semantic concept [21]. In [34], for example, the Markov clustering algorithm was applied to discover corpus-specific word senses in an unsupervised way. Specifically, a word association graph is first constructed in which related words would be connected with an edge. For a given word that potentially has multiple senses, we can then isolate the subgraph representing its neighbors. These neighbors are expected to form clusters according to different senses of the target word, thus by grouping together neighbors that are well connected with each other, we can discover word clusters that characterize different senses of the target word. In [21], an n-gram class language model was proposed to cluster words based on minimizing the loss of mutual information between adjacent words, which can achieve the effect of grouping together words that share similar context in natural language text.

4.3 Co-clustering Words and Documents

In many cases, it is desirable to simultaneously cluster the rows and columns of the contingency table, and explore the interplay between word clusters and document clusters during the clustering process. Since the clusters among words and documents are clearly related, it is often desirable to cluster both simultaneously when when it is desirable to find clusters along one of the two dimensions. Such an approach is referred to as *co-clustering* [30, 31]. Co-clustering is defined as a pair of maps from rows to row-cluster indices and columns to column-cluster indices. These maps are determined *simultaneously* by the algorithm in order to optimize the corresponding cluster representations.

We further note that the matrix factorization approach [58] discussed earlier in this chapter can be naturally used for co-clustering because it discovers word clusters and document clusters simultaneously. In that section, we have also discussed how matrix factorization can be viewed as a co-clustering technique. While matrix factorization has not widely been used as a technique for co-clustering, we point out this natural connection, as possible exploration for future comparison with other co-clustering methods. Some recent work [60] has shown how matrix factorization can be used in order to transform knowledge from word space to document space in the context of document clustering techniques.

The problem of co-clustering is also closely related to the problem of *subspace clustering* [7] or *projected clustering* [5] in quantitative data in the database literature. In this problem, the data is clustered by simultaneously associating it with a set of points and subspaces in multi-dimensional space. The concept of co-clustering is a natural application of this broad idea to data domains which can be represented as **sparse** high dimensional matrices in which most of the entries are 0. Therefore, traditional methods for subspace clustering can also be extended to the problem of co-clustering. For example, an adaptive iterative subspace clustering method for documents was proposed in [59].

We note that subspace clustering or co-clustering can be considered a form of *local feature selection*, in which the features selected are specific to each cluster. A natural question arises, as to whether the features can be selected as a linear combination of dimensions as in the case of traditional dimensionality reduction techniques such as PCA [53]. This is also known as *local dimensionality reduction* [22] or *generalized projected clustering* [6] in the traditional database literature. In this method, PCA-based techniques are used in order to generate subspace representations which are *specific to each cluster*, and are leveraged in order to achieve a better clustering process. In particular, such an approach has recently been designed [32], which has been shown to work well with document data.

In this section, we will study two well known methods for document co-clustering, which are commonly used in the document clustering literature. One of these methods uses graph-based term-document representations [30] and the other uses information theory [31]. We will discuss both of these methods below.

4.3.1 Co-clustering with graph partitioning. The core idea in this approach [30] is to represent the term-document matrix as a bipartite graph $G = (V_1 \cup V_2, E)$, where V_1 and V_2 represent the vertex sets in the two bipartite portions of this graph, and E represents the edge set. Each node in V_1 corresponds to one of the n documents, and each node in V_2 corresponds to one of the d terms. An undirected edge exists between node $i \in V_1$ and node $j \in V_2$ if document i contains the term j. We note that there are no edges in E directly between terms, or directly between documents. Therefore, the graph is bipartite. The weight of each edge is the corresponding normalized term-frequency.

We note that a word partitioning in this bipartite graph induces a document partitioning and vice-versa. Given a partitioning of the documents in this graph, we can associate each word with the document cluster to which it is connected with the most weight of edges. Note that this criterion also minimizes the weight of the edges across the partitions. Similarly, given a word partitioning, we can associate each document with the word partition to which it is connected with the greatest weight of edges. Therefore, a natural solution to this problem would

A Survey of Text Clustering Algorithms

be *simultaneously* perform the k-way partitioning of this graph which minimizes the total weight of the edges across the partitions. This is of course a classical problem in the graph partitioning literature. In [30], it has been shown how a spectral partitioning algorithm can be used effectively for this purpose. Another method discussed in [75] uses an isometric bipartite graph-partitioning approach for the clustering process.

4.3.2 Information-Theoretic Co-clustering.

In [31], the optimal clustering has been defined to be one which maximizes the mutual information between the clustered random variables. The normalized non-negative contingency table is treated as a joint probability distribution between two discrete random variables which take values over rows and columns. Let $X = x_1 \ldots x_n$ be the random variables corresponding to the rows, and let $Y = y_1 \ldots y_d$ be the random variables corresponding to the columns. We would like to partition X into k clusters, and Y into l clusters. Let the clusters be denoted by $\hat{X} = \hat{x}_1 \ldots \hat{x}_k$ and $\hat{Y} = \hat{y}_1 \ldots \hat{y}_l$. In other words, we wish to find the maps C_X and C_Y, which define the clustering:

$$C_X : x_1 \ldots x_n \Rightarrow \hat{x}_1 \ldots \hat{x}_k$$
$$C_Y : y_1 \ldots y_d \Rightarrow \hat{y}_1 \ldots \hat{y}_l$$

The partition functions C_X and C_Y are allowed to depend on the joint probability distribution $p(X, Y)$. We note that since $\hat{X}$ and $\hat{Y}$ are higher level clusters of X and Y, there is loss in mutual information in the higher level representations. In other words, the distribution $p(\hat{X}, \hat{Y})$ contains less information than $p(X, Y)$, and the mutual information $I(\hat{X}, \hat{Y})$ is lower than the mutual information $I(X, Y)$. Therefore, the optimal co-clustering problem is to determine the mapping which minimizes the loss in mutual information. In other words, we wish to find a co-clustering for which $I(X, Y) - I(\hat{X}, \hat{Y})$ is as small as possible. An iterative algorithm for finding a co-clustering which minimizes mutual information loss is proposed in [29].

4.4 Clustering with Frequent Phrases

One of the key differences of this method from other text clustering methods is that it treats a document as a string as opposed to a bag of words. Specifically, each document is treated as a string of *words*, rather than characters. The main difference between the string representation and the bag-of-words representation is that the former also retains ordering information for the clustering process. As is the case with many clustering methods, it uses an indexing method in order to organize the phrases in the document collection, and then uses this organization to create the clusters [103, 104]. Several steps are used in order to create the clusters:

(1) The first step is to perform the cleaning of the strings representing the documents. A light stemming algorithm is used by deleting word prefixes and suffixes and reducing plural to singular. Sentence boundaries are marked and non-word tokens are stripped.

(2) The second step is the identification of base clusters. These are defined by the frequent phases in the collection which are represented in the form of a *suffix tree*. A suffix tree [45] is essentially a trie which contains all the suffixes of the entire collection. Each node of the suffix tree represents a group of documents, and a phrase which is common to all these documents. Since each node of the suffix-tree also corresponds to a group of documents, it also corresponds to a base clustering. Each base cluster is given a score which is essentially the product of the number of documents in that cluster and a non-decreasing function of the length of the underlying phrase. Therefore, clusters containing a large number of documents, and which are defined by a relatively long phrase are more desirable.

(3) An important characteristic of the base clusters created by the suffix tree is that they do not define a strict partitioning and have overlaps with one another. For example, the same document may contain multiple phrases in different parts of the suffix tree, and will therefore be included in the corresponding document groups. The third step of the algorithm merges the clusters based on the similarity of their underlying document sets. Let P and Q be the document sets corresponding to two clusters. The base similarity $BS(P,Q)$ is defined as follows:

$$BS(P,Q) = \left\lfloor \frac{|P \cap Q|}{\max\{|P|,|Q|\}} + 0.5 \right\rfloor \quad (4.11)$$

This base similarity is either 0 or 1, depending upon whether the two groups have at least 50% of their documents in common. Then, we construct a graph structure in which the nodes represent the base clusters, and an edge exists between two cluster nodes, if the corresponding base similarity between that pair of groups is 1. The connected components in this graph define the final clusters. Specifically, the union of the groups of documents in each connected component is used as the final set of clusters. We note that the final set of clusters have much less overlap with one another, but they still do not define a strict partitioning. This is sometimes the case with clustering algorithms in which modest overlaps are allowed to enable better clustering quality.

5. Probabilistic Document Clustering and Topic Models

A popular method for probabilistic document clustering is that of *topic modeling*. The idea of topic modeling is to create a *probabilistic generative model* for the text documents in the corpus. The main approach is to represent a corpus as a function of hidden random variables, the parameters of which are estimated using a particular document collection. The primary assumptions in any topic modeling approach (together with the corresponding random variables) are as follows:

- The n documents in the corpus are assumed to have a probability of belonging to one of k topics. Thus, a given document may have a probability of belonging to multiple topics, and this reflects the fact that the same document may contain a multitude of subjects. For a given document D_i, and a set of topics $T_1 \ldots T_k$, the probability that the document D_i belongs to the topic T_j is given by $P(T_j|D_i)$. We note that the the topics are essentially analogous to clusters, and the value of $P(T_j|D_i)$ provides a probability of cluster membership of the ith document to the jth cluster. In non-probabilistic clustering methods, the membership of documents to clusters is deterministic in nature, and therefore the clustering is typically a clean partitioning of the document collection. However, this often creates challenges, when there are overlaps in document subject matter across multiple clusters. The use of a *soft cluster membership in terms of probabilities* is an elegant solution to this dilemma. In this scenario, the determination of the membership of the documents to clusters is a secondary goal to that of finding the *latent topical clusters* in the underlying text collection. Therefore, this area of research is referred to as *topic modeling*, and while it is related to the clustering problem, it is often studied as a distinct area of research from clustering.

  The value of $P(T_j|D_i)$ is estimated using the topic modeling approach, and is one of the primary outputs of the algorithm. The value of k is one of the inputs to the algorithm and is analogous to the number of clusters.

- Each topic is associated with a probability vector, which quantifies the probability of the different terms in the lexicon for that topic. Let $t_1 \ldots t_d$ be the d terms in the lexicon. Then, for a document that belongs completely to topic T_j, the probability that the term t_l occurs in it is given by $P(t_l|T_j)$. The value of $P(t_l|T_j)$ is another important parameter which needs to be estimated by the topic modeling approach.

Note that the number of documents is denoted by n, topics by k and lexicon size (terms) by d. Most topic modeling methods attempt to learn the above parameters using maximum likelihood methods, so that the probabilistic fit to the given corpus of documents is as large as possible. There are two basic methods which are used for topic modeling, which are *Probabilistic Latent Semantic Indexing (PLSI)* [49] and *Latent Dirichlet Allocation (LDA)*[20] respectively.

In this section, we will focus on the probabilistic latent semantic indexing method. Note that the above set of random variables $P(T_j|D_i)$ and $P(t_l|T_j)$ allow us to model the probability of a term t_l occurring in any document D_i. Specifically, the probability $P(t_l|D_i)$ of the term t_l occurring document D_i can be expressed in terms of afore-mentioned parameters as follows:

$$P(t_l|D_i) = \sum_{j=1}^{k} p(t_l|T_j) \cdot P(T_j|D_i) \qquad (4.12)$$

Thus, for each term t_l and document D_i, we can generate a $n \times d$ matrix of probabilities in terms of these parameters, where n is the number of documents and d is the number of terms. For a given corpus, we also have the $n \times d$ term-document occurrence matrix X, which tells us which term *actually* occurs in each document, and how many times the term occurs in the document. In other words, $X(i,l)$ is the number of times that term t_l occurs in document D_i. Therefore, we can use a maximum likelihood estimation algorithm which maximizes the product of the probabilities of terms that are observed in each document in the entire collection. The logarithm of this can be expressed as a weighted sum of the logarithm of the terms in Equation 4.12, where the weight of the (i,l)th term is its frequency count $X(i,l)$. This is a constrained optimization problem which optimizes the value of the log likelihood probability $\sum_{i,l} X(i,l) \cdot \log(P(t_l|D_i))$ subject to the constraints that the probability values over each of the topic-document and term-topic spaces must sum to 1:

$$\sum_{l} P(t_l|T_j) = 1 \quad \forall T_j \qquad (4.13)$$

$$\sum_{j} P(T_j|D_i) = 1 \quad \forall D_i \qquad (4.14)$$

The value of $P(t_l|D_i)$ in the objective function is expanded and expressed in terms of the model parameters with the use of Equation 4.12. We note that a Lagrangian method can be used to solve this constrained problem. This is quite similar to the approach that we discussed for the non-negative matrix factorization problem in this chapter. The Lagrangian solution essentially leads to a set of iterative update equations for the corresponding parameters which need to be estimated. It can be shown that these parameters can be estimated [49] with the iterative update of two matrices $[P_1]_{k \times n}$ and $[P_2]_{d \times k}$ containing the topic-document probabilities and term-topic probabilities respectively. We start off by initializing these matrices randomly, and normalize each of them so that the probability values in their columns sum to one. Then, we iteratively perform the following steps on each of P_1 and P_2 respectively:

for each entry (j, i) in P_1 **do update**
$$P_1(j, i) \leftarrow P_1(j, i) \cdot \sum_{r=1}^{d} P_2(r, j) \cdot \frac{X(i,r)}{\sum_{v=1}^{k} P_1(v,i) \cdot P_2(r,v)}$$
Normalize each column of P_1 to sum to 1;
for each entry (l, j) in P_2 **do update**
$$P_2(l, j) \leftarrow P_2(l, j) \cdot \sum_{q=1}^{n} P_1(j, q) \cdot \frac{X(q,l)}{\sum_{v=1}^{k} P_1(v,q) \cdot P_2(l,v)}$$
Normalize each column of P_2 to sum to 1;

The process is iterated to convergence. The output of this approach are the two matrices P_1 and P_2, the entries of which provide the topic-document and term-topic probabilities respectively.

The second well known method for topic modeling is that of *Latent Dirichlet Allocation*. In this method, the term-topic probabilities and topic-document probabilities are modeled with a Dirichlet distribution as a prior. Thus, the LDA method is the Bayesian version of the PLSI technique. It can also be shown the the PLSI method is equivalent to the LDA technique, when applied with a uniform Dirichlet prior [42].

The method of LDA was first introduced in [20]. Subsequently, it has generally been used much more extensively as compared to the PLSI method. Its main advantage over the PLSI method is that it is not quite as susceptible to overfitting. This is generally true of Bayesian methods which reduce the number of model parameters to be estimated, and therefore work much better for smaller data sets. Even for larger data sets, PLSI has the disadvantage that the number of model parameters grows linearly with the size of the collection. It has been argued [20] that the PLSI model is not a fully generative model, because there is no accurate way to model the topical distribution of a document which is not included in the current data set. For example, one can use the current set of topical distributions to perform the modeling of a new document, but it is likely to be much more inaccurate because of the overfitting inherent in PLSI. A Bayesian model, which uses a small number of parameters in the form of a well-chosen prior distribution, such as a *Dirichlet*, is likely to be much more robust in modeling new documents. Thus, the LDA method can also be used in order to model the topic distribution of a new document more robustly, even if it is not present in the original data set. Despite the theoretical advantages of LDA over PLSA, a recent study has shown that their task performances in clustering, categorization and retrieval tend to be similar [63]. The area of topic models is quite vast, and will be treated in more depth in Chapter 5 and Chapter 8 of this book; the purpose of this section is to simply acquaint the reader with the basics of this area and its natural connection to clustering.

We note that the EM-concepts which are used for topic modeling are quite general, and can be used for different variations on the text clustering tasks, such as text classification [72] or incorporating user feedback into clustering [46]. For example, the work in [72] uses an EM-approach in order to perform supervised clustering (and classification) of the documents, when a mixture of labeled and unlabeled data is available. A more detailed discussion is provided in Chapter 6 on text classification.

6. Online Clustering with Text Streams

The problem of streaming text clustering is particularly challenging in the context of text data because of the fact that the clusters need to be continuously maintained in real time. One of the earliest methods for streaming text clustering was proposed in [112]. This technique is referred to as the *Online Spherical k-Means Algorithm (OSKM)*, which reflects the broad approach used by the methodology. This technique divides up the incoming stream into small segments, each of which can be processed effectively in main memory. A set of k-means iterations are applied to each such data segment in order to cluster them. The advantage of using a segment-wise approach for clustering is that since each segment can be held in main memory, we can process each data point multiple times as long as it is held in main memory. In addition, the centroids from the previous segment are used in the next iteration for clustering purposes. A decay factor is introduced in order to age-out the old documents, so that the new documents are considered more important from a clustering perspective. This approach has been shown to be extremely effective in clustering massive text streams in [112].

A different method for clustering massive text and categorical data streams is discussed in [3]. The method discussed in [3] uses an approach which examines the relationship between outliers, emerging trends, and clusters in the underlying data. Old clusters may become inactive, and eventually get replaced by new clusters. Similarly, when newly arriving data points do not naturally fit in any particular cluster, these need to be initially classified as outliers. However, as time progresses, these new points may create a distinctive pattern of activity which can be recognized as a new cluster. The temporal locality of the data stream is manifested by these new clusters. For example, the first web page belonging to a particular category in a crawl may be recognized as an outlier, but may later form a cluster of documents of its own. On the other hand, the new outliers may not necessarily result in the formation of new clusters. Such outliers are true short-term abnormalities in the data since they do not result in the emergence of sustainable patterns. The approach discussed in [3] recognizes new clusters by first recognizing them as outliers. This approach works with the use of a summarization methodology, in which we use the concept of *condensed droplets* [3] in order to create concise representations of the underlying clusters.

As in the case of the OSKM algorithm, we ensure that recent data points are given greater importance than older data points. This is achieved by creating a time-sensitive weight for each data point. It is assumed that each data point has a time-dependent weight defined by the function $f(t)$. The function $f(t)$ is also referred to as the *fading function*. The fading function $f(t)$ is a non-monotonic decreasing function which decays uniformly with time t. The aim of defining a half life is to quantify the rate of decay of the importance of each data point in the stream clustering process. The *decay-rate* is defined as the inverse of the half life of the data stream. We denote the decay rate by $\lambda = 1/t_0$. We denote the weight function of each point in the data stream by $f(t) = 2^{-\lambda \cdot t}$. From the perspective of the clustering process, the weight of each data point is $f(t)$. It is easy to see that this decay function creates a half life of $1/\lambda$. It is also evident that by changing the value of λ, it is possible to change the rate at which the importance of the historical information in the data stream decays.

When a cluster is created during the streaming process by a newly arriving data point, it is allowed to remain as a trend-setting outlier for at least one half-life. During that period, if at least one more data point arrives, then the cluster becomes an active and mature cluster. On the other hand, if no new points arrive during a half-life, then the trend-setting outlier is recognized as a true anomaly in the data stream. At this point, this anomaly is removed from the list of current clusters. We refer to the process of removal as *cluster death*. Thus, a new cluster containing one data point dies when the (weighted) number of points in the cluster is 0.5. The same criterion is used to define the death of mature clusters. A necessary condition for this criterion to be met is that the inactivity period in the cluster has exceeded the half life $1/\lambda$. The greater the number of points in the cluster, the greater the level by which the inactivity period would need to exceed its half life in order to meet the criterion. This is a natural solution, since it is intuitively desirable to have stronger requirements (a longer inactivity period) for the death of a cluster containing a larger number of points.

The statistics of the data points are captured in summary statistics, which are referred to as *condensed droplets*. These represent the word distributions within a cluster, and can be used in order to compute the similarity of an incoming data point to the cluster. The overall algorithm proceeds as follows. At the beginning of algorithmic execution, we start with an empty set of clusters. As new data points arrive, unit clusters containing individual data points are created. Once a maximum number k of such clusters have been created, we can begin the process of online cluster maintenance. Thus, we initially start off with a trivial set of k clusters. These clusters are updated over time with the arrival of new data points.

When a new data point $\overline{X}$ arrives, its similarity to each cluster droplet is computed. In the case of text data sets, the cosine similarity measure between $\overline{DF1}$ and $\overline{X}$ is used. The similarity value $S(\overline{X}, \mathcal{C}_j)$ is computed from the incoming document $\overline{X}$ to every cluster $\mathcal{C}_j$. The cluster with the maximum value of $S(\overline{X}, \mathcal{C}_j)$ is chosen as the relevant cluster for data insertion. Let us assume that this cluster is $\mathcal{C}_{mindex}$. We use a threshold denoted by *thresh* in order to determine whether the incoming data point is an outlier. If the value of $S(\overline{X}, \mathcal{C}_{mindex})$ is larger than the threshold *thresh*, then the point $\overline{X}$ is assigned to the cluster $\mathcal{C}_{mindex}$. Otherwise, we check if some inactive cluster exists in the current set of cluster droplets. If no such inactive cluster exists, then the data point $\overline{X}$ is added to $\mathcal{C}_{mindex}$. On the other hand, when an inactive cluster does exist, a new cluster is created containing the solitary data point $\overline{X}$. This newly created cluster replaces the inactive cluster. We note that this new cluster is a potential true outlier or the beginning of a new trend of data points. Further understanding of this new cluster may only be obtained with the progress of the data stream.

In the event that $\overline{X}$ is inserted into the cluster $\mathcal{C}_{mindex}$, we update the statistics of the cluster in order to reflect the insertion of the data point and temporal decay statistics. Otherwise, we replace the most inactive cluster by a new cluster containing the solitary data point $\overline{X}$. In particular, the replaced cluster is the least recently updated cluster among all inactive clusters. This process is continuously performed over the life of the data stream, as new documents arrive over time. The work in [3] also presents a variety of other applications of the stream clustering technique such as evolution and correlation analysis.

A different way of utilizing the temporal evolution of text documents in the clustering process is described in [48]. Specifically, the work in [48] uses *bursty features* as markers of new topic occurrences in the data stream. This is because the semantics of an up-and-coming topic are often reflected in the frequent presence of a few distinctive words in the text stream. At a given period in time, the nature of relevant topics could lead to bursts in specific features of the data stream. Clearly, such features are extremely important from a clustering perspective. Therefore, the method discussed in [48] uses a new representation, which is referred to as the *bursty feature representation* for mining text streams. In this representation, a time-varying weight is associated with the features depending upon its burstiness. This also reflects the varying importance of the feature to the clustering process. Thus, it is important to remember that a particular document representation is dependent upon the particular instant in time at which it is constructed.

Another issue which is handled effectively in this approach is an implicit reduction in dimensionality of the underlying collection. Text is inherently a high dimensional data domain, and the pre-selection of some of the features on the basis of their burstiness can be a natural way to reduce the dimensionality of document representation. This can help in both the effectiveness and efficiency of the underlying algorithm.

The first step in the process is to identify the bursty features in the data stream. In order to achieve this goal, the approach uses Kleinberg's 2-state finite automaton model [57]. Once these features have been identified, the bursty features are associated with weights which depend upon their level of burstiness. Subsequently, a bursty feature representation is defined in order to reflect the underlying weight of the feature. Both the identification and the weight of the bursty feature are dependent upon its underlying frequency. A standard k-means approach is applied to the new representation in order to construct the clustering. It was shown in [48] that the approach of using burstiness improves the cluster quality. Once criticism of the work in [48] is that it is mostly focused on the issue of improving effectiveness with the use of temporal characteristics of the data stream, and does not address the issue of efficient clustering of the underlying data stream.

In general, it is evident that feature extraction is important for all clustering algorithms. While the work in [48] focuses on using temporal characteristics of the stream for feature extraction, the work in [61] focuses on using *phrase extraction* for effective feature selection. This work is also related to the concept of topic-modeling, which will be discussed in detail in the next section. This is because the different topics in a collection can be related to the clusters in a collection. The work in [61] uses topic-modeling techniques for clustering. The core idea in the work of [61] is that individual words are not very effective for a clustering algorithm because they miss the context in which the word is used. For example, the word "star" may either refer to a celestial body or to an entertainer. On the other hand, when the phrase "fixed star" is used, it becomes evident that the word "star" refers to a celestial body. The phrases which are extracted from the collection are also referred to as *topic signatures*.

The use of such phrasal clarification for improving the quality of the clustering is referred to as *semantic smoothing* because it reduces the noise which is associated with semantic ambiguity. Therefore, a key part of the approach is to extract phrases from the underlying data stream. After phrase extraction, the training process determines a translation probability of the phrase to terms in the vocabulary. For example, the word "planet" may have high probability of association with the phrase "fixed star", because both refer to celestial bodies. Therefore, for a given document, a rational probability count may also be assigned to all terms. For each document, it is assumed that all terms in it are generated either by a topic-signature model, or a background collection model.

The approach in [61] works by modeling the soft probability $p(w|C_j)$ for word w and cluster C_j. The probability $p(w|C_j)$ is modeled as a linear combination of two factors; (a) A maximum likelihood model which computes the probabilities of generating specific words for each cluster (b) An indirect (translated) word-membership probability which first determines the maximum likelihood probability for each topic-signature, and then multiplying with the conditional probability of each word, given the topic-signature. We note that we can use $p(w|C_j)$ in order to estimate $p(d|C_j)$ by using the product of the constituent words in the document. For this purpose, we use the frequency $f(w,d)$ of word w in document d.

$$p(d|C_j) = \prod_{w \in d} p(w|C_j)^{f(w,d)} \qquad (4.15)$$

We note that in the static case, it is also possible to add a background model in order to improve the robustness of the estimation process. This is however not possible in a data stream because of the fact that the background collection model may require multiple passes in order to build effectively. The work in [61] maintains these probabilities in online fashion with the use of a *cluster profile*, that weights the probabilities with the use of a fading function. We note that the concept of cluster profile is analogous to the concept of condensed droplet introduced in [3]. The key algorithm (denoted by OCTS) is to maintain a dynamic set of clusters into which documents are progressively assigned with the use of similarity computations. It has been shown in [61] how the cluster profile can be used in order to efficiently compute $p(d|C_j)$ for each incoming document. This value is then used in order to determine the similarity of the documents to the different clusters. This is used in order to assign the documents to their closest cluster. We note that the methods in [3, 61] share a number of similarities in terms of (a) maintenance of cluster profiles, (b) use of cluster profiles (or condensed droplets) to compute similarity and assignment of documents to most similar clusters, and (c) the rules used to decide when a new singleton cluster should be created, or one of the older clusters should be replaced.

The main difference between the two algorithms is the technique which is used in order to compute cluster similarity. The OCTS algorithm uses the probabilistic computation $p(d|C_j)$ to compute cluster similarity, which takes the phrasal information into account during the computation process. One observation about OCTS is that it may allow for very similar clusters to co-exist in the current set. This reduces the space available for distinct cluster profiles. A second algorithm called OCTSM is also proposed in [61], which allows for merging of very similar clusters. Before each assignment, it checks whether pairs of similar clusters can be merged on the basis of similarity. If this is the case, then we allow the merging of the similar clusters and their corresponding cluster profiles. Detailed experimental results on the different clustering algorithms and their effectiveness are presented in [61].

A closely related area to clustering is that of topic modeling, which we discussed in an earlier section. Recently, the topic modeling method has also been extended to the *dynamic* case which is helpful for topic modeling of text streams [107].

7. Clustering Text in Networks

Many social networks contain both text content in the nodes, as well as links between the different nodes. Clearly, the links provide useful cues in understanding the related nodes in the network. The impact of different link types on the quality of the clustering has been studied in [109], and it has been shown that many forms of implicit and explicit links improve clustering quality, because they encode human knowledge. Therefore, a natural choice is to combine these two factors in the process of clustering the different nodes. In this section, we will discuss a number of such techniques.

In general, links may be considered as a kind of side-information, which can be represented in the form of attributes. A general approach for incorporating side attributes into the clustering process has been proposed in [1]. This algorithm uses a combination of a k-means approach on the text attributes, and Bayesian probability estimations on the side attributes for the clustering process. The idea is to identify those attributes, which are helpful for the clustering process, and use them in order to enhance the quality of the clustering. However, this approach is really designed for general attributes of any kind, rather than link-based attributes, in which an underlying graph structure is implied by the document-to-document linkages. In spite of this, it has been shown in [1], that it is possible to significantly enhance the quality of clustering by treating linkage information as side-attributes. Many other techniques, which will be discussed in this section, have been proposed specifically for the case of text documents, which are linked together in a network structure.

The earliest methods for combining text and link information for the clustering process are proposed in [12]. Two different methods were proposed in this paper for the clustering process. The first method uses the link information in the neighbors of a node in order to bias the term weights in a document. Term weights which are common between a document and its neighbors are given more importance in the clustering process. One advantage of such an approach is that we can use any of the existing clustering algorithms for this purpose, because the link information is implicitly encoded in the modified term weights. The second method proposed in [12] is a graph-based approach which directly uses the links in the clustering process. In this case, the approach attempts to model the probability that a particular document belongs to a given cluster for a particular set of links and content. This is essentially a *soft-clustering*, in which a probability of assignment is determined for each cluster. The cluster with the largest probability of assignment is considered the most relevant cluster. A Markov Random Field (MRF) technique is used in order to perform the clustering. An iterative technique called relaxation labeling is used in order to compute the maximum likelihood parameters of this MRF. More details of this approach may be found in [12].

A recent method to perform clustering with both structural and attribute similarities is proposed in [113]. The techniques of this paper can be applied to both relational and text attributes. This paper integrates structural and attribute-based clustering by adding attribute vertices to the network in addition to the original structural vertices. In the context of text data, this implies that a vertex exists for each word in the lexicon. Therefore, in addition to the original set of vertices V in the graph $G = (V, E)$, we now have the augmented vertex set $V \cup V_1$, such that V_1 contains one vertex for each nodes. We also augment the edge set, in order to add to the original set of structural edges E. We add an edge between a structural vertex $i \in V$ and an attribute vertex $j \in V_1$, if word j is contained in the node i. This new set of edges added is denoted by E_1. Therefore, we now have an augmented graph $G_1 = (V \cup V_1, E \cup E_1)$ which is semi-bipartite. A neighborhood random walk model is used in order to determine the closeness of vertices. This closeness measure is used in order to perform the clustering. The main challenge in the algorithm is to determine the relative importance of structural and attribute components in the clustering process. In the context of the random walk model, this translates to determining the appropriate weights of different edges during the random walk process. A learning model has been proposed in [113] in order to learn these weights, and leverage them for an effective clustering process.

The problem of clustering network content is often encountered in the context of community detection in social networks. The text content in the social network graph may be attached to either the nodes [101] of the network, or to the edges [74]. The node-based approach is generally more common, and most of the afore-mentioned techniques in this paper can be modeled in terms of content attached to the nodes. In the method proposed in [101], the following link-based and content-based steps are combined for effective community detection:

- A conditional model is proposed for link analysis, in which the conditional probability for the destination of given link is modeled. A hidden variable is introduced in order to capture the popularity of a node in terms of the likelihood of that node being cited by other nodes.

- A discriminative content model is introduced in order to reduce the impact of noisy content attributes. In this model, the attributes are weighed by their ability to discriminate between the different communities.

- The two models are combined into a unified framework with the use of a two-stage optimization algorithm for maximum likelihood inference. One interesting characteristic of this broad framework is that it can also be used in the context of other complementary approaches.

The details of the algorithm are discussed in [101].

For the case of edge-based community detection, it is assumed that the text content in the network is attached to the edges [74]. This is common in applications which involve extensive communication between the different nodes. For example, in email networks, or online chat networks, the text in the network is associated with the communications between the different entities. In such cases, the text is associated with an edge in the underlying network. The presence of content associated with edges allows for a much more nuanced approach in community detection, because a given node may participate in communities of different kinds. The presence of content associated with edges helps in separating out these different associations of the same individual to different communities. The work in [74] uses a matrix-factorization methodology in order to jointly model the content and structure for the community detection process. The matrix factorization method is used to transform the representation into multi-dimensional representation, which can be easily clustered by a simple algorithm such as the k-means algorithm. It was shown in [74], that the use of such an approach can provide much more effective results than a pure content- or link-based clustering methodology.

A closely related area to clustering is that of *topic modeling*, in which we attempt to model the probability of a document belonging to a particular cluster. A natural approach to network-based topic modeling is to add a network-based regularization constraint to traditional topic models such as NetPLSA [65]. The relational topic model (RTM) proposed in [23] tries to model the generation of documents and links sequentially. The first step for generating the documents is the same as LDA. Subsequently, the model predicts links based on the similarity of the topic mixture used in two documents. Thus, this method can be used both for topic modeling and predicting missing links. A more unified model is proposed in the *iTopicModel* [91] framework which creates a Markov Random Field model in order to create a generative model which simultaneously captures both text and links. Experimental results have shown this approach to be more general and superior to previously existing methods. A number of other methods for incorporating network information into topic modeling are discussed in the next chapter on dimensionality reduction.

8. Semi-Supervised Clustering

In some applications, prior knowledge may be available about the kinds of clusters that are available in the underlying data. This prior knowledge may take on the form of labels attached with the documents, which indicate its underlying topic. For example, if we wish to use the broad distribution of topics in the *Yahoo!* taxonomy in order to supervise the clustering process of a new web collection, one way to performing supervision would be add some labeled pages from the *Yahoo!* taxonomy to the collection. Typically such pages would contain labels of the form *@Science@Astronomy* or *@Arts@Painting*, which indicate the subject area of the added pages. Such knowledge can be very useful in creating significantly more coherent clusters, especially when the total number of clusters is large. The process of using such labels to guide the clustering process is referred to as *semi-supervised clustering*. This form of learning is a bridge between the clustering and classification problem, because it uses the underlying class structure, but it is not completely tied down by the specific structure. As a result, such an approach finds applicability both to the clustering and classification scenarios.

The most natural method for incorporating supervision into the clustering process is to do so in partitional clustering methods such as k-means. This is because the supervision can be easily incorporated by changing the seeds in the clustering process. For example, the work in [4] uses the initial seeds in the k-means clustering process as the centroids of the original classes in the underlying data. A similar approach has also been used in [15], except a wider variation of how the seeds may be selected has been explored.

A number of probabilistic frameworks have also been designed for semi-supervised clustering [72, 14]. The work in [72] uses an iterative EM-approach in which the unlabeled documents are assigned labels using a naive Bayes approach on the currently labeled documents. These newly labeled documents are then again used for re-training a Bayes classifier. This process is iterated to convergence. The iterative labeling approach in [72] can be considered a partially supervised approach for clustering the unlabeled documents. The work in [14] uses a Heterogeneous Markov Random Field (HMRF) model for the clustering process.

A graph-based method for incorporating prior knowledge into the clustering process has been proposed in [52]. In this method, the documents are modeled as a graph, in which nodes represent documents and edges represent the similarity among them. New edges may also be added to this graph, which correspond to the prior knowledge. Specifically, an edge is added to the graph, when it is known on the basis of prior knowledge that these two documents are similar. A normalized cut algorithm [84] is then applied to this graph in order to create the final clustering. This approach implicitly uses the prior knowledge because of the augmented graph representation which is used for the clustering.

Since semi-supervised clustering forms a natural bridge between the clustering and classification problems, it is natural that semi-supervised methods can be used for classification as well [68]. This is also referred to as *co-training*, because it involves the use of unsupervised document clustering in order to assist the training process. Since semi-supervised methods use both the clustering structure in the feature space and the class information, they are sometimes more robust in classification scenarios, especially in cases where the amount of available labeled data is small. It has been shown in [72], how a partially supervised co-training approach which mixes supervised and unsupervised data may yield more effective classification results, when the amount of training data available is small. The work in [72] uses a partially supervised EM-algorithm which iteratively assigns labels to the unlabeled documents and refines them over time as convergence is achieved. A number of similar methods along this spirit are proposed in [4, 14, 35, 47, 89] with varying levels of supervision in the clustering process. Partially supervised clustering methods are also used feature transformation in classification using the methods as discussed in [17, 18, 88]. The idea is that the clustering structure provides a compressed feature space, which capture the relevant classification structure very well, and can therefore be helpful for classification.

Partially supervised methods can also be used in conjunction with pre-existing categorical hierarchies (or prototype hierarchies) [4, 56, 67]. A typical example of a prototype hierarchy would be the *Yahoo!* taxonomy, the *Open Directory Project*, or the *Reuters collection*. The idea is that such hierarchies provide a good general idea of the clustering structure, but also have considerable noise and overlaps in them because of their typical manual origins. The partial supervision is able to correct the noise and overlaps, and this results in a relatively clean and coherent clustering structure.

An unusual kind of supervision for document clustering is the method of use of a *universum* of documents which are known *not* to belong to a cluster [106]. This is essentially, the background distribution which cannot be naturally clustered into any particular group. The intuition is that the universum of examples provide an effective way of avoiding mistakes in the clustering process, since it provides a background of examples to compare a cluster with.

9. Conclusions and Summary

In this chapter, we presented a survey of clustering algorithms for text data. A good clustering of text requires effective feature selection and a proper choice of the algorithm for the task at hand. Among the different classes of algorithms, the distance-based methods are among the most popular in a wide variety of applications.

In recent years, the main trend in research in this area has been in the context of two kinds of text data:

- **Dynamic Applications:** The large amounts of text data being created by dynamic applications such as social networks or online chat applications has created an immense need for streaming text clustering applications. Such streaming applications need to be applicable in the case of text which is not very clean, as is often the case for applications such as social networks.

- **Heterogeneous Applications:** Text applications increasingly arise in heterogeneous applications in which the text is available in the context of links, and other heterogeneous multimedia data. For example, in social networks such as *Flickr* the clustering often needs to be applied in such scenario. Therefore, it is critical to effectively adapt text-based algorithms to heterogeneous multimedia scenarios.

We note that the field of text clustering is too vast to cover comprehensively in a single chapter. Some methods such as *committee-based clustering* [73] cannot even be neatly incorporated into any class of methods, since they use a combination of the different clustering methods in order to create a final clustering result. The main purpose of this chapter is to provide a comprehensive overview of the main algorithms which are often used in the area, as a starting point for further study.

References

[1] C. C. Aggarwal, Y. Zhao, P. S. Yu. On Text Clustering with Side Information, *ICDE Conference*, 2012.

[2] C. C. Aggarwal, P. S. Yu. On Effective Conceptual Indexing and Similarity Search in Text, *ICDM Conference*, 2001.

[3] C. C. Aggarwal, P. S. Yu. A Framework for Clustering Massive Text and Categorical Data Streams, *SIAM Conference on Data Mining*, 2006.

[4] C. C. Aggarwal, S. C. Gates, P. S. Yu. On Using Partial Supervision for Text Categorization, *IEEE Transactions on Knowledge and Data Engineering*, 16(2), 245–255, 2004.

[5] C. C. Aggarwal, C. Procopiuc, J. Wolf, P. S. Yu, J.-S. Park. Fast Algorithms for Projected Clustering, *ACM SIGMOD Conference*, 1999.

[6] C. C. Aggarwal, P. S. Yu. Finding Generalized Projected Clusters in High Dimensional Spaces, *ACM SIGMOD Conference*, 2000.

[7] R. Agrawal, J. Gehrke, P. Raghavan. D. Gunopulos. Automatic Subspace Clustering of High Dimensional Data for Data Mining Applications, *ACM SIGMOD Conference*, 1999.

[8] R. Agrawal, R. Srikant. Fast Algorithms for Mining Association Rules in Large Databases, *VLDB Conference*, 1994.

[9] J. Allan, R. Papka, V. Lavrenko. Online new event detection and tracking. *ACM SIGIR Conference*, 1998.

[10] P. Andritsos, P. Tsaparas, R. Miller, K. Sevcik. LIMBO: Scalable Clustering of Categorical Data. *EDBT Conference*, 2004.

[11] P. Anick, S. Vaithyanathan. Exploiting Clustering and Phrases for Context-Based Information Retrieval. *ACM SIGIR Conference*, 1997.

[12] R. Angelova, S. Siersdorfer. A neighborhood-based approach for clustering of linked document collections. *CIKM Conference*, 2006.

[13] R. A. Baeza-Yates, B. A. Ribeiro-Neto, *Modern Information Retrieval - the concepts and technology behind search, Second edition*, Pearson Education Ltd., Harlow, England, 2011.

[14] S. Basu, M. Bilenko, R. J. Mooney. A probabilistic framework for semi-supervised clustering. *ACM KDD Conference*, 2004.

[15] S. Basu, A. Banerjee, R. J. Mooney. Semi-supervised Clustering by Seeding. *ICML Conference*, 2002.

[16] F. Beil, M. Ester, X. Xu. Frequent term-based text clustering, *ACM KDD Conference*, 2002.

[17] L. Baker, A. McCallum. Distributional Clustering of Words for Text Classification, *ACM SIGIR Conference*, 1998.

[18] R. Bekkerman, R. El-Yaniv, Y. Winter, N. Tishby. On Feature Distributional Clustering for Text Categorization. *ACM SIGIR Conference*, 2001.

[19] D. Blei, J. Lafferty. Dynamic topic models. *ICML Conference*, 2006.

[20] D. Blei, A. Ng, M. Jordan. Latent Dirichlet allocation, *Journal of Machine Learning Research*, 3: pp. 993–1022, 2003.

[21] P. F. Brown, P. V. deSouza, R. L. Mercer, V. J. Della Pietra, and J/ C. Lai. Class-based n-gram models of natural language, *Computational Linguistics*, 18, 4 (December 1992), 467-479.

[22] K. Chakrabarti, S. Mehrotra. Local Dimension reduction: A new Approach to Indexing High Dimensional Spaces, *VLDB Conference*, 2000.

[23] J. Chang, D. Blei. Topic Models for Document Networks. *AISTA-SIS*, 2009.

[24] W. B. Croft. Clustering large files of documents using the single-link method. *Journal of the American Society of Information Science*, 28: pp. 341–344, 1977.

[25] D. Cutting, D. Karger, J. Pedersen, J. Tukey. Scatter/Gather: A Cluster-based Approach to Browsing Large Document Collections. *ACM SIGIR Conference*, 1992.

[26] D. Cutting, D. Karger, J. Pederson. Constant Interaction-time Scatter/Gather Browsing of Large Document Collections, *ACM SIGIR Conference*, 1993.

[27] M. Dash, H. Liu. Feature Selection for Clustering, *PAKDD Conference*, pp. 110–121, 1997.

[28] S. Deerwester, S. Dumais, T. Landauer, G. Furnas, R. Harshman. Indexing by Latent Semantic Analysis. *JASIS*, 41(6), pp. 391–407, 1990.

[29] I. Dhillon, D. Modha. Concept Decompositions for Large Sparse Data using Clustering, 42(1), pp. 143–175, 2001.

[30] I. Dhillon. Co-clustering Documents and Words using bipartite spectral graph partitioning, *ACM KDD Conference*, 2001.

[31] I. Dhillon, S. Mallela, D. Modha. Information-theoretic Co-Clustering, *ACM KDD Conference*, 2003.

[32] C. Ding, X. He, H. Zha, H. D. Simon. Adaptive Dimension Reduction for Clustering High Dimensional Data, *ICDM Conference*, 2002.

[33] C. Ding, X. He, H. Simon. On the equivalence of nonnegative matrix factorization and spectral clustering. *SDM Conference*, 2005.

[34] B. Dorow, D. Widdows. Discovering corpus-specific word senses, *Proceedings of the tenth conference on European chapter of the Association for Computational Linguistics - Volume 2 (EACL '03)*, pages 79-82, 2003.

[35] R. El-Yaniv, O. Souroujon. Iterative Double Clustering for Unsupervised and Semi-supervised Learning. *NIPS Conference*, 2002.

[36] H. Fang, T. Tao, C. Zhai, A formal study of information retrieval heuristics, *Proceedings of ACM SIGIR 2004*, 2004.

[37] D. Fisher. Knowledge Acquisition via incremental conceptual clustering. *Machine Learning*, 2: pp. 139–172, 1987.

[38] M. Franz, T. Ward, J. McCarley, W.-J. Zhu. Unsupervised and supervised clustering for topic tracking. *ACM SIGIR Conference*, 2001.

[39] G. P. C. Fung, J. X. Yu, P. Yu, H. Lu. Parameter Free Bursty Events Detection in Text Streams, *VLDB Conference*, 2005.

[40] J. H. Gennari, P. Langley, D. Fisher. Models of incremental concept formation. *Journal of Artificial Intelligence*, 40 pp. 11–61, 1989.

[41] D. Gibson, J. Kleinberg, P. Raghavan. Clustering Categorical Data: An Approach Based on Dynamical Systems, *VLDB Conference*, 1998.

[42] M. Girolami, A Kaban. On the Equivalance between PLSI and LDA, *SIGIR Conference*, pp. 433–434, 2003.

[43] S. Guha, R. Rastogi, K. Shim. ROCK: a robust clustering algorithm for categorical attributes, *International Conference on Data Engineering*, 1999.

[44] S. Guha, R. Rastogi, K. Shim. CURE: An Efficient Clustering Algorithm for Large Databases. *ACM SIGMOD Conference*, 1998.

[45] D. Gusfield. Algorithms for strings, trees and sequences, *Cambridge University Press*, 1997.

[46] Y. Huang, T. Mitchell. Text clustering with extended user feedback. *ACM SIGIR Conference*, 2006.

[47] H. Li, K. Yamanishi. Document classification using a finite mixture model. *Annual Meeting of the Association for Computational Linguistics*, 1997.

[48] Q. He, K. Chang, E.-P. Lim, J. Zhang. Bursty feature representation for clustering text streams. *SDM Conference*, 2007.

[49] T. Hofmann. Probabilistic Latent Semantic Indexing. *ACM SIGIR Conference*, 1999.

[50] A. Jain, R. C. Dubes. Algorithms for Clustering Data, *Prentice Hall*, Englewood Cliffs, NJ, 1998.

[51] N. Jardine, C. J.van Rijsbergen. The use of hierarchical clustering in information retrieval, *Information Storage and Retrieval*, 7: pp. 217–240, 1971.

[52] X. Ji, W. Xu. Document clustering with prior knowledge. *ACM SIGIR Conference*, 2006.

[53] I. T. Jolliffee. Principal Component Analysis. *Springer*, 2002.

[54] L. Kaufman, P. J. Rousseeuw. Finding Groups in Data: An Introduction to Cluster Analysis, *Wiley Interscience*, 1990.

[55] W. Ke, C. Sugimoto, J. Mostafa. Dynamicity vs. effectiveness: studying online clustering for scatter/gather. *ACM SIGIR Conference*, 2009.

[56] H. Kim, S. Lee. A Semi-supervised document clustering technique for information organization, *CIKM Conference*, 2000.

[57] J. Kleinberg, Bursty and hierarchical structure in streams, *ACM KDD Conference*, pp. 91–101, 2002.

[58] D. D. Lee, H. S. Seung. Learning the parts of objects by non-negative matrix factorization, *Nature*, 401: pp. 788–791, 1999.

[59] T. Li, S. Ma, M. Ogihara, Document Clustering via Adaptive Subspace Iteration, *ACM SIGIR Conference*, 2004.

[60] T. Li, C. Ding, Y. Zhang, B. Shao. Knowledge transformation from word space to document space. *ACM SIGIR Conference*, 2008.

[61] Y.-B. Liu, J.-R. Cai, J. Yin, A. W.-C. Fu. Clustering Text Data Streams, *Journal of Computer Science and Technology*, Vol. 23(1), pp. 112–128, 2008.

[62] T. Liu, S. Lin, Z. Chen, W.-Y. Ma. An Evaluation on Feature Selection for Text Clustering, *ICML Conference*, 2003.

[63] Y. Lu, Q. Mei, C. Zhai. Investigating task performance of probabilistic topic models: an empirical study of PLSA and LDA, *Information Retrieval*, 14(2): 178-203 (2011).

[64] A. McCallum. Bow: A toolkit for statistical language modeling, text retrieval, classification and clustering. http://www.cs.cmu.edu/~mccallum/bow, 1996.

[65] Q. Mei, D. Cai, D. Zhang, C.-X. Zhai. Topic Modeling with Network Regularization. *WWW Conference*, 2008.

[66] D. Metzler, S. T. Dumais, C. Meek, Similarity Measures for Short Segments of Text, *Proceedings of ECIR 2007*, 2007.

[67] Z. Ming, K. Wang, T.-S. Chua. Prototype hierarchy-based clustering for the categorization and navigation of web collections. *ACM SIGIR Conference*, 2010.

[68] T. M. Mitchell. The role of unlabeled data in supervised learning. *Proceedings of the Sixth International Colloquium on Cognitive Science*, 1999.

[69] F. Murtagh. A Survey of Recent Advances in Hierarchical Clustering Algorithms, *The Computer Journal*, 26(4), pp. 354–359, 1983.

[70] F. Murtagh. Complexities of Hierarchical Clustering Algorithms: State of the Art, *Computational Statistics Quarterly*, 1(2), pp. 101–113, 1984.

[71] R. Ng, J. Han. Efficient and Effective Clustering Methods for Spatial Data Mining. *VLDB Conference*, 1994.

[72] K. Nigam, A. McCallum, S. Thrun, T. Mitchell. Learning to classify text from labeled and unlabeled documents. *AAAI Conference*, 1998.

[73] P. Pantel, D. Lin. Document Clustering with Committees, *ACM SIGIR Conference*, 2002.

[74] G. Qi, C. Aggarwal, T. Huang. Community Detection with Edge Content in Social Media Networks, *ICDE Conference*, 2012.

[75] M. Rege, M. Dong, F. Fotouhi. Co-clustering Documents and Words Using Bipartite Isoperimetric Graph Partitioning. *ICDM Conference*, pp. 532–541, 2006.

[76] C. J. van Rijsbergen. *Information Retrieval*, Butterworths, 1975.

[77] C. J.van Rijsbergen, W. B. Croft. Document Clustering: An Evaluation of some experiments with the Cranfield 1400 collection, *Information Processing and Management*, 11, pp. 171–182, 1975.

[78] S. E. Robertson and S. Walker. Some simple effective approximations to the 2-poisson model for probabilistic weighted retrieval. In *SIGIR*, pages 232–241, 1994.

[79] M. Sahami, T. D. Heilman, A web-based kernel function for measuring the similarity of short text snippets, *Proceedings of WWW 2006*, pages 377-386, 2006.

[80] N. Sahoo, J. Callan, R. Krishnan, G. Duncan, R. Padman. Incremental Hierarchical Clustering of Text Documents, *ACM CIKM Conference*, 2006.

[81] G. Salton. An Introduction to Modern Information Retrieval, *Mc Graw Hill*, 1983.

[82] G. Salton, C. Buckley. Term Weighting Approaches in Automatic Text Retrieval, *Information Processing and Management*, 24(5), pp. 513–523, 1988.

[83] H. Schutze, C. Silverstein. Projections for Efficient Document Clustering, *ACM SIGIR Conference*, 1997.

[84] J. Shi, J. Malik. Normalized cuts and image segmentation. *IEEE Transaction on Pattern Analysis and Machine Intelligence*, 2000.

[85] C. Silverstein, J. Pedersen. Almost-constant time clustering of arbitrary corpus subsets. *ACM SIGIR Conference*, pp. 60–66, 1997.

[86] A. Singhal, C. Buckley, M. Mitra. Pivoted Document Length Normalization. *ACM SIGIR Conference*, pp. 21–29, 1996.

[87] N. Slonim, N. Tishby. Document Clustering using word clusters via the information bottleneck method, *ACM SIGIR Conference*, 2000.

[88] N. Slonim, N. Tishby. The power of word clusters for text classification. *European Colloquium on Information Retrieval Research (ECIR)*, 2001.

[89] N. Slonim, N. Friedman, N. Tishby. Unsupervised document classification using sequential information maximization. *ACM SIGIR Conference*, 2002.

[90] M. Steinbach, G. Karypis, V. Kumar. A Comparison of Document Clustering Techniques, *KDD Workshop on text mining*, 2000.

[91] Y. Sun, J. Han, J. Gao, Y. Yu. iTopicModel: Information Network Integrated Topic Modeling, *ICDM Conference*, 2009.

[92] E. M. Voorhees. Implementing Agglomerative Hierarchical Clustering for use in Information Retrieval, *Technical Report TR86-765, Cornell University, Ithaca, NY*, July 1986.

[93] F. Wang, C. Zhang, T. Li. Regularized clustering for documents. *ACM SIGIR Conference*, 2007.

[94] J. Wilbur, K. Sirotkin. The automatic identification of stopwords, *J. Inf. Sci.*, 18: pp. 45–55, 1992.

[95] P. Willett. Document Clustering using an inverted file approach. *Journal of Information Sciences*, 2: pp. 223–231, 1980.

[96] P. Willett. Recent Trends in Hierarchical Document Clustering: A Critical Review. *Information Processing and Management*, 24(5): pp. 577–597, 1988.

[97] W. Xu, X. Liu, Y. Gong. Document Clustering based on non-negative matrix factorization, *ACM SIGIR Conference*, 2003.

[98] W. Xu, Y. Gong. Document clustering by concept factorization. *ACM SIGIR Conference*, 2004.

[99] Y. Yang, J. O. Pederson. A comparative study on feature selection in text categorization, *ACM SIGIR Conference*, 1995.

[100] Y. Yang. Noise Reduction in a Statistical Approach to Text Categorization, *ACM SIGIR Conference*, 1995.

[101] T. Yang, R. Jin, Y. Chi, S. Zhu. Combining link and content for community detection: a discriminative approach. *ACM KDD Conference*, 2009.

[102] L. Yao, D. Mimno, A. McCallum. Efficient methods for topic model inference on streaming document collections, *ACM KDD Conference*, 2009.

[103] O. Zamir, O. Etzioni. Web Document Clustering: A Feasibility Demonstration, *ACM SIGIR Conference*, 1998.

[104] O. Zamir, O. Etzioni, O. Madani, R. M. Karp. Fast and Intuitive Clustering of Web Documents, *ACM KDD Conference*, 1997.

[105] C. Zhai, *Statistical Language Models for Information Retrieval (Synthesis Lectures on Human Language Technologies)*, Morgan & Claypool Publishers, 2008.

[106] D. Zhang, J. Wang, L. Si. Document clustering with universum. *ACM SIGIR Conference*, 2011.

[107] J. Zhang, Z. Ghahramani, Y. Yang. A probabilistic model for online document clustering with application to novelty detection. In *Saul L., Weiss Y., Bottou L. (eds) Advances in Neural Information Processing Letters*, 17, 2005.

[108] T. Zhang, R. Ramakrishnan, M. Livny. BIRCH: An Efficient Data Clustering Method for Very Large Databases. *ACM SIGMOD Conference*, 1996.

[109] X. Zhang, X. Hu, X. Zhou. A comparative evaluation of different link types on enhancing document clustering. *ACM SIGIR Conference*, 2008.

[110] Y. Zhao, G. Karypis. Evaluation of hierarchical clustering algorithms for document data set, *CIKM Conference*, 2002.

[111] Y. Zhao, G. Karypis. Empirical and Theoretical comparisons of selected criterion functions for document clustering, *Machine Learning*, 55(3), pp. 311–331, 2004.

[112] S. Zhong. Efficient Streaming Text Clustering. *Neural Networks*, Volume 18, Issue 5–6, 2005.

[113] Y. Zhou, H. Cheng, J. X. Yu. Graph Clustering based on Structural/Attribute Similarities, *VLDB Conference*, 2009.

[114] http://www.lemurproject.org/

Chapter 5

DIMENSIONALITY REDUCTION AND TOPIC MODELING:
FROM LATENT SEMANTIC INDEXING TO LATENT DIRICHLET ALLOCATION AND BEYOND

Steven P. Crain
School of Computational Science and Engineering
College of Computing
Georgia Institute of Technology
s.crain@gatech.edu

Ke Zhou
School of Computational Science and Engineering
College of Computing
Georgia Institute of Technology
kzhou@gatech.edu

Shuang-Hong Yang
School of Computational Science and Engineering
College of Computing
Georgia Institute of Technology
shy@gatech.edu

Hongyuan Zha
School of Computational Science and Engineering
College of Computing
Georgia Institute of Technology
zha@cc.gatech.edu

Abstract   The bag-of-words representation commonly used in text analysis can be analyzed very efficiently and retains a great deal of useful information, but it is also troublesome because the same thought can be expressed using many different terms or one term can have very different meanings. Dimension reduction can collapse together terms that have the same semantics, to identify and disambiguate terms with multiple meanings and to provide a lower-dimensional representation of documents that reflects concepts instead of raw terms. In this chapter, we survey two influential forms of dimension reduction. Latent semantic indexing uses spectral decomposition to identify a lower-dimensional representation that maintains semantic properties of the documents. Topic modeling, including probabilistic latent semantic indexing and latent Dirichlet allocation, is a form of dimension reduction that uses a probabilistic model to find the co-occurrence patterns of terms that correspond to semantic topics in a collection of documents. We describe the basic technologies in detail and expose the underlying mechanism. We also discuss recent advances that have made it possible to apply these techniques to very large and evolving text collections and to incorporate network structure or other contextual information.

Keywords: Dimension reduction, Latent semantic indexing, Topic modeling, Latent Dirichlet allocation.

1.   Introduction

In 1958, Lisowsky completed an index of the Hebrew scriptures to help scholars identify the meanings of terms that had long since become unfamiliar [42]. Through a tedious manual process, he collected together all of the contexts in which every term occurred. As he did this, he needed to suppress differences in word form that were not significant while preserving differences that might affect the semantics. He hoped by this undertaking to enable other researchers to analyze the different passages and understand the semantics of each term in context.

The core task of automated text mining shares many of the same challenges that Lisowsky faced. The same concept can be expressed using any number of different terms (*synonymy*) and conversely the apparently same term can have very different meanings in different contexts (*polysemy*). Automated text mining must leverage clues from the context to identify different ways of expressing the same concept and to identify and disambiguate terms that are polysemous. It must also present the data in a form that enables human analysts to identify the semantics involved when they are not known *a priori*.

It is common to represent documents as a *bag of words* (BOW), accounting for the number of occurrences of each term but ignoring the order. This representation balances computational efficiency with the need to retain the document content. It also results in a vector representation that can be analyzed with techniques from applied mathematics and machine learning, notably dimension reduction, a technique that is used to identify a lower-dimensional representation of a set of vectors that preserves important properties.

BOW vectors have a very high dimensionality — each dimension corresponding to one term from the language. However, for the task of analyzing the concepts present in documents, a lower-dimensional semantic space is ideal — each dimension corresponding to one concept or one topic. Dimension reduction can be applied to find the semantic space and its relationship to the BOW representation. The new representation in semantic space reveals the topical structure of the corpus more clearly than the original representation.

Two of the many dimension reduction techniques that have been applied to text mining stand out. *Latent semantic indexing*, discussed in Section 2, uses a standard matrix factorization technique (singular vector decomposition) to find a latent semantic space. *Topic models*, on the other hand, provide a probabilistic framework for the dimension reduction task. We describe topic modeling in Section 3, including probabilistic latent semantic indexing (PLSI) and latent Dirichlet allocation (LDA). In Section 4, we describe the techniques that are used to interpret and evaluate the latent semantic space that results from dimension reduction. Many recent advances have made it possible to apply dimension reduction and topic modeling to large and dynamic datasets. Other advances incorporate network structures like social networks or other contextual information. We highlight these extensions in Section 5 before concluding in Section 6.

1.1 The Relationship Between Clustering, Dimension Reduction and Topic Modeling

Clustering, dimension reduction and topic modeling have interesting relationships. For text mining, these techniques represent documents in a new way that reveals their internal structure and interrelations, yet there are subtle distinctions. *Clustering* uses information on the similarity (or dissimilarity) between documents to place documents into natural groupings, so that similar documents are in the same cluster. *Soft clustering* associates each document with multiple clusters. By viewing each cluster as a dimension, clustering induces a low-dimensional representation for documents. However, it is often difficult to characterize a cluster in terms of meaningful features because the clustering is independent of the document representation, given the computed similarity.

On the other hand, dimension reduction starts with a feature representation of documents (typically a BOW model) and looks for a lower-dimensional representation that is faithful to the original representation. Although this close coupling with the original features results in a more coherent representation that maintains more of the original information than clustering, interpretation of the compressed dimensions is still difficult. Specifically, each new dimension is usually a function of all the original features, so that generally a document can only be fully understood by considering all of the dimensions together.

Topic modeling essentially integrates soft clustering with dimension reduction. Documents are associated with a number of latent topics, which correspond to both document clusters and compact representations identified from a corpus. Each document is assigned to the topics with different weights, which specify both the degree of membership in the clusters as well as the coordinates of the document in the reduced dimension space. The original feature representation plays a key role in defining the topics and in identifying which topics are present in each document. The result is an understandable representation of documents that is useful for analyzing the themes in documents.

1.2 Notation and Concepts

Documents. We use the following notation to consistently describe the documents used for training or evaluation. D is a corpus of M documents, indexed by d. There are W distinct terms in the vocabulary, indexed by v. The term-document matrix X is a $W \times M$ matrix encoding the occurrences of each term in each document. The LDA model has K topics, indexed by i. The number of tokens in any set is given by N, with a subscript to specify the set. For example, N_i is the number of tokens assigned to topic i. A bar indicates set complement, as for example $\bar{z}_{dn} \equiv \{z_{d'n'} : d' \neq d \text{ or } n' \neq n\}$.

Multinomial distribution. A commonly used probabilistic model for texts is the multinomial distribution,

$$\mathcal{M}(\boldsymbol{X}|\boldsymbol{\Psi}) \propto \prod_{v=1}^{W} \psi_v^{x_v},$$

which captures the relative frequency of terms in a document and is essentially equivalent to the BOW-vector with ℓ_1-norm standardization as $\sum_{v=1}^{W} \psi_v = 1$.

Dirichlet distribution. Dirichlet distribution is the conjugate distribution to multinomial distribution and therefore commonly used as prior for multinomial models:

$$\mathcal{D}(\mathbf{\Psi}|\mathbf{\Xi}) = \frac{\Gamma\left(\sum_{i=1}^{K} \xi_i\right)}{\prod_{i=1}^{K} \Gamma(\xi_i)} \prod_{i=1}^{K} \psi_i^{\xi_i-1}.$$

This distributions favors imbalanced multinomial distributions, where most of the probability mass is concentrated on a small number of values. As a result, it is well suited for models that reflect commonly observed power law distributions in human language.

Generative process. A generative process is an algorithm describing how an outcome was selected. For example, one could describe the generative process of rolling a die: one side is selected from a multinomial distribution with 1/6 probability on each of the six sides. For topic modeling, a random generative process is valuable even though choosing the terms in a document is not random, because they capture real statistical correlations between topics and terms.

2. Latent Semantic Indexing

LSI is an automatic indexing method that projects both documents and terms into a low dimensional space which, by intent, represents the semantic concepts in the document. By projecting documents into the semantic space, LSI enables the analysis of documents at a *conceptual* level, purportedly overcoming the drawbacks of purely term-based analysis. For example, in information retrieval, users may use many different queries to describe the same information need, and likewise, many of the relevant documents may not contain the exact terms used in the particular query. In this case, projecting documents into the semantic space enables the search engine to find documents containing the same concepts but different terms. The projection also helps to resolve terms that are associated with multiple concepts. In this sense, LSI overcomes the issues of *synonymy* and *polysemy* that plague term-based information retrieval.

LSI was applied to text data in the 1980s and later used for indexing in information retrieval systems [23]. It has also been used for a variety of tasks, including assigning papers to reviewers [28] and cross-lingual retrieval.

LSI is based on the singular value decomposition (SVD) of the term-document matrix, which constructs a low rank approximation of the original matrix while preserving the similarity between the documents. LSI is meant to interpret the dimensions of the low-rank approximation as semantic concepts although it is surpassed in this regard by later improvements such as PLSI. We now describe the basic steps for performing LSI. Then, we will discuss the implementation issues and analyze the underlying mechanisms for LSI.

2.1 The Procedure of Latent Semantic Indexing

Given the term-document matrix X of a corpus, the d-th column $\boldsymbol{X}_d$ represents a document d in the corpus and the v-th row of the matrix X, denoted by $\boldsymbol{T}_v$, represents a term v. Several possibilities for the encoding are discussed in the implementation issues section.

Let the singular value decomposition of X be

$$X = U\Sigma V^T,$$

where the matrices U and V are orthonormal and Σ is diagonal—

$$\Sigma = \begin{bmatrix} \sigma_1 & & \\ & \ddots & \\ & & \sigma_{\min\{W,M\}} \end{bmatrix}.$$

The values $\sigma_1, \sigma_2, \ldots, \sigma_{\min\{W,M\}}$ are the singular values of the matrix X. Without loss of generality, we assume that the singular values are arranged in descending order, $\sigma_1 \geq \sigma_2 \geq \cdots \geq \sigma_{\min\{W,M\}}$.

For dimension reduction, we approximate the term-document matrix X by a rank-K approximation $\hat{X}$. This is done with a partial SVD using the singular vectors corresponding to the K largest singular values.

$$\begin{aligned} \hat{X} &= \hat{U}\hat{\Sigma}\hat{V}^T \\ &= \begin{bmatrix} \boldsymbol{U}_1 & \cdots & \boldsymbol{U}_K \end{bmatrix} \begin{bmatrix} \sigma_1 & & \\ & \ddots & \\ & & \sigma_K \end{bmatrix} \begin{bmatrix} \boldsymbol{V}_1^T \\ \vdots \\ \boldsymbol{V}_K^T \end{bmatrix}. \end{aligned} \quad (5.1)$$

SVD produces the rank-K matrix $\hat{X}$ that minimizes the distance from X in terms of the spectral norm and the Frobenius norm. Although X is typically sparse, $\hat{X}$ is generally not sparse. Thus, $\hat{X}$ can be viewed as a smoothed version of X, obtained by propagating the co-occurring terms in the document corpus. This smoothing effect is achieved by discovering a latent semantic space formed by the documents. Specifically, we can

Dimensionality Reduction and Topic Modeling 135

observe from Eqn. (5.1) that each document d can be represented by a K-dimensional vector $\hat{\boldsymbol{X}}_d$, which is the d-th row of the matrix $\hat{V}$. The relation between the representation of document d in term space $\boldsymbol{X}_d$ and the latent semantic space $\hat{\boldsymbol{X}}_d$ is given by

$$\boldsymbol{X}_d = \hat{U}\hat{\Sigma}\hat{\boldsymbol{X}}_d.$$

Similarly, each term v can be represented by the K-dimensional vector $\hat{\boldsymbol{T}}_v$ given by

$$\boldsymbol{T}_v = \hat{V}\hat{\Sigma}\hat{\boldsymbol{T}}_v.$$

Thus, LSI projects both terms and documents into a K-dimensional latent semantic space. We can utilize these projections into latent semantic space to perform several tasks.

Information retrieval. In information retrieval, we are given a query $\boldsymbol{q}$ which contains several key terms that describe the information need. The goal is to return documents that are related to the query. In this case, we can view the query as a short document and project it into the latent semantic space using

$$\hat{\boldsymbol{q}} = \hat{\Sigma}^{-1}\hat{U}^T\boldsymbol{q}.$$

Then, the similarity between the query and document can be measured in the latent semantic space. For example, we can use the inner product $\hat{\boldsymbol{V}}_d^T\hat{\boldsymbol{q}}$. By using the smoothed latent semantic space for the comparison, we mitigate the problems with synonymy and polysemy.

Document similarity. The similarity between document d and d' can be measured using their representations in the latent semantic space, for example, using the inner product of $\hat{\boldsymbol{X}}_d$ and $\hat{\boldsymbol{X}}_{d'}$. This can be used to cluster or classify documents. Additional regularization may be necessary to resolve the non-identifiability of the SVD [63].

Term similarity. Analogous to the document similarity, term similarities can be measured in the latent semantic space, so as to identify terms with similar meanings.

2.2 Implementation Issues

2.2.1 Term-Document Matrix Representation.
LSI utilizes the term-document matrix X for a document corpus, which represents the occurrences of terms in documents. In practice, the term-document matrix can be constructed in several ways. For example, each

entry x_{vd} can represent the number of times that the term v occurs in document d. However, Zipf's law shows that real documents tend to be *bursty*—a globally uncommon term is likely to occur multiple times in a document if it occurs at all [19]. As a result, simply using the term frequency tends to exaggerate the contribution of the term. This problem can be directly addressed by using a binary representation, which only indicates whether a term occurs in a particular document and ignores its frequency. Global term-weight methods, such as term frequency weighted with inverse document frequency (IDF) [44], provide a good compromise for most document corpora. Besides these BOW representations, the language pyramid model [70] provides a multi-resolution matrix representation for documents, encoding not only the semantic information of term occurrence but also the spatial information such as term proximity, ordering, long distance dependence and so on.

2.2.2 Computation.
LSI relies on a partial SVD of the term-document matrix, which can be computed using the Lanczos algorithm [7, 30, 73]. The Lanczos algorithm is an iterative algorithm that computes the eigenvalues and eigenvectors of a large and sparse matrix X using the matrix vector multiplication. This process can be accelerated by exploiting any special structure of the term-document matrix. For example, Zha and Zhang [75] provide an efficient algorithm when the matrix has a low-rank-plus-shift structure, which arises when regularization is added. Numerous implementations that use the Lanczos algorithm are available, including SVDPACK (http://www.netlib.org/svdpack).

2.2.3 Handling Changes.
In real world applications, the corpus often changes rapidly. As a result, it is impractical to apply LSI to the corpus every time a document is added, removed or changed. There are two strategies for efficiently handling these changes.

Fold-in. One method for updating LSI is called *fold-in*, where we compute the projection of the new documents and terms into the latent semantic space based on the projection for original documents and terms. In order to fold in a document represented by vector $\boldsymbol{d} \in \mathbb{R}^W$ into a existing latent semantic indexing, we can project the document into the latent semantic space based on the SVD decomposition obtained from the original corpus.

$$\hat{\boldsymbol{d}} = \hat{\Sigma}^{-1} \hat{U}^T \boldsymbol{d}.$$

Fold-in is very efficient because the SVD does not need to be recomputed. Because the term vector $\boldsymbol{d}$ is typically sparse, the fold-in process can be computed in $O(KN)$ time, where N is the number of unique terms in $\boldsymbol{d}$.

Dimensionality Reduction and Topic Modeling 137

Updating the semantic space. Although the fold-in process is efficient and maintains a consistent indexing, there is no longer any guarantee that the indexing provides the best rank-K approximation of the modified corpus. Over time, the outdated model becomes increasingly less useful. Several methods for updating the LSI model have been proposed that are both efficient and accurate [8, 52, 74]. For example, Zha and Simon [74] provide an updating algorithm based on performing LSI on $[\hat{X}\ X']$ instead of $[X\ X']$, where X' is the term-document matrix for new documents. Specifically, the low-rank approximate $\hat{X}$ is used to replace the document-term matrix X of the original corpus. Assume that the QR decomposition of the matrix $(I - \hat{U}\hat{U}^T)X'$ is

$$(I - \hat{U}\hat{U})X' = U'R,$$

where R is a triangular matrix and $\hat{X} = \hat{U}\hat{\Sigma}\hat{V}$ is the partial SVD of the matrix X. Then we have

$$[X\ X'] = [\hat{U}\ U'] \begin{bmatrix} \hat{\Sigma} & \hat{U}^T X \\ 0 & R \end{bmatrix} \begin{bmatrix} \hat{V}^T & 0 \\ 0 & I \end{bmatrix}.$$

Now we can compute the best rank-K approximation of $\begin{bmatrix} \hat{\Sigma} & \hat{U}^T X \\ 0 & R \end{bmatrix}$ by SVD:

$$\begin{bmatrix} \hat{\Sigma} & \hat{U}^T X \\ 0 & R \end{bmatrix} = \hat{P}\Sigma'\hat{Q}^T.$$

Then, the partial SVD for $[\hat{X}\ X']$ can be expressed as

$$\left([\hat{U}\ U']\hat{P}\right) \Sigma' \left(\begin{bmatrix} \hat{V} & 0 \\ 0 & I \end{bmatrix} \hat{Q}\right)^T,$$

which provides an approximation of the partial SVD for $[X\ X']$. A theoretical analysis by Zha and Simon shows that this approximation will not introduce unacceptable errors into LSI [74].

2.3 Analysis

Due to the popularity of LSI, there has been considerable research into the underlying mechanism of LSI.

Term context. LSI improves the performance of information retrieval by discovering the latent concepts in a document corpus and thus solving the problems of synonymy and polysemy. Bast and Majumdar [5] demonstrate this point by considering the projections of a query q

and document d into the latent semantic space by the mapping
$$f(x) = \hat{U}^T x.$$
The cosine similarity of the query and document in the latent semantic space is
$$S_{qd} = \frac{q^T \hat{U} \hat{U}^T d}{\|\hat{U}^T q\| \|\hat{U}^T d\|}.$$
Since the factor $\|\hat{U}^T q\|$ does not depend on documents, it can be neglected without affecting the ranking. Note that
$$\|\hat{U}^T d\| = \|\hat{U} \hat{U}^T d\|,$$
so the cosine similarity S_{qd} can expressed by
$$S_{qd} = \frac{q^T \hat{U} \hat{U}^T d}{\|\hat{U} \hat{U}^T d\|} = \frac{q^T T d}{\|T d\|},$$
where $T \equiv \hat{U} \hat{U}^T$.

The similarity S_{qd} between query q and document d can be expressed by the cosine similarity of query q and the transformed document Td. In term-based information retrieval, the transformation $T = I$, so the original document is used to calculate the similarity. In LSI, however, the transformation T is not the identity matrix. Intuitively, the entry $t_{vv'}$ represents the relationship between terms v and v'. Specifically, the occurrence of term v in document has an equivalent impact on the similarity to $t_{vv'}$ times the occurrence of term v in the same document. In this sense, LSI enriches the document by introducing similar terms that may not occur in the original document.

Bast and Majumdar [5] also analyze LSI from the view of identifying terms that appear in similar contexts in the documents. Consider the sequence of the similarities between a pair of terms with respect to the dimension of latent semantic space, $K_{vv'}(k) = \sum_{i=1}^{k} U_v^{(i)} U_{v'}^{(i)T}$, where $U^{(i)}$ is from the rank-i partial SVD. The trend of the sequence can be categorized into three different types: increasing steadily (A); first increasing and then decreasing (B); or, no clear trend (C). If terms v and v' are related, the sequence is usually of Type A or B. Otherwise, the sequence is of Type C. This result is closely related to global special structures in the term-document matrix X that arise from similar contexts for similar terms. Thus, the sequence $K_{vv'}$ of similar terms have the specific shapes described above.

Since LSI captures the contexts of terms in documents, it is able to deal with the problems of synonymy and polysemy: synonymy can be captured since terms with the same meaning usually occur in similar context; polysemy can be addressed since terms with different meaning can be distinguished by their occurrences in different context. Landauer [40] also provides intuition for LSI by showing that it captures several important aspects of human languages.

Dimension of the latent semantic space. Dupert [29] studies how to determine the optimal number of latent factors for finding the most similar terms of a query. In particular, he shows how LSI can deal with the problem of synonymy in the context of Correlation method. He also provides an upper bound for the dimension of latent semantic space in order to present the corpus correctly.

Probabilistic analysis. Kubato Ando and Lee [38] explore the relationship between the performance of LSI and the uniformity of the underlying distribution. When the topic-documents distribution is quite uniform, LSI can recover the optimal representation precisely. Papadimitriou et al. [53] and Ding [26] analyze LSI from a probabilistic perspective which is related to probabilistic latent semantic indexing [36], which we discuss next.

3. Topic Models and Dimension Reduction

Taboo® (a registered trademark of Hasbro) is a game where one player must help a teammate guess a word from a game card without using any of the taboo words listed on the card. The surprising difficulty of the game highlights that certain terms are very likely to be present based on the topic of a document. *Latent topic models* capture this idea by modeling the conditional probability that an author will use a term given the topic the author is writing about.

LSI reduced the dimensionality of documents by projecting the BOW vectors into a semantic space constructed from the SVD of the term-document matrix. By providing a mechanism to explicitly reason about latent topics, probabilistic topic models can achieve a similar yet more meaningful latent semantic space. The results are presented in familiar probabilistic terms, and thus can be directly incorporated into other probabilistic models and analyzed with standard statistical techniques. Moreover, Bayesian methods can be used to make the models robust to parameter selection. Finally, one of the most useful advantages is that the models can be easily extended by modifying the structure to solve interesting related problems.

3.1 Probabilistic Latent Semantic Indexing

PLSI, proposed by Hofmann [36], provides a crucial step in topic modeling by extending LSI in a probabilistic context. PLSI has seen widespread use in text document retrieval, clustering and related areas; it builds on the same conceptual assumptions as LSI, but uses a radically different *probabilistic* generative process for generating the terms in the documents of a text corpus.

PLSI is based on the following generative process for (w, d), a word w in document d:

- Sample a document d from multinomial distribution $p(d)$.

- Sample a topic $i \in \{1, \ldots, K\}$ based on the topic distribution $\theta_{di} = p(z = i|d)$.

- Sample a term v for token w based on $\Phi_{iv} = p(w = v|z = i)$.

In other words, an unobservable topic variable z is associated with each observation (v, d) in PLSI. The joint probability distribution for $p(v, d)$ can be expressed as

$$p(v, d) = p(d)p(v|d), \quad \text{where} \quad p(v|d) = \sum_{i=1}^{K} p(v|z = i)p(z = i|d).$$

This equation has the geometric interpretation that the distribution of terms conditioned on documents $p(z = i|d)$ is a convex combination of the topic-specific term distributions $p(v|z = i)$.

Connection to LSI. An alternative way to express the joint probability is given by

$$p(v, d) = \sum_{i=1}^{K} p(z = i)p(d|z = i)p(v|z = i).$$

This formulation is sometimes called the *symmetric formulation* because it models the documents and terms in a symmetric manner. This formulation has a nice connection to LSI: the probability distributions $p(d|z = i)$ and $p(w|z = i)$ can be viewed as the projections of documents and terms into the latent semantic spaces, just like the matrices $\hat{V}$ and $\hat{U}$ in LSI. Also, the distribution $p(z = i)$ is similar to the diagonal matrix $\hat{\Sigma}$ in LSI. This is the sense in which PLSI is a probabilistic version of LSI.

3.1.1 Algorithms.

The maximal likelihood method is used to estimate the parameters $p(d)$, $p(z|d)$ and $p(v|z)$. Given the term-document matrix X, the log-likelihood of observed data can be expressed as

$$\mathcal{L} = \sum_{d=1}^{M} \sum_{v=1}^{W} x_{vd} \log p(w = v, d)$$

$$= \sum_{d=1}^{M} \sum_{v=1}^{W} x_{vd} \log \sum_{i=1}^{K} p(w = v|z = i) p(z = i|d) p(d). \quad (5.2)$$

Maximizing the log-likelihood function is equivalent to minimizing the Kullback-Leibler divergence (KL) [39] between the measured empirical distribution $\hat{p}(v|d)$ and the model distribution $p(w|d) = \sum_{i=1}^{K} p(w|z = i) p(z = i|d)$. Since this is non-convex, expectation-maximization (EM) [24] is used to seek a locally optimal solution. The log-likelihood value (Eqn. (5.2)) increases on each iteration and converges to a local maximum.

Expectation. The E-step computes the posterior of the latent variable z based on the current estimation of the parameters.

$$p'(z = i|d, v) = \frac{p(d) p(z = i|d) p(v|z = i)}{\sum_{i'=1}^{K} p(d) p(z = i'|d) p(v|z = i')},$$

where the prime on p indicates the new estimate of the probability for the next step.

Maximization. The M-step updates the parameters once the latent variables are known using the posterior estimated in the previous E-step:

$$p'(w = v|z) \propto \sum_{d=1}^{M} x_{vd}\, p'(z = i|d, w = v);$$

$$p'(z = i|d) \propto \sum_{v=1}^{W} x_{vd}\, p'(z = i|d, w = v);$$

$$p'(d) \propto \sum_{v=1}^{W} x_{vd}.$$

3.1.2 Updating.

Given a new document d, the fold-in process can be applied to obtain its representation in the latent semantic space, much like for LSI. Specifically, an EM algorithm similar to parameter estimation can be used to obtain $p(z|d)$ [37]. $p(w|z)$ and $p(z)$ are not updated in the M-step during fold-in.

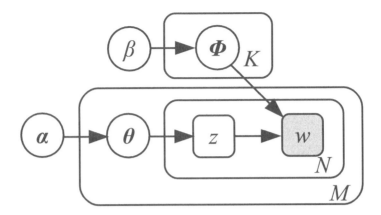

Figure 5.1. Diagram of the LDA graphical model

3.2 Latent Dirichlet Allocation

PLSI provides a good basis for text analysis, but it has two problems. First, it contains a large number of parameters that grows linearly with the number of documents so that it tends to overfit the training data. Second, there is no natural way to compute the probability of a document that was not in the training data. LDA includes a process for generating the topics in each document, thus greatly reducing the number of parameters to be learned and providing a clearly-defined probability for arbitrary documents. Because LDA has a rich generative model, it is also readily adapted to specific application requirements, which we describe in Section 5.

3.2.1 Model. Like PLSI, LDA is based on a hypothetical generative process for a corpus. A diagram of the graphical model showing how the different random variables are related is shown in Fig. 5.1. In the diagram, each random variable is represented by a circle (continuous) or square (discrete). A variable that is *observed* (its outcome is known) is shaded. An arrow is drawn from one random variable to another if the the outcome of the second variable depends on the value of the first variable. A rectangular plate is drawn around a set of variables to show that the set is repeated multiple times, as for example for each document or each token.

- CHOOSE THE TERM PROBABILITIES FOR EACH TOPIC. The distribution of terms for each topic i is represented as a multinomial distribution $\mathbf{\Phi}_i$, which is drawn from a symmetric Dirichlet distribution with parameter β.

$$\mathbf{\Phi}_i \sim \mathcal{D}(\beta); \qquad p(\mathbf{\Phi}_i|\beta) = \frac{\Gamma(W\beta)}{[\Gamma(\beta)]^W} \prod_{v=1}^{W} \phi_{iv}^{\beta-1}.$$

- CHOOSE THE TOPICS OF THE DOCUMENT. The topic distribution for document d is represented as a multinomial distribution $\boldsymbol{\theta}_d$, which is drawn from a Dirichlet distribution with parameters $\boldsymbol{\alpha}$. The Dirichlet distribution captures the document-independent popularity and the within-document burstiness of each topic.

$$\boldsymbol{\theta}_d \sim \mathcal{D}(\boldsymbol{\alpha}); \qquad p(\boldsymbol{\theta}_d|\boldsymbol{\alpha}) = \frac{\Gamma(\sum_{i=1}^{K} \alpha_i)}{\prod_{i=1}^{K} \Gamma(\alpha_i)} \prod_{i=1}^{K} \theta_{di}^{\alpha_i-1}.$$

- CHOOSE THE TOPIC OF EACH TOKEN. The topic z_{dn} for each token index n is chosen from the document topic distribution.

$$z_{dn} \sim \mathcal{M}(\boldsymbol{\theta}_d); \qquad p(z_{dn} = i|\boldsymbol{\theta}_d) = \theta_{di}.$$

- CHOOSE EACH TOKEN. Each token w at each index is chosen from the multinomial distribution associated with the selected topic.

$$w_{dn} \sim \mathcal{M}(\boldsymbol{\phi}_{z_{dn}}); \qquad p(w_{dn} = v|z_{dn} = i, \boldsymbol{\phi}_i) = \phi_{iv}.$$

Mechanism. LDA provides the mechanism for finding patterns of term co-occurrence and using those patterns to identify coherent topics. Suppose that we have used LDA to learn a topic i and that for term v, $p(w = v|z = i)$ is high. As a result of the LDA generative process, any document d that contains term v has an elevated probability for topic i, that is, $p(z_{dn'} = i|w_{dn} = v) > p(z_{dn'} = i)$. This in turn means that all terms that co-occur with term v are more likely to have been generated by topic i, especially as the number of co-occurrences increases. Thus, LDA results in topics in which the terms that are most probable frequently co-occur with each other in documents.

Moreover, LDA also helps with polysemy. Consider a term v with two distinct meanings in topics i and i'. Considering only this term, the model places equal probability on topics i and i'. However, if the other words in the context place a 90% probability on i and only a 9% probability on i', then LDA will be able to use the context to disambiguate the topic: it is topic i with 90% probability.

Wallach et al. [60] show that the symmetry or asymmetry of the Dirichlet priors strongly influences the mechanism. For the topic-specific term distributions, a symmetric Dirichlet prior provides smoothing so that unseen terms will have non-zero probability. However, an asymmetric prior would equally affect all topics, making them less distinctive.

In contrast, they showed that an asymmetric prior for the document-specific topic distributions made LDA more robust to stop words and less sensitive to the selection of the number of topics. The stop words were mainly relegated to a small number of highly probable topics that influence most documents uniformly. The asymmetric prior also results in more stable topics, which means that additional topics will make small improvements in the model instead of radically altering the topic structure. This is similar to the situation of LSI, where performance is optimal when Σ scales the contribution of each dimension according to its eigenvalue. In the same way, LDA will perform best if α is non-uniform and corresponds to some natural values characteristic of the dataset.

One disadvantage of LDA is that it tends to learn broad topics. Consider the case where a concept has a number of aspects to it. Each of the aspects co-occurs frequently with the main concept, and so LDA will favor a topic that includes the concept and all of its aspects. It will further favor adding other concepts to the same topic if they share the same aspects. As this process continues, the topics become more diffuse. When sharper topics are desired, a hierarchical topic model may be more appropriate.

Likelihood. Training an LDA model involves finding the optimal set of parameters, under which the probability of generating the training documents is maximized. The probability of the training documents under a given LDA model is called the *empirical likelihood* $\mathcal{L}$. It can also be used to identify the optimal model configuration using Bayesian model selection.

$$\mathcal{L} = \prod_{d=1}^{M} \prod_{n=1}^{N} p(w_{dn}|z_{dn}, \boldsymbol{\Phi}) p(z_{dn}|\boldsymbol{\theta}_d) p(\boldsymbol{\theta}_d|\boldsymbol{\alpha}) p(\boldsymbol{\Phi}|\boldsymbol{\beta})$$

$$= \phi_{zw}\, \theta_{dz}\, \frac{\Gamma\left(\sum_{i=1}^{K} \alpha_i\right)}{\prod_{i=1}^{K} \Gamma(\alpha_i)} \prod_{i=1}^{K} \theta_{di}^{\alpha_i-1} \frac{\Gamma(W\beta)}{[\Gamma(\beta)]^W} \prod_{v=1}^{W} \phi_v^{\beta-1}.$$

Unfortunately, the direct optimization of the likelihood is problematic because the topic assignments z_{dn} are not directly observed. Even

Dimensionality Reduction and Topic Modeling 145

inference for a single document is intractable. We describe two different approximations for LDA. Collapsed Gibbs sampling samples a value for each z_{dn} in turn, conditioned on the topic assignments for the other tokens. Variational Bayes approximates the model with a series of simpler models that bound the likelihood but neglect the troublesome dependencies.

3.2.2 Collapsed Gibbs.

Gibbs sampling is commonly used to estimate the distribution of values for a probability model when exact inference is intractable. First, values are assigned to each variable in the model, either randomly or using a heuristic. Each variable is then sampled in turn, conditioned on the values of the other variables. In the limit of the number of iterations, this process explores all configurations and yields unbiased estimates of the underlying distributions. In practice, Gibbs sampling is implemented by rejecting a large number of samples during an initial burn-in period and then averaging the assignments during an additional large number of samples.

In *collapsed* Gibbs sampling, certain variables are marginalized out of the model. Griffiths and Steyvers [32] propose collapsed Gibbs sampling for LDA, with both $\boldsymbol{\theta}$ and $\boldsymbol{\Phi}$ marginalized. Only z_{dn} is sampled, and the sampling is done conditioned on $\boldsymbol{\alpha}$, β and the topic assignments of other words $\bar{z}_{dn}$.

$$p(z_{dn}|\bar{z}_{dn}) \propto (N_{dz} + \alpha_z)(N_{zw} + \beta).$$

The N statistics do not include the contribution from the word being sampled, and must be updated after each sampling.

The equation makes intuitive sense. A topic that is used frequently in the document has a higher probability in $\boldsymbol{\theta}$ and so is more likely for the current token also. This characteristic corresponds to the burstiness observed in documents [19]. Similarly, a topic that is frequently assigned for the same term corpus-wide is more likely to be correct here also.

After burn-in, the implementation can keep statistics of the number of times each topic is selected for each word. These statistics can then be aggregated and normalized to estimate the topic distributions for each document or word. To apply a trained model to additional documents, the only change is that the N_{zw} statistic is not updated.

3.2.3 Variational Approximation.

Variational approximation provides an alternative algorithm for training an LDA model. We will first consider the case of inferring the topics of a document given an existing LDA model, before we explain how the model is trained. A

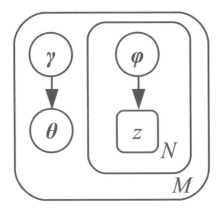

Figure 5.2. Diagram of the LDA variational model

direct approach for topic inference is to apply Bayes' rule:

$$p(\boldsymbol{\theta}|w_d) = \frac{p(\boldsymbol{\theta}, w_d)}{p(w_d)} = \frac{\int_{\boldsymbol{Z}} p(d, \boldsymbol{\theta}, \boldsymbol{Z}|\boldsymbol{\alpha}, \beta) d\boldsymbol{Z}}{\int_{\boldsymbol{Z}, \boldsymbol{\theta}} p(d, \boldsymbol{\theta}, \boldsymbol{Z}|\boldsymbol{\alpha}, \beta) d\boldsymbol{Z} d\boldsymbol{\theta}},$$

where $\boldsymbol{Z} = \{z_1, z_2, \ldots, z_N\}$. However, the marginalization in both numerator and denominator is intractable. The *Variational Bayesian* approach provides an approximate solution; instead of inferring the latent variables by directly marginalizing the joint distribution $p(w_d, \boldsymbol{\theta}, \boldsymbol{Z}|\boldsymbol{\alpha}, \beta)$, it uses a much simpler distribution as a proxy and performs the inference through optimization.

Variational inference approximates the true posterior distribution of the latent variables by a fully-factorized distribution—this proxy is usually referred to as the variational model, which assumes all the latent variables are independent of each other. For LDA,

$$q(\boldsymbol{Z}, \boldsymbol{\theta}|\boldsymbol{\gamma}, \boldsymbol{\phi}) = q(\boldsymbol{\theta}|\boldsymbol{\gamma}) \prod_{n=1}^{N} q(z_n|\boldsymbol{\phi}_n) = \mathcal{D}(\boldsymbol{\theta}|\boldsymbol{\gamma}) \prod_{n=1}^{N} \mathcal{M}(z_n|\boldsymbol{\phi}_n).$$

Essentially, this variational distribution is a simplification of the original LDA graphical model by removing the edges between the nodes $\boldsymbol{\theta}$ and $\boldsymbol{Z}$ (Figure 5.2). The optimal approximation is achieved by optimizing the distance (for example, the KL divergence) between the true model and the variational model:

$$\min_{\boldsymbol{\gamma}, \boldsymbol{\phi}} \; KL[q(\boldsymbol{\theta}, \boldsymbol{Z}|\boldsymbol{\gamma}, \boldsymbol{\phi}) \; || \; p(\boldsymbol{\theta}, \boldsymbol{Z}|\boldsymbol{\alpha}, \beta)].$$

Dimensionality Reduction and Topic Modeling

It can be shown that the above KL-divergence is the discrepancy between the true log-likelihood and its variational lower-bound that is used in the variational EM algorithm (described later in this section) for estimating the LDA hyperparameters $\boldsymbol{\alpha}$ and β.

The optimization has no close-form solution but can be implemented through iterative updates,

$$\gamma_i = \alpha_i + \sum_{n=1}^{N} \phi_{ni}, \quad \phi_{ni} \propto \beta_{iw_n} \exp[\Psi(\gamma_i)],$$

where $\Psi(\cdot)$ is the bi-gamma function.

Variational EM for parameter estimation. We can learn a LDA topic model by maximizing the likelihood of the corpus.

$$\max_{\boldsymbol{\alpha},\beta} \sum_{d=1}^{M} \log p(w_d|\boldsymbol{\alpha},\beta)$$

$$= \max_{\boldsymbol{\alpha},\beta} \sum_{d=1}^{M} \log \int_{\boldsymbol{\theta}_d, \boldsymbol{Z}_d} p(w_d, \boldsymbol{\theta}_d, \boldsymbol{Z}_d|\boldsymbol{\alpha},\beta) d\boldsymbol{\theta}_d d\boldsymbol{Z}_d.$$

Again, it involves intractable computation of the marginal distribution and we therefore resort to variational approximation, which provides a tractable lower bound,

$$\mathcal{L}(\boldsymbol{\gamma}, \boldsymbol{\phi}) = \log p(w_d|\boldsymbol{\alpha},\beta) - KL(q(\boldsymbol{Z},\boldsymbol{\theta}|\boldsymbol{\gamma},\boldsymbol{\phi})||p(\boldsymbol{Z},\boldsymbol{\theta}|\boldsymbol{\alpha},\beta))$$
$$\leq \log p(w_d|\boldsymbol{\alpha},\beta),$$

where $\mathcal{L}(\boldsymbol{\gamma}, \boldsymbol{\phi}) = \mathbf{E}_q[\log p(w_d, \boldsymbol{\theta}, \boldsymbol{Z}) - \log q]$ is the variational lower bound for the log-likelihood. The maximum likelihood estimation therefore involves a two-layer optimization,

$$\max_{\boldsymbol{\alpha},\beta} \sum_{d=1}^{M} \max_{\boldsymbol{\gamma}_d, \boldsymbol{\phi}_d} \mathcal{L}(\boldsymbol{\gamma}_d, \boldsymbol{\phi}_d).$$

The inner-loop (the optimization with respect to $\boldsymbol{\gamma}$ and $\boldsymbol{\phi}$, referred to as the Variational E-step) goes through the whole corpus and performs variational approximation for each of the documents, which ends up with a tight lower bound for the log-likelihood. Then the M-step updates the model parameters ($\boldsymbol{\alpha}$ and β) by optimizing this lower-bound approximation of the log-likelihood. The E- and M-steps are alternated in an outer loop until convergence.

In the E-step, γ and ϕ are alternately optimized for each document—in practice, 20 iterations is adequate for a good fit. The outer loop may need to be repeated hundreds of times for full convergence. For best results, the likelihood of a separate validation corpus controls early stopping.

3.2.4 Implementations. There have been substantial efforts in developing efficient and effective implementations of LDA, especially for parallel or distributed architectures. In order to provide a quick hands-on experience, we list a few implementations that are open-source or publicly accessible in Table 5.1.

Table 5.1. Publicly-accessible implementations of LDA.

Name	Language	Algorithm	Reference
LDA-C	C	Var. EM	www.cs.princeton.edu/~blei/lda-c
Mallet	Java	Gibbs	mallet.cs.umass.edu
GibbsLDA++	C++	Gibbs	gibbslda.sourceforge.net
Gensim	Python	Gibbs	nlp.fi.muni.cz/projekty/gensim
Matlab-LDA	Matlab	Gibbs	psiexp.ss.uci.edu/programs_data

4. Interpretation and Evaluation

We have looked at three methods for dimension reduction of textual data. These methods have much in common: they identify the relationships of terms and documents to the dimensions of a latent semantic space. Intuitively, the latent dimensions correspond to concepts or topics that are meaningful to the authors. In this section, we discuss how to jump from the mathematical representations to meaningful topics, how to evaluate the resulting models and how to apply them to applications.

4.1 Interpretation

The common way to interpret the topic models that are discovered by dimension reduction is through inspection of the term-topic associations. Typically, practitioners examine the five to twenty terms that are most strongly associated with each topic, and attempt to discern the commonality. For LSI, the terms can be sorted according to the coefficient corresponding to the given feature in the semantic space. For the probabilistic models, the terms are sorted by the probability of generating the term conditioned on the topic. This approach was popularized following Blei et al. [13], and is generally used to report qualitative topic model results even though it has many disadvantages. The chief problem is

Dimensionality Reduction and Topic Modeling 149

that the top terms are often dominated by globally probable terms that may not be representative of the topic. Stop word removal and variations on IDF weighting both help substantially, but the characterization is sensitive to the precise method used to order terms. Mei et al. [47] provide an alternative approach that automatically selects a portion of a document to use as a label for each topic. Buntine and Jakulin [16] provide a more general framework for interpreting topic models.

4.2 Evaluation

There are three main approaches to evaluating the models resulting from dimension reduction. The fit of the models to test data is important for understanding how well the models generalize to new data, but application-driven metrics are also essential if the model is to be useful. When it is necessary for a human to interact with the model, interpretability should also be evaluated.

Fit of test data. A very common approach is to train a model on a portion of the data and to evaluate the fit of the model on another portion of the data. For LSI, the test documents can be projected into the latent semantic space and then the ℓ_2 error introduced by the approximation can be calculated. The probabilistic models can be evaluated by computing the probability of generating the test documents given the model.

Perplexity [4] is the most common way to report this probability. Computed as

$$\exp\left(-\frac{1}{N}\sum_{d=1}^{M}\sum_{n=1}^{N_d}\log p(w_{dn}|\text{model})\right),$$

the perplexity corresponds to the effective size of the vocabulary. For example, a value of 100 indicates that the probabilities resulting from the model are equivalent to randomly picking each word from a vocabulary of 100 words. This means that smaller values indicate that the model fits the test data better.

Wallach et al. [61], evaluate several different ways to compute this probability and recommended the *left-to-right* method, in which the probability of generating each token in a document is conditioned on all previous tokens in the document so that the interaction between the tokens in the document are properly accounted for.

Application performance. Another common approach is to measure the utility of topic models in some application. Whenever the dimension reduction is being carried out with a specific application in mind, this in an important evaluation. For example, Wei and Croft [65] discuss the evaluation of LDA models for document search using standard information retrieval metrics.

Interpretability. For text mining, the ability to use the discovered models to better understand the documents is essential. Unfortunately, the fit of test data and application performance metrics completely ignore the topical structure. In fact, models with better perplexity are often harder to interpret [18]. This is not surprising, because the task of finding a meaningful model that fits well is more constrained than the task of finding any model that fits well, so the best fit is likely to be found using a less meaningful model.

Chang et al. [18] propose a new evaluation protocol based on a user study. Starting with a list of top terms for each topic that has been tainted with an additional term, users are asked to identify the spurious term. User performance on this task is higher when the topic is coherent so that the extra term stands out. They also conducted a similar experiment to measure the appropriateness of topic assignments to test documents.

4.3 Parameter Selection

Asuncion et al. [3] compare a variety of different algorithms for the LDA model. They found that with careful selection of the regularization hyperparameters α and β, all of the algorithms had similar perplexity. A grid search over possible values yields the best performance, but interleaving optimization of the hyperparameters with iterations of the algorithm is almost as good with much less computational cost.

4.4 Dimension Reduction

Latent topic models, including LSI, PLSI and LDA, are commonly used as dimension reduction tools for texts. After the training process, the document d can be represented by its topic distribution $p(z|d)$, where z can be viewed as a K-dimensional representation of the original document. The similarity between documents can then be measured by their similarity in the topic space.

$$S_{dd'} = \sum_{z=1}^{K} p(z|d)p(z|d').$$

Through this equation, documents are projected into a low dimensional space. The terms are projected into a K-dimensional space in the same way. For probabilistic topic models, KL divergence can be used for an alternative comparison.

Handling of synonymy is a natural result of dimension reduction. Multiple terms associated with the same concept are projected into the same place in the latent semantic space. Polysemy presents a more difficult challenge. Griffiths and Steyvers [31] found that LSI was able to detect polysemy: a term that was projected onto multiple latent dimensions generally had multiple meanings. LDA can resolve polysemy provided that one of the topics associated with a polysemous term is associated with additional tokens in the document.

5. Beyond Latent Dirichlet Allocation

LDA has many advantages for topic modeling, including its relative simplicity to implement and the useful topics that it unearths. However, with additional effort topic modeling can be adapted to the characteristics of a particular problem. In this section, we survey recent advances that make it practical to apply topic modeling to very large text corpora, dynamic data, data that is embedded in a network and other problems with special characteristics.

5.1 Scalability

Standard LDA learning algorithms read the documents in the training corpus numerous times and are inherently serial. In practice, this means that LDA models are trained on only a small fraction of the available data. However, recent advances in online and parallel algorithms make it reasonable to train and apply models at very large scale.

Efficient parallel implementations are available based on either collapsed Gibbs sampling or variational approximation. Smola et al. [56] perform Gibbs sampling based on slightly outdated term and topic statistics in parallel with threads that globally update the statistics. Using variational approximation, Asuncion et al. [3] interleave an inference step on all documents with a parallel aggregation of the term and topic statistics. Both of these methods achieve scalability through approximations that have no known convergence guarantee. In contrast, Yan et al. [66] and Liu et al. [43], use careful scheduling to achieve strong parallelization without approximation. However, the approximate methods are easier to implement correctly and work very well in practice.

5.2 Dynamic Data

Numerous approaches are possible when the corpus of documents is changing over time or must be processed as a stream. One common approach is to augment the corpus with the time of each document and incorporate time into the model. Wang and Agichtein [64] model the revision history of documents by considering the temporal dimension and extend LSI to tensor factorization. PLSI can be similarly augmented to model the temporal patterns of activities in videos [59]. Mølgaard et al. [50] study temporal PLSI for music retrieval, which can be viewed as a probabilistic model for tensor factorization. Blei and Lafferty modeled time evolution of topic models [10] to analyze how the topics used in a corpus changed over time.

For streaming data, Yao et al. [72] present a time and space efficient algorithm for applying an existing topic model to a stream of documents, using a modification of Gibbs sampling. Hoffman et al. [35] developed an online algorithm for LDA that retrains the model for each document in turn. Interestingly, they found that this approach did not sacrifice any quality in the learned model as measured using perplexity. They further show that the online algorithm corresponds to stochastic gradient descent on the variational objective function, and so converges to a stationary point of that function.

The online training process for LDA optimizes each document in turn. First, it uses standard variational approximation to estimate the probability distribution for the topic of each word ϕ_j. Next, topic models $\tilde{\Phi}_i$ are estimated as if the corpus consisted of M copies of this document, based on ϕ_j.

$$\tilde{\Phi}_{ij} = \beta + MN_j\phi_{ij}.$$

The estimate of the topic models Φ_i are then updated to include the contribution from this document by

$$\Phi_{ij} \leftarrow (\Phi_{ij} + \rho\tilde{\Phi}_{ij})/(1+\rho),$$

where $\rho = (t_0+t)^{-\kappa}$ when processing the t-th document. t_0 is a parameter that slows the algorithm during the early iterations and $0.5 < \kappa \leq 1$ is a parameter that controls the rate of learning. This algorithm is essentially the variational algorithm applied to a different single document on each iteration, with appropriate changes to how the topic models are updated. For very large datasets, this is many times faster than other algorithms and yet yields very excellent results.

5.3 Networked Data

Networks play an important role in many text mining problems. Email messages are linked to the senders and recipients. Publications are also linked by citations. Many documents are related to a social network.

The analysis of these documents can reveal more interesting structure if the network graph can be incorporated.

LSI has been applied to analysis network data. Ng et al. analyze the connection between the LSI and HIST [51], which is a widely used algorithm for network data. Other approaches learn low-dimensional representations of documents based on both their contents and the citation graph between them through learning from multiple relationships between different types of entities [69, 76].

PLSI has also been applied to analyze the network data. Cohn and Chang apply PLSI to model the citation graph and identity authoritative document based on the latent factors [21]. Citations between documents can be modeled together with the contents of the documents in a joint probabilistic model [20], through the probability of generating a citation given a latent topic. Guo et al. [34] model the interaction of topics between linked documents. Intuitively, the topics of a document are borrowed from the documents to which it links. Deng et al. [25] propose the two frameworks based on random walk and regularization to propagate the topics of documents according to the links between them.

We describe the work of Mei et al. [46] in detail since it is representative of combining PLSI and network analysis. This work utilizes the network structure as the regularization for PLSI through assuming that the topic distributions are similar for documents connected to each other. The regularization term induced from the structure of the network is optimized together with the log-likelihood function of PLSI. The model is applied to several applications such as author-topic analysis and spatial topic analysis, where network structures are constructed from co-authorships and adjacency of locations, respectively.

Much research has explored various ways to integrate network information into topic models. Rosen-Zvi et al. incorporate authorship information through author-specific topic mixtures [55]. Supervised topic models allow the per-topic term distributions to depend on a document label [12]. Chang and Blei incorporate relational information between documents [17]. It is also possible to integrate general first order logic [2]. McCallum et al. extend LDA so that it can identify topic models that are conditioned on the author and the audience of the communication [45]. This is useful for analyzing the social dynamics of communication in a network.

Relational topic models (RTM) extend LDA to jointly model the generation of documents and the generation of links between documents [17]. The model predicts links based on the similarity of the topic mixture used in two documents, which adds the capability of predicting missing links in the graph structure. Because the links influence the selection of topics, the model can more accurately predict links than a similar prediction based on topics from LDA.

The generative process of documents is the same for RTM as for LDA. Once the documents are generated, the link λ_{de} between documents d and e is generated from exponential regression on the empirical topic mixtures $\bar{Z}_d$ and $\bar{Z}_e$,

$$\lambda_{de} = \exp\left(\boldsymbol{\eta}^T(\bar{Z}_d \circ \bar{Z}_e) + \boldsymbol{\nu}\right),$$

where $\bar{Z}_d \equiv 1/N_d \sum_{n=1}^{N} Z_{dn}$ and $a \circ b$ is the element-wise product of vectors a and b. Typically the link is taken to be binary, in which case $\boldsymbol{\nu}$ is used to control the threshold. $\boldsymbol{\eta}$ is a parameter that must be learned which controls the importance of each topic in establishing the link. We generally expect it to have positive values, although a negative value in a social network would reflect the adage that opposites attract.

5.4 Adapting Topic Models to Applications

The graphical model of LDA can be easily extended to match the characteristics of a specific application. Here we survey some of the fruitful approaches.

One important class of extensions to LDA has been the introduction of richer priors for document topic and term distributions. Instead of using a fixed, global Dirichlet hyperparameter $\boldsymbol{\alpha}$ for all the documents in a corpus, Mimno and McCallum use regression from document features to establish a document-specific $\boldsymbol{\alpha}$ [48]. This is a valuable enhancement when other meta-features are available that are expected to influence the selected topics, as, for example, the identity of the author, the publication venue and the dates.

The Bayesian hierarchy of LDA provides a useful modeling pipeline for data with complex structure. The hierarchy can model web-like interconnections and uncertain labels [67, 71]. The *mixed membership stochastic block model* coupled two LDA hierarchies to model inter-connected entities [1], which provides a flexible model for network graphs and has proven useful for a variety of applications ranging from role discovery to community detection in social, biological and information networks.

Hierarchical topic models (hLDA) are used to identify subtopics that are increasingly more specific [9]. The hLDA model automatically learns a tree structure hierarchy for topics while they are discovered from the documents. For additional flexibility, hierarchical Dirichlet processes [57] can automatically discover an appropriate number of topics and subtopics. There are also principled ways to learn correlations between topics [11, 41]. Other extensions support richer document representations and contextual information, including bigrams [62], syntactic relationships [15, 33] and product aspects [58].

Multinomial distributions for term occurrences usually have a difficult time modeling the word burstiness in language — if a word appears in a document once, it will likely appear again in the same document. This effect is commonly referred to as Zipf's law, a profound characteristic of language. To discount this impact, Doyle and Elkan replace the per-topic Multinomial distribution with a Dirichlet-Compound Multinomial (also called the multivariate Pòlya) distribution) [27]. Reisinger et al. substitutes spherical admixture models [54], which not only incorporate negative correlations among term occurrence but also admit the natural use of cosine similarity to compare topics or documents.

Standard topic models are not appropriate for identifying consistent topics across multiple languages, because the multiple languages do not co-occur in documents frequently enough to be assigned into the same topics. Mimno et al. developed an extension that works with loosely *aligned* documents [49]—pairs of documents in different languages that have nearly the same mixture of topics. Boyd-Graber and Blei explore various strategies for discovering multilingual topics from unaligned documents [14]. Similar issues arise with documents in multiple dialects. Crain et al. [22] and Yang et al. [68] discuss extensions of LDA that find shared topics between consumer and technical medical documents.

6. Conclusion

Using a BOW representation results in very efficient text mining because more complex factors like grammar and word order can be neglected. However, working directly with individual terms has a number of strong limitations, because multiple documents can discuss the same ideas using very different words, and likewise, the same word can have very different meanings. Dimension reduction is able to lift the BOW representation to a more abstract level that better reflects the needs of a human analyst, where the new dimensions correspond to concepts or topics. In this way, alternative ways of expressing the same content can be reduced to a common representation and terms with multiple meanings can be identified.

LSI is based on a spectral analysis of the term-document matrix. This approach identifies common generalizations that are guaranteed to provide the best lower-dimensional representation of the original data. This representation is not necessarily easy to interpret, but is very useful for performing a conceptual match between two documents that may use different terms for the same concepts.

Probabilistic topic models provide an intuitive, probabilistic foundation for dimension reduction. They allow us to reason about the topics present in a document and expose the probability of seeing each word in any given topic. This makes it much easier to interpret what the topics mean. It also makes it easier to extend the models in interesting ways. Many extensions to PLSI and LDA have been developed, both to allow them to be applied to large scale data and to incorporate special structure for a particular application.

Acknowledgment

Part of the work is supported by NSF grants IIS-1049694, IIS-1116886, a Yahoo! Faculty Research and Engagement Grant and a Department of Homeland Security Career Development Grant.

References

[1] E. Airoldi, D. Blei, S. Fienberg, and E. Xing. Mixed membership stochastic blockmodels. *J. Mach. Learn. Res.*, 9:1981–2014, June 2008.

[2] D. Andrzejewski, X. Zhu, M. Craven, and B. Recht. A framework for incorporating general domain knowledge into latent Dirichlet allocation using first-order logic. In *IJCAI*, 2011.

[3] A. Asuncion, M. Welling, P. Smyth, and Y. Teh. On smoothing and inference for topic models. In *UAI*, pages 27–34, 2009.

[4] L. Bahl, J. Baker, E. Jelinek, and R. Mercer. Perplexity—a measure of the difficulty of speech recognition tasks. In *Program, 94th Meeting of the Acoustical Society of America*, volume 62, page S63, 1977.

[5] H. Bast and D. Majumdar. Why spectral retrieval works. In *SIGIR*, page 11, 2005.

[6] J.-P. Benzecri. *L'Analyse des Donnees. Volume II*. 1973.

[7] M. Berry. Large-scale sparse singular value computations. *The International Journal Of Supercomputer Applications*, 6(1):13–49, 1992.

[8] M. Berry, S. Dumais, and G. O'Brien. Using linear algebra for intelligent information retrieval. *SIAM review*, 37(4):573–595, 1995.

[9] D. Blei, T. Griffiths, M. Jordan, and J. Tenenbaum. Hierarchical topic models and the nested chinese restaurant process. In *NIPS*, 2003.

[10] D. Blei and J. Lafferty. Dynamic topic models. In *ICML*, pages 113–120, 2006.

[11] D. Blei and J. Lafferty. A correlated topic model of science. *AAS*, 1(1):17–35, 2007.

[12] D. Blei and J. McAuliffe. Supervised topic models. In *NIPS*, 2007.

[13] D. Blei, A. Ng, and M. Jordan. Latent Dirichlet allocation. *J. Mach. Learn. Res.*, 3:993–1022, 2003.

[14] J. Boyd-Graber and D. Blei. Multilingual topic models for unaligned text. In *UAI*, pages 75–82, 2009.

[15] J. Boyd-Graber and D. Blei. Syntactic topic models. In *NIPS*, pages 185–192. 2009.

[16] W. Buntine and A. Jakulin. Discrete component analysis. In Craig Saunders, Marko Grobelnik, Steve Gunn, and John Shawe-Taylor, editors, *Subspace, Latent Structure and Feature Selection*, volume 3940 of *Lecture Notes in Computer Science*, pages 1–33. Springer Berlin / Heidelberg, 2006.

[17] J. Chang and D. Blei. Relational topic models for document networks. In *AIStats*, 2009.

[18] J. Chang, J. Boyd-Graber, S. Gerrish, C. Wang, and D. Blei. Reading tea leaves: How humans interpret topic models. In *NIPS*, pages 288–296. 2009.

[19] K. Church and W. Gale. Poisson mixtures. *Natural Language Engineering*, 1:163–190, 1995.

[20] D. Cohn. The missing link-a probabilistic model of document content and hypertext connectivity. In *NIPS*, 2001.

[21] D. Cohn and H. Chang. Learning to probabilistically identify authoritative documents. In *ICML*, pages 167–174, 2001.

[22] S. Crain, S.-H. Yang, Y. Jiao, and H. Zha. Dialect topic modeling for improved consumer medical search. In *AMIA Annual Symposium*, 2010.

[23] S. Deerwester, S. Dumais, G. Furnas, T. Landauer, and R. Harshman. Indexing by latent semantic analysis. *Journal of the American Society for Information Science*, 41(6):391–407, September 1990.

[24] A. Dempster, N. Laird, and D. Rubin. Maximum Likelihood from Incomplete Data via the EM Algorithm. *Journal of the Royal Statistical Society*, 39(1):1–38, 1977.

[25] H. Deng, J. Han, B. Zhao, Y. Yu, and C. Lin. Probabilistic Topic Models with Biased Propagation on Heterogeneous Information Networks. In *KDD*, pages 1271—-1279, San Diego, 2011. ACM.

[26] C. Ding. A similarity-based probability model for latent semantic indexing. In *SIGIR*, pages 58–65, 1999.

[27] G. Doyle and C. Elkan. Accounting for burstiness in topic models. In *ICML*, 2009.

[28] S. Dumais and J. Nielsen. Automating the assignment of submitted manuscripts to reviewers. In *SIGIR*, pages 233–244, 1992.

[29] G. Dupret. Latent concepts and the number orthogonal factors in latent semantic analysis. *SIGIR*, pages 221–226, 2003.

[30] G. Golub and C. Van Loan. *Matrix computations (3rd ed.)*. Johns Hopkins University Press, Baltimore, MD, USA, 1996.

[31] T. Griffiths and M. Steyvers. *Latent Semantic Analysis: A Road to Meaning*, chapter Probabilistic topic models. 2006.

[32] T. Griffiths and M. Steyvers. Finding scientific topics. In *Proceedings of the National Academy of Sciences of the United States of America*, volume 101, pages 5228–5235, 2004.

[33] T. Griffiths, M. Steyvers, D. Blei, and J. Tenenbaum. Integrating topics and syntax. In *NIPS*, pages 537–544, 2005.

[34] Z, Guo, S. Zhu, Y. Chi, Z. Zhang, and Y. Gong. A latent topic model for linked documents. In *SIGIR*, page 720, 2009.

[35] M. Hoffman, D. Blei, and F. Bach. Online learning for latent Dirichlet allocation. In *NIPS*, pages 856–864, 2010.

[36] T. Hofmann. Probabilistic latent semantic analysis. In *UAI*, page 21, 1999.

[37] T. Hofmann. Probabilistic latent semantic indexing. In *SIGIR*, pages 50–57, 1999.

[38] R. Kubota Ando and L. Lee. Iterative residual rescaling: An analysis and generalization of LSI. In *SIGIR*, pages 154–162, 2001.

[39] S. Kullback and R. Leibler. On information and sufficiency. *The Annals of Mathematical Statistics*, 22(1):79–86, March 1951.

[40] T. Landauer. On the computational basis of learning and cognition: Arguments from LSA. *Psychology of learning and motivation*, (1):1–63, 2002.

[41] W. Li, D. Blei, and A. McCallum. Nonparametric Bayes Pachinko allocation. In *UAI*, 2007.

[42] G. Lisowsky and L. Rost. *Konkordanz zum hebräischen Alten Testament: nach dem von Paul Kahle in der Biblia Hebraica edidit Rudolf Kittel besorgten Masoretischen Text*. Deutsche Bibelgesellschaft, 1958.

[43] Z. Liu, Y. Zhang, E.Y. Chang, and M. Sun. PLDA+: Parallel latent Dirichlet allocation with data placement and pipeline processing. *ACM Trans. Intell. Syst. Technol.*, 2:26:1–26:18, May 2011.

[44] C. Manning, P. Raghavan, and H. Schutze. *Introduction to Information Retrieval*. Cambridge University Press, New York, NY, USA, 2008.

[45] A. McCallum, A. Corrada-Emmanuel, and X. Wang. Topic and role discovery in social networks. In *Proceedings of the 19th international joint conference on Artificial intelligence*, pages 786–791, 2005.

[46] Q. Mei, D. Cai, D. Zhang, and C. Zhai. Topic modeling with network regularization. In *WWW*, page 101, 2008.

[47] Q. Mei, X. Shen, and C. Zhai. Automatic labeling of multinomial topic models. In *KDD*, pages 490–499, 2007.

[48] D. Mimno and A. McCallum. Topic models conditioned on arbitrary features with dirichlet-multinomial regression. In *UAI*, 2008.

[49] D. Mimno, H. Wallach, J. Naradowsky, D. Smith, and A. McCallum. Polylingual topic models. In *Proceedings of the 2009 Conference on Empirical Methods in Natural Language Processing*, pages 880–889, 2009.

[50] L. Mølgaard, J. Larsen, and D. Lyngby. Temporal analysis of text data using latent variable models. *2009 IEEE International Workshop on Machine Learning for Signal Processing*, 2009.

[51] A. Ng, A. Zheng, and M. Jordan. Link analysis, eigenvectors and stability. In *International Joint Conference on Artificial Intelligence*, volume 17, pages 903–910, 2001.

[52] G. O'Brien. Information management tools for updating an SVD-encoded indexing scheme. *Master's thesis, The University of Knoxville, Tennessee*, (October), 1994.

[53] C. Papadimitriou, P. Raghavan, H. Tamaki, and S. Vempala. Latent semantic indexing: A probabilistic analysis. In *Proceedings of the seventeenth ACM SIGACT-SIGMOD-SIGART symposium on Principles of database systems*, pages 159–168, 1998.

[54] J. Reisinger, A. Waters, B. Silverthorn, and R. Mooney. Spherical topic models. In *ICML*, pages 903–910, 2010.

[55] M. Rosen-Zvi, T. Griffiths, M. Steyvers, and P. Smyth. The author-topic model for authors and documents. In *UAI*, 2004.

[56] A. Smola and S. Narayanamurthy. An architecture for parallel topic models. *Proc. VLDB Endow.*, 3:703–710, September 2010.

[57] Y. Teh, M. Jordan, M. Beal, and D. Blei. Hierarchical Dirichlet processes. *JASA*, 101, 2006.

[58] I. Titov and R. McDonald. Modeling online reviews with multi-grain topic models. In *WWW*, pages 111–120, 2008.

[59] J. Varadarajan, R. Emonet, and J. Odobez. Probabilistic latent sequential motifs: Discovering temporal activity patterns in video scenes. In *BMVC 2010*, volume 42, pages 177–196, 2010.

[60] H. Wallach, D. Mimno, and A. McCallum. Rethinking LDA: Why priors matter. In *NIPS*, pages 1973–1981, 2009.

[61] H. Wallach, I. Murray, R. Salakhutdinov and D. Mimno. Evaluation methods for topic models In *ICML*, pages 1105–1112, 2009.

[62] H. Wallach. Topic modeling: beyond bag-of-words. In *ICML*, 2006.

[63] Q. Wang, J. Xu, and H. Li. Regularized latent semantic indexing. In *SIGIR*, 2011.

[64] Y. Wang and E. Agichtein. Temporal latent semantic analysis for collaboratively generated content: preliminary results. In *SIGIR*, pages 1145—-1146, 2011.

[65] X. Wei and W. Bruce Croft. LDA-based document models for ad-hoc retrieval. In *SIGIR*, pages 178–185, 2006.

[66] F. Yan, N. Xu, and Y. Qi. Parallel inference for latent Dirichlet allocation on graphics processing units. In *NIPS*, pages 2134–2142. 2009.

[67] S. Yang, J. Bian, and H. Zha. Hybrid generative/discriminative learning for automatic image annotation. In *UAI*, 2010.

[68] S. Yang, S. Crain, and H. Zha. Briding the language gap: topic-level adaptation for cross-domain knowledge transfer. In *AIStat*, 2011.

[69] S. Yang, B. Long, A. Smola, N. Sadagopan, Z. Zheng, and H. Zha. Like like alike – joint friendship and interest propagation in social networks. In *WWW*, 2011.

[70] S. Yang and H. Zha. Language pyramid and multi-scale text analysis. In *CIKM*, pages 639–648, 2010.

[71] S. Yang, H. Zha, and B. Hu. Dirichlet-bernoulli alignment: A generative model for multi-class multi-label multi-instance corpora. In *NIPS*, 2009.

[72] L. Yao, D. Mimno, and A. McCallum. Efficient methods for topic model inference on streaming document collections. In *KDD*, pages 937–946, 2009.

[73] Y. Saad. *Numerical Methods for Large Eigenvalue Problems.* Manchester University Press ND, 1992.

[74] H. Zha and H. Simon. On updating problems in latent semantic indexing. *SIAM Journal on Scientific Computing*, 21(2):782, 1999.

[75] H. Zha and Z. Zhang. On matrices with low-rank-plus-shift structures: Partial SVD and latent semantic indexing. *SIAM Journal Matrix Analysis and Applications*, 21:522–536, 1999.

[76] D. Zhou, S. Zhu, K. Yu, X. Song, B. Tseng, H. Zha, and C. Lee Giles. Learning multiple graphs for document recommendations. In *WWW*, page 141, 2008.

Chapter 6

A SURVEY OF TEXT CLASSIFICATION ALGORITHMS

Charu C. Aggarwal
IBM T. J. Watson Research Center
Yorktown Heights, NY
charu@us.ibm.com

ChengXiang Zhai
University of Illinois at Urbana-Champaign
Urbana, IL
czhai@cs.uiuc.edu

Abstract The problem of classification has been widely studied in the data mining, machine learning, database, and information retrieval communities with applications in a number of diverse domains, such as target marketing, medical diagnosis, news group filtering, and document organization. In this paper we will provide a survey of a wide variety of text classification algorithms.

Keywords: Text Classification

1. Introduction

The problem of classification has been widely studied in the database, data mining, and information retrieval communities. The problem of classification is defined as follows. We have a set of training records $\mathcal{D} = \{X_1, \ldots, X_N\}$, such that each record is labeled with a class value drawn from a set of k different discrete values indexed by $\{1 \ldots k\}$. The training data is used in order to construct a *classification model*, which relates the features in the underlying record to one of the class labels. For a given *test instance* for which the class is unknown, the training model is used to predict a class label for this instance. In the *hard version* of the classification problem, a particular label is explicitly assigned to the instance, whereas in the *soft version* of the classification problem, a probability value is assigned to the test instance. Other variations of the classification problem allow ranking of different class choices for a test instance, or allow the assignment of multiple labels [52] to a test instance.

The classification problem assumes categorical values for the labels, though it is also possible to use continuous values as labels. The latter is referred to as the regression modeling problem. The problem of text classification is closely related to that of classification of records with set-valued features [28]; however, this model assumes that only information about the presence or absence of words is used in a document. In reality, the frequency of words also plays a helpful role in the classification process, and the typical domain-size of text data (the entire lexicon size) is much greater than a typical set-valued classification problem. A broad survey of a wide variety of classification methods may be found in [42, 62], and a survey which is specific to the text domain may be found in [111]. A relative evaluation of different kinds of text classification methods may be found in [132]. A number of the techniques discussed in this chapter have also been converted into software and are publicly available through multiple toolkits such as the *BOW* toolkit [93], Mallot [96], WEKA [1], and LingPipe [2].

The problem of text classification finds applications in a wide variety of domains in text mining. Some examples of domains in which text classification is commonly used are as follows:

- **News filtering and Organization:** Most of the news services today are electronic in nature in which a large volume of news articles are created very single day by the organizations. In such cases, it is difficult to organize the news articles manually. Therefore, automated methods can be very useful for news categorization in a variety of web portals [78]. This application is also referred to as *text filtering*.

- **Document Organization and Retrieval:** The above application is generally useful for many applications beyond news filtering and organization. A variety of supervised methods may be used for document organization in many domains. These include large digital libraries of documents, web collections, scientific literature,

---
[1] http://www.cs.waikato.ac.nz/ml/weka/
[2] http://alias-i.com/lingpipe/

or even social feeds. Hierarchically organized document collections can be particularly useful for browsing and retrieval [19].

- **Opinion Mining:** Customer reviews or opinions are often short text documents which can be mined to determine useful information from the review. Details on how classification can be used in order to perform opinion mining are discussed in [89] and Chapter 13 in this book.

- **Email Classification and Spam Filtering:** It is often desirable to classify email [23, 27, 85] in order to determine either the subject or to determine junk email [113] in an automated way. This is also referred to as *spam filtering* or *email filtering*.

A wide variety of techniques have been designed for text classification. In this chapter, we will discuss the broad classes of techniques, and their uses for classification tasks. We note that these classes of techniques also generally exist for other data domains such as quantitative or categorical data. Since text may be modeled as quantitative data with frequencies on the word attributes, it is possible to use most of the methods for quantitative data directly on text. However, text is a particular kind of data in which the word attributes are sparse, and high dimensional, with low frequencies on most of the words. Therefore, it is critical to design classification methods which effectively account for these characteristics of text. In this chapter, we will focus on the specific changes which are applicable to the text domain. Some key methods, which are commonly used for text classification are as follows:

- **Decision Trees:** Decision trees are designed with the use of a hierarchical division of the underlying data space with the use of different text features. The hierarchical division of the data space is designed in order to create class partitions which are more skewed in terms of their class distribution. For a given text instance, we determine the partition that it is most likely to belong to, and use it for the purposes of classification.

- **Pattern (Rule)-based Classifiers:** In rule-based classifiers we determine the word patterns which are most likely to be related to the different classes. We construct a set of rules, in which the left-hand side corresponds to a word pattern, and the right-hand side corresponds to a class label. These rules are used for the purposes of classification.

- **SVM Classifiers:** SVM Classifiers attempt to partition the data space with the use of linear or non-linear delineations between the different classes. The key in such classifiers is to determine the optimal boundaries between the different classes and use them for the purposes of classification.

- **Neural Network Classifiers:** Neural networks are used in a wide variety of domains for the purposes of classification. In the context of text data, the main difference for neural network classifiers is to adapt these classifiers with the use of word features. We note that neural network classifiers are related to SVM classifiers; indeed, they both are in the category of discriminative classifiers, which are in contrast with the *generative classifiers* [102].

- **Bayesian (Generative) Classifiers:** In Bayesian classifiers (also called generative classifiers), we attempt to build a probabilistic classifier based on modeling the underlying word features in different classes. The idea is then to classify text based on the posterior probability of the documents belonging to the different classes on the basis of the word presence in the documents.

- **Other Classifiers:** Almost all classifiers can be adapted to the case of text data. Some of the other classifiers include nearest neighbor classifiers, and genetic algorithm-based classifiers. We will discuss some of these different classifiers in some detail and their use for the case of text data.

The area of text categorization is so vast that it is impossible to cover all the different algorithms in detail in a single chapter. Therefore, our goal is to provide the reader with an overview of the most important techniques, and also the pointers to the different variations of these techniques.

Feature selection is an important problem for text classification. In feature selection, we attempt to determine the features which are most relevant to the classification process. This is because some of the words are much more likely to be correlated to the class distribution than others. Therefore, a wide variety of methods have been proposed in the literature in order to determine the most important features for the purpose of classification. These include measures such as the gini-index or the entropy, which determine the level of which the presence of a particular feature skews the class distribution in the underlying data. We will also discuss the different feature selection methods which are commonly used for text classification.

The rest of this chapter is organized as follows. In the next section, we will discuss methods for feature selection in text classification. In section

3, we will describe decision tree methods for text classification. Rule-based classifiers are described in detail in section 4. We discuss naive Bayes classifiers in section 5. The nearest neighbor classifier is discussed in section 7. In section 7, we will discuss a number of linear classifiers, such as the SVM classifier, direct regression modeling and the neural network classifier. A discussion of how the classification methods can be adapted to text and web data containing hyperlinks is discussed in section 8. In section 9, we discuss a number of different meta-algorithms for classification such as boosting, bagging and ensemble learning. Section 10 contains the conclusions and summary.

2. Feature Selection for Text Classification

Before any classification task, one of the most fundamental tasks that needs to be accomplished is that of document representation and feature selection. While feature selection is also desirable in other classification tasks, it is especially important in text classification due to the high dimensionality of text features and the existence of irrelevant (noisy) features. In general, text can be represented in two separate ways. The first is as a bag of words, in which a document is represented as a set of words, together with their associated frequency in the document. Such a representation is essentially independent of the sequence of words in the collection. The second method is to represent text directly as *strings*, in which each document is a sequence of words. Most text classification methods use the bag-of-words representation because of its simplicity for classification purposes. In this section, we will discuss some of the methods which are used for feature selection in text classification.

The most common feature selection which is used in both supervised and unsupervised applications is that of stop-word removal and stemming. In stop-word removal, we determine the common words in the documents which are not specific or discriminatory to the different classes. In stemming, different forms of the same word are consolidated into a single word. For example, singular, plural and different tenses are consolidated into a single word. We note that these methods are not specific to the case of the classification problem, and are often used in a variety of unsupervised applications such as clustering and indexing. In the case of the classification problem, it makes sense to supervise the feature selection process with the use of the class labels. This kind of selection process ensures that those features which are highly skewed towards the presence of a particular class label are picked for the learning process. A wide variety of feature selection methods are discussed in [133, 135]. Many of these feature selection methods have been compared with one another, and the experimental results are presented in [133]. We will discuss each of these feature selection methods in this section.

2.1 Gini Index

One of the most common methods for quantifying the discrimination level of a feature is the use of a measure known as the *gini-index*. Let $p_1(w)\ldots p_k(w)$ be the fraction of class-label presence of the k different classes for the word w. In other words, $p_i(w)$ is the conditional probability that a document belongs to class i, given the fact that it contains the word w. Therefore, we have:

$$\sum_{i=1}^{k} p_i(w) = 1 \qquad (6.1)$$

Then, the gini-index for the word w, denoted by $G(w)$ is defined[3] as follows:

$$G(w) = \sum_{i=1}^{k} p_i(w)^2 \qquad (6.2)$$

The value of the gini-index $G(w)$ always lies in the range $(1/k, 1)$. Higher values of the gini-index $G(w)$ represent indicate a greater discriminative power of the word w. For example, when all documents which contain word w belong to a particular class, the value of $G(w)$ is 1. On the other hand, when documents containing word w are evenly distributed among the k different classes, the value of $G(w)$ is $1/k$.

One criticism with this approach is that the global class distribution may be skewed to begin with, and therefore the above measure may sometimes not accurately reflect the discriminative power of the underlying attributes. Therefore, it is possible to construct a normalized gini-index in order to reflect the discriminative power of the attributes more accurately. Let $P_1\ldots P_k$ represent the global distributions of the documents in the different classes. Then, we determine the normalized probability value $p'_i(w)$ as follows:

$$p'_i(w) = \frac{p_i(w)/P_i}{\sum_{j=1}^{k} p_j(w)/P_j} \qquad (6.3)$$

---

[3]The gini-index is also sometimes defined as $1 - \sum_{i=1}^{k} p_i(w)^2$, with lower values indicating greater discriminative power of the feature w.

Then, the gini-index is computed in terms of these normalized probability values.

$$G(w) = \sum_{i=1}^{k} p'_i(w)^2 \qquad (6.4)$$

The use of the global probabilities P_i ensures that the gini-index more accurately reflects class-discrimination in the case of biased class distributions in the whole document collection. For a document corpus containing n documents, d words, and k classes, the complexity of the information gain computation is $O(n \cdot d \cdot k)$. This is because the computation of the term $p_i(w)$ for all the different words and the classes requires $O(n \cdot d \cdot k)$ time.

2.2 Information Gain

Another related measure which is commonly used for text feature selection is that of information gain or entropy. Let P_i be the global probability of class i, and $p_i(w)$ be the probability of class i, given that the document contains the word w. Let $F(w)$ be the fraction of the documents containing the word w. The information gain measure $I(w)$ for a given word w is defined as follows:

$$I(w) = -\sum_{i=1}^{k} P_i \cdot \log(P_i) + F(w) \cdot \sum_{i=1}^{k} p_i(w) \cdot \log(p_i(w)) +$$
$$+(1 - F(w)) \cdot \sum_{i=1}^{k}(1 - p_i(w)) \cdot \log(1 - p_i(w))$$

The greater the value of the information gain $I(w)$, the greater the discriminatory power of the word w. For a document corpus containing n documents and d words, the complexity of the information gain computation is $O(n \cdot d \cdot k)$.

2.3 Mutual Information

This *mutual information measure* is derived from information theory [31], and provides a formal way to model the mutual information between the features and the classes. The pointwise mutual information $M_i(w)$ between the word w and the class i is defined on the basis of the level of co-occurrence between the class i and word w. We note that the expected co-occurrence of class i and word w on the basis of mutual independence is given by $P_i \cdot F(w)$. The true co-occurrence is of course given by $F(w) \cdot p_i(w)$. In practice, the value of $F(w) \cdot p_i(w)$ may be much larger or smaller than $P_i \cdot F(w)$, depending upon the level of correlation between the class i and word w. The mutual information is defined in terms of the ratio between these two values. Specifically, we have:

$$M_i(w) = \log\left(\frac{F(w) \cdot p_i(w)}{F(w) \cdot P_i}\right) = \log\left(\frac{p_i(w)}{P_i}\right) \qquad (6.5)$$

Clearly, the word w is positively correlated to the class i, when $M_i(w) > 0$, and the word w is negatively correlated to class i, when $M_i(w) < 0$. We note that $M_i(w)$ is specific to a particular class i. We need to compute the overall mutual information as a function of the mutual information of the word w with the different classes. These are defined with the use of the average and maximum values of $M_i(w)$ over the different classes.

$$M_{avg}(w) = \sum_{i=1}^{k} P_i \cdot M_i(w)$$
$$M_{max}(w) = \max_i \{M_i(w)\}$$

Either of these measures may be used in order to determine the relevance of the word w. The second measure is particularly useful, when it is more important to determine high levels of positive correlation of the word w with any of the classes.

2.4 χ^2-Statistic

The χ^2 statistic is a different way to compute the lack of independence between the word w and a particular class i. Let n be the total number of documents in the collection, $p_i(w)$ be the conditional probability of class i for documents which contain w, P_i be the global fraction of documents containing the class i, and $F(w)$ be the global fraction of documents which contain the word w. The χ^2-statistic of the word between word w and class i is defined as follows:

$$\chi_i^2(w) = \frac{n \cdot F(w)^2 \cdot (p_i(w) - P_i)^2}{F(w) \cdot (1 - F(w)) \cdot P_i \cdot (1 - P_i))} \qquad (6.6)$$

As in the case of the mutual information, we can compute a global χ^2 statistic from the class-specific values. We can use either the average of maximum values in order to create the composite value:

$$\chi_{avg}^2(w) = \sum_{i=1}^{k} P_i \cdot \chi_i^2(w)$$
$$\chi_{max}^2(w) = \max_i \chi_i^2(w)$$

We note that the χ^2-statistic and mutual information are different ways of measuring the the correlation between terms and categories. One major advantage of the χ^2-statistic over the mutual information measure is that it is a normalized value, and therefore these values are more comparable across terms in the same category.

2.5 Feature Transformation Methods: Supervised LSI

While feature selection attempts to reduce the dimensionality of the data by picking from the original set of attributes, feature transformation methods create a new (and smaller) set of features as a function of the original set of features. A typical example of such a feature transformation method is Latent Semantic Indexing (LSI) [38], and its probabilistic variant PLSA [57]. The LSI method transforms the text space of a few hundred thousand word features to a new axis system (of size about a few hundred) which are a linear combination of the original word features. In order to achieve this goal, Principal Component Analysis techniques [69] are used to determine the axis-system which retains the greatest level of information about the variations in the underlying attribute values. The main disadvantage of using techniques such as LSI is that these are unsupervised techniques which are blind to the underlying class-distribution. Thus, the features found by LSI are not necessarily the directions along which the *class-distribution* of the underlying documents can be best separated. A modest level of success has been obtained in improving classification accuracy by using boosting techniques in conjunction with the conceptual features obtained from unsupervised pLSA method [17]. A more recent study has systematically compared pLSA and LDA (which is a Bayesian version of pLSA) in terms of their effectiveness in transforming features for text categorization and drawn a similar conclusion and found that pLSA and LDA tend to perform similarly.

A number of techniques have also been proposed to perform the feature transformation methods by using the class labels for effective supervision. The most natural method is to adapt LSI in order to make it work more effectively for the supervised case. A number of different methods have been proposed in this direction. One common approach is to perform local LSI on the subsets of data representing the individual classes, and identify the discriminative eigenvectors from the different reductions with the use of an iterative approach [123]. This method is known as SLSI (Supervised Latent Semantic Indexing), and the advantages of the method seem to be relatively limited, because the experiments in [123] show that the improvements over a standard SVM classifier, which did not use a dimensionality reduction process, are relatively limited. The work in [129] uses a combination of class-specific LSI and global analysis. As in the case of [123], class-specific LSI representations are created. Test documents are compared against each LSI representation in order to create the most discriminative reduced space. One problem with this approach is that the different local LSI representations use a different subspace, and therefore it is difficult to compare the similarities of the different documents across the different subspaces. Furthermore, both the methods in [123, 129] tend to be computationally expensive.

A method called *sprinkling* is proposed in [21], in which artificial terms are added to (or "sprinkled" into) the documents, which correspond to the class labels. In other words, we create a term corresponding to the class label, and add it to the document. LSI is then performed on the document collection with these added terms. The sprinkled terms can then be removed from the representation, once the eigenvectors have been determined. The sprinkled terms help in making the LSI more sensitive to the class distribution during the reduction process. It has also been proposed in [21] that it can be generally useful to make the sprinkling process *adaptive*, in which all classes are not necessarily treated equally, but the relationships between the classes are used in order to regulate the sprinkling process. Furthermore, methods have also been proposed in [21] to make the sprinkling process adaptive to the use of a particular kind of classifier.

2.6 Supervised Clustering for Dimensionality Reduction

One technique which is commonly used for feature transformation is that of text clustering [7, 71, 83, 121]. In these techniques, the clusters are constructed from the underlying text collection, with the use of supervision from the class distribution. The exception is [83] in which supervision is not used. In the simplest case, each class can be treated as a separate cluster, though better results may be obtained by using the classes for supervision of the clustering process. The frequently occurring words in these supervised clusters can be used in order to create the new set of dimensions. The classification can be performed with respect to this new feature representation. One advantage of such an approach is that it retains interpretability with respect to the original words of the document collection. The disadvantage is that the optimum directions of separation may not necessarily be represented in the form of clusters of words. Furthermore, the underlying axes are not necessarily orthonormal to one another. The use of supervised methods [1, 7, 71, 121] has generally led to good results either in terms of improved classification accuracy, or significant performance gains at the expense of a small reduction in accuracy. The results with the use of unsupervised clustering [83, 87] are mixed. For example, the work in [83] suggests that the use of unsupervised term-clusters and phrases is generally not helpful [83] for the classification process. The key observation in [83] is that the loss of granularity associated with the use of phrases and term clusters is not necessarily advantageous for the classification process. The work in [8] has shown that the use of the information bottleneck method for feature distributional clustering can create clustered pseudo-word representations which are quite effective for text classification.

2.7 Linear Discriminant Analysis

Another method for feature transformation is the use of linear discriminants, which explicitly try to construct directions in the feature space, along which there is best separation of the different classes. A common method is the *Fisher's linear discriminant* [46]. The main idea in the Fisher's discriminant method is to determine the directions in the data along which the points are as well separated as possible. The subspace of lower dimensionality is constructed by iteratively finding such unit vectors α_i in the data, where α_i is determined in the ith iteration. We would also like to ensure that the different values of α_i are orthonormal to one another. In each step, we determine this vector α_i by discriminant analysis, and project the data onto the remaining orthonormal subspace. The next vector α_{i+1} is found in this orthonormal subspace. The quality of vector α_i is measured by an objective function which measures the separation of the different classes. This objective function reduces in each iteration, since the value of α_i in a given iteration is the optimum discriminant in that subspace, and the vector found in the next iteration is the optimal one from a smaller search space. The process of finding linear discriminants is continued until the class separation, as measured by an objective function, reduces below a given threshold for the vector determined in the current iteration. The power of such a dimensionality reduction approach has been illustrated in [18], in which it has been shown that a simple decision tree classifier can perform much more effectively on this transformed data, as compared to more sophisticated classifiers.

Next, we discuss how the Fisher's discriminant is actually constructed. First, we will set up the objective function $J(\overline{\alpha})$ which determines the level of separation of the different classes along a given direction (unit-vector) $\overline{\alpha}$. This sets up the crisp optimization problem of determining the value of $\overline{\alpha}$ which maximizes $J(\overline{\alpha})$. For simplicity, let us assume the case of binary classes. Let D_1 and D_2 be the two sets of documents belonging to the two classes. Then, the projection of a document $\overline{X} \in D_1 \cup D_2$ along $\overline{\alpha}$ is given by $\overline{X} \cdot \overline{\alpha}$. Therefore, the squared class separation $S(D_1, D_2, \overline{\alpha})$ along the direction $\overline{\alpha}$ is given by:

$$S(D_1, D_2, \overline{\alpha}) = \left(\frac{\sum_{\overline{X} \in D_1} \overline{\alpha} \cdot \overline{X}}{|D_1|} - \frac{\sum_{\overline{X} \in D_2} \overline{\alpha} \cdot \overline{X}}{|D_2|} \right)^2 \qquad (6.7)$$

In addition, we need to normalize this absolute class separation with the use of the underlying class variances. Let $Var(D_1, \overline{\alpha})$ and $Var(D_2, \overline{\alpha})$ be the individual class variances along the direction α. In other words, we have:

$$Var(D_1, \overline{\alpha}) = \frac{\sum_{\overline{X} \in D_1} (\overline{X} \cdot \overline{\alpha})^2}{|D_1|} - \left(\frac{\sum_{\overline{X} \in D_1} \overline{X} \cdot \overline{\alpha}}{|D_1|} \right)^2 \qquad (6.8)$$

The value of $Var(D_2, \overline{\alpha})$ can be defined in a similar way. Then, the normalized class-separation $J(\overline{\alpha})$ is defined as follows:

$$J(\overline{\alpha}) = \frac{S(D_1, D_2, \overline{\alpha})}{Var(D_1, \overline{\alpha}) + Var(D_2, \overline{\alpha})} \qquad (6.9)$$

The optimal value of α needs to be determined subject to the constraint that $\overline{\alpha}$ is a unit vector. Let μ_1 and μ_2 be the means of the two data sets D_1 and D_2, and C_1 and C_2 be the corresponding covariance matrices. It can be shown that the optimal (unscaled) direction $\overline{\alpha} = \overline{\alpha^*}$ can be expressed in closed form, and is given by the following:

$$\overline{\alpha^*} = \left(\frac{C_1 + C_2}{2} \right)^{-1} (\mu_1 - \mu_2) \qquad (6.10)$$

The main difficulty in computing the above equation is that this computation requires the inversion of the covariance matrix, which is sparse and computationally difficult in the high-dimensional text domain. Therefore, a gradient descent approach can be used in order to determine the value of $\overline{\alpha}$ in a more computationally effective way. Details of the approach are presented in [18].

Another related method which attempts to determine projection directions that maximize the topical differences between different classes is the *Topical Difference Factor Analysis* method proposed in [72]. The problem has been shown to be solvable as a generalized eigenvalue problem. The method was used in conjunction with a k-nearest neighbor classifier, and it was shown that the use of this approach significantly improves the accuracy over a classifier which uses the original set of features.

2.8 Generalized Singular Value Decomposition

While the method discussed above finds one vector $\overline{\alpha_i}$ at a time in order to determine the relevant dimension transformation, it is possible to be much more direct in finding the optimal subspaces simultaneously by using a generalized version of dimensionality reduction [58, 59]. It is important to note that this method has really been proposed in [58, 59] as an *unsupervised* method which preserves the underlying *clustering* structure, assuming the data has already been clustered in a pre-processing phase. Thus, the generalized dimensionality reduction method has been proposed as a much more aggressive dimensionality reduction technique, which preserves the underlying clustering structure rather than the individual points. This method can however also be used as a *supervised* technique in which the different classes are used as input to the dimensionality reduction algorithm, instead of the clusters constructed in the pre-processing phase [131]. This method is known as the *Optimal Orthogonal Centroid Feature Selection Algorithm (OCFS)*, and it directly targets at the maximization of inter-class scatter. The algorithm is shown to have promising results for supervised feature selection in [131].

2.9 Interaction of Feature Selection with Classification

Since the classification and feature selection processes are dependent upon one another, it is interesting to test how the feature selection process interacts with the underlying classification algorithms. In this context, two questions are relevant:

- Can the feature-specific insights obtained from the intermediate results of some of the classification algorithms be used for creating feature selection methods that can be used more generally by other classification algorithms?

- Do the different feature selection methods work better or worse with different kinds of classifiers?

Both these issues were explored in some detail in [99]. In regard to the first question, it was shown in [99] that feature selection which was derived from *linear* classifiers, provided very effective results. In regard to the second question, it was shown in [99] that the sophistication of the feature selection process itself was more important than the specific pairing between the feature selection process and the classifier.

Linear Classifiers are those for which the output of the linear predictor is defined to be $p = \overline{A} \cdot \overline{X} + b$, where $\overline{X} = (x_1 \ldots x_n)$ is the normalized document word frequency vector, $\overline{A} = (a_1 \ldots a_n)$ is a vector of linear coefficients with the same dimensionality as the feature space, and b is a scalar. Both the basic neural network and basic SVM classifiers [65] (which will be discussed later in this chapter) belong to this category. The idea here is that if the coefficient a_i is close to zero, then the corresponding feature does not have a significant effect on the classification process. On the other hand, since large absolute values of a_j may significantly influence the classification process, such features should be selected for classification. In the context of the SVM method, which attempts to determine linear planes of separation between the different classes, the vector $\overline{A}$ is essentially the normal vector to the corresponding plane of separation between the different classes. This intuitively explains the choice of selecting features with large values of $|a_j|$. It was shown in [99] that this class of feature selection methods was quite robust, and performed well even for classifiers such as the Naive Bayes method, which were unrelated to the linear classifiers from which these features were derived. Further discussions on how SVM and maximum margin techniques can be used for feature selection may be found in [51, 56].

3. Decision Tree Classifiers

A decision tree [106] is essentially a hierarchical decomposition of the (training) data space, in which a *predicate* or a condition on the attribute value is used in order to divide the data space hierarchically. In the context of text data, such predicates are typically conditions on the presence or absence of one or more words in the document. The division of the data space is performed recursively in the decision tree, until the leaf nodes contain a certain minimum number of records, or some conditions on class purity. The majority class label (or cost-weighted majority label) in the leaf node is used for the purposes of classification. For a given test instance, we apply the sequence of predicates at the nodes, in order to traverse a path of the tree in top-down fashion and determine the relevant leaf node. In order to further reduce the overfitting, some of the nodes may be be pruned by holding out a part of the data, which are not used to construct the tree. The portion of the data which is held out is used in order to determine whether or not the constructed leaf node should be pruned or not. In particular, if the class distribution in the training data (for decision tree construction) is very different from the class distribution in the training data which is used for pruning, then it is assumed that the node overfits the training data. Such a node can be pruned. A detailed discussion of decision tree methods may be found in [15, 42, 62, 106].

In the particular case of text data, the predicates for the decision tree nodes are typically defined in terms of the terms in the underlying text collection. For example, a node may be partitioned into its children nodes depending upon the presence or absence of a particular term in the document. We note that different nodes at the same level of the tree may use different terms for the partitioning process.

Many other kinds of predicates are possible. It may not be necessary to use individual terms for partitioning, but one may measure the similarity of documents to correlated sets of terms. These correlated sets of terms may be used to further partition the document collection, based on the similarity of the document to them. The different kinds of splits are as follows:

- **Single Attribute Splits:** In this case, we use the presence or absence of particular words (or even phrases) at a particular node in the tree in order to perform the split. At any given level, we pick the word which provides the maximum discrimination between the different classes. Measures such as the gini-index or information gain can be used in order to determine the level of entropy. For example, the DT-min10 algorithm [81] is based on this approach.

- **Similarity-based multi-attribute split:** In this case, we use documents (or meta-documents such as frequent word clusters), and use the similarity of the documents to these words clusters in order to perform the split. For the selected word cluster, the documents are further partitioned into groups by rank ordering the documents by similarity value, and splitting at a particular threshold. We select the word-cluster for which rank-ordering by similarity provides the best separation between the different classes.

- **Discriminant-based multi-attribute split:** For the multi-attribute case, a natural choice for performing the split is to use discriminants such as the Fisher discriminant for performing the split. Such discriminants provide the directions in the data along which the classes are best separated. The documents are projected on this discriminant vector for rank ordering, and then split at a particular coordinate. The choice of split point is picked in order to maximize the discrimination between the different classes.

The work in [18] uses a discriminant-based split, though this is done indirectly because of the use of a feature transformation to the discriminant representation, before building the classifier.

Some of the earliest implementation of classifiers may be found in [80, 81, 87, 127]. The last of these is really a rule-based classifier, which can be interpreted either as a decision tree or a rule-based classifier. Most of the decision tree implementations in the text literature tend to be small variations on standard packages such as ID3 and C4.5, in order to adapt the model to text classification. Many of these classifiers are typically designed as baselines for comparison with other learning models [65].

A well known implementation of the decision tree classifier is based on the C4.5 taxonomy of algorithms [106] is presented in [87]. More specifically, the work in [87] uses the successor to the C4.5 algorithm, which is also known as the C5 algorithm. This algorithm uses single-attribute splits at each node, where the feature with the highest information gain [31] is used for the purpose of the split. Decision trees have also been used in conjunction with boosting techniques. An adaptive boosting technique [48] is used in order to improve the accuracy of classification. In this technique, we use n different classifiers. The ith classifier is constructed by examining the errors of the $(i-1)$th classifier. A voting scheme is applied among these classifiers in order to report the final label. Other boosting techniques for improving decision tree classification accuracy are proposed in [116].

The work in [43] presents a decision tree algorithm based on the Bayesian approach developed in [22]. In this classifier, the decision tree is grown by recursive greedy splits, where the splits are chosen using Bayesian posterior probability of model structure. The structural prior penalizes additional model parameters at each node. The output of the process is a class probability rather than a deterministic class label for the test instance.

4. Rule-based Classifiers

Decision trees are also generally related to *rule-based classifiers*. In rule-based classifiers, the data space is modeled with a set of rules, in which the left hand side is a condition on the underlying feature set, and the right hand side is the class label. The rule set is essentially the model which is generated from the training data. For a given test instance, we determine the set of rules for which the test instance satisfies the condition on the left hand side of the rule. We determine the predicted class label as a function of the class labels of the rules which are satisfied by the test instance. We will discuss more on this issue slightly later.

In its most general form, the left hand side of the rule is a boolean condition, which is expressed in Disjunctive Normal Form (DNF). However, in most cases, the condition on the left hand side is much simpler and represents a set of terms, all of which must be present in the document for the condition to be satisfied. The *absence* of terms is rarely used, because such rules are not likely to be very informative for sparse text data, in which most words in the lexicon will typically not be present in it by default (sparseness property). Also, while the *set intersection* of conditions on term presence is used often, the union of such conditions is rarely used in a single rule. This is because such rules can be split into two separate rules, each of which is more informative on its own. For example, the rule $Honda \cup Toyota \Rightarrow Cars$ can be replaced by two separate rules $Honda \Rightarrow Cars$ and $Toyota \Rightarrow Cars$ without any loss of information. In fact, since the confidence of each of the two rules can now be measured separately, this can be more useful. On the other hand, the rule $Honda \cap Toyota \Rightarrow Cars$ is certainly much more informative than the individual rules. Thus, in practice, for sparse data sets such as text, rules are much more likely to be expressed as a simple conjunction of conditions on term presence.

We note that decision trees and decision rules both tend to encode rules on the feature space, except that the decision tree tends to achieve this goal with a hierarchical approach. In fact, the original work on decision tree construction in C4.5 [106] studied the decision tree problem and decision rule problem within a single framework. This is because a particular path in the decision tree can be considered a rule for classification of the text instance. The main difference is that the decision tree framework is a strict hierarchical partitioning of the data space, whereas rule-based classifiers allow for overlaps in the decision space. The general principle is to create a rule set, such that all points in the decision space are covered by *at least* one rule. In most cases, this is achieved by generating a set of targeted rules which are related to the different classes, and one default *catch-all* rule, which can cover all the remaining instances.

A number of criteria can be used in order to generate the rules from the training data. Two of the most common conditions which are used for rule generation are those of *support* and *confidence*. These conditions are common to all rule-based pattern classifiers [88] and may be defined as follows:

- **Support:** This quantifies the **absolute number** of instances in the training data set which are relevant to the rule. For example, in a corpus containing 100,000 documents, a rule in which **both** the left-hand set and right-hand side are satisfied by 50,000 documents is more important than a rule which is satisfied by 20 documents. Essentially, this quantifies the statistical *volume* which is associated with the rule. However, it does not encode the strength of the rule.

- **Confidence:** This quantifies the **conditional probability** that the right hand side of the rule is satisfied, if the left-hand side is satisfied. This is a more direct measure of the strength of the underlying rule.

We note that the afore-mentioned measures are not the only measures which are possible, but are widely used in the data mining and machine learning literature [88] for both textual and non-textual data, because of their intuitive nature and simplicity of interpretation. One criticism of the above measures is that they do not normalize for the a-priori presence of different terms and features, and are therefore prone to misinterpretation, when the feature distribution or class-distribution in the underlying data set is skewed.

The training phase constructs all the rules, which are based on measures such as the above. For a given test instance, we determine all the rules which are relevant to the test instance. Since we allow overlaps, it is possible that more than one rule may be relevant to the test instance. If the class labels on the right hand sides of all these rules are the same, then it is easy to pick this class as the relevant label for the test instance. On the other hand, the problem becomes more challenging when there are conflicts between these different rules. A variety of different methods are used to rank-order the different rules [88], and report the most relevant rule as a function of these different rules. For example, a common approach is to rank-order the rules by their confidence, and pick the top-k rules as the most relevant. The class label on the right-hand side of the most number of these rules is reported as the relevant one.

Am interesting rule-based classifier for the case of text data has been proposed in [5]. This technique uses an iterative methodology, which was first proposed in [128] for generating rules. Specifically, the method determines the single best rule related to any particular class in the training data. The best rule is defined in terms of the confidence of the rule, as defined above. This rule along with its corresponding instances are removed from the training data set. This approach is continuously repeated, until it is no longer possible to find strong rules in the training data, and complete predictive value is achieved.

The transformation of decision trees to rule-based classifiers is discussed generally in [106], and for the particular case of text data in [68]. For each path in the decision tree a rule can be generated, which represents the conjunction of the predicates along that path. One advantage of the rule-based classifier over a decision tree is that it is not restricted to a strict hierarchical partitioning of the feature space, and it allows for overlaps and inconsistencies among the different rules. Therefore, if a new set of training examples are encountered, which are related to a new class or new part of the feature space, then it is relatively easy to modify the rule set for these new examples. Furthermore, rule-based classifiers also allow for a tremendous interpretability of the underlying decision space. In cases in which domain-specific expert knowledge is known, it is possible to encode this into the classification process by manual addition of rules. In many practical scenarios, rule-based techniques are more commonly used because of their ease of maintenance and interpretability.

One of the most common rule-based techniques is the *RIPPER* technique discussed in [26–28]. The *RIPPER* technique essentially determines frequent combinations of words which are related to a particular class. The *RIPPER* method has been shown to be especially effective in scenarios where the number of training examples is relatively small [25]. Another method called *sleeping experts* [26, 49] generates rules which take the placement of the words in the documents into account. Most of the classifiers such as *RIPPER* [26–28] treat documents as set-valued objects, and generate rules based on the co-presence of the words in the documents. The rules in *sleeping experts* are different from most of the other classifiers in this respect. In this case [49, 26], the left hand side of the rule consists of a *sparse phrase*, which is a group of words close to one another in the document (though not necessarily completely sequential). Each such rule has a weight, which depends upon its classification specificity in the training data. For a given test example, we determine the sparse phrases which are present in it, and perform the classification by combining the weights of the different rules that are fired. The *sleeping experts* and *RIPPER* systems have been compared in [26], and have been shown to have excellent performance on a variety of text collections.

5. Probabilistic and Naive Bayes Classifiers

Probabilistic classifiers are designed to use an implicit mixture model for generation of the underlying documents. This mixture model typically assumes that each class is a component of the mixture. Each mixture component is essentially a generative model, which provides the probability of sampling a particular term for that component or class. This is why this kind of classifiers are often also called generative classifier. The naive Bayes classifier is perhaps the simplest and also the most commonly used generative classifers. It models the distribution of the documents in each class using a probabilistic model with independence assumptions about the distributions of different terms. Two classes of models are commonly used for naive Bayes classification. Both models essentially compute the posterior probability of a class, based on the distribution of the words in the document. These models ignore the actual position of the words in the document, and work with the "bag of words" assumption. The major difference between these two models is the assumption in terms of taking (or not taking) word frequencies into account, and the corresponding approach for sampling the probability space:

- **Multivariate Bernoulli Model:** In this model, we use the presence or absence of words in a text document as features to represent a document. Thus, the frequencies of the words are not used for the modeling a document, and the word features in the text are assumed to be binary, with the two values indicating presence or absence of a word in text. Since the features to be modeled are binary, the model for documents in each class is a multivariate Bernoulli model.

- **Multinomial Model:** In this model, we captuer the frequencies of terms in a document by representing a document with a bag of words. The documents in each class can then be modeled as samples drawn from a multinomial word distribution. As a result, the conditional probability of a document given a class is simply a product of the probability of each observed word in the corresponding class.

No matter how we model the documents in each class (be it a multivariate Bernoulli model or a multinomial model), the component class models (i.e., generative models for documents in each class) can be used in conjunction with the Bayes rule to compute the posterior probability of the class for a given document, and the class with the highest posterior probability can then be assigned to the document.

There has been considerable confusion in the literature on the differences between the multivariate Bernoulli model and the multinomial model. A good exposition of the differences between these two models may be found in [94]. In the following, we describe these two models in more detail.

5.1 Bernoulli Multivariate Model

This class of techniques treats a document as a set of distinct words with no frequency information, in which an element (term) may be either present or absent. The seminal work on this approach may be found in [82].

Let us assume that the lexicon from which the terms are drawn are denoted by $V = \{t_1 \ldots t_n\}$. Let us assume that the bag-of-words (or text document) in question contains the terms $Q = \{t_{i_1} \ldots t_{i_m}\}$, and the class is drawn from $\{1 \ldots k\}$. Then, our goal is to model the posterior probability that the document (which is assumed to be generated from the term distributions of one of the classes) belongs to class i, given that it contains the terms $Q = \{t_{i_1} \ldots t_{i_m}\}$. The best way to understand the Bayes method is by understanding it as a sampling/generative process from the underlying mixture model of classes. The Bayes probability of class i can be modeled by sampling a set of terms T from the term distribution of the classes:

If we sampled a term set T of any size **from the term distribution of one of the randomly chosen classes**, *and the final outcome is the set Q, then what is the posterior probability that we had originally picked class i for sampling? The a-priori probability of picking class i is equal to its fractional presence in the collection.*

We denote the class of the sampled set T by C^T and the corresponding posterior probability by $P(C^T = i | T = Q)$. This is essentially what we are trying to find. It is important to note that since we do not allow replacement, we are essentially picking a subset of terms from V with no frequencies attached to the picked terms. Therefore, the set Q may not contain duplicate elements. Under the naive Bayes assumption of independence between terms, this is essentially equivalent to either selecting or not selecting each term with a probability that depends upon the underlying term distribution. Furthermore, it is also important to note that this model has no restriction on the number of terms picked. As we will see later, these assumptions are the key differences with the multinomial Bayes model. The Bayes approach classifies a given set Q based on the posterior probability that Q is a sample from the data distribution of class i, i.e., $P(C^T = i | T = Q)$, and it requires us to compute the following two probabilities in order to achieve this:

1. What is the prior probability that a set T is a sample from the term distribution of class i? This probability is denoted by $P(C^T = i)$.

2 If we sampled a set T of any size *from the term distribution of class* i, then what is the probability that our sample is the set Q? This probability is denoted by $P(T = Q|C^T = i)$.

We will now provide a more mathematical description of Bayes modeling. In other words, we wish to model $P(C^T = i|Q \text{ is sampled})$. We can use the Bayes rule in order to write this conditional probability in a way that can be *estimated* more easily from the underlying corpus. In other words, we can simplify as follows:

$$P(C^T = i|T = Q) = \frac{P(C^T = i) \cdot P(T = Q|C^T = i)}{P(T = Q)}$$

$$= \frac{P(C^T = i) \cdot \prod_{t_j \in Q} P(t_j \in T|C^T = i) \cdot \prod_{t_j \notin Q}(1 - P(t_j \in T|C^T = i))}{P(T = Q)}$$

We note that the last condition of the above sequence uses the *naive independence assumption*, because we are assuming that the probabilities of occurrence of the different terms are independent of one another. This is practically necessary, in order to transform the probability equations to a form which can be estimated from the underlying data.

The class assigned to Q is the one with the highest posterior probability given Q. It is easy to see that this decision is not affected by the denominator, which is the marginal probability of observing Q. That is, we will assign the following class to Q:

$$\hat{i} = \arg\max_i P(C^T = i|T = Q)$$

$$= \arg\max_i P(C^T = i) \cdot$$

$$\prod_{t_j \in Q} P(t_j \in T|C^T = i) \cdot \prod_{t_j \notin Q}(1 - P(t_j \in T|C^T = i)).$$

It is important to note that all terms in the right hand side of the last equation can be estimated from the training corpus. The value of $P(C^T = i)$ is estimated as the global fraction of documents belonging to class i, the value of $P(t_j \in T|C^T = i)$ is the fraction of documents in the ith class which contain term t_j, and the value of $P(t_j \in T)$ is the fraction of documents (in the whole corpus) containing the term t_j. We note that all of the above are maximum likelihood estimates of the corresponding probabilities. In practice, Laplacian smoothing [124] is used, in which small values are added to the frequencies of terms in order to avoid zero probabilities of sparsely present terms.

In most applications of the Bayes classifier, we only care about the *identity of the* class with the highest probability value, rather than the actual probability value associated with it, which is why we do not need to compute the normalizer $P(T = Q)$. In fact, in the case of **binary** classes, a number of simplifications are possible in computing these Bayes "probability" values by using the logarithm of the Bayes expression, and removing a number of terms which do not affect the ordering of class probabilities. We refer the reader to [108] for details.

Although for classification, we do not need to compute $P(T = Q)$, some applications necessitate the exact computation of the posterior probability $P(C^T = i|T = Q)$. For example, in the case of supervised anomaly detection (or rare class detection), the exact posterior probability value $P(C^T = i|T = Q)$ is needed in order to fairly compare the probability value over different test instances, and rank them for their anomalous nature. In such cases, we would need to compute $P(T = Q)$. One way to achieve this is simply to take a sum over all the classes:

$$P(T = Q) = \sum_i P(T = Q|C^T = i)P(C^T = i).$$

This is based on the conditional independence of features for each class. Since the parameter values are estimated for each class separately, we may face the problem of data sparseness. An alternative way of computing it, which may alleviate the data sparseness problem, is to further make the assumption of (global) independence of terms, and compute it as:

$$P(T = Q) = \prod_{j \in Q} P(t_j \in T) \cdot \prod_{t_j \notin Q} (1 - P(t_j \in T))$$

where the term probabilities are based on global term distributions in *all* the classes.

A natural question arises, as to whether it is possible to design a Bayes classifier which does not use the naive assumption, and models the dependencies between the terms during the classification process. Methods which generalize the naive Bayes classifier by not using the independence assumption do not work well because of the higher computational costs and the inability to estimate the parameters accurately and robustly in the presence of limited data. The most interesting line of work in relaxing the independence assumption is provided in [112]. In this work, the tradeoffs in spectrum of allowing different levels of dependence among the terms have been explored. On the one extreme, an assumption of complete dependence results in a Bayesian network model which turns out to be computationally very expensive. On the other hand, it has been shown that allowing limited levels of dependence can provide good tradeoffs between accuracy and computational costs. We note that while the independence assumption is a practical approximation, it has been shown in [29, 39] that the approach does have some theoretical merit. Indeed, extensive experimental tests have tended to show that the naive classifier works quite well in practice.

A number of papers [19, 64, 74, 79, 108, 113] have used the naive Bayes approach for classification in a number of different application domains. The classifier has also been extended to modeling temporally aware training data, in which the importance of a document may decay with time [114]. As in the case of other statistical classifiers, the naive Bayes classifier [113] can easily incorporate *domain-specific knowledge* into the classification process. The particular domain that the work in [113] addresses is that of filtering junk email. Thus, for such a problem, we often have a lot of additional domain knowledge which helps us determine whether a particular email message is junk or not. For example, some common characteristics of the email which would make an email to be more or less likely to be junk are as follows:

- The domain of the sender such as *.edu* or *.com* can make an email to be more or less likely to be junk.

- Phrases such as *"Free Money"* or over emphasized punctuation such as *"!!!"* can make an email more likely to be junk.

- Whether the recipient of the message was a particular user, or a mailing list.

The Bayes method provides a natural way to incorporate such additional information into the classification process, by creating new features for each of these characteristics. The standard Bayes technique is then used in conjunction with this augmented representation for classification. The Bayes technique has also been used in conjunction with the incorporation of other kinds of domain knowledge, such as the incorporation of hyperlink information into the classification process [20, 104].

The Bayes method is also suited to hierarchical classification, when the training data is arranged in a taxonomy of topics. For example, the Open Directory Project (ODP), *Yahoo!* Taxonomy, and a variety of news sites have vast collections of documents which are arranged into hierarchical groups. The hierarchical structure of the topics can be exploited to perform more effective classification [19, 74], because it has been observed that context-sensitive feature selection can provide more useful classification results. In hierarchical classification, a Bayes classifier is built at *each node*, which then provides us with the next branch to follow for classification purposes. Two such methods are proposed in [19, 74], in which node specific features are used for the classification process. Clearly, much fewer features are required at a particular node

A Survey of Text Classification Algorithms 187

in the hierarchy, because the features which are picked are relevant to that branch. An example in [74] suggests that a branch of the taxonomy which is related to *Computer* may have no relationship with the word "cow". These node-specific features are referred to as *signatures* in [19]. Furthermore, it has been observed in [19] that in a given node, the most discriminative features for a given class may be different from their parent nodes. For example, the word "health" may be discriminative for the *Yahoo*! category @*Health*, but the word "baby" may be much more discriminative for the category @*Health*@*Nursing*. Thus, it is critical to have an appropriate feature selection process at each node of the classification tree. The methods in [19, 74] use different methods for this purpose.

- The work in [74] uses an information-theoretic approach [31] for feature selection which takes into account the dependencies between the attributes [112]. The algorithm greedily eliminates the features one-by-one so as the least disrupt the conditional class distribution at that node.

- The node-specific features are referred to as *signatures* in [19]. These node-specific signatures are computed by calculating the ratio of intra-class variance to inter-class variance for the different words at each node of the tree. We note that this measure is the same as that optimized by the Fisher's discriminant, except that it is applied to the original set of words, rather than solved as a general optimization problem in which arbitrary directions in the data are picked.

A Bayesian classifier is constructed at each node in order to determine the appropriate branch. A small number of context-sensitive features provide One advantage of these methods is that Bayesian classifiers work much more effectively with a much smaller number of features. Another major difference between the two methods is that the work in [74] uses the Bernoulli model, whereas that in [19] uses the multinomial model, which will be discussed in the next subsection. This approach in [74] is referred to as the *Pachinko Machine classifier* and that in [19] is known as *TAPER (Taxonomy and Path Enhanced Retrieval System)*.

Other noteworthy methods for hierarchical classification are proposed in [11, 130, 95]. The work [11] addresses two common problems associated with hierarchical text classification: (1) error propagation; (2) non-linear decision surfaces. The problem of error propagation occurs when the classification mistakes made at a parent node are propagated to its children node. This problem was solved in [11] by using cross validation to obtain a training data set for a child node that is more similar to the actual test data passed to the child node from its parent node than the training data set normally used for training a classifier at the child node. The problem of non-linear decision surfaces refers to that the decision boundary of a category at a higher level is often non-linear (since its members are the union of the members of its children nodes). This problem is addressed by using the tentative class labels obtained at the children nodes as features for use at a parent node. These are general strategies that can be applied to any base classifier, and the experimental results in [11] show that both strategies are effective.

5.2 Multinomial Distribution

This class of techniques treats a document as a set of words with frequencies attached to each word. Thus, the set of words is allowed to have duplicate elements.

As in the previous case, we assume that the set of words in document is denoted by Q, drawn from the vocabulary set V. The set Q contains the distinct terms $\{t_{i_1} \ldots t_{i_m}\}$ with associated frequencies $F = \{F_{i_1} \ldots F_{i_m}\}$. We denote the terms and their frequencies by $[Q, F]$. The total number of terms in the document (or document length) is denoted by $L = \sum_{j=1}^{m} F(i_j)$. Then, our goal is to model the posterior probability that the document T belongs to class i, given that it contains the terms in Q *with the associated frequencies* F. The Bayes probability of class i can be modeled by using the following sampling process:

If we sampled L terms sequentially **from the term distribution of one of the randomly chosen classes** *(allowing repetitions) to create the term set T, and the final outcome for sampled set T is the set Q with the corresponding frequencies F, then what is the posterior probability that we had originally picked class i for sampling? The a-priori probability of picking class i is equal to its fractional presence in the collection.*

The aforementioned probability is denoted by $P(C^T = i | T = [Q, F])$. An assumption which is commonly used in these models is that the length of the document is independent of the class label. While it is easily possible to generalize the method, so that the document length is used as a prior, independence is usually assumed for simplicity. As in the previous case, we need to estimate two values in order to compute the Bayes posterior.

1. What is the prior probability that a set T is a sample from the term distribution of class i? This probability is denoted by $P(C^T = i)$.

2 If we sampled L terms *from the term distribution of class i* (with repetitions), then what is the probability that our sampled set T is the set Q with associated frequencies F? This probability is denoted by $P(T = [Q, F]|C^T = i)$.

Then, the Bayes rule can be applied to this case as follows:

$$P(C^T = i|T = [Q, F]) = \frac{P(C^T = i) \cdot P(T = [Q, F]|C^T = i)}{P(T = [Q, F])}$$

$$\propto P(C^T = i) \cdot P(T = [Q, F]|C^T = i) \quad (6.11)$$

As in the previous case, it is not necessary to compute the denominator, $P(T = [Q, F])$, for the purpose of deciding the class label for Q. The value of the probability $P(C^T = i)$ can be estimated as the fraction of documents belonging to class i. The computation of $P([Q, F]|C^T = i)$ is much more complicated. When we consider the sequential order of the L different samples, the number of possible ways to sample the different terms so as to result in the outcome $[Q, F]$ is given by $\frac{L!}{\prod_{i=1}^{m} F_i!}$. The probability of *each* of these sequences is given by $\prod_{t_j \in Q} P(t_j \in T)^{F_j}$, by using the naive independence assumption. Therefore, we have:

$$P(T = [Q, F]|C^T = i) = \frac{L!}{\prod_{i=1}^{m} F_i!} \cdot \prod_{t_j \in Q} P(t_j \in T|C^T = i)^{F_j} \quad (6.12)$$

We can substitute Equations 6.12 in Equation 6.11 to obtain the class with the highest Bayes posterior probability, where the class priors are computed as in the previous case, and the probabilities $P(t_j \in T|C^T = i)$ can also be easily estimated as previously with Laplacian smoothing [124]. We note that the probabilities of class absence are not present in the above equations because of the way in which the sampling is performed.

A number of different variations of the multinomial model have been proposed in [53, 70, 84, 95, 97, 103]. In the work [95], it is shown that a category hierarchy can be leveraged to improve the estimate of multinomial parameters in the naive Bayes classifier to significantly improve classification accuracy. The key idea is to apply shrinkage techniques to smooth the parameters for data-sparse child categories with their common parent nodes. As a result, the training data of related categories are essentially "shared" with each other in a weighted manner, which helps improving the robustness and accuracy of parameter estimation when there are insufficient training data for each individual child category. The work in [94] has performed an extensive comparison between the bernoulli and the multinomial models on different corpora, and the following conclusions were presented:

- The multi-variate Bernoulli model can sometimes perform better than the multinomial model at small vocabulary sizes.

- The multinomial model outperforms the multi-variate Bernoulli model for large vocabulary sizes, and almost always beats the multi-variate Bernoulli when vocabulary size is chosen optimally for both. On the average a 27% reduction in error was reported in [94].

The afore-mentioned results seem to suggest that the two models may have different strengths, and may therefore be useful in different scenarios.

5.3 Mixture Modeling for Text Classification

We note that the afore-mentioned Bayes methods simply assume that each component of the mixture corresponds to the documents belonging to a class. A more general interpretation is one in which the components of the mixture are created by a clustering process, and the class membership probabilities are modeled in terms of this mixture. Mixture modeling is typically used for unsupervised (probabilistic) clustering or topic modeling, though the use of clustering can also help in enhancing the effectiveness of probabilistic classifiers [86, 103]. These methods are particularly useful in cases where the amount of training data is limited. In particular, clustering can help in the following ways:

- The Bayes method implicitly estimates the word probabilities $P(t_i \in T | C^T = i)$ of a large number of terms in terms of their fractional presence in the corresponding component. This is clearly noisy. By treating the clusters as separate entities from the classes, we now only need to relate (a much smaller number of) cluster membership probabilities to class probabilities. This reduces the number of parameters and greatly improves classification accuracy [86].

- The use of clustering can help in incorporating unlabeled documents into the training data for classification. The premise is that unlabeled data is much more copiously available than labeled data, and when labeled data is sparse, it should be used in order to assist the classification process. While such unlabeled documents do not contain class-specific information, they do contain a lot of information about the clustering behavior of the underlying data. This can be very useful for more robust modeling [103], when the amount of training data is low. This general approach is also referred to as *co-training* [9, 13, 37].

The common characteristic of both the methods [86, 103] is that they both use a form of supervised clustering for the classification process. While the goal is quite similar (limited training data), the approach used for this purpose is quite different. We will discuss both of these methods in this section.

In the method discussed in [86], the document corpus is modeled with the use of supervised word clusters. In this case, the k mixture components are clusters which are correlated to, but are distinct from the k groups of documents belonging to the different classes. The main difference from the Bayes method is that the term probabilities are computed indirectly by using clustering as an intermediate step. For a sampled document T, we denote its class label by $C^T \in \{1 \ldots k\}$, and its mixture component by $M^T \in \{1 \ldots k\}$. The k different mixture components are essentially word-clusters whose frequencies are generated by using the frequencies of the terms in the k different classes. This ensures that the word clusters for the mixture components are correlated to the classes, but they are not assumed to be drawn from the same distribution. As in the previous case, let us assume that the a document contains the set of words Q. Then, we would like to estimate the probability $P(T = Q | C^T = i)$ for each class i. An interesting variation of the work in [86] from the Bayes approach is that it does not attempt to determine the posterior probability $P(C^T = i | T = Q)$. Rather, it simply reports the class with the highest likelihood $P(T = Q | C^T = i)$. This is essentially equivalent to assuming in the Bayes approach, that the prior distribution of each class is the same.

The other difference of the approach is in terms of how the value of $P(T = Q | C^T = i)$ is computed. As before, we need to estimate the value of $P(t_j \in T | C^T = i)$, according to the naive Bayes rule. However, unlike the standard Bayes classifier, this is done very indirectly with the use of mixture modeling. Since the mixture components do not directly correspond to the class, this term can only be estimated by summing up the expected value over all the mixture components:

$$P(t_j \in T | C^T = i) = \sum_{s=1}^{k} P(t_j \in T | M^T = s) \cdot P(M^T = s | C^T = i) \quad (6.13)$$

The value of $P(t_j \in T | M^T = s)$ is easy to estimate by using the fractional presence of term t_j in the sth mixture component. The main unknown here are the set of model parameters $P(M^T = s | C^T = i)$.

Since a total of k classes and k mixture-components are used, this requires the estimation of only k^2 model parameters, which is typically quite modest for a small number of classes. An EM-approach has been used in [86] in order to estimate this small number of model parameters in a robust way. It is important to understand that the work in [86] is an interesting combination of supervised topic modeling (dimensionality reduction) and Bayes classification after reducing the effective dimensionality of the feature space to a much smaller value by clustering. The scheme works well because of the use of supervision in the topic modeling process, which ensures that the use of an intermediate clustering approach does not lose information for classification. We also note that in this model, the number of mixtures can be made to vary from the number of classes. While the work in [86] does not explore this direction, there is really no reason to assume that the number of mixture components is the same as the number of classes. Such an assumption can be particularly useful for data sets in which the classes may not be contiguous in the feature space, and a natural clustering may contain far more components than the number of classes.

Next, we will discuss the second method [103] which uses unlabeled data. The approach is [103] uses the unlabeled data in order to improve the training model. Why should unlabeled data help in classification at all? In order to understand this point, recall that the Bayes classification process effectively uses k mixture components, which are assumed to be the k different classes. If we had an infinite amount of training data, it would be possible to create the mixture components, but it would not be possible to assign labels to these components. However, the most data-intensive part of modeling the mixture, is that of determining the shape of the mixture components. The actual assignment of mixture components to class labels can be achieved with a relatively small number of class labels. It has been shown in [24] that the accuracy of assigning components to classes increases exponentially with the number of labeled samples available. Therefore, the work in [103] designs an EM-approach [36] to simultaneously determine the relevant mixture model and its class assignment.

It turns out that the EM-approach, as applied to this problem is quite simple to implement. It has been shown in [103] that the EM-approach is equivalent to the following iterative methodology. First, a naive Bayes classifier is constructed by estimating the model parameters from the labeled documents only. This is used in order to assign probabilistically-weighted class labels to the unlabeled documents. Then, the Bayes classifier is re-constructed, except that we also use the newly labeled documents in the estimation of the underlying model parameters. We again use this classifier to re-classify the (originally unlabeled) documents. The process is continually repeated till convergence is achieved.

The ability to significantly improve the quality of text classification with a small amount of labeled data, and the use of clustering on a large amount of unlabeled data has been a recurring theme in the text mining literature. For example, the method in [122] performs purely unsupervised clustering (with no knowledge of class labels), and then as a final step assigns all documents in the cluster to the *dominant* class label of that cluster (as an evaluation step for the unsupervised clustering process in terms of its ability in matching clusters to known topics).[4] It has been shown that this approach is able to achieve a comparable accuracy of matching clusters to topics as a supervised Naive Bayes classifier trained over a small data set of about 1000 documents. Similar results were obtained in [47] where the quality of the unsupervised clustering process were shown to comparable to an SVM classifier which was trained over a small data set.

6.   Linear Classifiers

Linear Classifiers are those for which the output of the linear predictor is defined to be $p = \overline{A} \cdot \overline{X} + b$, where $\overline{X} = (x_1 \ldots x_n)$ is the normalized document word frequency vector, $\overline{A} = (a_1 \ldots a_n)$ is a vector of linear coefficients with the same dimensionality as the feature space, and b is a scalar. A natural interpretation of the predictor $p = \overline{A} \cdot \overline{X} + b$ in the *discrete scenario* (categorical class labels) would be as a *separating hyperplane* between the different classes. *Support Vector Machines* [30, 125] are a form of classifiers which attempt to determine "good" linear separators between the different classes. One characteristic of linear classifiers is that they are closely related to many *feature transformation methods* (such as the Fisher discriminant), which attempt to use these directions in order to transform the feature space, and then use other classifiers on this transformed feature space [51, 56, 99]. Thus, linear classifiers are intimately related to linear feature transformation methods as well.

Regression modeling (such as the least squares method) is a more direct and traditional statistical method for text classification. However, it is generally used in cases where the target variable to be learned is numerical rather than categorical. A number of methods have been

---

[4] In a supervised application, the last step would require only a small number of class labels in the cluster to be known to determine the dominant label very accurately.

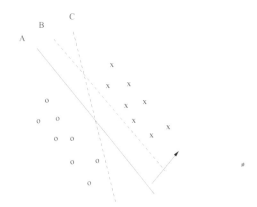

Figure 6.1. What is the Best Separating Hyperplane?

proposed in the literature for adapting such methods to the case of text data classification [134]. A comparison of different linear regression techniques for classificationm including SVM, may be found in [138].

Finally, simple neural networks are also a form of linear classifiers, since the function computed by a set of neurons is essentially linear. The simplest form of neural network, known as the *perceptron* (or single layer network) are essentially designed for linear separation, and work well for text. However, by using multiple layers of neurons, it is also possible to generalize the approach for non-linear separation. In this section, we will discuss the different linear methods for text classification.

6.1   SVM Classifiers

Support-vector machines were first proposed in [30, 124] for numerical data. The main principle of SVMs is to determine separators in the search space which can best separate the different classes. For example, consider the example illustrated in Figure 6.1, in which we have two classes denoted by 'x' and 'o' respectively. We have denoted three different separating hyperplanes, which are denoted by A, B, and C respectively. It is evident that the hyperplane A provides the best separation between the different classes, because the *normal distance* of any of the data points from it is the largest. Therefore, the hyperplane A represents the maximum *margin of separation*. We note that the normal vector to this hyperplane (represented by the arrow in the figure) is a direction in the feature space along which we have the maximum discrimination. One advantage of the SVM method is that since it attempts to determine the optimum direction of discrimination in the feature space by examining the appropriate combination of features, it is quite robust to high dimensionality. It has been noted in [64] that text data is ideally suited for SVM classification because of the sparse high-dimensional nature of text, in which few features are irrelevant, but they tend to be correlated with one another and generally organized into linearly separable categories. We note that it is not necessary to use a linear function for the SVM classifier. Rather, with the kernel trick [6], SVM can construct a non-linear *decision surface* in the original feature space by mapping the data instances non-linearly to an inner product space where the classes can be separated linearly with a hyperplane. However, in practice, linear SVM is used most often because of their simplicity and ease of interpretability. The first set of SVM classifiers, as adapted to the text domain were proposed in [64–66]. A deeper theoretical study of the SVM method has been provided in [67]. In particular, it has been shown why the SVM classifier is expected to work well under a wide variety of circumstances. This has also been demonstrated experimentally in a few different scenarios. For example, the work in [41] applied the method to email data for classifying it as spam or non-spam data. It was shown that the SVM method provides much more robust performance as compared to many other techniques such as boosting decision trees, the rule based RIPPER method, and the Rocchio method. The SVM method is flexible and can easily be combined with interactive user-feedback methods [107].

We note that the problem of finding the best separator is essentially an optimization problem, which can typically be reduced to a Quadratic Programming problem. For example, many of these methods use Newton's method for iterative minimization of a convex function. This can sometimes be slow, especially for high dimensional domains such as text data. It has been shown [43] that by breaking a large Quadratic Programming problem (QP problem) to a set of smaller problems, an efficient solution can be derived for the task. The SVM approach has also been used successfully [44] in the context of a hierarchical organization of the classes, as often occurs in web data. In this approach, a different classifier is built at different positions of the hierarchy.

The SVM classifier has also been shown to be useful in large scale scenarios in which a large amount of unlabeled data and a small amount of labeled data is available [120]. This is essentially a semi-supervised approach because of its use of unlabeled data in the classification process. This techniques is also quite scalable because of its use of a number of modified quasi-newton techniques, which tend to be efficient in practice.

6.2 Regression-Based Classifiers

Regression modeling is a method which is commonly used in order to learn the relationships between real-valued attributes. Typically, these methods are designed for real valued attributes, as opposed to binary attributes. This is however not an impediment to its use in classification, because the binary value of a class may be treated as a rudimentary special case of a real value, and some regression methods such as logistic regression can also naturally model discrete response variables.

An early application of regression to text classification is the Linear Least Squares Fit (LLSF) method [134], which works as follows. Suppose the predicted class label be $p_i = \overline{A} \cdot \overline{X_i} + b$, and y_i is known to be the true class label, then our aim is to learn the values of A and b, such that the *Linear Least Squares Fit (LLSF)* $\sum_{i=1}^{n}(p_i - y_i)^2$ is minimized. In practice, the value of b is set to 0 for the learning process. let P be $1 \times n$ vector of binary values indicating the binary class to which the corresponding class belongs. Thus, if X be the the $n \times d$ term-matrix, then we wish to determine the $1 \times d$ vector of regression coefficients A for which $||A \cdot X^T - P||$ is minimized, where $||\cdot||$ represents the Froebinus norm. The problem can be easily generalized from the binary class scenario to the multi-class scenario with k classes, by using P as a $k \times n$ matrix of binary values. In this matrix, exactly one value in each column is 1, and the corresponding row identifier represents the class to which that instance belongs. Similarly, the set A is a $k \times d$ vector in the multi-class scenario. The LLSF method has been compared to a variety of other methods [132, 134, 138], and has been shown to be very robust in practice.

A more natural way of modeling the classification problem with regression is the logistic regression classifier [102], which differs from the LLSF method in that the objective function to be optimized is the likelihood function. Specifically, instead of using $p_i = \overline{A} \cdot \overline{X_i} + b$ directly to fit the true label y_i, we assume that the probability of observing label y_i is:

$$p(C = y_i|X_i) = \frac{exp(\overline{A} \cdot \overline{X_i} + b)}{1 + exp(\overline{A} \cdot \overline{X_i} + b)}.$$

This gives us a conditional generative model for y_i given X_i. Putting it in another way, we assume that the logit transformation of $p(C = y_i|X_i)$ can be modeled by the linear combination of features of the instance X_i, i.e.,

$$\log \frac{p(C = y_i|X_i)}{1 - p(C = y_i|X_i)} = \overline{A} \cdot \overline{X_i} + b.$$

Thus logistic regression is also a linear classifier as the decision boundary is determined by a linear function of the features. In the case of binary classification, $p(C = y_i|X_i)$ can be used to determine the class label (e.g., using a threshold of 0.5). In the case of multi-class classification, we have $p(C = y_i|X_i) \propto exp(\overline{A} \cdot \overline{X_i} + b)$, and the class label with the highest value according to $p(C = y_i|X_i)$ would be assigned to X_i. Given a set of training data points $\{(X_1, y_i), ...(X_n, y_n)\}$, the logistic regression classifier can be trained by choosing parameters $\overline{A}$ to maximize the conditional likelihood $\prod_{i=1}^{n} p(y_i|X_i)$.

In some cases, the domain knowledge may be of the form, where some sets of words are more important than others for a classification problem. For example, in a classification application, we may know that certain domain-words (*Knowledge Words (KW)*) may be more important to classification of a particular target category than other words. In such cases, it has been shown [35] that it may be possible to encode such domain knowledge into the logistic regression model in the form of prior on the model parameters and use Bayesian estimation of model parameters.

It is clear that the regression classifiers are extremely similar to the SVM model for classification. Indeed, since LLSF, Logistic Regression, and SVM are all linear classifiers, they are thus identical at a conceptual level; the main difference among them lies in the details of the optimization formulation and implementation. As in the case of SVM classifiers, training a regression classifier also requires an expensive optimization process. For example, fitting LLSF requires expensive matrix computations in the form of a singular value decomposition process.

6.3 Neural Network Classifiers

The basic unit in a neural network is a *neuron* or *unit*. Each unit receives a set of inputs, which are denoted by the vector $\overline{X_i}$, which in this case, correspond to the term frequencies in the ith document. Each neuron is also associated with a set of weights A, which are used in order to compute a function $f(\cdot)$ of its inputs. A typical function which is often used in the neural network is the linear function as follows:

$$p_i = A \cdot \overline{X_i} \qquad (6.14)$$

Thus, for a vector $\overline{X_i}$ drawn from a lexicon of d words, the weight vector A should also contain d elements. Now consider a binary classification problem, in which all labels are drawn from $\{+1, -1\}$. We assume that the class label of $\overline{X_i}$ is denoted by y_i. In that case, the sign of the predicted function p_i yields the class label.

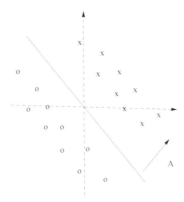

Figure 6.2. The sign of the projection onto the weight vector A yields the class label

In order to illustrate this point, let us consider a simple example in a 2-dimensional feature space, as illustrated in Figure 6.2. In this case, we have illustrated two different classes, and the plane corresponding to $Ax = 0$ is illustrated in the same figure. It is evident that the sign of the function $A \cdot \overline{X_i}$ yields the class label. Thus, the goal of the approach is to *learn* the set of weights A with the use of the training data. The idea is that we start off with random weights and gradually update them when a mistake is made by applying the current function on the training example. The magnitude of the update is regulated by a learning rate μ. This forms the core idea of the *perceptron algorithm*, which is as follows:

Perceptron Algorithm
Inputs: Learning Rate: μ
　　　　Training Data $(\overline{X_i}, y_i) \; \forall i \in \{1 \ldots n\}$
Initialize weight vectors in A to 0 or small random numbers
repeat
Apply each training data to the neural network to check if the
　　　sign of $A \cdot \overline{X_i}$ matches y_i;
if sign of $A \cdot \overline{X_i}$ does **not** match y_i, then
　　　update weights A based on learning rate μ
until weights in A converge

The weights in A are typically updated (increased or decreased) proportionally to $\mu \cdot \overline{X_i}$, so as to reduce the direction of the error of the neuron. We further note that many different update rules have been proposed in the literature. For example, one may simply update each weight by μ, rather than by $\mu \cdot \overline{X_i}$. This is particularly possible in domains such

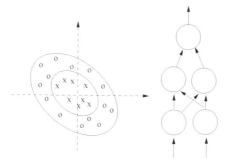

Figure 6.3. Multi-Layered Neural Networks for Nonlinear Separation

as text, in which all feature values take on small non-negative values of relatively similar magnitude. A number of implementations of neural network methods for text data have been studied in [34, 90, 101, 117, 129].

A natural question arises, as to how a neural network may be used, if all the classes may not be neatly separated from one another with a linear separator, as illustrated in Figure 6.2. For example, in Figure 6.3, we have illustrated an example in which the classes may not be separated with the use of a single linear separator. The use of *multiple layers of neurons* can be used in order to induce such non-linear classification boundaries. The effect of such multiple layers is to induce multiple piece-wise linear boundaries, which can be used to approximate enclosed regions belonging to a particular class. In such a network, the outputs of the neurons in the earlier layers feed into the neurons in the later layers. The training process of such networks is more complex, as the errors need to be back-propagated over different layers. Some examples of such classifiers include those discussed in [75, 110, 126, 132]. However, the general observation [117, 129] for text has been that linear classifiers generally provide comparable results to non-linear data, and the improvements of non-linear classification methods are relatively small. This suggests that the additional complexity of building more involved non-linear models does not pay for itself in terms of significantly better classification.

6.4 Some Observations about Linear Classifiers

While the different linear classifiers have been developed independently from one another in the research literature, they are surprisingly similar at a basic conceptual level. Interestingly, these different lines of work have also resulted in a number of similar conclusions in terms of the effectiveness of the different classifiers. We note that the main difference between the different classifiers is in terms of the details of the objective function which is optimized, and the iterative approach used in order to determine the optimum direction of separation. For example, the SVM method uses a Quadratic Programming (QP) formulation, whereas the LLSF method uses a closed-form least-squares formulation. On the other hand, the perceptron method does not try to formulate a closed-form objective function, but works with a softer iterative hill climbing approach. This technique is essentially inherited from the iterative learning approach used by neural network algorithms. However, its goal remains quite similar to the other two methods. Thus, the differences between these methods are really at a detailed level, rather than a conceptual level, in spite of their very different research origins.

Another general observation about these methods is that all of them can be implemented with non-linear versions of their classifiers. For example, it is possible to create non-linear decision surfaces with the SVM classifier, just as it is possible to create non-linear separation boundaries by using layered neurons in a neural network [132]. However, the general consensus has been that the linear versions of these methods work very well, and the additional complexity of non-linear classification does not tend to pay for itself, except for some special data sets. The reason for this is perhaps because text is a high dimensional domain with highly correlated features and small non-negative values on sparse features. For example, it is hard to easily create class structures such as that indicated in Figure 6.3 for a sparse domain such as text containing only small non-negative values on the features. On the other hand, the high dimensional nature of correlated text dimensions is especially suited to classifiers which can exploit the redundancies and relationships between the different features in separating out the different classes. Common text applications have generally resulted in class structures which are linearly separable over this high dimensional domain of data. This is one of the reasons that linear classifiers have shown an unprecedented success in text classification.

7. Proximity-based Classifiers

Proximity-based classifiers essentially use distance-based measures in order to perform the classification. The main thesis is that documents which belong to the same class are likely to be close to one another based on similarity measures such as the dot product or the cosine metric [115]. In order to perform the classification for a given test instance, two possible methods can be used:

- We determine the k-nearest neighbors in the training data to the test instance. The majority (or most abundant) class from these k neighbors are reported as the class label. Some examples of such methods are discussed in [25, 54, 134]. The choice of k typically ranges between 20 and 40 in most of the afore-mentioned work, depending upon the size of the underlying corpus.

- We perform training data aggregation during pre-processing, in which clusters or groups of documents belonging to the same class are created. A representative meta-document is created from each group. The same k-nearest neighbor approach is applied as discussed above, except that it is applied to this new set of meta-documents (or *generalized instances* [76]) rather than to the original documents in the collection. A pre-processing phase of summarization is useful in improving the efficiency of the classifier, because it significantly reduces the number of distance computations. In some cases, it may also boost the accuracy of the technique, especially when the data set contains a large number of outliers. Some examples of such methods are discussed in [55, 76, 109].

A method for performing nearest neighbor classification in text data is the *WHIRL* method discussed in [25]. The *WHIRL* method is essentially a method for performing soft similarity joins on the basis of text attributes. By *soft* similarity joins, we refer to the fact that the two records may not be exactly the same on the joined attribute, but a notion of similarity used for this purpose. It has been observed in [25] that any method for performing a similarity-join can be adapted as a nearest neighbor classifier, by using the relevant text documents as the joined attributes.

One observation in [134] about nearest neighbor classifiers was that feature selection and document representation play an important part in the effectiveness of the classification process. This is because most terms in large corpora may not be related to the category of interest. Therefore, a number of techniques were proposed in [134] in order to learn the associations between the words and the categories. These are then used to create a feature representation of the document, so that the nearest neighbor classifier is more sensitive to the classes in the document collection. A similar observation has been made in [54], in which it has been shown that the addition of weights to the terms (based on their class-sensitivity) significantly improves the underlying classifier performance. The nearest neighbor classifier has also been extended to the temporally-aware scenario [114], in which the timeliness of a training document plays a role in the model construction process.

In order to incorporate such factors, a temporal weighting function has been introduced in [114], which allows the importance of a document to gracefully decay with time.

For the case of classifiers which use grouping techniques, the most basic among such methods is that proposed by Rocchio in [109]. In this method, a *single* representative meta-document is constructed from each of the representative classes. For a given class, the weight of the term t_k is the normalized frequency of the term t_k in documents belonging to that class, minus the normalized frequency of the term in documents which do not belong to that class. Specifically, let f_p^k be the expected weight of term t_k in a randomly picked document belonging to the positive class, and f_n^k be the expected weight of term t_k in a randomly picked document belonging to the negative class. Then, for weighting parameters α_p and α_n, the weight $f_{rocchio}^k$ is defined as follows:

$$f_{rocchio}^k = \alpha_p \cdot f_p^k - \alpha_n \cdot f_n^k \qquad (6.15)$$

The weighting parameters α_p and α_n are picked so that the positive class has much greater weight as compared to the negative class. For the relevant class, we now have a vector representation of the terms $(f_{rocchio}^1, f_{rocchio}^2 \ldots f_{rocchio}^n)$. This approach is applied separately to each of the classes, in order to create a separate meta-document for each class. For a given test document, the closest meta-document to the test document can be determined by using a vector-based dot product or other similarity metric. The corresponding class is then reported as the relevant label. The main distinguishing characteristic of the Rocchio method is that it creates a single profile of the entire class. This class of methods is also referred to as the *Rocchio framework*. The main disadvantage of this method is that if a single class occurs in multiple disjoint clusters which are not very well connected in the data, then the centroid of these examples may not represent the class behavior very well. This is likely to be a source of inaccuracy for the classifier. The main advantage of this method is its extreme simplicity and efficiency; the training phase is linear in the corpus size, and the number of computations in the testing phase are linear to the number of classes, since all the documents have already been aggregated into a small number of classes. An analysis of the Rocchio algorithm, along with a number of different variations may be found in [64].

In order to handle the shortcomings of the Rocchio method, a number of classifiers have also been proposed [1, 14, 55, 76], which explicitly perform the clustering of each of the classes in the document collection. These clusters are used in order to generate class-specific profiles. These profiles are also referred to as *generalized instances* in [76]. For a given test instance, the label of the closest generalized instance is reported by the algorithm. The method in [14] is also a centroid-based classifier, but is specifically designed for the case of text documents. The work in [55] shows that there are some advantages in designing schemes in which the similarity computations take account of the dependencies between the terms of the different classes.

We note that the nearest neighbor classifier can be used in order to generate a ranked list of categories for each document. In cases, where a document is related to multiple categories, these can be reported for the document, as long as a thresholding method is available. The work in [136] studies a number of thresholding strategies for the k-nearest neighbor classifier. It has also been suggested in [136] that these thresholding strategies can be used to understand the thresholding strategies of other classifiers which use ranking classifiers.

8. Classification of Linked and Web Data

In recent years, the proliferation of the web and social network technologies has lead to a tremendous amount of document data, which is expressed in the form of linked networks. The simplest example of this is the web, in which the documents are linked to one another with the use of hyper-links. Social networks can also be considered a noisy example of such data, because the comments and text profiles of different users are connected to one another through a variety of links. Linkage information is quite relevant to the classification process, because documents of similar subjects are often linked together. This observation has been used widely in the *collective classification literature* [12], in which a subset of network nodes are labeled, and the remaining nodes are classified on the basis of the linkages among the nodes.

In general, a content-based network may be denoted by $G = (N, A, C)$, where N is the set of nodes, A is the set of edges between the nodes, and C is a set of text documents. Each node in N corresponds to a text document in C, and it is possible for a document to be the empty, when the corresponding node does not contain any content. A subset of the nodes in N are labeled. This corresponds to the training data. The classification problem in this scenario is to determine the labels of the remaining nodes with the use of the training data. It is clear that both the content and structure can play a useful and complementary role in the classification process.

An interesting method for combining linkage and content information for classification was discussed in [20]. In this paper, a hypertext categorization method was proposed, which uses the content and labels of neighboring web pages for the classification process. When the labels of all the nearest neighbors are available, then a Bayesian method can be adapted easily for classification purposes. Just as the presence of a word in a document can be considered a Bayesian feature for a text classifier, the presence of a link between the target page, and a page (for which the label is known) can be considered a feature for the classifier. The real challenge arises when the labels of all the nearest neighbors are not available. In such cases, a relaxation labeling method was proposed in order to perform the classification. Two methods have been proposed in this work:

- **Fully Supervised Case of Radius one Enhanced Linkage Analysis:** In this case, it is assumed that all the neighboring class labels are known. In such a case, a Bayesian approach is utilized in order to treat the labels on the nearest neighbors as features for classification purposes. In this case, the linkage information is the sole information which is used for classification purposes.

- **When the class labels of the nearest neighbors are not known:** In this case, an iterative approach is used for combining text and linkage based classification. Rather than using the pre-defined labels (which are not available), we perform a first labeling of the neighboring documents with the use of document content. These labels are then used to classify the label of the target document, with the use of *both* the local text and the class labels of the neighbors. This approach is used iteratively for re-defining the labels of both the target document and its neighbors until convergence is achieved.

The conclusion from the work in [20] is that a combination of text and linkage based classification always improves the accuracy of a text classifier. Even when none of the neighbors of the document have known classes, it seemed to be always beneficial to add link information to the classification process. When the class labels of all the neighbors are known, then the advantages of using the scheme seem to be quite significant.

An additional idea in the paper is that of the use of *bridges* in order to further improve the classification accuracy. The core idea in the use of a bridge is the use of *2-hop* propagation for link-based classification. The results with the use of such an approach are somewhat mixed, as the accuracy seems to reduce with an increasing number of hops. The work in [20] shows results on a number of different kinds of data sets such as the *Reuters database*, *US patent database*, and *Yahoo!*. Since the *Reuters database* contains the least amount of noise, and pure text classifiers were able to do a good job. On the other hand, the *US patent database* and the *Yahoo! database* contain an increasing amount of noise which reduces the accuracy of text classifiers. An interesting observation in [20] was that a scheme which simply absorbed the neighbor text into the current document performed *significantly worse* than a scheme which was based on pure text-based classification. This is because there are often significant cross-boundary linkages between topics, and such linkages are able to confuse the classifier. A publicly available implementation of this algorithm may be found in the *NetKit* tool kit available in [92].

Another relaxation labeling method for graph-based document classification is proposed in [4]. In this technique, the probability that the end points of a link take on a particular pair of class labels is quantified. We refer to this as the *link-class pair probability*. The posterior probability of classification of a node T into class i is expressed as sum of the probabilities of pairing all possible class labels of the neighbors of T with class label i. We note a significant percentage of these (exponential number of) possibilities are pruned, since only the currently most probable[5] labelings are used in this approach. For this purpose, it is assumed that the class labels of the different neighbors of T (while dependent on T) are independent of each other. This is similar to the naive assumption, which is often used in Bayes classifiers. Therefore, the probability for a particular combination of labels on the neighbors can be expressed as the product of the corresponding link-class pair probabilities. The approach starts off with the use of a standard content-based Bayes or SVM classifier in order to assign the initial labels to the nodes. Then, an iterative approach is used to refine the labels, by using the most probably label estimations from the previous iteration in order to refine the labels in the current iteration. We note that the link-class pair probabilities can be estimated as the smoothed fraction of edges in the last iteration which contain a particular pair of classes as the end points (hard labeling), or it can also be estimated as the average product of node probabilities over all edges which take on that particular class pair (soft labeling). This approach is repeated to convergence.

Another method which uses a naive Bayes classifier to enhance link-based classification is proposed in [104]. This method incrementally assigns class labels, starting off with a temporary assignment and then gradually making them permanent. The initial class assignment is based on a simple Bayes expression based on both the terms and links in the

---

[5]In the case of *hard labeling*, the single most likely labeling is used, whereas in the case of *soft labeling*, a small set of possibilities is used.

document. In the final categorization, the method changes the term weights for Bayesian classification of the target document with the terms in the neighbor of the current document. This method uses a broad framework which is similar to that in [20], except that it differentiates between the classes in the neighborhood of a document in terms of their influence on the class label of the current document. For example, documents for which the class label was either already available in the training data, or for which the algorithm has performed a final assignment, have a different confidence weighting factor than those documents for which the class label is currently temporarily assigned. Similarly, documents which belong to a completely different subject (based on content) are also removed from consideration from the assignment. Then, the Bayesian classification is performed with the re-computed weights, so that the document can be assigned a final class label. By using this approach the technique is able to compensate for the noise and inconsistencies in the link structures among different documents.

One major difference between the work in [20] and [104], is that the former is focussed on using link information in order to propagate the labels, whereas the latter attempts to use the content of the neighboring pages. Another work along this direction, which uses the content of the neighboring pages more explicitly is proposed in [105]. In this case, the content of the neighboring pages is broken up into different fields such as titles, anchor text, and general text. The different fields are given different levels of importance, which is learned during the classification process. It was shown in [105] that the use of title fields and anchor fields is much more relevant than the general text. This accounts for much of the accuracy improvements demonstrated in [105].

The work in [2] proposes a method for dynamic classification in text networks with the use of a random-walk method. The key idea in the work is to *transform* the combination of structure and content in the network into a pure network containing only content. Thus, we transform the original network $G = (N, A, C)$ into an *augmented* network $G^A = (N \cup N_c, A \cup A_c)$, where N_c and A_c are an additional set of nodes and edges added to the original network. Each node in N_c corresponds to a distinct word in the lexicon. Thus, the augmented network contains the original structural nodes N, and a new set of word nodes N_c. The added edges in A_c are undirected edges added between the structural nodes N and the word nodes N_c. Specifically, an edge (i, j) is added to A_c, if the word $i \in N_c$ occurs in the text content corresponding to the node $j \in N$. Thus, this network is *semi-bipartite*, in that there are no edges between the different word nodes. An illustration of the semi-bipartite content-structure transformation is provided in Figure 6.4.

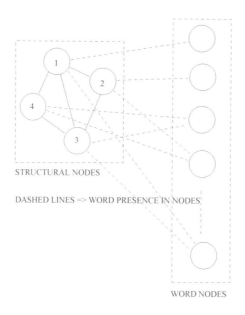

Figure 6.4. The Semi-bipartite Transformation

It is important to note that once such a transformation has been performed, any of the collective classification methods [12] can be applied to the structural nodes. In the work in [2], a random-walk method has been used in order to perform the collective classification of the underlying nodes. In this method, repeated random walks are performed starting at the unlabeled nodes which need to be classified. The random walks are defined only on the structural nodes, and each hop may either be a *structural* hop or a *content* hop. We perform l different random walks, each of which contains h nodes. Thus, a total of $l \cdot h$ nodes are encountered in the different walks. The class label of this node is predicted to be the label with the highest frequency of presence in the different $l \cdot h$ nodes encountered in the different walks. The error of this random walk-based sampling process has been bounded in [12]. In addition, the method in [12] can be adapted to dynamic content-based networks, in which the nodes, edges and their underlying content continuously evolve over time. The method in [2] has been compared to that proposed in [18] (based on the implementation in [92]), and it has been shown that the classification methods of [12] are significantly superior.

Another method for classification of linked text data is discussed in [139]. This method designs two separate regularization conditions; one is for the text-only classifier (also referred to as the *local* classifier), and the other is for the link information in the network structure. These regularizers are expressed in the terms of the underlying kernels; the link regularizer is related to the standard graph regularizer used in the machine learning literature, and the text regularizer is expressed in terms of the kernel gram matrix. These two regularization conditions are combined in two possible ways. One can either use linear combinations of the regularizers, or linear combinations of the associated kernels. It was shown in [139] that both combination methods perform better than either pure structure-based or pure text-based methods. The method using a linear combination of regularizers was slightly more accurate and robust than the method which used a linear combination of the kernels.

A method in [32] designs a classifier which combines a Naive Bayes classifier (on the text domain), and a rule-based classifier (on the structural domain). The idea is to invent a set of predicates, which are defined in the space of links, pages and words. A variety of predicates (or relations) are defined depending upon the presence of the word in a page, linkages of pages to each other, the nature of the anchor text of the hyperlink, and the neighborhood words of the hyperlink. These essentially encode the graph structure of the documents in the form of boolean predicates, and can also be used to construct relational learners. The main contribution in [32] is to combine the relational learners on the structural domain with the Naive Bayes approach in the text domain. We refer the reader to [32, 33] for the details of the algorithm, and the general philosophy of such relational learners.

One of the interesting methods for collective classification in the context of email networks was proposed in [23]. The technique in [23] is designed to classify *speech acts* in email. Speech acts essentially characterize, whether an email refers to a particular kind of action (such as scheduling a meeting). It has been shown in [23] that the use of sequential thread-based information from the email is very useful for the classification process. An email system can be modeled as a network in several ways, one of which is to treat an email as a node, and the edges as the thread relationships between the different emails. In this sense, the work in [23] devises a network-based mining procedure which uses both the content and the structure of the email network. However, this work is rather specific to the case of email networks, and it is not clear whether the technique can be adapted (effectively) to more general networks.

A different line of solutions to such problems, which are defined on a heterogeneous feature space is to use latent space methods in order to simultaneously homogenize the feature space, and also determine the latent factors in the underlying data. The resulting representation can be used in conjunction with any of the text classifiers which are designed for latent space representations. A method in [140] uses a matrix factorization approach in order to construct a latent space from the underlying data. Both supervised and unsupervised methods were proposed for constructing the latent space from the underlying data. It was then shown in [140] that this feature representation provides more accurate results, when used in conjunction with an SVM-classifier.

Finally, a method for web page classification is proposed in [119]. This method is designed for using intelligent agents in web page categorization. The overall approach relies on the design of two functions which correspond to scoring web pages and links respectively. An advice language is created, and a method is proposed for mapping advice to neural networks. It is has been shown in [119] how this general purpose system may be used in order to find home pages on the web.

9. Meta-Algorithms for Text Classification

Meta-algorithms play an important role in classification strategies because of their ability to enhance the accuracy of existing classification algorithms by combining them, or making a general change in the different algorithms to achieve a specific goal. Typical examples of classifier meta-algorithms include *bagging*, *stacking* and *boosting* [42]. Some of these methods change the underlying distribution of the training data, others combine classifiers, and yet others change the algorithms in order to satisfy specific classification criteria. We will discuss these different classes of methods in this section.

9.1 Classifier Ensemble Learning

In this method, we use *combinations* of classifiers in conjunction with a voting mechanism in order to perform the classification. The idea is that since different classifiers are susceptible to different kinds of overtraining and errors, a combination classifier is likely to yield much more robust results. This technique is also sometimes referred to as *stacking* or *classifier committee construction*.

Ensemble learning has been used quite frequently in text categorization. Most methods simply use weighted combinations of classifier outputs (either in terms of scores or ranks) in order to provide the final classification result. For example, the work by Larkey and Croft [79] used weighted linear combinations of the classifier scores or ranks. The work by Hull [60] used linear combinations of probabilities for the same goal. A linear combination of the normalized scores was used for classification [137]. The work in [87] used classifier selection techniques and voting in order to provide the final classification result. Some examples of such voting and selection techniques are as follows:

- In a binary-class application, the class label which obtains the majority vote is reported as the final result.

- For a given test instance, a specific classifier is selected, depending upon the performance of the classifier which are closest to that test instance.

- A weighted combination of the results from the different classifiers are used, where the weight is regulated by the performance of the classifier on validation instances which are most similar to the current test instance.

The last two methods above try to select the final classification in a smarter way by discriminating between the performances of the classifiers in different scenarios. The work by [77] used category-averaged features in order to construct a different classifier for each category.

The major challenge in ensemble learning is to provide the appropriate combination of classifiers for a particular scenario. Clearly, this combination can significantly vary with the scenario and the data set. In order to achieve this goal, the method in [10] proposes a method for probabilistic combination of text classifiers. The work introduces a number of variables known as *reliability variables* in order to regulate the importance of the different classifiers. These reliability variables are learned dynamically for each situation, so as to provide the best classification.

9.2 Data Centered Methods: Boosting and Bagging

While ensemble techniques focus on combining different classifiers, data-centered methods such as boosting and bagging typically focus on training the same classifier on different parts of the training data in order to create different models. For a given test instance, a combination of the results obtained from the use of these different models is reported. Another major difference between ensemble-methods and boosting methods is that the training models in a boosting method are not constructed independently, but are constructed sequentially. Specifically, after i classifiers are constructed, the $(i+1)$th classifier is constructed on those parts of the training data which the first i classifiers are unable to accurately classify. The results of these different classifiers are combined together carefully, where the weight of each classifier is typically a function of its error rate. The most well known meta-algorithm for boosting is the

AdaBoost algorithm [48]. Such boosting algorithms have been applied to a variety of scenarios such as decision tree learners, rule-based systems, and Bayesian classifiers [49, 61, 73, 100, 116, 118].

We note that boosting is also a kind of ensemble learning methodology, except that we train the same model on different subsets of the data in order to create the ensemble. One major criticism of boosting is that in many data sets, some of the training records are noisy, and a classification model should be resistant to overtraining on the data. Since the boosting model tends to weight the error-prone examples more heavily in successive rounds, this can cause the classification process to be more prone to overfitting. This is particularly noticeable in the case of noisy data sets. Some recent results have suggested that all convex boosting algorithms may perform poorly in the presence of noise [91]. These results tend to suggest that the choice of boosting algorithm may be critical for a successful outcome, depending upon the underlying data set.

Bagging methods [16] are generally designed to reduce the model overfitting error which arises during the learning process. The idea in bagging is to pick *bootstrap samples* (samples with replacement) from the underlying collection, and train the classifiers in these samples. The classification results from these different samples are then combined together in order to yield the final result. Bagging methods are generally used in conjunction with decision trees, though these methods can be used in principle with any kind of classifier. The main criticism of the bagging method is that it can sometimes lead to a reduction in accuracy because of the smaller size of each individual training sample. Bagging is useful only if the model is unstable to small details of the training algorithm, because it reduces the overfitting error. An example of such an algorithm would be the decision tree model, which is highly sensitive to how the higher levels of the tree are constructed in a high dimensional feature space such as text. Bagging methods have not been used frequently in text classification.

9.3 Optimizing Specific Measures of Accuracy

We note that the use of the absolute classification accuracy is not the only measure which is relevant to classification algorithms. For example, in skewed-class scenarios, as often arise in the context of applications such as fraud detection, and spam filtering, it is more costly to misclassify examples of one class than another. For example, while it may be tolerable to misclassify a few spam emails (thereby allowing them into the inbox), it is much more undesirable to incorrectly mark a legitimate email as spam. Cost-sensitive classification problems also naturally arise in cases in which one class is more rare than the other, and it is therefore more desirable to identify the rare examples. In such cases, it is desirable to optimize the *cost-weighted accuracy* of the classification process. We note that many of the broad techniques which have been designed for non-textual data [40, 42, 45] are also applicable to text data, because the specific feature representation is not material to how standard algorithms for modified to the cost-sensitive case. A good understanding of cost-sensitive classification both for the textual and non-textual case may be found in [40, 45, 3]. Some examples of how classification algorithms may be modified in straightforward ways to incorporate cost-sensitivity are as follows:

- In a decision-tree, the split condition at a given node tries to maximize the accuracy of its children nodes. In the cost-sensitive case, the split is engineered to maximize the cost-sensitive accuracy.

- In rule-based classifiers, the rules are typically quantified and ordered by measures corresponding to their predictive accuracy. In the cost-sensitive case, the rules are quantified and ordered by their *cost-weighted accuracy*.

- In Bayesian classifiers, the posterior probabilities are weighted by the cost of the class for which the prediction is made.

- In linear classifiers, the optimum hyperplane separating the classes is determined in a cost-weighted sense. Such costs can typically be incorporated in the underlying objective function. For example, the least-square error in the objective function of the LLSF method can be weighted by the underlying costs of the different classes.

- In a k-nearest neighbor classifier, we report the cost-weighted majority class among the k nearest neighbors of the test instance.

We note that the use of a cost-sensitive approach is essentially a change of the objective function of classification, which can also be formulated as an optimization problem. While the standard classification problem generally tries to optimize accuracy, the cost-sensitive version tries to optimize a cost-weighted objective function. A more general approach was proposed in [50] in which a meta-algorithm was proposed for optimizing a specific figure of merit such as the accuracy, precision, recall, or F_1-measure. Thus, this approach generalizes this class of methods to *any arbitrary objective function*, making it essentially an *objective-centered classification method*. A generalized probabilistic descent algorithm (with the desired objective function) is used in conjunction with the classifier of interest in order to derive the class labels of the test instance. The work in [50] shows the advantages of using the technique over a standard SVM-based classifier.

10. Conclusions and Summary

The classification problem is one of the most fundamental problems in the machine learning and data mining literature. In the context of text data, the problem can also be considered similar to that of classification of *discrete set-valued* attributes, when the frequencies of the words are ignored. The domains of these sets are rather large, as it comprises the entire lexicon. Therefore, text mining techniques need to be designed to effectively manage large numbers of elements with varying frequencies. Almost all the known techniques for classification such as decision trees, rules, Bayes methods, nearest neighbor classifiers, SVM classifiers, and neural networks have been extended to the case of text data. Recently, a considerable amount of emphasis has been placed on linear classifiers such as neural networks and SVM classifiers, with the latter being particularly suited to the characteristics of text data. In recent years, the advancement of web and social network technologies have lead to a tremendous interest in the classification of text documents containing links or other meta-information. Recent research has shown that the incorporation of linkage information into the classification process can significantly improve the quality of the underlying results.

References

[1] C. C. Aggarwal, S. C. Gates, P. S. Yu. On Using Partial Supervision for Text Categorization, *IEEE Transactions on Knowledge and Data Engineering*, 16(2), 245–255, 2004.

[2] C. C. Aggarwal, N. Li. On Node Classification in Dynamic Content-based Networks, *SDM Conference*, 2011.

[3] I. Androutsopoulos, J. Koutsias, K. Chandrinos, G. Paliouras, C. Spyropoulos. An Evaluation of Naive Bayesian Anti-Spam Filtering. Workshop on *Machine Learning in the New Information Age*, in conjunction with *ECML Conference*, 2000.
http://arxiv.org/PS_cache/cs/pdf/0006/0006013v1.pdf

[4] R. Angelova, G. Weikum. Graph-based text classification: learn from your neighbors. *ACM SIGIR Conference*, 2006.

[5] C. Apte, F. Damerau, S. Weiss. Automated Learning of Decision Rules for Text Categorization, *ACM Transactions on Information Systems*, 12(3), pp. 233–251, 1994.

[6] M. Aizerman, E. Braverman, L. Rozonoer. Theoretical foundations of the potential function method in pattern recognition learning, *Automation and Remote Control*, 25: pp. 821–837, 1964.

[7] L. Baker, A. McCallum. Distributional Clustering of Words for Text Classification, *ACM SIGIR Conference*, 1998.

[8] R. Bekkerman, R. El-Yaniv, Y. Winter, N. Tishby. On Feature Distributional Clustering for Text Categorization. *ACM SIGIR Conference*, 2001.

[9] S. Basu, A. Banerjee, R. J. Mooney. Semi-supervised Clustering by Seeding. *ICML Conference*, 2002.

[10] P. Bennett, S. Dumais, E. Horvitz. Probabilistic Combination of Text Classifiers using Reliability Indicators: Models and Results. *ACM SIGIR Conference*, 2002.

[11] P. Bennett, N. Nguyen. Refined experts: improving classification in large taxonomies. *ACM SIGIR Conference*, 2009.

[12] S. Bhagat, G. Cormode, S. Muthukrishnan. Node Classification in Social Networks, Book Chapter in *Social Network Data Analytics, Ed. Charu Aggarwal, Springer*, 2011.

[13] A. Blum, T. Mitchell. Combining labeled and unlabeled data with co-training. *COLT*, 1998.

[14] D. Boley, M. Gini, R. Gross, E.-H. Han, K. Hastings, G. Karypis, V. Kumar, B. Mobasher, J. Moore. Partitioning-based clustering for web document categorization. *Decision Support Systems*, Vol. 27, pp. 329–341, 1999.

[15] L. Brieman, J. Friedman, R. Olshen, C. Stone. *Classification and Regression Trees*, Wadsworth Advanced Books and Software, CA, 1984.

[16] L. Breiman. Bagging Predictors. *Machine Learning*, 24(2), pp. 123–140, 1996.

[17] L. Cai, T. Hofmann. Text categorization by boosting automatically extracted concepts. *ACM SIGIR Conference*, 2003.

[18] S. Chakrabarti, S. Roy, M. Soundalgekar. Fast and Accurate Text Classification via Multiple Linear Discriminant Projections, *VLDB Journal*, 12(2), pp. 172–185, 2003.

[19] S. Chakrabarti, B. Dom. R. Agrawal, P. Raghavan. Using taxonomy, discriminants and signatures for navigating in text databases, *VLDB Conference*, 1997.

[20] S. Chakrabarti, B. Dom, P. Indyk. Enhanced hypertext categorization using hyperlinks. *ACM SIGMOD Conference*, 1998.

[21] S. Chakraborti, R. Mukras, R. Lothian, N. Wiratunga, S. Watt, D. Harper. Supervised Latent Semantic Indexing using Adaptive Sprinkling, *IJCAI*, 2007.

[22] D. Chickering, D. Heckerman, C. Meek. A Bayesian approach for learning Bayesian networks with local structure. *Thirteenth Conference on Uncertainty in Artificial Intelligence*, 1997.

[23] V. R. de Carvalho, W. Cohen. On the collective classification of email "speech acts", *ACM SIGIR Conference*, 2005.

[24] V. Castelli, T. M. Cover. On the exponential value of labeled samples. *Pattern Recognition Letters*, 16(1), pp. 105–111, 1995.

[25] W. Cohen, H. Hirsh. Joins that generalize: text classification using Whirl. *ACM KDD Conference*, 1998.

[26] W. Cohen, Y. Singer. Context-sensitive learning methods for text categorization. *ACM Transactions on Information Systems*, 17(2), pp. 141–173, 1999.

[27] W. Cohen. Learning rules that classify e-mail. *AAAI Conference*, 1996.

[28] W. Cohen. Learning with set-valued features. *AAAI Conference*, 1996.

[29] W. Cooper. Some inconsistencies and misnomers in probabilistic information retrieval. *ACM Transactions on Information Systems*, 13(1), pp. 100–111, 1995.

[30] C. Cortes, V. Vapnik. Support-vector networks. *Machine Learning*, 20: pp. 273–297, 1995.

[31] T. M. Cover, J. A. Thomas. *Elements of information theory.* New York: John Wiley and Sons, 1991.

[32] M. Craven, S. Slattery. Relational learning with statistical predicate invention: Better models for hypertext. *Machine Learning*, 43: pp. 97–119, 2001.

[33] M. Craven, D. DiPasquo, D. Freitag, A. McCallum, T. Mitchell, K. Nigam, S. Slattery. Learning to Extract Symbolic Knowledge from the Worldwide Web. *AAAI Conference*, 1998.

[34] I. Dagan, Y. Karov, D. Roth. Mistake-driven Learning in Text Categorization, *Proceedings of EMNLP*, 1997.

[35] A. Dayanik, D. Lewis, D. Madigan, V. Menkov, A. Genkin. Constructing informative prior distributions from domain knowledge in text classification. *ACM SIGIR Conference*, 2006.

[36] A. P. Dempster, N.M. Laird, D.B. Rubin. Maximum likelihood from incomplete data via the em algorithm. *Journal of the Royal Statistical Society, Series B*, 39(1): pp. 1–38, 1977.

[37] F. Denis, A. Laurent. Text Classification and Co-Training from Positive and Unlabeled Examples, *ICML 2003 Workshop: The Continuum from Labeled to Unlabeled Data.* http://www.grappa.univ-lille3.fr/ftp/reports/icmlws03.pdf.

[38] S. Deerwester, S. Dumais, T. Landauer, G. Furnas, R. Harshman. Indexing by Latent Semantic Analysis. *JASIS*, 41(6), pp. 391–407, 1990.

[39] P. Domingos, M. J. Pazzani. On the the optimality of the simple Bayesian classifier under zero-one loss. *Machine Learning*, 29(2–3), pp. 103–130, 1997.

[40] P. Domingos. MetaCost: A General Method for making Classifiers Cost-Sensitive. *ACM KDD Conference*, 1999.

[41] H. Drucker, D. Wu, V. Vapnik. Support Vector Machines for Spam Categorization. *IEEE Transactions on Neural Networks*, 10(5), pp. 1048–1054, 1999.

[42] R. Duda, P. Hart, W. Stork. *Pattern Classification*, Wiley Interscience, 2000.

[43] S. Dumais, J. Platt, D. Heckerman, M. Sahami. Inductive learning algorithms and representations for text categorization. *CIKM Conference*, 1998.

[44] S. Dumais, H. Chen. Hierarchical Classification of Web Content. *ACM SIGIR Conference*, 2000.

[45] C. Elkan. The foundations of cost-sensitive learning, *IJCAI Conference*, 2001.

[46] R. Fisher. The Use of Multiple Measurements in Taxonomic Problems. *Annals of Eugenics*, 7, pp. 179–188, 1936.

[47] R. El-Yaniv, O. Souroujon. Iterative Double Clustering for Unsupervised and Semi-supervised Learning. *NIPS Conference*, 2002.

[48] Y. Freund, R. Schapire. A decision-theoretic generalization of on-line learning and an application to boosting. In *Proc. Second European Conference on Computational Learning Theory*, pp. 23–37, 1995.

[49] Y. Freund, R. Schapire, Y. Singer, M. Warmuth. Using and combining predictors that specialize. *Proceedings of the 29th Annual ACM Symposium on Theory of Computing*, pp. 334–343, 1997.

[50] S. Gao, W. Wu, C.-H. Lee, T.-S. Chua. A maximal figure-of-merit learning approach to text categorization. *SIGIR Conference*, 2003.

[51] R. Gilad-Bachrach, A. Navot, N. Tishby. Margin based feature selection – theory and algorithms. *ICML Conference*, 2004.

[52] S. Gopal, Y. Yang. Multilabel classification with meta-level features. *ACM SIGIR Conference*, 2010.

[53] L. Guthrie, E. Walker. Document Classification by Machine: Theory and Practice. *COLING*, 1994.

[54] E.-H. Han, G. Karypis, V. Kumar. Text Categorization using Weighted-Adjusted k-nearest neighbor classification, *PAKDD Conference*, 2001.

[55] E.-H. Han, G. Karypis. Centroid-based Document Classification: Analysis and Experimental Results, *PKDD Conference*, 2000.

[56] D. Hardin, I. Tsamardinos, C. Aliferis. A theoretical characterization of linear SVM-based feature selection. *ICML Conference*, 2004.

[57] T. Hofmann. Probabilistic latent semantic indexing. *ACM SIGIR Conference*, 1999.

[58] P. Howland, M. Jeon, H. Park. Structure Preserving Dimension Reduction for Clustered Text Data based on the Generalized Singular Value Decomposition. *SIAM Journal of Matrix Analysis and Applications*, 25(1): pp. 165–179, 2003.

[59] P. Howland, H. Park. Generalizing discriminant analysis using the generalized singular value decomposition, *IEEE Transactions on Pattern Analysis and Machine Intelligence*, 26(8), pp. 995–1006, 2004.

[60] D. Hull, J. Pedersen, H. Schutze. Method combination for document filtering. *ACM SIGIR Conference*, 1996.

[61] R. Iyer, D. Lewis, R. Schapire, Y. Singer, A. Singhal. Boosting for document routing. *CIKM Conference*, 2000.

[62] M. James. *Classification Algorithms*, Wiley Interscience, 1985.

[63] D. Jensen, J. Neville, B. Gallagher. Why collective inference improves relational classification. *ACM KDD Conference*, 2004.

[64] T. Joachims. A Probabilistic Analysis of the Rocchio Algorithm with TFIDF for Text Categorization. *ICML Conference*, 1997.

[65] T. Joachims. Text categorization with support vector machines: learning with many relevant features. *ECML Conference*, 1998.

[66] T. Joachims. Transductive inference for text classification using support vector machines. *ICML Conference*, 1999.

[67] T. Joachims. A Statistical Learning Model of Text Classification for Support Vector Machines. *ACM SIGIR Conference*, 2001.

[68] D. Johnson, F. Oles, T. Zhang, T. Goetz. A Decision Tree-based Symbolic Rule Induction System for Text Categorization, *IBM Systems Journal*, 41(3), pp. 428–437, 2002.

[69] I. T. Jolliffee. Principal Component Analysis. *Springer*, 2002.

[70] T. Kalt, W. B. Croft. A new probabilistic model of text classification and retrieval. *Technical Report IR-78, University of Massachusetts Center for Intelligent Information Retrieval*, 1996. http://ciir.cs.umass.edu/publications/index.shtml

[71] G. Karypis, E.-H. Han. Fast Supervised Dimensionality Reduction with Applications to Document Categorization and Retrieval, *ACM CIKM Conference*, 2000.

[72] T. Kawatani. Topic difference factor extraction between two document sets and its application to text categorization. *ACM SIGIR Conference*, 2002.

[73] Y.-H. Kim, S.-Y. Hahn, B.-T. Zhang. Text filtering by boosting naive Bayes classifiers. *ACM SIGIR Conference*, 2000.

[74] D. Koller, M. Sahami. Hierarchically classifying documents with very few words, *ICML Conference*, 2007.

[75] S. Lam, D. Lee. Feature reduction for neural network based text categorization. *DASFAA Conference*, 1999.

[76] W. Lam, C. Y. Ho. Using a generalized instance set for automatic text categorization. *ACM SIGIR Conference*, 1998.

[77] W. Lam, K.-Y. Lai. A meta-learning approach for text categorization. *ACM SIGIR Conference*, 2001.

[78] K. Lang. Newsweeder: Learning to filter netnews. *ICML Conference*, 1995.

[79] L. S. Larkey, W. B. Croft. Combining Classifiers in text categorization. *ACM SIGIR Conference*, 1996.

[80] D. Lewis, J. Catlett. Heterogeneous uncertainty sampling for supervised learning. *ICML Conference*, 1994.

[81] D. Lewis, M. Ringuette. A comparison of two learning algorithms for text categorization. *SDAIR*, 1994.

[82] D. Lewis. Naive (Bayes) at forty: The independence assumption in information retrieval. *ECML Conference*, 1998.

[83] D. Lewis. An Evaluation of Phrasal and Clustered Representations for the Text Categorization Task, *ACM SIGIR Conference*, 1992.

[84] D. Lewis, W. Gale. A sequential algorithm for training text classifiers, *SIGIR Conference*, 1994.

[85] D. Lewis, K. Knowles. Threading electronic mail: A preliminary study. *Information Processing and Management*, 33(2), pp. 209–217, 1997.

[86] H. Li, K. Yamanishi. Document classification using a finite mixture model. *Annual Meeting of the Association for Computational Linguistics*, 1997.

[87] Y. Li, A. Jain. Classification of text documents. *The Computer Journal*, 41(8), pp. 537–546, 1998.

[88] B. Liu, W. Hsu, Y. Ma. Integrating Classification and Association Rule Mining. *ACM KDD Conference*, 1998.

[89] B. Liu, L. Zhang. A Survey of Opinion Mining and Sentiment Analysis. Book Chapter in *Mining Text Data, Ed. C. Aggarwal, C. Zhai, Springer*, 2011.

[90] N. Littlestone. Learning quickly when irrelevant attributes abound: A new linear-threshold algorithm. *Machine Learning*, 2: pp. 285–318, 1988.

[91] P. Long, R. Servedio. Random Classification Noise defeats all Convex Potential Boosters. *ICML Conference*, 2008.

[92] S. A. Macskassy, F. Provost. Classification in Networked Data: A Toolkit and a Univariate Case Study, *Journal of Machine Learning Research*, Vol. 8, pp. 935–983, 2007.

[93] A. McCallum. Bow: A toolkit for statistical language modeling, text retrieval, classification and clustering. http://www.cs.cmu.edu/~mccallum/bow, 1996.

[94] A. McCallum, K. Nigam. A Comparison of Event Models for Naive Bayes Text Classification. *AAAI Workshop on Learning for Text Categorization*, 1998.

[95] A. McCallum, R. Rosenfeld, T. Mitchell, A. Ng. Improving text classification by shrinkage in a hierarchy of classes. *ICML Conference*, 1998.

[96] McCallum, Andrew Kachites. "MALLET: A Machine Learning for Language Toolkit." http://mallet.cs.umass.edu. 2002.

[97] T. M. Mitchell. *Machine Learning*. WCB/McGraw-Hill, 1997.

[98] T. M. Mitchell. The role of unlabeled data in supervised learning. *Proceedings of the Sixth International Colloquium on Cognitive Science*, 1999.

[99] D. Mladenic, J. Brank, M. Grobelnik, N. Milic-Frayling. Feature selection using linear classifier weights: interaction with classification models. *ACM SIGIR Conference*, 2004.

[100] K. Myers, M. Kearns, S. Singh, M. Walker. A boosting approach to topic spotting on subdialogues. *ICML Conference*, 2000.

[101] H. T. Ng, W. Goh, K. Low. Feature selection, perceptron learning, and a usability case study for text categorization. *ACM SIGIR Conference*, 1997.

[102] A. Y. Ng, M. I. Jordan. On discriminative vs. generative classifiers: a comparison of logistic regression and naive Bayes. *NIPS*. pp. 841-848, 2001.

[103] K. Nigam, A. McCallum, S. Thrun, T. Mitchell. Learning to classify text from labeled and unlabeled documents. *AAAI Conference*, 1998.

[104] H.-J. Oh, S.-H. Myaeng, M.-H. Lee. A practical hypertext categorization method using links and incrementally available class information. *ACM SIGIR Conference*, 2000.

[105] X. Qi, B. Davison. Classifiers without borders: incorporating fielded text from neighboring web pages. *ACM SIGIR Conference*, 2008.

[106] J. R. Quinlan, Induction of Decision Trees, *Machine Learning*, 1(1), pp 81–106, 1986.

[107] H. Raghavan, J. Allan. An interactive algorithm for asking and incorporating feature feedback into support vector machines. *ACM SIGIR Conference*, 2007.

[108] S. E. Robertson, K. Sparck-Jones. Relevance weighting of search terms. *Journal of the American Society for Information Science*, 27: pp. 129–146, 1976.

[109] J. Rocchio. Relevance feedback information retrieval. *The Smart Retrieval System- Experiments in Automatic Document Processing*, G. Salton, Ed. Prentice Hall, Englewood Cliffs, NJ, pp 313–323, 1971.

[110] M. Ruiz, P. Srinivasan. Hierarchical neural networks for text categorization. *ACM SIGIR Conference*, 1999.

[111] F. Sebastiani. Machine Learning in Automated Text Categorization, *ACM Computing Surveys*, 34(1), 2002.

[112] M. Sahami. Learning limited dependence Bayesian classifiers, *ACM KDD Conference*, 1996.

[113] M. Sahami, S. Dumais, D. Heckerman, E. Horvitz. A Bayesian approach to filtering junk e-mail. *AAAI Workshop on Learning for Text Categorization. Tech. Rep. WS-98-05, AAAI Press.* http://robotics.stanford.edu/users/sahami/papers.html

[114] T. Salles, L. Rocha, G. Pappa, G. Mourao, W. Meira Jr., M. Goncalves. Temporally-aware algorithms for document classification. *ACM SIGIR Conference*, 2010.

[115] G. Salton. An Introduction to Modern Information Retrieval, *Mc Graw Hill*, 1983.

[116] R. Schapire, Y. Singer. BOOSTEXTER: A Boosting-based System for Text Categorization, *Machine Learning*, 39(2/3), pp. 135–168, 2000.

[117] H. Schutze, D. Hull, J. Pedersen. A comparison of classifiers and document representations for the routing problem. *ACM SIGIR Conference*, 1995.

[118] R. Shapire, Y. Singer, A. Singhal. Boosting and Rocchio applied to text filtering. *ACM SIGIR Conference*, 1998.

[119] J. Shavlik, T. Eliassi-Rad. Intelligent agents for web-based tasks: An advice-taking approach. *AAAI-98 Workshop on Learning for Text Categorization. Tech. Rep. WS-98-05, AAAI Press*, 1998. http://www.cs.wisc.edu/~shavlik/mlrg/publications.html

[120] V. Sindhwani, S. S. Keerthi. Large scale semi-supervised linear SVMs. *ACM SIGIR Conference*, 2006.

[121] N. Slonim, N. Tishby. The power of word clusters for text classification. *European Colloquium on Information Retrieval Research (ECIR)*, 2001.

[122] N. Slonim, N. Friedman, N. Tishby. Unsupervised document classification using sequential information maximization. *ACM SIGIR Conference*, 2002.

[123] J.-T. Sun, Z. Chen, H.-J. Zeng, Y. Lu, C.-Y. Shi, W.-Y. Ma. Supervised Latent Semantic Indexing for Document Categorization. *ICDM Conference*, 2004.

[124] V. Vapnik. Estimations of dependencies based on statistical data, *Springer*, 1982.

[125] V. Vapnik. *The Nature of Statistical Learning Theory*, Springer, New York, 1995.

[126] A. Weigand, E. Weiner, J. Pedersen. Exploiting hierarchy in text catagorization. *Information Retrieval*, 1(3), pp. 193–216, 1999.

[127] S, M. Weiss, C. Apte, F. Damerau, D. Johnson, F. Oles, T. Goetz, T. Hampp. Maximizing text-mining performance. *IEEE Intelligent Systems*, 14(4), pp. 63–69, 1999.

[128] S. M. Weiss, N. Indurkhya. Optimized Rule Induction, *IEEE Exp.*, 8(6), pp. 61–69, 1993.

[129] E. Wiener, J. O. Pedersen, A. S. Weigend. A Neural Network Approach to Topic Spotting. *SDAIR*, pp. 317–332, 1995.

[130] G.-R. Xue, D. Xing, Q. Yang, Y. Yu. Deep classification in large-scale text hierarchies. *ACM SIGIR Conference*, 2008.

[131] J. Yan, N. Liu, B. Zhang, S. Yan, Z. Chen, Q. Cheng, W. Fan, W.-Y. Ma. OCFS: optimal orthogonal centroid feature selection for text categorization. *ACM SIGIR Conference*, 2005.

[132] Y. Yang, L. Liu. A re-examination of text categorization methods, *ACM SIGIR Conference*, 1999.

[133] Y. Yang, J. O. Pederson. A comparative study on feature selection in text categorization, *ACM SIGIR Conference*, 1995.

[134] Y. Yang, C.G. Chute. An example-based mapping method for text categorization and retrieval. *ACM Transactions on Information Systems*, 12(3), 1994.

[135] Y. Yang. Noise Reduction in a Statistical Approach to Text Categorization, *ACM SIGIR Conference*, 1995.

[136] Y. Yang. A Study on Thresholding Strategies for Text Categorization. *ACM SIGIR Conference*, 2001.

[137] Y. Yang, T. Ault, T. Pierce. Combining multiple learning strategies for effective cross-validation. *ICML Conference*, 2000.

[138] J. Zhang, Y. Yang. Robustness of regularized linear classification methods in text categorization. *ACM SIGIR Conference*, 2003.

[139] T. Zhang, A. Popescul, B. Dom. Linear prediction models with graph regularization for web-page categorization, *ACM KDD Conference*, 2006.

[140] S. Zhu, K. Yu, Y. Chi, Y. Gong. Combining content and link for classification using matrix factorization. *ACM SIGIR Conference*, 2007.

Chapter 7

TRANSFER LEARNING FOR TEXT MINING

Weike Pan
Hong Kong University of Science and Technology
Clearwater Bay, Kowloon, Hong Kong
weikep@cse.ust.hk

Erheng Zhong
Hong Kong University of Science and Technology
Clearwater Bay, Kowloon, Hong Kong
ezhong@cse.ust.hk

Qiang Yang
Hong Kong University of Science and Technology
Clearwater Bay, Kowloon, Hong Kong
qyang@cse.ust.hk

Abstract Over the years, transfer learning has received much attention in machine learning research and practice. Researchers have found that a major bottleneck associated with machine learning and text mining is the lack of high-quality annotated examples to help train a model. In response, transfer learning offers an attractive solution for this problem. Various transfer learning methods are designed to extract the useful knowledge from different but related auxiliary domains. In its connection to text mining, transfer learning has found novel and useful applications. In this chapter, we will review some most recent developments in transfer learning for text mining, explain related algorithms in detail, and project future developments of this field. We focus on two important topics: cross-domain text document classification and heterogeneous transfer learning that uses labeled text documents to help classify images.

Keywords: Transfer learning, text mining, classification, clustering, learning-to-rank.

1. Introduction

Transfer learning refers to the machine learning framework in which one extracts knowledge from some auxiliary domains to help boost the learning performance in a target domain. Transfer learning as a new paradigm of machine learning has achieved great success in various areas over the last two decades [17, 67], e.g. text mining [8, 26, 23], speech recognition [95, 52], computer vision (e.g. image [75] and video [100] analysis), and ubiquitous computing [108, 93].

For text mining, transfer learning can be found in many application scenarios, e.g., knowledge transfer from Wikipedia documents (auxiliary) to Twitter text (target), from WWW webpages to Flick images, from English documents to Chinese documents in search engine, etc. One fundamental motivation of transfer learning in text mining is the so-called *data sparsity* problem in a target domain, where data sparsity can be defined by a lack of useful labels or sufficient data in the training set. For example, Twitter messages are short documents that are generated by users. These documents are often unlabeled, which are difficult to classify. Thus, it would be useful for us to transfer the supervised knowledge from another fully labeled text corpus to help classify Twitter messages. When data sparsity happens, overfitting can easily happen when we train a model. In the past, many traditional machine learning methods have been proposed for addressing the *data sparsity* problem, including semi-supervised learning [111, 18], co-training [9] and active learning [91]. However, in many practical situations, we still have to look elsewhere for additional knowledge for learning in our domain of interest.

We can take the following two views on knowledge transfer,

1 *In theory*, transfer learning can be considered as a new *learning paradigm*, where most non-transfer learning methods are considered as a special case when learning happens within a single target domain only, e.g., text classification in Twitter, and

2 *In applications*, transfer learning can be considered as a new cross-domain *learning technique*, since it explicitly addresses the various aspects of domain differences, e.g. data distribution, feature and label space, noise in the auxiliary data, relevance of auxiliary and target domains, etc. For example, we have to address most of the above issues when we transfer knowledge from Wikipedia documents to Twitter text.

Machine learning algorithms such as classification and regression (e.g. discriminative learning, ensemble learning) have been widely adopted in various text mining applications, e.g. text classification [42], sentiment analysis [68], named entity recognition (NER) [106], part-of-speech (POS) tagging [77], relation extraction (RE) [104], etc. In this chapter, we will survey some recent transfer learning extensions in aforementioned machine learning and data mining techniques and their applications for text mining. The organization of the chapter is as follows. We first give an overview of the scope of text-mining problems that we consider, and motivate the need for transfer learning in text classification. We then describe some typical approaches in transfer learning, such that we can subsequently categorize various text-mining approaches under these transfer-learning categories. This is followed by an overview of transfer learning approaches that extracts knowledge from labeled text data for the benefit of image classification and processing. This latter approach is known as heterogeneous transfer learning (HTL). Finally, we conclude the chapter with a summary and discussion of future work.

2. Transfer Learning in Text Classification

We first review the problem formulation in cross-domain text classification problems. In the next section, we first look at some typical benchmark data examples where the cross domain classification methods are needed. We then consider the nature of these problems, their differences from a traditional text classification problem, as well as how to formulate these problems into a machine learning problem.

2.1 Cross Domain Text Classification

2.1.1 Support Vector Machines for Text Classification.
Text classification [42] aims to categorize a document to some predefined categories $\mathcal{Y}$, where a document is usually represented in the form of bag of words $\mathcal{X}$, denoted as a vector $x \in \mathbb{R}^{d \times 1}$ with d unique words. The entries in the feature vector x can be 1/0 indicating whether the corresponding word appears or not or TF-IDF (term frequency inverse-document frequency).

There are enormous user-generated contents in online products and services on social media forums, blogs and microblogs, social networks, etc. It is very important to be able to summarize consumers' opinions on existing products and services. Sentiment analysis (or opinion mining) [68] addresses this problem, by classifying the reviews or sentiments into positive and negative categories. Similar to text classification, reviews or sentiments can be represented as a feature vector $x \in \mathbb{R}^{d \times 1}$, and the label space is $\mathcal{Y} = \{\pm 1\}$.

Extension of text classification has also been done in sequence classification areas. For example, POS tagging [77] aims to assign a tag to each word in a text, or equivalently classify each word in a text to some specified categories such as norm, verb, adjective, etc. POS tagging is very important for language pre-processing, speech synthesis, word sense disambiguation, information retrieval, etc. POS tagging can be considered as a structure prediction problem, and can be reduced to multiple binary-classification problems.

As support vector machines (SVM) [42] have been recognized as a state-of-the-art model in text mining, below, we will use SVM as a representative base model among various discriminative models to illustrate how the labeled data in auxiliary domains can be used to achieve knowledge transfer from auxiliary domains to the target domain. We first consider the text data representation.

In text mining, we assume that the data are represented as a bag-of-words $\mathcal{X} = \mathbb{R}^{d \times 1}$ with the same feature space for both auxiliary and target learning domains. For notational simplicity, we consider binary classification problems, $\mathcal{Y} = \{\pm 1\}$, which can be extended to multi-class classification via common tricks of one-vs-one or one-vs-rest pre-processing. We generally assume the same feature space and label space in both auxiliary and target domains, but in Section 3, we mention some recent works on heterogeneous feature space and/or heterogeneous label space learning. We use X and Y to denote variables for feature space and label space, respectively, and we use x, y, $\tilde{x}$, $\tilde{y}$ to denote the corresponding instantiations of variables in target and auxiliary domains, respectively.

For each word in a text, we can extract a feature vector based on the context information that is represented as $x \in \mathbb{R}^{d \times 1}$. Many text mining problems can be modeled this way. In POS tagging, for example, the learning problem is basically a classification problem by assigning a label y to x. In Named Entity Recognition (NER) problems [106], the aim is to classify each word in a text to some pre-defined categories, e.g. location, time, organization, etc. Another interesting problem is relation extraction [104], where each pair of entities in a sentence is represented as a feature vector x, which is assigned to a certain type of relation, e.g. family, user-owner, employer-executive, etc.

Text classification can be addressed by discriminative learning methods, which explicitly model the conditional distribution $P_r(Y|X)$. We can find many text mining formulations as variants of this formulation, e.g., maximum entropy (MaxEnt) [5], logistic regression (LR) [36], conditional random field (CRF) [47]. With this in mind, we consider the following basic SVM algorithm for text classification.

Basic SVM for Text Classification Given ℓ labeled data points $\{(\boldsymbol{x}_i, y_i)\}_{i=1}^{\ell}$ with $\boldsymbol{x}_i \in \mathbb{R}^{d \times 1}$ and $y_i \in \{\pm 1\}$ in the target domain, we have the following optimization problem for the linear SVM with soft margin [82],

$$\min_{\boldsymbol{w},\boldsymbol{\xi}} \quad \frac{1}{2}||\boldsymbol{w}||_2^2 + \lambda \sum_{i=1}^{\ell} \xi_i \qquad (7.1)$$
$$\text{s.t.} \quad y_i \boldsymbol{w}^T \boldsymbol{x}_i \geq 1 - \xi_i, \ \xi_i \geq 0, \ i=1,\ldots,\ell$$

where $\boldsymbol{w} \in \mathbb{R}^{d \times 1}$ is the model parameter, $\boldsymbol{\xi} \in \mathbb{R}^{\ell \times 1}$ are the slack variables, and $\lambda > 0$ is the tradeoff parameter to balance the model complexity $||\boldsymbol{w}||_2^2$ and loss function $\sum_{i=1}^{\ell} \xi_i$. Solving the convex optimization problem in Eq.(7.1), we have a decision function

$$f(\boldsymbol{x}) = \boldsymbol{w}^T \boldsymbol{x} = \sum_{k=1}^{d} w_k x_k. \qquad (7.2)$$

In this section, we will consider how to extend this formulation to include transfer learning capabilities.

2.1.2 Cross Domain Text Classification Problems.

With the above baseline algorithm in mind, we now consider several problem domains where we show examples of cross domain text classification. These examples illustrates some of the benchmark data often used in transfer learning experiments. They also help demonstrate why transfer learning is needed when the domain difference is large between the auxiliary and target learning domains.

20 Newsgroups First, we consider the well-known 20-newsgroup data. The 20-newsgroup [48] is a text collection of approximately 20,000 newsgroup documents, which are partitioned across 20 different newsgroups nearly evenly. This data collection provides an ideal benchmark for evaluating and comparing different transfer learning algorithms for text classification. A typical method is to generate six different data sets from the 20-newsgroup data for evaluating cross-domain classification algorithms. For each data set, two top categories[1] are chosen, one as positive and the other as negative. Then, we can split the data based on

---
[1] Three top categories, `misc`, `soc` and `alt` are removed, because they are too small.

Data Set	$\tilde{D}$	D
comp vs sci	comp.graphics comp.os.ms-windows.misc sci.crypt sci.electronics	comp.sys.ibm.pc.hardware comp.sys.mac.hardware comp.windows.x sci.med sci.space
rec vs talk	rec.autos rec.motorcycles talk.politics.guns talk.politics.misc	rec.sport.baseball rec.sport.hockey talk.politics.mideast talk.religion.misc
rec vs sci	rec.autos rec.sport.baseball sci.med sci.space	rec.motorcycles rec.sport.hockey sci.crypt sci.electronics
sci vs talk	sci.electronics sci.med talk.politics.misc talk.religion.misc	sci.crypt sci.space talk.politics.guns talk.politics.mideast
comp vs rec	comp.graphics comp.sys.ibm.pc.hardware comp.sys.mac.hardware rec.motorcycles rec.sport.hockey	comp.os.ms-windows.misc comp.windows.x rec.autos rec.sport.baseball
comp vs talk	comp.graphics comp.sys.mac.hardware comp.windows.x talk.politics.mideast talk.religion.misc	comp.os.ms-windows.misc comp.sys.ibm.pc.hardware talk.politics.guns talk.politics.misc

Table 7.1. A description of 20-newsgroup data sets for cross-domain classification.

sub-categories. Different sub-categories can be considered as different domains, while the task is defined as top category classification. The splitting strategy ensures the domains of labeled and unlabeled data related, since they are under the same top categories. Table 7.1 shows details of this data.

SRAA SRAA [61] is a UseNet data set for document classification that describes documents in Simulated/Real/Aviation/Auto classes. 73,218 UseNet articles are collected from four discussion groups about simulated autos (`sim-auto`), simulated aviation (`sim-aviation`), real autos (`real-auto`) and real aviation (`real-aviation`).

For a task to predict labels of instances between *real* and *simulated*, we can use the documents in `real-auto` and `sim-auto` as auxiliary domain data, while `real-aviation` and `sim-aviation` as target domain data.

Data Set	$\tilde{D}$	D
auto vs aviation	sim-auto & sim-aviation	real-auto & real-aviation
real vs simulated	real-aviation & sim-aviation	real-auto & sim-auto

Table 7.2. The description of SRAA data sets for cross-domain classification.

Data Set	$KL(\tilde{D}\|\|D)$	Documents $\|\tilde{X}\|$	$\|X\|$	SVM $\tilde{D} \to D$	D+CV
real vs simulated	1.161	8,000	8,000	0.266	0.032
auto vs aviation	1.126	8,000	8,000	0.228	0.033
rec vs talk	1.102	3,669	3,561	0.233	0.003
rec vs sci	1.021	3,961	3,965	0.212	0.007
comp vs talk	0.967	4,482	3,652	0.103	0.005
comp vs sci	0.874	3,930	4,900	0.317	0.012
comp vs rec	0.866	4,904	3,949	0.165	0.008
sci vs talk	0.854	3,374	3,828	0.226	0.009
orgs vs places	0.329	1,079	1,080	0.454	0.085
people vs places	0.307	1,239	1,210	0.266	0.113
orgs vs people	0.303	1,016	1,046	0.297	0.106

Table 7.3. Description of the data sets for cross-domain text classification, including errors given by SVM. "$\tilde{D} \to D$" means training on the auxiliary domain $\tilde{D}$ and testing on the target domain D; "D+CV" means 10-fold cross-validation using target domain data only. The performances are in test error rate. The table is quoted from [22].

Then, the data set `real vs simulated` is generated as shown in Table 7.2. As a result, all the data in the auxiliary domain data set are about autos, while all the data in the target domain set are about aviation. The `auto vs aviation` data set is generated in the similar way as shown in Table 7.2.

Reuters-21578 Reuters-21578 [49] is a well known test collections for evaluating text classification techniques. This dataset contains 5 top categories, among which `orgs`, `people` and `places` are three large ones. There is also a hierarchical structure which allows us to generate different data sets such as `orgs vs people`, `orgs vs places`, and `people vs places` for cross-domain classification in a similar way as what we have done on the 20-newsgroup and SRAA corpora.

Properties of the Data Sets Table 7.3 gives an overview of applying the basic SVM algorithm to the above data sets. The first three columns of the table show the statistical properties of the data sets. The first two data sets are from SRAA corpus. The next six are generated using

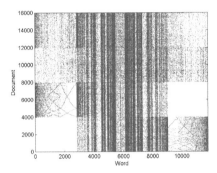

Figure 7.1. Document-word co-occurrence distribution on the `auto vs aviation` data set (quoted from [22]).

20-newsgroup data set. The last three are from Reuters-21578 test collection. To show the distribution differences between the training and testing data, KL-divergence values are calculated by $\text{KL}(\tilde{D}||D)$ on all the data set and are presented in the second column in the table, sorted in decreasing order from top down. Note that the Kullback-Leibler (KL) divergence [45] of two distributions of $\{p_i\}_{i=1}^{\ell}$ and $\{q_i\}_{i=1}^{\ell}$ is defined as

$$\text{KL}(\{p_i\}_{i=1}^{\ell}||\{q_i\}_{i=1}^{\ell}) = \sum_{i=1}^{\ell} p_i \ln(p_i/q_i) + (1-p_i)\ln((1-p_i)/(1-q_i)) \quad (7.3)$$

Here $\tilde{D}$ is the auxiliary domain data and D is the target domain data. It can be seen that the KL-divergence values for all the data sets are much larger than the identical-distribution case which has a KL value of nearly zero. The next column titled "Documents" shows the size of the data sets used.

Under the column titled "SVM", we show two groups of classification results in two sub-columns. First, "$\tilde{D} \to D$" denotes the test error rate obtained when a classifier trained based on the auxiliary domain data set $\tilde{D}$ is applied to the target domain data set D. The column titled "D+CV" denotes the best-case obtained by the corresponding classifier, where the best case is to conduct a 10-fold cross-validation on the target domain data set D using that classifier. Note that in obtaining the best case for each classifier, the training part is labeled data from D and the test part is also D, according to different folds, which gives the best result for that classifier. It can be found that the test error rates, given by SVM, in the case of "$\tilde{D} \to D$" is much worse than those in the case of "D+CV". This indicates that for these data sets, it is not suitable to apply traditional supervised classification algorithms.

Figure 7.1 shows the document-word co-occurrence distribution on the `auto vs aviation` data set. In this figure, documents 1 to 8000 are from target domain D, while documents 8001 to 16000 are from auxiliary domain $\tilde{D}$. The documents are order first by their domains ($\tilde{D}$ or D), and second by their categories (positive or negative). The words are sorted by $n_+(w)/n_-(w)$, where $n_+(w)$ and $n_-(w)$ represent the number of word positions w appears in positive and negative document, respectively. From Figure 7.1, it can be found that the distributions of auxiliary domain and target domain data are somewhat different, but almost consistent. That is, in general, the probabilities of a word belongs to a category in two domains do not differ very much.

2.2   Instance-based Transfer

One of the most intuitive methods is to transfer the knowledge between the domains by identifying a subset of source instances and insert them into the training set of the target domain data. We can observe that some instances in auxiliary domains are helpful for training the target domain model, while others may do harm to the target learning task. Thus, we need to select those that are useful and kick out those that are not. One effective way to achieve this is to perform instance weighting on the source domain data according to their importance to learning in the target domain. Taking SVM as an example, suppose that we have $\tilde{\ell}$ labeled data in the auxiliary domain, $\{(\tilde{x}_i, \tilde{y}_i)\}_{i=1}^{\tilde{\ell}}$ with $\tilde{x}_i \in \mathbb{R}^{d \times 1}$ and $\tilde{y}_i \in \{\pm 1\}$, which can be incorporated into the standard SVM in Eq.(7.1) as follows [96, 54],

$$\min_{w,\xi,\tilde{\xi}} \quad \frac{1}{2}\|w\|_2^2 + \lambda \sum_{i=1}^{\ell} \xi_i + \lambda \sum_{i=1}^{\tilde{\ell}} \tilde{\rho}_i \tilde{\xi}_i \qquad (7.4)$$
$$\text{s.t.} \quad y_i w^T x_i \geq 1 - \xi_i, \; \xi_i \geq 0, \; i=1,\ldots,\ell$$
$$\tilde{y}_i w^T \tilde{x}_i \geq 1 - \tilde{\xi}_i, \; \tilde{\xi}_i \geq 0, \; i=1,\ldots,\tilde{\ell}$$

where $\tilde{\rho}_i \in \mathbb{R}$ is the weight on the data point $(\tilde{x}_i, \tilde{y}_i)$ in the auxiliary domain, which can be estimated via some heuristics [54, 40] or optimization techniques [55]. We can see that the only difference between the standard SVM in Eq.(7.1) and SVM with instance-based transfer in Eq.(7.4) is from the loss function $\lambda \sum_{i=1}^{\tilde{\ell}} \tilde{\rho}_i \tilde{\xi}_i$ and its corresponding constraints defined on the labeled data in the auxiliary domain. The auxiliary data $\{(\tilde{x}_i, \tilde{y}_i)\}_{i=1}^{\tilde{\ell}}$ can be the support vectors of a trained SVM in the auxiliary domain [54, 40] or the whole auxiliary data set [96, 55]. Note that the approach in [96] uses a slightly different base model of linear programming SVM (LP-SVM) [59] instead of the standard SVM in Eq.(7.1). Similar techniques are also developed in the context of *incremental learning* [80], where support vectors of a learned SVM in the auxiliary domain are combined with labeled data in the target domain with different weight.

Research works have also been done in sample selection bias [35, 103] with $\tilde{P}_r(X) \neq P_r(X)$, $\tilde{P}_r(Y|X) \neq P_r(Y|X)$, and covariate shift [88] with $\tilde{P}_r(X) \neq P_r(X)$, $\tilde{P}_r(Y|X) = P_r(Y|X)$. For example, Bickel et al. [6] explicitly consider the difference of conditional distributions, $\tilde{P}_r(Y|X) \neq P_r(Y|X)$, and propose an alternative gradient descent algorithm to automatically learn the weight of the instances besides the model parameter of Logistic regression. Jiang and Zhai [39] propose a general instance weighting framework from a distribution view considering differences from both marginal distributions, $\tilde{P}_r(X) \neq P_r(X)$, and conditional distributions, $\tilde{P}_r(Y|X) \neq P_r(Y|X)$.

Xiang et al. proposed an algorithm known as BIG (Bridging Information Gap) [97], which is a framework to make use of a wolrdwide knowledge base (e.g. Wikipedia) as a bridge to achieve knowledge transfer from an auxiliary domain with labeled data to a target domain with test data. Specifically, Xiang et al. [97] study the *information gap* between the target domain and auxiliary domain, and propose a margin related criteria to sample unlabeled data from Wikipedia to fill the *information gap*, which enables more effective knowledge transfer. Transductive SVM [41] is then trained using the improved data pool of labeled data in the auxiliary domain, unlabeled data from Wikipedia, and test data in the target domain. The proposed framework is studied in cross-domain text classification, sentiment analysis and query classification [97].

2.3 Cross-Domain Ensemble Learning

It is well known in text mining that ensemble methods are very effective in gaining top performance. AdaBoost [31] and Bagging [11] are two of the most popular ensemble learning algorithms in machine learning. In this section, we show how to use AdaBoost [31] as a representative base algorithm to be extended for transfer learning.

The AdaBoost [31] algorithm, as shown in Figure 7.2, starts with a uniform distribution of instance weights. It then gradually *increases* the weights of misclassified instances and *decreases* the weights of correctly classified instances, in order to concentrate more on "hard-to-learn" instances to improve overall classification performance. AdaBoost [31] finally generates a set of weighted weak learners $\{(\alpha^t, \boldsymbol{w}^t)\}_{t=1}^{\Gamma}$, which

---

Input: labeled data in the target domain $\{(\boldsymbol{x}_i, y_i)\}_{i=1}^{\ell}$
Initialization: initialize instance weight $\{\rho_i^1\}_{i=1}^{\ell}$
For $t = 1 \ldots \Gamma$ Step 1. Train a model $\boldsymbol{w}^t$ using $\{(\boldsymbol{x}_i, y_i, \rho_i^t)\}_{i=1}^{\ell}$
Step 2. Calculate the error ϵ^t of $\boldsymbol{w}^t$ on $\{(\boldsymbol{x}_i, y_i, \rho_i^t)\}_{i=1}^{\ell}$
Step 3. Calculate the weight α^t from ϵ^t
Step 4. Update instance weight $\{\rho_i^{t+1}\}_{i=1}^{\ell}$ using α^t: *decrease* ρ_i^{t+1} for correct predictions in the target domain *increase* ρ_i^{t+1} for incorrect predictions in the target domain
Output: learned weight and weak models $\{(\alpha^t, \boldsymbol{w}^t)\}_{t=1}^{\Gamma}$.

Figure 7.2. The AdaBoost algorithm [31].

Input: labeled data in the target domain $\{(\boldsymbol{x}_i, y_i)\}_{i=1}^{\ell}$, labeled data in the auxiliary domain $\{(\tilde{\boldsymbol{x}}_i, \tilde{y}_i)\}_{i=1}^{\tilde{\ell}}$
Initialization: initialize instance weight $\{\rho_i^1\}_{i=1}^{\ell}$, $\{\tilde{\rho}_i^1\}_{i=1}^{\tilde{\ell}}$
For $t = 1 \ldots \Gamma$ Step 1. Train a model $\boldsymbol{w}^t$ using $\{(\boldsymbol{x}_i, y_i, \rho_i^t)\}_{i=1}^{\ell}$ and $\{(\tilde{\boldsymbol{x}}_i, \tilde{y}_i, \tilde{\rho}_i^t)\}_{i=1}^{\tilde{\ell}}$, which minimizes the weighted error only on labeled target data.
Step 2. Calculate the error ϵ^t of $\boldsymbol{w}^t$ on $\{(\boldsymbol{x}_i, y_i, \rho_i^t)\}_{i=1}^{\ell}$
Step 3. Calculate the weight α^t from ϵ^t
Step 4. Update instance weight $\{\rho_i^{t+1}\}_{i=1}^{\ell}$ and $\{\tilde{\rho}_i^{t+1}\}_{i=1}^{\tilde{\ell}}$ using α^t: *decrease* $\tilde{\rho}_i^{t+1}$ for incorrect predictions in the auxiliary domain *increase* ρ_i^{t+1} for incorrect predictions in the target domain
Output: learned weight and weak models $\{(\alpha^t, \boldsymbol{w}^t)\}_{t=\lceil \Gamma/2 \rceil}^{\Gamma}$.

Figure 7.3. The TrAdaBoost algorithm [23].

can be used to predict the label of an incoming instance $\boldsymbol{x}$,

$$f(\boldsymbol{x}) = \sum_{t=1}^{\Gamma} \alpha^t \boldsymbol{w}^{t^T} \boldsymbol{x}. \qquad (7.5)$$

TrAdaBoost In order to leverage auxiliary instances, various ensemble learning based transfer learning algorithms are proposed. TrAdaBoost [23] is a well-known instance-based transfer learning algorithm, which is shown in Figure 7.3. The idea behind this algorithm is to pick those auxilary instances which are similar to the target domain and ignore others. One observation is that we can integrate some unlabeled data from the target domain, if there are any [23]. Although the detailed implementations of "Steps 1, 2, 3" in TrAdaBoost [23] are all different from that of AdaBoost [31], an interesting part of TrAdaBoost [23] is in "Step 4", which has a different instance weight update strategy. TrAdaBoost [23] aims at transferring the most useful instances from the auxiliary domain. Thus it *decreases* the weight of misclassified instances in the auxiliary domain. Furthermore, as in transfer learning, we care more about the prediction performance on labeled data in the target domain, thus, TrAdaBoost [23] *increases* the weights of misclassified instances in the target domain.

TransferBoost [28] extends TrAdaBoost [23] by considering both an instance level and set-of-instances level weights of an auxiliary data. By doing so it allows the model to be more robust.

TrAdaBoost.R2 [69] studies the regression problem based on TrAdaBoost [23] and AdaBoost.R2 [27]. It achieves knowledge transfer from weighted instances from the auxiliary domain. An additional feature is that TrAdaBoost.R2 [69] proposes a two-stage instance weight update strategy in order to avoid model overfitting.

MultiSourceTrAdaBoost [102] extends TrAdaBoost [23] for *multiple* auxiliary data sources, aiming at alleviating negative transfer that may happen if we only have a single auxiliary data source. MultiSourceTrAdaBoost [102] replaces "Step 1" in the TrAdaBoost algorithm in Figure 7.3 as follows,

" Step 1. Train a model using $\{(\boldsymbol{x}_i, y_i, \rho_i^t)\}_{i=1}^{\ell}$ and labeled data from one of the n_a auxiliary data sources. Select one model from those n_a trained models that minimizes the weighted error on labeled data in the target domain. The selected model is denoted as $\boldsymbol{w}^t$. "

MultiSourceTrAdaBoost [102] combines the instance update strategy of TrAdaBoost [23] for auxiliary data and that of AdaBoost [31] for the target data.

TrAdaBoost [23] is further extended in [94] by adding an additional feature selection step. In [94], the authors replace "Step 1" of TrAdaBoost in Figure 7.3 with the following step, in order to select the most *discriminative* feature in each iteration:

"Step 1. Select a *single-feature* and train a *single-feature* model $\boldsymbol{w}^t$ using $\{(\boldsymbol{x}_i, y_i, \rho_i^t)\}_{i=1}^{\ell}$ and $\{(\tilde{\boldsymbol{x}}_i, \tilde{y}_i, \tilde{\rho}_i^t)\}_{i=1}^{\tilde{\ell}}$, which minimizes the weighted error on the labeled data in the target domain."

This feature selection approach based on transfer learning models achieves very promising results in lunar crater discovery applications, as reported in [94], which is quite general and can be adapted for text classification and ranking.

2.4 Feature-based Transfer Learning for Document Classification

Feature-based transfer is another main transfer learning paradigm, where algorithms are designed from the perspective of feature space transformation. Examples include feature replication [37, 46], feature projection [8, 7, 64], dimensionality reduction [63, 65, 66, 89, 21], feature correlation [76, 44, 107], feature subsetting [81], feature weighting [2], etc.

Feature Replication The feature replication or feature augmentation approach [37] is basically a pre-processing step on the labeled data $\{(\tilde{x}_i, \tilde{y}_i)\}_{i=1}^{\tilde{\ell}}$ in the auxiliary domain and labeled data $\{(x_i, y_i)\}_{i=1}^{\ell}$ in the target domain,

$$(\tilde{x}_i, \tilde{y}_i) \to ([\tilde{x}_i^T \, \tilde{x}_i^T \, \mathbf{0}^T]^T, \tilde{y}_i), \; i = 1, \ldots, \tilde{\ell}$$
$$(x_i, y_i) \to ([x_i^T \, \mathbf{0}^T \, x_i^T]^T, y_i), \; i = 1, \ldots, \ell$$

where the feature dimensionality is expanded from $\mathbb{R}^{d \times 1}$ to $\mathbb{R}^{3d \times 1}$, and standard supervised learning methods can then be used, e.g. SVM in Eq.(7.1).

As a follow-up work, Kumar et al. [46] further generalize the idea of *feature replication* via incorporating unlabeled data $\{x_i\}_{i=\ell+1}^{n}$ in the target domain,

$$x_i \to ([\mathbf{0}^T \, x^T \, -x^T]^T, +1), \; i = \ell+1, \ldots, n$$
$$x_i \to ([\mathbf{0}^T \, x^T \, -x^T]^T, -1), \; i = \ell+1, \ldots, n$$

where the processed data points are all with labels now.

The relationship of the feature replication method and the model-based transfer is discussed in [37] and some theoretical results of generalization bound are given in [46]. Feature replication approach have been successfully applied in cross-domain named entity recognition [37], part-of-speech tagging [37] and sentiment analysis [46].

Feature Projection Structured correspondence learning (SCL) [8] introduces the concept of *pivot features*, which possess high frequency and similar meaning in both auxiliary and target domains. Non-pivot features can be mapped to each other via the pivot features from the unlabeled data of both auxiliary and target domains. Learning in SCL [8] is based on the alternating structure optimization (ASO) algorithm [1]. Typically, SCL [8] goes through the following steps. First, it selects n_p pivot features. Then, for each pivot feature, SCL trains an SVM model in Eq.(7.1) using unlabeled data instances from both domains with labels indicating whether the pivot feature appears in the data instance. In this step it obtains n_p models such that $\mathbf{W} = [\boldsymbol{w}_j]_{j=1}^{n_p} \in \mathbb{R}^{d \times n_p}$. Third, SCL applies Singular Value Decomposition (SVD) to the model parameters $\mathbf{W}$, $[\mathbf{U}\,\Sigma\,\mathbf{V}^T] = \text{svd}(\mathbf{W})$, and it takes the top k columns of $\mathbf{U}$ as the projection matrix $\boldsymbol{\theta} \in \mathbb{R}^{d \times k}$. Finally, it obtains the following transformation for each labeled data point in the auxiliary domain,

$$(\tilde{\boldsymbol{x}}_i, \tilde{y}_i) \to ([\tilde{\boldsymbol{x}}_i^T\,\lambda\,(\boldsymbol{\theta}^T \tilde{\boldsymbol{x}}_i)^T]^T, \tilde{y}_i),\ i = 1, \ldots, \tilde{\ell} \qquad (7.6)$$

In the above equation, $\lambda > 0$ is a tradeoff parameter. The transformed data points is augmented with k additional features encoded with *structural correspondence* information between the features from auxiliary and target domains. With the transformed labeled data in the auxiliary domain, SCL can train a discriminative model, e.g. SVM in Eq.(7.1). For any future data instance $\boldsymbol{x}$, it is transformed via $\boldsymbol{x} \to [\boldsymbol{x}^T\,\lambda\,(\boldsymbol{\theta}^T\boldsymbol{x})^T]^T$ before $\boldsymbol{x}$ is classified by the learned model according to Eq.(7.2).

Blitzer et al. [7] successfully apply SCL [8] to cross-domain sentiment classification, and Prettenhofer and Stein [70, 71] extend SCL [8] with an additional cross-language translator to achieve knowledge transfer from English to German, French and Japanese for text classification and sentiment analysis. Pan et al. [64] propose a spectral learning algorithm for cross-domain sentiment classification using co-occurrence information from auxiliary-domain-specific, target-domain-specific and domain-independent features. They then align domain-specific features from both domains in a latent space via a learned projection matrix $\boldsymbol{\theta} \in \mathbb{R}^{k \times d}$. In some practical cases, the cross-domain sentiment and review classification performance of [64] is empirically shown to be superior to SCL [8] and other baselines.

Dimensionality Reduction In order to bridge two domains to enable knowledge transfer, Pan et al. [63] introduce *maximum mean discrepancy* (MMD) [10] as a distribution measurement of unlabeled data from auxiliary and target domains,

$$\|\frac{1}{\tilde{\ell}}\sum_{i=1}^{\tilde{\ell}} \phi(\tilde{\boldsymbol{x}}_i) - \frac{1}{n-\ell}\sum_{i=\ell+1}^{n} \phi(\boldsymbol{x})_i\|_2^2 \qquad (7.7)$$

which is used to minimize the distribution distance in a latent space. The MMD measurement is formulated as a kernel learning problem [63], which can be solved by SDP (semi-definite programming) by learning a kernel matrix $\mathbf{K} \in \mathbb{R}^{(\tilde{\ell}+n-\ell) \times (\tilde{\ell}+n-\ell)}$. Principal Component Analysis (PCA) is then applied on the learned kernel matrix $\mathbf{K}$ to obtain a low-dimensional representation,

$$[\mathbf{U} \Sigma \mathbf{U}^T] = \text{PCA}(\mathbf{K}), \ \mathbf{U} \in \mathbb{R}^{(\tilde{\ell}+n-\ell) \times k} \qquad (7.8)$$

As a result of the transformation, the original data can now be represented with a reduced dimensionality of $\mathbb{R}^{k \times 1}$ in the corresponding rows of $\mathbf{U}$. Standard supervised discriminative method such as SVM in Eq.(7.1) can be used to train a model using the transformed labeled data in the auxiliary domain.

Note that as a transductive learning method, the algorithm in [63] cannot be directly used to classify out-of-sample test data, which problem is addressed in [65, 66] by learning a projection matrix to minimize the MMD [10] criteria. Si et al. [89] introduce the Bregman divergence measurement as an additional regularization term in traditional dimensionality reduction techniques to bring two domains together in the latent space.

The EigenTransfer framework [21] introduces a novel approach to integrate co-occurrence information of instance-feature, instance-label from both auxiliary and target domains in a single graph. Normalized cut [85] is then adopted to learn a low-dimensional representation from the graph to replace original data in both target and auxiliary domains. Finally, standard supervised discriminative model, e.g. SVM in Eq.(7.1) is trained using the transformed labeled data in the auxiliary domain. An advantage of EigenTrasnfer is its ability to unify almost all available information in auxiliary and target domains, allowing the consideration of heterogenous feature and label space.

Feature Correlation Transferring *feature correlation* from auxiliary domains to a target domain is introduced in [76, 44, 107], where a feature-feature covariance matrix $\Sigma_0 \in \mathbb{R}^{d \times d}$ estimated from some auxiliary data is taken as an additional regularization term,

$$\lambda \boldsymbol{w}^T \Sigma_0^{-1} \boldsymbol{w} \qquad (7.9)$$

In this equation, the feature-feature correlation information is encoded in the covariance matrix Σ_0, which can be estimated from labeled or unlabeled data in auxiliary domains. Σ_0 will constrain the model parameters w_i and w_j of two high-correlated features i and j to be similar, and constrain the low-correlated features to be dissimilar. Such a regularization term is quite general and can be considered in various regularization based learning frameworks to incorporate the feature-feature correlation knowledge. Feature correlation is quite intuitive, and thus it has attracted several practical applications. For example, Raina et al. [76] transfer the feature-feature correlation knowledge from a newsgroups domain to a webpage domain for text classification, and Zhang et al. [107] study text classification with different time periods.

Feature Subsetting Feature selection via feature subsetting has been proposed for named entity recognition in CRF [81], which makes use of labeled data in auxiliary domains and the unlabeled data in the target domain. To illustrate the idea more clearly, we consider a simplified case of binary classification, where $y \in \{\pm 1\}$, instead of sequence labeling [81]. We re-write the optimization problem as follows,

$$\min_{\tilde{w},\tilde{\xi}} \quad \frac{1}{2}||\tilde{w}||_2^2 + \lambda \sum_{i=1}^{\tilde{\ell}} \tilde{\xi}_i \tag{7.10}$$

$$\text{s.t.} \quad \tilde{w}^T \phi(\tilde{x}_i, \tilde{y}_i) \geq 1 - \tilde{\xi}_i, \ \tilde{\xi}_i \geq 0, \ i = 1, \ldots, \tilde{\ell}$$

$$\sum_{k=1}^{d} |\tilde{w}_k|^\gamma dist(\tilde{E}_k, E_k) \leq \epsilon$$

Here we have:

$$E_k = \frac{1}{n-\ell} \sum_{i=\ell+1}^{n} (\phi_k(x_i, +1) P_r(+1|x_i, \tilde{w}) + \phi_k(x_i, -1) P_r(-1|x_i, \tilde{w}))$$

Furthermore, $\tilde{E}_k = \frac{1}{\tilde{\ell}} \sum_{i=1}^{\tilde{\ell}} \phi_k(\tilde{x}_i, \tilde{y}_i)$ are expected values of the kth feature of the joint feature mapping function $\phi(X, Y)$ in the target and auxiliary data, respectively, and $P_r(+1|x_i, \tilde{w}))$ and $P_r(-1|x_i, \tilde{w}))$ are the posterior probabilities of instance x_i belonging to classes $+1$ and -1, respectively. The parameter γ is used to control the sparsity of the model parameter $\tilde{w}$, which produces a subset of non-zeros; this is why it is called feature subsetting. The distance $dist(\tilde{E}_k, E_k)$ can be square distance $(\tilde{E}_k - E_k)^2$ for optimization simplicity [81], which is used to punish *highly distorted features* in order to bring two domains closer. The trained model $\tilde{w}$ will have better prediction performance in the target domain, especially when some features distort seriously in two domains.

Feature Weighting Arnold et al. [2] propose a feature weighting (or rescaling) approach to bridge two domains with labeled data in the auxiliary domain and test data in the target domain. Specifically, the kth feature of instance $\tilde{x}_j$ in the auxiliary domain is weighted as follows,

$$\tilde{x}_{j,k} \rightarrow \tilde{x}_{j,k} \frac{E_k(\tilde{y}_j|\mathbf{X}_U, \tilde{w})}{\tilde{E}_k(\tilde{y}_j|\tilde{D}_L)} \qquad (7.11)$$

where $E_k(\tilde{y}_j|\mathbf{X}_U, \tilde{w}) = \frac{1}{n-\ell} \sum_{i=\ell+1}^{n} x_{i,k} P_r(\tilde{y}_j|x_i, \tilde{w})$ is the expected value of kth feature (belonging to class $\tilde{y}_j$) in the target domain using the trained MaxEnt model $\tilde{w}$ from auxiliary domain. The value $\tilde{E}_k(\tilde{y}_j|\tilde{D}_L) = \frac{1}{\tilde{\ell}} \sum_{i=1}^{\tilde{\ell}} \tilde{x}_{i,k} \delta(\tilde{y}_j, \tilde{y}_i)$ represents the expected value of kth feature (belonging to class $\tilde{y}_j$) in the auxiliary domain. The weighted data (feature) in the auxiliary domain then have the same expected values of joint distribution about kth feature and class label y, $\tilde{E}_k(y|\tilde{D}_L) = E_k(y|\mathbf{X}_U, \tilde{w}), y \in \mathcal{Y}$. As a result, the two domains are brought closer together. Note that the learning procedure can be iterated with (a) learning $\tilde{w}$ and (b) weighting the feature, and that is the reason the model is called IFT (iterative feature transformation) [2]. Since $E_k(\tilde{y}_j|\mathbf{X}_U, \tilde{w})$ is only an estimated value, [2] adopts a common trick to preserve the original feature, which works quite well in NER problems. In particular,

$$\tilde{x}_{j,k} \rightarrow \lambda \tilde{x}_{j,k} + (1-\lambda) \tilde{x}_{j,k} \frac{E_k(\tilde{y}_j|\mathbf{X}_U, \tilde{w})}{\tilde{E}_k(\tilde{y}_j|\tilde{D}_L)} \qquad (7.12)$$

where $0 \leq \lambda \leq 1$ is a tradeoff parameter.

In the same spirit, other feature-based transfer methods have also been proposed, such as distance minimization [4], feature clustering [22, 57], kernel mapping [109], etc.

3. Heterogeneous Transfer Learning

Above we have surveyed transfer learning tasks where both the source and target domains are text documents in English. Recently, researchers in transfer learning area have started to consider transfer learning across heterogeneous feature and/or label space, namely heterogeneous transfer learning (HTL) [101]. HTL can be roughly categorized into two branches, (1) heterogeneous feature space, e.g. text and image space [20, 101, 87, 112, 72], English and Chinese vocabulary space [56, 105], and (2) heterogeneous label space, e.g. label space of Open Directory Project (ODP) [2] and query categories in KDDCUP 2005 [3] [84, 83], label space

---
[2]http://dmoz.org/
[3]http://www.sigkdd.org/kdd2005/kddcup.html

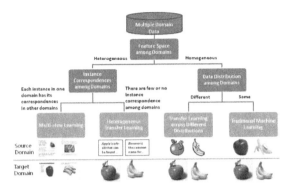

Figure 7.4. An intuitive illustration of heterogeneous transfer learning via classification of the images of `apple` and `banana` (quoted from [101]).

in Yahoo! Directory [4] and ODP [62], "head" (frequent) categories and "tail" (infrequent) categories in label-frequency distribution, and document categories in Newsgroup and categories in Wikipedia [98].

In Figure 7.4, we show different kinds of transfer learning and their relations to heterogeneous transfer learning. When features (or labels) are different between different domains, as shown on the left side of the figure, we have heterogeneous transfer learning when the instances in different domains lack a direct correspondence.

In general, recent works of heterogeneous transfer learning (HTL) can be classified into the following categories:

HTL for Image Classification An example is heterogeneous transfer learning for image classification [112]). In this work Zhu et al. consider how to use unlabeled text documents that we find on the Web to help boost the performance of image classification, by exploiting their semantic level similarity when the labeled images are in short supply.

HTL for Image Clustering An example of this direction is heterogeneous transfer learning for image clustering, where Yang et al. proposed a heterogenous transfer learning algorithm for image clustering by levering auxiliary annotated images ([101]).

HTL Across Different label Space An example is the cross-category learning in [73]. In this work, it adapts Adaboost with learning a feature correlation matrix to transfer knowledge from frequent categories to infrequent categories.

---
[4]http://dir.yahoo.com/

3.1 Heterogeneous Feature Space

Dai et al. [20] propose a novel approach named translated learning via risk minimization (TLRisk) to achieve knowledge transfer from text to image for image classification. The key idea is to bridge heterogeneous feature space in two domains via the co-occurrence information of image-feature and text-feature (or feature-level translator [20]) contained in the annotated auxiliary images, e.g. annotated images in Flickr. The knowledge in an auxiliary domain is then transferred along the path,

auxiliary-label → auxiliary-feature → target-feature → target-label

The TLRisk model is formulated in the risk minimization framework combining the feature translator and nearest neighbor learning, and is empirically studied for both image classification and cross-lingual (from English to German) text classification.

Yang et al. [101] proposed a probabilistic approach named annotation-based probabilistic latent semantic analysis (aPLSA) to achieve knowledge transfer from text to image for image clustering. Some multi-view auxiliary data of images and text is first transformed to a new representation of correlations between image-feature and text-feature. The aPLSA model [101] then discovers latent topics of image features of both multi-view data and target image data, which are shared as a bridge to bring two domains together.

Zhu et al. [112] propose a matrix-factorization based approach named heterogeneous transfer learning for image classification (HTLIC), in order to achieve knowledge transfer from text to image for image classification. To enable classification for out-of-sample images, HTLIC adopts collective matrix factorization [90] to learn an image-feature projection matrix from the auxiliary data of documents and the multi-view data, which is then used to obtain a new representation of the target images. Finally, a classifier (e.g. support vector machine) is trained using the newly projected target images.

Given a set of images to classify, we often need to have high-quality labeled images to train a classification model. However, obtaining the labeled image data is difficult and costly. In ([112]), the following question is addressed: is it possible for us to make use of some auxiliary labeled images and large quantities of unlabeled text to help us build a classifier? Suppose that we are given a few labeled image instances $\mathbf{X} = \{\mathbf{x}_i, y_i\}_{i=1}^n$ where $\mathbf{x}_i \in \mathbb{R}^d$ is an input vector of image features and y_i is the corresponding label of image i. We assume that the labeled images are not sufficient to build a high quality image classifier. In addition, we are also given a set of auxiliary annotated images $\mathbf{I} = \{\mathbf{z}_i, \mathbf{t}_i\}_{i=1}^{l}$ and a set of text documents $\mathbf{D} = \{\mathbf{d}_i\}_{i=1}^{k}$, where $\mathbf{z}_i \in \mathbb{R}^d$ is an image represented by a feature vector as $\mathbf{x}_i$, $\mathbf{t}_i \in \mathbb{R}^h$ is its corresponding vector of tags, and h is the number of tags. $\mathbf{d}_i \in \mathbb{R}^m$ is a document represented by a vector of bag-of-words, and l and k are the numbers of auxiliary images and documents respectively. The goal is to learn an accurate image classifier $f(\cdot)$ from $\mathbf{X}$, $\mathbf{I}$ and $\mathbf{D}$ to make predictions on $\mathbf{X}^*$, $f(\mathbf{X}^*)$.

We can make use of a set of auxiliary images $\mathbf{Z} \in \mathbb{R}^{l \times d}$ with their corresponding tags $\mathbf{T} \in \mathbb{R}^{l \times h}$ from Web resources such as Flickr. We can also easily obtain a set of unlabeled text documents $\mathbf{D} \in \mathbb{R}^{k \times m}$ via a search engine. To help build an image classifier, we need to first build some connection between image features and text features. To do this, we construct a two-layer bipartite graph based on images, tags and text documents. The top layer of the bipartite graph is used to represent the relationship between images and tags. Each image can be annotated by some tags, and some images may share one or multiple tags. If two images are annotated by some common tags, they tend to be related to each other semantically. Similarly, if two tags co-occur in annotations of shared images, they tend to be related to each other. This image-tag bipartite graph is represented by a tag matrix $\mathbf{T}$. The bottom layer bipartite graph is used to represent the relationship between tags and documents. If a tag occurs in a text document, there is an edge connecting the tag and the document.

Based on the bipartite graph, we can then learn semantic features for images by exploiting the relationship between images and text from the auxiliary sources. We first define a new matrix $\mathbf{G} = \mathbf{Z}^\top \mathbf{T} \in \mathbb{R}^{d \times h}$ to denote the correlation between low-level image features and annotations which can be referred to as high-level concepts. We then apply the Latent Semantic Analysis (LSA) as described in ([25]). Finally, we apply matrix factorization to decompose $\mathbf{G}$ into latent factor matrices as $\mathbf{G} = \mathbf{U}\mathbf{V}_1^\top$, where $\mathbf{U} \in \mathbb{R}^{d \times g}$, $\mathbf{V}_1 \in \mathbb{R}^{h \times g}$, and g is the number of latent factors. Then $\mathbf{u}_i$ can be treated as a latent semantic representation of the ith image low-level feature, and $\mathbf{v}_{1j}$ can be treated as a latent semantic representation of jth tag.

Zhu et al. [112] describe a method to learn the best decomposition via collective matrix factorization, as follows.

$$\min_{\mathbf{U},\mathbf{V},\mathbf{W}} \lambda \left\| \mathbf{G} - \mathbf{U}\mathbf{V}^\top \right\|_F^2 + (1-\lambda) \left\| \mathbf{F} - \mathbf{W}\mathbf{V}^\top \right\|_F^2 + R(\mathbf{U},\mathbf{V},\mathbf{W}), \quad (7.13)$$

where $0 \leq \lambda \leq 1$ is a tradeoff parameter to control the decomposition error between the two matrix factorizations, $||\cdot||_F$ denotes the Frobenius norm of matrix, and $R(\mathbf{U}, \mathbf{V}, \mathbf{W})$ is the regularization function to control the complexity of the latent matrices $\mathbf{U}$, $\mathbf{V}$ and $\mathbf{W}$. The optimization problem is an unconstrained non-convex optimization problem with three matrix variables **U**, **V** and **W**, thus it only has local optimal solutions. However, (7.13) is convex with respect to any one of the three matrices while fixing the other two. Thus a common technique to solve this kind of optimization problem is to fix two matrices and optimize the left one iteratively until the results converge.

Qi et al. [72] adopt Singular value thresholding (SVT) [14] and support vector machine to learn a low-rank feature-level correlation matrix (or translator) using multi-view data (text and images), and then the labels of text can be propagated (or transferred) to images through the feature-level translator. Note that both text and images are from the multi-view data, e.g. annotated images in Flickr. The problem setting of [72] is different from that of [112], where in [112] the multi-view data is considered as a bridge to transfer knowledge from auxiliary documents to target images, while in [72] the multi-view data is considered as a two-domain data sources in which knowledge is transferred from text to image.

3.2 Heterogeneous Label Space

Heterogeneous transfer learning may be needed when there is label mismatch between the auxiliary and target learning domains. The problem has attracted increasing attention in transfer learning, both in text mining and image understanding. One of the earliest works in matching labels across different classification domains is on the KDDCUP 2005 dataset, which task is to classify short, ambiguous and unlabeled search queries from a search engine log into a set of predefined categories. In [84, 83], Shen et al. considered the problem of quickly adapting a query categorization classifier when the target domain label taxonomy changes in the target learning domain. Their approach was to make the use of a large intermediate taxonomy to compile a collection of classifiers, and then adapt these classifiers to the new target label taxonomy in real time.

Shi et al. presented an approach to solving the label mismatch problem by a risk-sensitive spectral partition (RSP) algorithm [86]. A multi-task learning with mutual information (MTL-MI) is developed in [74] for learning the label correspondence.

Qi et al. [73] use quadratic programming (QP) to learn a diagonal feature-level correlation matrix on single-view data (e.g. image or video), and then use the AdaBoost framework to transfer knowledge from "head" (frequent) categories to "tail" (infrequent) categories, e.g.

from mountain images to castle images. In both [72] and [73], the decision function for a target instance is defined as a weighted linear combination of labels of auxiliary instances, where the *weight* is represented as the *similarity* of the target instance and any auxiliary instance estimated via learning a feature-level correlation matrix. The difference between [72] and [73] is that the former works on heterogeneous feature space (e.g. text and images) but same label space, while the latter focus on same feature space (e.g. images) but heterogeneous label space (e.g. semantically related categories of mountain and castle).

Rohrbach et al. [79] propose to automatically mine semantic relationships between class labels (or equivalently class attributes) from linguistic data (e.g. wikepedia, WordNet, Yahoo image, Flickr), which can be considered as a label-level translator. The trained classifiers of auxiliary classes can then be reused by target domain (different) classes through the label-level translator and Bayesian rules. The proposed approach allows different label space but assuming same feature space, and is empirically verified for image classification. A follow-up work [78] conducts extensive and in-depth study of transfer learning for image classification.

Xiang et al. [98] propose a novel approach named source-selection-free transfer learning (SSFTL) to achieve knowledge transfer from some large-scale auxiliary data set, e.g. Wikipedia, which does not require practitioners to manually select some particular part of auxiliary data to transfer from. The main idea is to bridge large-scale auxiliary label space and target label space via social tagging data, e.g. Flick. Specifically, each label (*scalar*) is represented as a *vector* in a latent space, where two vectors are similar if the corresponding labels are semantically correlated. An additional advantage of SSFTL is that the training procedure of auxiliary classifiers can be implemented offline, which makes the whole learning approach very efficient.

There are also some other heterogeneous transfer learning settings in different data domains and scenarios e.g. target domains with few instances [50], transfer from text to video [51], etc.

3.3 Summary

Heterogeneous transfer learning is mainly based on feature-level translator and label-level translator, which bridges heterogeneous feature space and heterogeneous label space of two domains. The techniques of heterogeneous transfer learning and transfer learning methods in previous sections are complementary, which enables knowledge transfer in a much wider application scope with very little limitation.

Table 7.4. Learning paradigms and techniques. The notation "req." means that the test data are *required* during model training, and "✓" means the corresponding data are available to the learner. $\tilde{D}_L$ and $\tilde{D}_U$ are labeled and unlabeled data in an auxiliary domain. D_L, D_U and D_T are labeled, unlabeled and test data in the target domain. Unsupervised and supervised transfer learning are categorized by the availability of labeled data in the target domain.

	Learning Paradigm	Auxiliary		Target			Learning Technique
		$\tilde{D}_L$	$\tilde{D}_U$	D_L	D_U	D_T	
ML	Unsupervised	N/A				req.	Spectral clustering [58], etc.
	Transductive			✓		req.	TSVM [41], etc.
	Supervised			✓			AdaBoost [31], etc.
	Semi-supervised			✓	✓		SSL [111], etc.
TL	*Unsupervised*		✓			req.	STC [24], etc.
		✓				req.	LWE [32], etc.
		✓			✓		SCL [8], etc.
	Supervised	✓		✓			MTL [30], etc.
		✓		✓	✓		TrAdaBoost [23], etc.
		✓	✓	✓	✓	req.	EigenTransfer [21], etc.
			✓	✓			STL [75], etc.
	Heterogeneous	across different feature space					Translated learning [20], aPLSA [101], TTI [72], HTLIC [112], etc.
		across different label space					RSP [86], CCTL [73], Semantic relatedness [79], SSFTL [98], etc.

4. Discussion

Above we have seen that there are several important applications of transfer learning. What insights can be gained from these applications and extensions on transfer learning? Below, we consider a few such issues.

What, How and When to Transfer As pointed out by Pan and Yang [67], there are three fundamental questions in transfer learning, namely "what to transfer", "how to transfer" and "when to transfer". We have answered the "what to transfer" question from two perspectives, (1) instance-based transfer and (2) feature-based transfer, where the corresponding knowledge are selected and weighted instances and

Table 7.5. Applications in text mining.

Application	Transfer learning work
Text classification	[29, 76, 22, 107, 63, 65, 66, 89, 21, 97, 70, 71, 57], etc.
Sentiment analysis	[46, 7, 71, 97, 64], etc.
Named entity recognition	[39, 2, 37, 81], etc.
Part-of-speech tagging	[8, 39, 4, 37], etc.
Relation extraction	[38], etc.

learned or transformed features. The "how to transfer" question [67] is quite related to "what to transfer", and we have surveyed *instance weighting*, *feature projection* and other various techniques adopted in different works to achieve knowledge transfer. The "when to transfer" question [67] is related to negative transfer, cross-domain validation and transfer bounds, where some works focus on empirical study to avoid negative transfer [28, 102, 16]. Some research works also focus on theoretical developments of transfer learning, such as [4, 53, 23, 60, 109, 46]. In addition, researchers have also proposed cross-domain cross-validation strategies [110, 12] for text mining and other learning tasks.

Learning Paradigms and Techniques Transfer learning can be considered as a new *learning paradigm*. One perspective is to consider transfer learning as an over-arching framework that includes the traditional learning as a special case, as shown in Table 7.4. Here we can see that traditional machine learning (ML) methods do not consider data from auxiliary domains; instead and they study the learning problems under the same data distribution $P_r(X,Y)$. In contrast, transfer learning goes beyond the learning paradigm via transferring knowledge from auxiliary domains with different distribution $\tilde{P}_r(X,Y) \neq P_r(X,Y)$.

Text Mining Applications As we surveyed so far, transfer learning have been wildly adopted in various text mining applications; a summary can be found in Table 7.5. Note that many transfer learning methods surveyed in previous sections have been applied to non-text mining applications as well; e.g. in speech recognition, in image and video analysis, etc.

5. Conclusions

In this chapter, we have focused on transfer learning approaches for text mining. Specifically, we have reviewed transfer learning techniques in text related classification tasks, including discriminative learning and ensemble learning, and heterogeneous transfer. We have considered these learning approaches from two perspectives, namely, (1) instance-based transfer and (2) feature-based transfer. Most of the surveyed transfer learning methods are proposed or can be applied in text mining applications, e.g. text classification, sentiment analysis, POS tagging, NER and relation extraction. In addition, the introduced heterogeneous transfer techniques can explore the knowledge in text to help the learning task in other domain, such as image classification.

A current research issue is how to apply transfer learning to the learning-to-rank framework [43, 13, 99], where the ranking model in the target domain may benefit from knowledge transferred from auxiliary domains. In this area, works include model-based transfer [34], instance-based transfer [19, 33, 15] and feature-based transfer [92, 19, 3], which extend the pairwise ranking algorithms of RankSVM [43], RankNet [13], or list-wise ranking model of AdaRank [99]. We expect to see much research progress in this new direction, e.g. generalizations of learning to rank to heterogeneous settings [101].

In the future, we expect to see more extensive applications of transfer learning in text mining, where the concept of "text" can be more general. For example, we expect to see transfer learning methods to be applied to analyzing microblogging contents and structure, in association with social network mining. We also expect to see more cross-domain transfer learning approaches, for knowledge transfer between very different domains, e.g., text and videos, etc.

References

[1] Rie Kubota Ando and Tong Zhang. A framework for learning predictive structures from multiple tasks and unlabeled data. *J. Mach. Learn. Res.*, 6:1817–1853, December 2005.

[2] Andrew Arnold, Ramesh Nallapati, and William W. Cohen. A comparative study of methods for transductive transfer learning. In *Proceedings of the Seventh IEEE International Conference on Data Mining Workshops*, ICDMW '07, pages 77–82, Washington, DC, USA, 2007. IEEE Computer Society.

[3] Jing Bai, Ke Zhou, Guirong Xue, Hongyuan Zha, Gordon Sun, Belle Tseng, Zhaohui Zheng, and Yi Chang. Multi-task learning for learning to rank in web search. In *Proceeding of the 18th ACM conference on Information and knowledge management*, CIKM '09, pages 1549–1552, New York, NY, USA, 2009. ACM.

[4] Shai Ben-David, John Blitzer, Koby Crammer, Fernando Pereira, and Artur Dubrawski. Analysis of representations for domain adaptation. In *NIPS*, 2006.

[5] Adam L. Berger, Vincent J. Della Pietra, and Stephen A. Della Pietra. A maximum entropy approach to natural language processing. *Comput. Linguist.*, 22:39–71, March 1996.

[6] Steffen Bickel, Michael Brückner, and Tobias Scheffer. Discriminative learning for differing training and test distributions. In *Proceedings of the 24th international conference on Machine learning*, ICML '07, pages 81–88, New York, NY, USA, 2007. ACM.

[7] John Blitzer, Mark Dredze, and Fernando Pereira. Biographies, bollywood, boom-boxes and blenders: Domain adaptation for sentiment classification. In *Association for Computational Linguistics*, Prague, Czech Republic.

[8] John Blitzer, Ryan McDonald, and Fernando Pereira. Domain adaptation with structural correspondence learning. In *Proceedings of the 2006 Conference on Empirical Methods in Natural Language Processing*, EMNLP '06, pages 120–128, Stroudsburg, PA, USA, 2006. Association for Computational Linguistics.

[9] Avrim Blum and Tom Mitchell. Combining labeled and unlabeled data with co-training. In *Proceedings of the eleventh annual conference on Computational learning theory*, COLT' 98, pages 92–100, New York, NY, USA, 1998. ACM.

[10] Karsten M. Borgwardt, Arthur Gretton, Malte J. Rasch, Hans-Peter Kriegel, Bernhard Schölkopf, and Alexander J. Smola. Integrating structured biological data by kernel maximum mean discrepancy. In *Proceedings of the 14th International Conference on Intelligent Systems for Molecular Biology*, pages 49–57, Fortaleza, Brazil, August 2006.

[11] Leo Breiman. Bagging predictors. *Mach. Learn.*, 24:123–140, August 1996.

[12] Lorenzo Bruzzone and Mattia Marconcini. Domain adaptation problems: A dasvm classification technique and a circular validation strategy. *IEEE Trans. Pattern Anal. Mach. Intell.*, 32(5):770–787, 2010.

[13] Chris Burges, Tal Shaked, Erin Renshaw, Ari Lazier, Matt Deeds, Nicole Hamilton, and Greg Hullender. Learning to rank using gradient descent. In *Proceedings of the 22nd international conference on Machine learning*, ICML '05, pages 89–96, New York, NY, USA, 2005. ACM.

[14] Jian-Feng Cai, Emmanuel J. Candès, and Zuowei Shen. A singular value thresholding algorithm for matrix completion. *SIAM J. on Optimization*, 20:1956–1982, March 2010.

[15] Peng Cai and Aoying Zhou. A novel framework for ranking model adaptation. *Web Information Systems and Applications Conference*, 0:149–154, 2010.

[16] Bin Cao, Sinno Jialin Pan, Yu Zhang, Dit-Yan Yeung, and Qiang Yang. Adaptive transfer learning. In *AAAI*, 2010.

[17] Rich Caruana. Multitask learning: A knowledge-based source of inductive bias. In *ICML*, pages 41–48, 1993.

[18] Olivier Chapelle, Bernhard Schölkopf, and Alexander Zien. *Semi-Supervised Learning (Adaptive Computation and Machine Learning)*. The MIT Press, 2006.

[19] Depin Chen, Yan Xiong, Jun Yan, Gui-Rong Xue, Gang Wang, and Zheng Chen. Knowledge transfer for cross domain learning to rank. *Inf. Retr.*, 13:236–253, June 2010.

[20] Wenyuan Dai, Yuqiang Chen, Gui-Rong Xue, Qiang Yang, and Yong Yu. Translated learning: Transfer learning across different feature spaces. In *NIPS*, pages 353–360, 2008.

[21] Wenyuan Dai, Ou Jin, Gui-Rong Xue, Qiang Yang, and Yong Yu. Eigentransfer: a unified framework for transfer learning. In *Proceedings of the 26th Annual International Conference on Machine Learning*, ICML '09, pages 193–200, New York, NY, USA, 2009. ACM.

[22] Wenyuan Dai, Gui-Rong Xue, Qiang Yang, and Yong Yu. Co-clustering based classification for out-of-domain documents. In *Proceedings of the 13th ACM SIGKDD international conference on Knowledge discovery and data mining*, KDD '07, pages 210–219, New York, NY, USA, 2007. ACM.

[23] Wenyuan Dai, Qiang Yang, Gui-Rong Xue, and Yong Yu. Boosting for transfer learning. In *Proceedings of the 24th international conference on Machine learning*, ICML '07, pages 193–200, New York, NY, USA, 2007. ACM.

[24] Wenyuan Dai, Qiang Yang, Gui-Rong Xue, and Yong Yu. Self-taught clustering. In *Machine Learning, Proceedings of the Twenty-Fifth International Conference (ICML 2008), Helsinki, Finland, June 5-9, 2008*, volume 307, pages 200–207. ACM, 2008.

[25] Scott Deerwester, Susan T. Dumais, George W. Furnas, Thomas K. Landauer, and Richard Harshman. Indexing by latent

semantic analysis. *Journal of the American Society for Information Science*, 41:391–407, 1990.

[26] Chuong B. Do and Andrew Y. Ng. Transfer learning for text classification. In *NIPS*, 2005.

[27] Harris Drucker. Improving regressors using boosting techniques. In *Proceedings of the Fourteenth International Conference on Machine Learning*, ICML '97, pages 107–115, San Francisco, CA, USA, 1997. Morgan Kaufmann Publishers Inc.

[28] Eric Eaton and Marie desJardins. Set-based boosting for instance-level transfer. In *Proceedings of the 2009 IEEE International Conference on Data Mining Workshops*, ICDMW '09, pages 422–428, Washington, DC, USA, 2009. IEEE Computer Society.

[29] Eric Eaton, Marie Desjardins, and Terran Lane. Modeling transfer relationships between learning tasks for improved inductive transfer. In *Proceedings of the 2008 European Conference on Machine Learning and Knowledge Discovery in Databases - Part I*, ECML PKDD '08, pages 317–332, Berlin, Heidelberg, 2008. Springer-Verlag.

[30] Theodoros Evgeniou and Massimiliano Pontil. Regularized multi-task learning. In *Proceedings of the tenth ACM SIGKDD international conference on Knowledge discovery and data mining*, KDD '04, pages 109–117, New York, NY, USA, 2004. ACM.

[31] Yoav Freund and Robert E. Schapire. A decision-theoretic generalization of on-line learning and an application to boosting. *J. Comput. Syst. Sci.*, 55(1):119–139, 1997.

[32] Jing Gao, Wei Fan, Jing Jiang, and Jiawei Han. Knowledge transfer via multiple model local structure mapping. In *Proceeding of the 14th ACM SIGKDD international conference on Knowledge discovery and data mining*, KDD '08, pages 283–291, New York, NY, USA, 2008. ACM.

[33] Wei Gao, Peng Cai, Kam-Fai Wong, and Aoying Zhou. Learning to rank only using training data from related domain. In *Proceeding of the 33rd international ACM SIGIR conference on Research and development in information retrieval*, SIGIR '10, pages 162–169, New York, NY, USA, 2010. ACM.

[34] Bo Geng, Linjun Yang, Chao Xu, and Xian-Sheng Hua. Ranking model adaptation for domain-specific search. In *Proceeding of the 18th ACM conference on Information and knowledge management*, CIKM '09, pages 197–206, New York, NY, USA, 2009. ACM.

[35] James J Heckman. Sample selection bias as a specification error. *Econometrica*, 47(1):153–61, January 1979.

[36] David W. Hosmer and Stanley Lemeshow. *Applied logistic regression*. Wiley-Interscience, 2 edition, 2000.

[37] Hal Daumé III. Frustratingly easy domain adaptation. In *ACL*, 2007.

[38] Jing Jiang. Multi-task transfer learning for weakly-supervised relation extraction. In *Proceedings of the Joint Conference of the 47th Annual Meeting of the ACL and the 4th International Joint Conference on Natural Language Processing of the AFNLP: Volume 2 - Volume 2*, ACL '09, pages 1012–1020, Stroudsburg, PA, USA, 2009. Association for Computational Linguistics.

[39] Jing Jiang and ChengXiang Zhai. Instance weighting for domain adaptation in nlp. In *ACL*, 2007.

[40] Wei Jiang, Eric Zavesky, Shih-Fu Chang, and Alexander C. Loui. Cross-domain learning methods for high-level visual concept classification. In *ICIP*, pages 161–164, 2008.

[41] Thorsten Joachims. Transductive inference for text classification using support vector machines. In *Proceedings of the Sixteenth International Conference on Machine Learning*, ICML '99, pages 200–209, San Francisco, CA, USA, 1999. Morgan Kaufmann Publishers Inc.

[42] Thorsten Joachims. *Learning to Classify Text Using Support Vector Machines: Methods, Theory and Algorithms*. Kluwer Academic Publishers, Norwell, MA, USA, 2002.

[43] Thorsten Joachims. Optimizing search engines using clickthrough data. In *Proceedings of the eighth ACM SIGKDD international conference on Knowledge discovery and data mining*, KDD '02, pages 133–142, New York, NY, USA, 2002. ACM.

[44] Eyal Krupka and Naftali Tishby. Incorporating prior knowledge on features into learning. In *Proceedings of the 11th International Conference on Artificial Intelligence and Statistics*, San Juan, Puerto Rico, 2007.

[45] S. Kullback and R. A. Leibler. On information and sufficiency. *Annals of Mathematical Statistics*, 22(1):79–86, 1951.

[46] Abhishek Kumar, Avishek Saha, and Hal Daumé III. A co-regularization based semi-supervised domain adaptation. In *Proceedings of the Conference on Neural Information Processing Systems (NIPS)*, Vancouver, Canada, 2010.

[47] John D. Lafferty, Andrew McCallum, and Fernando C. N. Pereira. Conditional random fields: Probabilistic models for segmenting and labeling sequence data. In *Proceedings of the Eighteenth International Conference on Machine Learning*, ICML '01, pages 282–289, San Francisco, CA, USA, 2001. Morgan Kaufmann Publishers Inc.

[48] Ken Lang. Newsweeder: Learning to filter netnews. In *Proceedings of the Twelfth International Conference on Machine Learning*, 1995.

[49] David Dolan Lewis. Reuters-21578 test collection. http://www.daviddlewis.com/.

[50] Fei-Fei Li, Fergus Rob, and Perona Pietro. One-shot learning of object categories. *IEEE Trans. Pattern Anal. Mach. Intell.*, 28:594–, April 2006.

[51] Liangda Li, Ke Zhou, Gui-Rong Xue, Hongyuan Zha, and Yong Yu. Video summarization via transferrable structured learning. In *Proceedings of the 20th international conference on World wide web*, WWW '11, pages 287–296, New York, NY, USA, 2011. ACM.

[52] Xiao Li and Jeff Bilmes. Regularized adaptation of discriminative classifiers. In *Proc. IEEE Intl. Conf. on Acoustics, Speech, and Signal Processing*, Toulouse, France, May 2006.

[53] Xiao Li and Jeff Bilmes. A bayesian divergence prior for classifier adaptation. In *Eleventh International Conference on Artificial Intelligence and Statistics (AISTATS-2007)*, March 2007.

[54] Xiao Li, Jeff Bilmes, and Joh Malkin. Maximum margin learning and adaptation of MLP classifers. September 2005.

[55] Xuejun Liao, Ya Xue, and Lawrence Carin. Logistic regression with an auxiliary data source. In *Proceedings of the 22nd international conference on Machine learning*, ICML '05, pages 505–512, New York, NY, USA, 2005. ACM.

[56] Xiao Ling, Gui-Rong Xue, Wenyuan Dai, Yun Jiang, Qiang Yang, and Yong Yu. Can chinese web pages be classified with english data source? In *WWW*, pages 969–978, 2008.

[57] Mingsheng Long, Wei Cheng, Xiaoming Jin, Jianmin Wang, and Dou Shen. Transfer learning via cluster correspondence inference. In *ICDM*, pages 917–922, 2010.

[58] Ulrike Luxburg. A tutorial on spectral clustering. *Statistics and Computing*, 17:395–416, December 2007.

[59] Olvi L. Mangasarian. Generalized support vector machines. Technical report, Computer Sciences Department, University of Wisconsin, 1998.

[60] Yishay Mansour, Mehryar Mohri, and Afshin Rostamizadeh. Domain adaptation with multiple sources. In *NIPS*, 2008.

[61] Andrew Kachites McCallum. Simulated/real/aviation/auto usenet data. http://www.cs.umass.edu/~mccallum/code-data.html.

[62] Tiberio Caetano S. V. N. Vishwanathan Novi Quadrianto, Alex Smola and James Petterson. Multitask learning without label correspondences. In *NIPS*, 2010.

[63] Sinno Jialin Pan, James T. Kwok, and Qiang Yang. Transfer learning via dimensionality reduction. In *AAAI*, pages 677–682, 2008.

[64] Sinno Jialin Pan, Xiaochuan Ni, Jian-Tao Sun, Qiang Yang, and Zheng Chen. Cross-domain sentiment classification via spectral feature alignment. In *WWW*, pages 751–760, 2010.

[65] Sinno Jialin Pan, Ivor W. Tsang, James T. Kwok, and Qiang Yang. Domain adaptation via transfer component analysis. In *IJCAI*, pages 1187–1192, 2009.

[66] Sinno Jialin Pan, Ivor W. Tsang, James T. Kwok, and Qiang Yang. Domain adaptation via transfer component analysis. *IEEE Transactions on Neural Networks*, 22(2):199–210, 2011.

[67] Sinno Jialin Pan and Qiang Yang. A survey on transfer learning. *IEEE Transactions on Knowledge and Data Engineering*, 22(10):1345–1359, October 2010.

[68] Bo Pang and Lillian Lee. Opinion mining and sentiment analysis. *Found. Trends Inf. Retr.*, 2:1–135, January 2008.

[69] David Pardoe and Peter Stone. Boosting for regression transfer. In *ICML*, pages 863–870, 2010.

[70] Peter Prettenhofer and Benno Stein. Cross-language text classification using structural correspondence learning. In *ACL*, pages 1118–1127, 2010.

[71] Peter Prettenhofer and Benno Stein. Cross-lingual adaptation using structural correspondence learning. *ACM TIST*, 3(1), 2012.

[72] Guo-Jun Qi, Charu Aggarwal, and Thomas Huang. Towards semantic knowledge propagation from text corpus to web images. In *Proceedings of the 20th international conference on World wide web*, WWW '11, pages 297–306, New York, NY, USA, 2011. ACM.

[73] Guo-Jun Qi, Charu Aggarwal, Yong Rui, Qi Tian, Shiyu Chang, and Thomas Huang. Towards cross-category knowledge propagation for learning visual concepts. In *CVPR*, 2011.

[74] Novi Quadrianto, Alex J. Smola, Tiberio S. Caetano, S.V.N. Vishwanathan, and James Petterson. Multitask learning without label correspondences. In *NIPS 23*, 2010.

[75] Rajat Raina, Alexis Battle, Honglak Lee, Benjamin Packer, and Andrew Y. Ng. Self-taught learning: transfer learning from unlabeled data. In *Proceedings of the 24th international conference on Machine learning*, ICML '07, pages 759–766, New York, NY, USA, 2007. ACM.

[76] Rajat Raina, Andrew Y. Ng, and Daphne Koller. Constructing informative priors using transfer learning. In *Proceedings of the 23rd international conference on Machine learning*, ICML '06, pages 713–720, New York, NY, USA, 2006. ACM.

[77] Adwait Ratnaparkhi. A maximum entropy model for part-of-speech tagging. In *Proceedings of the Conference on Empirical Methods in Natural Language Processing*, April 1996.

[78] Marcus Rohrbach, Michael Stark, and Bernt Schiele. Evaluating knowledge transfer and zero-shot learning in a large-scale setting. In *CVPR*, 2011.

[79] Marcus Rohrbach, Michael Stark, György Szarvas, Iryna Gurevych, and Bernt Schiele. What helps where - and why? semantic relatedness for knowledge transfer. In *CVPR*, pages 910–917, 2010.

[80] Stefan Rüping. Incremental learning with support vector machines. In *Proceedings of the 2001 IEEE International Conference on Data Mining*, ICDM '01, pages 641–642, Washington, DC, USA, 2001. IEEE Computer Society.

[81] Sandeepkumar Satpal and Sunita Sarawagi. Domain adaptation of conditional probability models via feature subsetting. In *Proceedings of the 11th European conference on Principles and Practice of Knowledge Discovery in Databases*, PKDD 2007, pages 224–235, Berlin, Heidelberg, 2007. Springer-Verlag.

[82] Bernhard Scholkopf and Alexander J. Smola. *Learning with Kernels: Support Vector Machines, Regularization, Optimization, and Beyond*. MIT Press, Cambridge, MA, USA, 2001.

[83] Dou Shen, Rong Pan, Jian-Tao Sun, Jeffrey Junfeng Pan, Kangheng Wu, Jie Yin, and Qiang Yang. Query enrichment for web-query classification. *ACM Trans. Inf. Syst.*, 24:320–352, July 2006.

[84] Dou Shen, Jian-Tao Sun, Qiang Yang, and Zheng Chen. Building bridges for web query classification. In *Proceedings of the 29th*

annual international ACM SIGIR conference on Research and development in information retrieval, SIGIR '06, pages 131–138, New York, NY, USA, 2006. ACM.

[85] Jianbo Shi and Jitendra Malik. Normalized cuts and image segmentation. *IEEE Trans. Pattern Anal. Mach. Intell.*, 22:888–905, August 2000.

[86] Xiaoxiao Shi, Wei Fan, Qiang Yang, and Jiangtao Ren. Relaxed transfer of different classes via spectral partition. In *ECML/PKDD*, 2009.

[87] Xiaoxiao Shi, Qi Liu, Wei Fan, Philip S. Yu, and Ruixin Zhu. Transfer learning on heterogenous feature spaces via spectral transformation. In *ICDM*, pages 1049–1054, 2010.

[88] Hidetoshi Shimodaira. Improving predictive inference under covariate shift by weighting the log-likelihood function. *Journal of Statistical Planning and Inference*, 90(2):227–244, 2000.

[89] Si Si, Dacheng Tao, and Bo Geng. Bregman divergence-based regularization for transfer subspace learning. *IEEE Trans. Knowl. Data Eng.*, 22(7):929–942, 2010.

[90] Ajit P. Singh and Geoffrey J. Gordon. Relational learning via collective matrix factorization. In *KDD*, pages 650–658, 2008.

[91] Simon Tong and Daphne Koller. Support vector machine active learning with applications to text classification. *J. Mach. Learn. Res.*, 2:45–66, March 2002.

[92] Bo Wang, Jie Tang, Wei Fan, Songcan Chen, Zi Yang, and Yanzhu Liu. Heterogeneous cross domain ranking in latent space. In *Proceeding of the 18th ACM conference on Information and knowledge management*, CIKM '09, pages 987–996, New York, NY, USA, 2009. ACM.

[93] Hua-Yan Wang, Vincent Wenchen Zheng, Junhui Zhao, and Qiang Yang. Indoor localization in multi-floor environments with reduced effort. In *PerCom*, pages 244–252, 2010.

[94] Yang Mu Lourenco Bandeira Ricardo Ricardo Youxi Wu Zhenyu Lu Tianyu Cao Xindong Wu Wei Ding, Tomasz F. Stepinski. Sub-kilometer crater discovery with boosting and transfer learning. *ACM TIST*, x(x), 201x.

[95] Philip C. Woodland. Speaker adaptation for continuous density hmms: a review. In *ISCA Tutorial and Research Workshop (ITRW) on Adaptation Methods for Speech Recognition*, ITRW' 01, pages 29–30, August 2001.

[96] Pengcheng Wu and Thomas G. Dietterich. Improving svm accuracy by training on auxiliary data sources. In *Proceedings of the twenty-first international conference on Machine learning*, ICML '04, pages 110–, New York, NY, USA, 2004. ACM.

[97] Evan Wei Xiang, Bin Cao, Derek Hao Hu, and Qiang Yang. Bridging domains using world wide knowledge for transfer learning. *IEEE Trans. Knowl. Data Eng.*, 22(6):770–783, 2010.

[98] Evan Wei Xiang, Sinno Jialin Pan, Weik Pan, Qiang Yang, and Jian Su. Source-free transfer learning. In *IJCAI*, 2011.

[99] Jun Xu and Hang Li. Adarank: a boosting algorithm for information retrieval. In *Proceedings of the 30th annual international ACM SIGIR conference on Research and development in information retrieval*, SIGIR '07, pages 391–398, New York, NY, USA, 2007. ACM.

[100] Jun Yang, Rong Yan, and Alexander G. Hauptmann. Cross-domain video concept detection using adaptive svms. In *Proceedings of the 15th international conference on Multimedia*, MULTIMEDIA '07, pages 188–197, New York, NY, USA, 2007. ACM.

[101] Qiang Yang, Yuqiang Chen, Gui-Rong Xue, Wenyuan Dai, and Yong Yu. Heterogeneous transfer learning for image clustering via the social web. In *Proceedings of the Joint Conference of the 47th Annual Meeting of the ACL and the 4th International Joint Conference on Natural Language Processing of the AFNLP: Volume 1 - Volume 1*, ACL '09, pages 1–9, Stroudsburg, PA, USA, 2009. Association for Computational Linguistics.

[102] Yi Yao and Gianfranco Doretto. Boosting for transfer learning with multiple sources. In *CVPR*, pages 1855–1862, 2010.

[103] Bianca Zadrozny. Learning and evaluating classifiers under sample selection bias. In *Proceedings of the twenty-first international conference on Machine learning*, ICML '04, pages 114–, New York, NY, USA, 2004. ACM.

[104] Dmitry Zelenko, Chinatsu Aone, and Anthony Richardella. Kernel methods for relation extraction. *J. Mach. Learn. Res.*, 3:1083–1106, March 2003.

[105] Duo Zhang, Qiaozhu Mei, and ChengXiang Zhai. Cross-lingual latent topic extraction. In *ACL*, pages 1128–1137, 2010.

[106] Tong Zhang and David Johnson. A robust risk minimization based named entity recognition system. In *Proceedings of the seventh conference on Natural language learning at HLT-NAACL 2003 -*

Volume 4, CONLL '03, pages 204–207, Stroudsburg, PA, USA, 2003. Association for Computational Linguistics.

[107] Yi Zhang, Jeff Schneider, and Artur Dubrawski. Learning the semantic correlation: An alternative way to gain from unlabeled text. In *NIPS*, pages 1945–1952, 2008.

[108] Vincent Wenchen Zheng, Derek Hao Hu, and Qiang Yang. Cross-domain activity recognition. In *Proceedings of the 11th international conference on Ubiquitous computing*, Ubicomp '09, pages 61–70, New York, NY, USA, 2009. ACM.

[109] Erheng Zhong, Wei Fan, Jing Peng, Kun Zhang, Jiangtao Ren, Deepak Turaga, and Olivier Verscheure. Cross domain distribution adaptation via kernel mapping. In *Proceedings of the 15th ACM SIGKDD international conference on Knowledge discovery and data mining*, KDD '09, pages 1027–1036, New York, NY, USA, 2009. ACM.

[110] Erheng Zhong, Wei Fan, Qiang Yang, Olivier Verscheure, and Jiangtao Ren. Cross validation framework to choose amongst models and datasets for transfer learning. In *ECML/PKDD (3)*, pages 547–562, 2010.

[111] Xiaojin Zhu, Andrew B. Goldberg, Ronald Brachman, and Thomas Dietterich. *Introduction to Semi-Supervised Learning*. Morgan and Claypool Publishers, 2009.

[112] Yin Zhu, Yuqiang Chen, Zhongqi Lu, Sinno Jialin Pan, Gui-Rong Xue, Yong Yu, and Qiang Yang. Heterogeneous transfer learning for image classification. In *AAAI*, 2011.

Chapter 8

PROBABILISTIC MODELS FOR TEXT MINING

Yizhou Sun
Department of Computer Science
University of Illinois at Urbana-Champaign
sun22@illinois.edu

Hongbo Deng
Department of Computer Science
University of Illinois at Urbana-Champaign
hbdeng@illinois.edu

Jiawei Han
Department of Computer Science
University of Illinois at Urbana-Champaign
hanj@illinois.edu

Abstract   A number of probabilistic methods such as LDA, hidden Markov models, Markov random fields have arisen in recent years for probabilistic analysis of text data. This chapter provides an overview of a variety of probabilistic models for text mining. The chapter focuses more on the fundamental probabilistic techniques, and also covers their various applications to different text mining problems. Some examples of such applications include topic modeling, language modeling, document classification, document clustering, and information extraction.

Keywords: Probabilistic models, mixture model, stochastic process, graphical model

1. Introduction

Probabilistic models are widely used in text mining nowadays, and applications range from topic modeling, language modeling, document classification and clustering to information extraction. For example, the well known topic modeling methods PLSA and LDA are special applications of mixture models.

A probabilistic model is a model that uses probability theory to model the uncertainty in the data. For example, terms in topics are modeled by multinomial distribution; and the observations for a random field are modeled by Gibbs distribution. A probabilistic model describes a set of possible probability distributions for a set of observed data, and the goal is to use the observed data to learn the distribution (usually associated with parameters) in the probabilistic model that can best describe the current data. In this chapter, we introduce several frequently used fundamental probabilistic models and their applications in text mining. For each probabilistic model, we will introduce their general framework of modeling, the probabilistic explanation, the standard algorithms to learn the model, and their applications in text mining.

The major probabilistic models covered in this chapter include:

- Mixture Models. Mixture models are used for clustering data points, where each component is a distribution for that cluster, and each data point belongs to one cluster with a certain probability. Finite mixture models require user to specify the number of clusters. The typical applications of mixture model in text mining include topic models, like PLSA and LDA.

- Bayesian Nonparametric Models. Beyesian nonparametric models refer to probabilistic models with infinite-dimensional parameters, which usually have a stochastic process that is infinite-dimensional as the prior distribution. Infinite mixture model is one type of non-parametric models, which can deal with the problem of selecting the number of clusters for clustering. Dirichlet process mixture model belongs to infinite mixture model, and can help to detect the number of topics in topic modeling.

- Bayesian Networks. A Bayesian network is a graphical model with directed acyclic links indicating the dependency relationship between random variables, which are represented as nodes in the network. A Bayesian network can be used to inference the unobserved node in the network, by learning parameters via training datasets.

- Hidden Markov Model. A hidden Markov model (HMM) is a simple case of dynamic Bayesian network, where the hidden states are forming a chain and only some possible value for each state can be observed. One goal of HMM is to infer the hidden states according to the observed values and their dependency relationships. A very important application of HMM is part-of-speech tagging in NLP.

- Markov Random Fields. A Markov random field (MRF) belongs to undirected graphical model, where the joint density of all the random variables in the network is modeled as a production of potential functions defined on cliques. An application of MRF is to model the dependency relationship between queries and documents, and thus to improve the performance of information retrieval.

- Conditional Random Fields. A conditional random field (CRF) is a special case of Markov random field, but each state of node is conditional on some observed values. CRFs can be considered as a type of discriminative classifiers, as they do not model the distribution over observations. Name entity recognition in information extraction is one of CRF's applications.

This chapter is organized as follows. In Section 2, mixture models that are frequently used in topic modeling and clustering is introduced, as well as its standard learning algorithms. In Section 3, we present several Bayesian nonparametric models, where stochastic processes are used as priors and can be used in modeling the uncertainty of the number of clusters in mixture models. In Section 4, several well-known graphical models that use nodes to represent random variables and use links in the graph to model the dependency relations between variables are introduced. Section 5 introduces several situations that constraints with domain knowledge can be integrated into probabilistic models. Section 6 is a brief introduction of parallel computing of probabilistic models for large scale datasets. The concluding remarks are given in Section 7.

2. Mixture Models

Mixture model [39] is a probabilistic model originally proposed to address the multi-modal problem in the data, and now is frequently used for the task of clustering in data mining, machine learning and statistics. Generally, a mixture model defines the distribution of a random variable, which contains multiple components and each component represents a different distribution following the same distribution family but with different parameters. The number of components are specified by users in this section, and these mixture models are called finite mixture models. Infinite mixture models that deal with how to learn the number of components in mixture models will be covered in Section 3. To learn the model, not only the probability membership for each observed data point but also the parameter set for each component need to be learned. In this section, we introduce the basic framework of mixture models, their variations and applications in text mining area, and the standard learning algorithms for them.

2.1 General Mixture Model Framework

In a mixture model, given a set of data points, e.g., the height of people in a region, they are treated as an instantiation of a set of random variables, which are following the mixture model. Then, according to the observed data points, the parameters in the mixture model can be learned. For example, we can learn the mean and standard deviation for female and male height distributions, if we model height of people as a mixture model of two Gaussian distributions. Formally, assume we have n i.i.d. random variables $X_1, X_2, \ldots, X_n$ with observations $x_1, x_2, \ldots, x_n$, following the mixture model with K components. Let each of the kth component be a distribution following a distribution family with parameters (θ_k) and have the form of $F(x|\theta_k)$, and let π_k ($\pi_k \geq 0$ and $\sum_k \pi_k = 1$) be the weight for kth component denoting the probability that an observation is generated from the component, then the probability of x_i can be written as:

$$p(x_i) = \sum_{k=1}^{K} \pi_k f(x_i|\theta_k)$$

where $f(x_i|\theta_k)$ is the density or mass function for $F(x|\theta_k)$. The joint probability of all the observations is then:

$$p(x_1, x_2, \ldots, x_n) = \prod_{i=1}^{n} \sum_{k=1}^{K} \pi_k f(x_i|\theta_k)$$

Let $Z_i \in \{1, 2, \ldots, K\}$ be the hidden cluster label for X_i, the probability function can be viewed as the summation over a complete joint distribution of both X_i and Z_i:

$$p(x_i) = \sum_{z_i} p(x_i, z_i) = \sum_{z_i} p(x_i|Z_i = z_i)p(z_i)$$

where $X_i|Z_i = z_i \sim F(x_i|\theta_{z_i})$ and $Z_i \sim \mathcal{M}_K(1; \pi_1, \ldots, \pi_K)$, the multinomial distribution of K dimensions with 1 observation. Z_i is also referred

Probabilistic Models for Text Mining 263

to missing variable or auxiliary variable, which identifies the cluster label of the observation x_i. From generative process point of view, each observed data x_i is generated by:

1. sample its hidden cluster label by $z_i|\pi \sim \mathcal{M}_K(1; \pi_1, \ldots, \pi_K)$
2. sample the data point in component z_i: $x_i|z_i, \{\theta_k\} \sim F(x_i|\theta_{z_i})$

The most well-known mixture model is the Gaussian mixture model, where each component is a Gaussian distribution. In this case, the parameter set for kth component is $\theta_k = (\mu_k, \sigma_k^2)$, where μ_k and σ_k^2 are the mean and variance of the Gaussian distribution.

Example: Mixture of Unigrams. The most common choice of the component distribution for terms in text mining is multinomial distribution, which can be considered as a unigram language model and determines the probability of a bag of terms. In Nigam et al. [50], a document d_i composed of a bag of words $\mathbf{w}_i = (c_{i,1}, c_{i,2}, \ldots, c_{i,m})$, where m is the size of the vocabulary and $c_{i,j}$ is the number of term w_j in document d_i, is considered as a mixture of unigram language models. That is, each component is a multinomial distribution over terms, with parameters $\beta_{k,j}$, denoting the probability of term w_j in cluster k, i.e., $p(w_j|\beta_k)$, for $k = 1, \ldots, K$ and $j = 1, \ldots, m$. The joint probability of observing the whole document collection is then:

$$p(\mathbf{w}_1, \mathbf{w}_2, \ldots, \mathbf{w}_n) = \prod_{i=1}^{n} \sum_{k=1}^{K} \pi_k \prod_{j=1}^{m} (\beta_{k,j})^{c_{i,j}}$$

where π_k is the proportion weight for cluster k. Note that, in mixture of unigrams, one document is modeled as being sampled from exactly one cluster, which is not typically true, since one document usually covers several topics.

2.2 Variations and Applications

Besides the mixture of unigrams, there are many other applications for mixture models in text mining, with some variations to the general framework. The most frequent variation to the framework of general mixture models is to adding all sorts of priors to the parameters, which are sometimes called Bayesian (finite) mixture models [33]. The topic models PLSA [29, 30] and LDA [11, 28] are among the most famous applications, which have been introduced in Chapter 5 in a dimension reduction view. In this section, we briefly describe them in the view of mixture models. Some other applications in text mining, such as comparative text mining, contextual text mining, and topic sentiment analysis, are introduced too.

2.2.1 Topic Models.

- PLSA. Probabilistic latent semantic analysis (PLSA) [29] is also known as probabilistic latent semantic indexing (PLSI) [30]. Different from the mixture unigram, where each document d_i connects to one latent variable Z_i, in PLSA, each observed term w_j in d_i corresponds to a different latent variable $Z_{i,j}$. The probability of observation term w_j in d_i is then defined by the mixture in the following:

$$p(w_j|d_i) = \sum_{k=1}^{K} p(k|d_i)p(w_j|\beta_k)$$

where $p(k|d_i) = p(z_{i,j} = k)$ is the mixing proportion of different topics for d_i, β_k is the parameter set for multinomial distribution over terms for topic k, and $p(w_j|\beta_k) = \beta_{k,j}$. $p(k|d_i)$ is usually denoted by the parameter $\theta_{i,k}$, and $Z_{i,j}$ is then following the discrete distribution with K-d vector parameter $\theta_i = (\theta_{i,1}, \ldots, \theta_{i,K})$. The joint probability of observing all the terms in document d_i is:

$$p(d_i, \mathbf{w}_i) = p(d_i) \prod_{j=1}^{m} p(w_j|d_i)^{c_{i,j}}$$

where $\mathbf{w}_i$ is the same defined as in the mixture of unigrams and $p(d_i)$ is the probability of generating d_i. And the joint probability of observing all the document corpus is $\prod_{i=1}^{n} p(d_i, \mathbf{w}_i)$.

- LDA. Latent Dirichlet allocation (LDA) [11] extends PLSA by further adding priors to the parameters. That is, $Z_{i,j} \sim \mathcal{M}_K(1; \theta_i)$ and $\theta_i \sim Dir(\alpha)$, where $\mathcal{M}_K$ is the K-dimensional multinomial distribution, θ_i is the K-d parameter vector denoting the mixing portion of different topics for document d_i, $Dir(\alpha)$ denotes a Dirichlet distribution with K-d parameter vector α, which is the conjugate prior of multinomial distribution. Usually, another Dirichlet prior $\beta \sim Dir(\eta)$ [11, 28] is added further to the multinomial distribution β over terms, which serves as a smoothing functionality over terms, where η is a m-d parameter vector and m is the size of the vocabulary. The probability of observing all the terms in document d_i is then:

$$p(\mathbf{w}_i|\alpha, \beta) = \int p(\mathbf{w}_i, \theta_i|\alpha, \beta)d\theta_i$$

Probabilistic Models for Text Mining

where
$$p(\mathbf{w}_i, \theta_i | \alpha, \beta) = p(\mathbf{w}_i | \theta_i, \beta) p(\theta_i | \alpha)$$
and
$$p(\mathbf{w}_i | \theta_i, \beta) = \prod_{j=1}^{m} (\sum_{k=1}^{K} p(z_{i,j} = k | \theta_i) p(w_j | \beta_k))^{c_{i,j}}$$

The probability of observing all the document corpus is:
$$p(\mathbf{w}_1, \ldots, \mathbf{w}_n | \alpha, \eta) = \prod_{i=1}^{n} \int p(\mathbf{w}_i | \alpha, \beta) p(\beta | \eta) d\beta$$

Notice that, compared with PLSA, LDA has stronger generative power, as it describes how to generate the topic distribution θ_i for an unseen document d_i.

2.2.2 Other Applications. Now, we briefly introduce some other applications of mixture models in text mining.

- Comparative Text Mining. Comparative text mining (CTM) is proposed in [71]. Given a set of comparable text collections (e.g., the reviews for different brands of laptops), the task of comparative text mining is to discover any latent common themes across all collections as well as special themes within one collection. The idea is to model each document as a mixture model of the background theme, common themes cross different collection, and specific themes within its collection, where a theme is a topic distribution over terms, the same as in topic models.

- Contextual Text Mining. Contextual text mining (CtxTM) is proposed in [43], which extracts topic models from a collection of text with context information (e.g., time and location) and models the variations of topics over different context. The idea is to model a document as a mixture model of themes, where the theme coverage in a document would be a mixture of the document-specific theme coverage and the context-specific theme coverage.

- Topic Sentiment Analysis. Topic Sentiment Mixture (TSM) is proposed in [42], which aims at modeling facets and opinions in weblogs. The idea is to model a blog article as a mixture model of a background language model, a set of topic language models, and two (positive and negative) sentiment language models. Therefore, not only the topics but their sentiments can be detected simultaneously for a collection of weblogs.

2.3 The Learning Algorithms

In this section, several frequently used algorithms for learning parameters in mixture models are introduced.

2.3.1 Overview. The general idea of learning parameters in mixture models (and other probabilistic models) is to find a set of "good" parameters θ that maximizes the probability of generating the observed data. Two estimation criteria are frequently used, one is maximum-likelihood estimation (MLE) and the other is maximum-a-posteriori-probability (MAP).

The likelihood (or likelihood function) of a set of parameters given the observed data is defined as the probability of all the observations under those parameter values. Formally, let $x_1, \ldots, x_n$ (assumed iid) be the observations, let the parameter set be θ, the likelihood of θ given the data set is defined as:

$$\mathcal{L}(\theta \mid x_1, \ldots, x_n) = p(x_1, x_2, \ldots, x_n \mid \theta) = \prod_{i=1}^{n} p(x_i \mid \theta)$$

In the general form of mixture models, the parameter set includes both the component distribution parameter θ_k for each component k, and the mixing proportion of each component π_k. MLE estimation is then to find the parameter values that maximizes the likelihood function. Most of the time, log-likelihood is optimized instead, as it converts products into summations and makes the computation easier:

$$\log \mathcal{L}(\theta \mid x_1, \ldots, x_n) = \sum_{i=1}^{n} \log p(x_i \mid \theta)$$

When priors are incorporated to the mixture models (such as in LDA), the MAP estimation is used instead, which is to find a set of parameters θ that maximizes the posterior density function of θ given the observed data:

$$p(\theta \mid x_1, \ldots, x_n) \propto p(x_1, \ldots, x_n \mid \theta) p(\theta)$$

where $p(\theta)$ is the prior distribution for θ and may involve some further hyper-parameters.

Several frequently used algorithms of finding MLE or MAP estimations for parameters in mixture models are introduced briefly in the following.

2.3.2 EM Algorithm. Expectation-Maximum (EM) [7, 22, 21, 12] algorithm is a method for learning MLE estimations for probabilistic

Probabilistic Models for Text Mining

models with latent variables, which is a standard learning algorithm for mixture models. For mixture models, the likelihood function can be further viewed as the marginal over the complete likelihood involving hidden variables:

$$\mathcal{L}(\theta \mid x_1, \ldots, x_n) = \sum_{\mathbf{Z}} p(x_1, \ldots, x_n, z_1, \ldots, z_n | \theta) = \prod_{i=1}^{n} \sum_{z_i} p(x_i, z_i | \theta)$$

The log-likelihood function is then:

$$\log \mathcal{L}(\theta \mid x_1, \ldots, x_n) = \sum_{i=1}^{n} \log \sum_{z_i} p(x_i | \theta, z_i) p(z_i)$$

which is difficult to maximize directly, as there is summation inside the logarithm operation. EM algorithm is an iterative algorithm involving two steps that maximizes the above log-likelihood, which can solve this problem. The two steps in each iteration are E-step and M-step respectively.

In **E-step (Expectation step)**, a tight lower bound for the log-likelihood called Q-function is calculated, which is the expectation of the complete log-likelihood function with respect to the conditional distribution of hidden variable **Z** given the observations of the data **X** and current estimation of parameters $\theta^{(t)}$:

$$Q(\theta|\theta^{(t)}) = \mathrm{E}_{\mathbf{Z}|\mathbf{X},\theta^{(t)}} [\log L(\theta; \mathbf{X}, \mathbf{Z})]$$

Note $L(\theta; \mathbf{X}, \mathbf{Z})$ is a complete likelihood function as it uses both the observed data **X** and the hidden cluster labels **Z**.

In **M-step (Maximization-step)**, a new $\theta = \theta^{(t+1)}$ is computed which maximizes the Q-function that is derived in E-step:

$$\theta^{(t+1)} = \arg\max_{\theta} Q(\theta|\theta^{(t)})$$

It is guaranteed that EM algorithm converges to a local maximum of the log-likelihood function, since Q-function is a tight lower bound and the M-step can always find a θ that increases the log-likelihood. The learning algorithm in PLSA is a typical application of EM algorithm. Notice that, in M-step there could exist no closed form solution for $\theta^{(t+1)}$ and requires iterative solutions via methods such as gradient descent or Newton's method (also called Newton-Raphson method) [34].

There are several variants for EM algorithm when the original EM algorithm is difficult to compute, and some of which are listed in the following:

- **Generalized EM.** For generalized EM (GEM) [12], it relaxes the requirement of finding the θ that maximizes Q-function in M-step to finding a θ that increases Q-function somehow. The convergence can still be guaranteed using GEM, and it is often used when maximization in M-step is difficult to compute.

- **Variational EM.** Variational EM is one of the approximate algorithms used in LDA [11]. The idea is to find a set of variational parameters with respective to the hidden variables that attempts to obtain the tightest possible lower bound in E-step, and to maximize the lower bound in M-step. The variational parameters are chosen in a way that simplifies the original probabilistic model and are thus easier to calculate.

2.3.3 Gibbs Sampling. Gibbs sampling is the simplest form of Markov chain Monte Carlo (MCMC) algorithm, which is a sampling-based approximation algorithm for model inference. The basic idea of Gibbs sampling is to generate samples that converge to the target distribution, which itself is difficult to obtain, and to estimate the parameters using the statistics of the distribution according to the samples.

In [28], a Gibbs sampling-based inference algorithm is proposed for LDA. The goal is to maximize the posterior distribution of hidden variables (MAP estimation) given the observations of the documents $p(\mathbf{Z}|\mathbf{w})$, which is a very complex density function with hyper-parameters α and η that are specified by users. As it is difficult to directly maximize the posterior, Gibbs sampling is then used to construct a Markov chain of $\mathbf{Z}$, which converges to the posterior distribution in the long run. The hidden cluster $z_{i,j}$ for term $w_{i,j}$, i.e., the term w_j in document d_i, is sampled according to the conditional distribution of $z_{i,j}$, given the observations of all the terms as well as the their hidden cluster labels except for $w_{i,j}$ in the corpus:

$$p(z_{i,j}|\mathbf{Z}_{-i,j},\mathbf{w}) \propto p(z_{i,j}, w_{i,j}|\mathbf{Z}_{-i,j}, \mathbf{W}_{-i,j}) = p(w_{i,j}|\mathbf{Z}, \mathbf{W}_{-i,j})p(z_{i,j}|\mathbf{Z}_{-i,j})$$

which turns out to be easy to calculate, where $\mathbf{z}_{-i,j}$ denotes the hidden variables of all the terms except for $w_{i,j}$ and $\mathbf{w}_{-i,j}$ denotes the all the terms except $w_{i,j}$ in the corpus. Note that the conditional probability is also involving the hyper-parameters α and η, which are not shown explicitly. After thousands of iterations (called burning period), the Markov chain is considered to be stable and converges to the target posterior distribution. Then the parameters of θ and β can be estimated according to the sampled hidden cluster labels from the chain as well as the given observations and the hyper-parameters. Please refer to [28] and

[53] for more details of Gibbs sampling in LDA, and more fundamental introductions for Gibbs sampling and other MCMC algorithms in [3].

3. Stochastic Processes in Bayesian Nonparametric Models

Priors are frequently used in probabilistic models. For example, in LDA, Dirichlet priors are added for topic distributions and term distributions, which are both multinomial distributions. A special type of priors that are stochastic processes, which emerges recently in text related probabilistic models, is introduced in this section. Different from previous methods, with the introduction of priors of stochastic processes, the parameters in such models become infinite-dimensional. These models belong to the category of *Bayesian nonparametric models* [51].

Different from the traditional priors as static distributions, stochastic process priors can model more complex structures for the probabilistic models, such as the number of the components in the mixture model, the hierarchical structures and evolution structure for topic models, and the power law distribution for terms in language models. For example, it is always a difficult task for users to determine the number of topics when applying topic models for a collection of documents, and a Dirichlet process prior can model infinite number of topics and finally determine the best number of topics.

3.1 Chinese Restaurant Process

The Chinese Restaurant Process (CRP) [33, 9, 67] is a discrete-time stochastic process, which defines a distribution on the partitions of the first n integers, for each discrete time index n. As for each n, CRP defines the distribution of the partitions over the n integers, it can be used as the prior for the sizes of clusters in the mixture model-based clustering, and thus provides a way to guide the selection of K, which is the number of clusters, in the clustering process.

Chinese restaurant process can be described using a random process as a metaphor of costumers choosing tables in a Chinese restaurant. Suppose there are countably infinite tables in a restaurant, and the nth costumer walks in the restaurant and sits down at some table with the following probabilities:

1 The first customer sits at the first table (with probability 1).

2 The nth customer either sits at an occupied table k with probability $\frac{m_k}{n-1+\alpha}$, or sits at the first unoccupied table with probability

$\frac{\alpha}{n-1+\alpha}$, where m_k is the number of existing customers sitting at table k and α is a parameter of the process.

It is easy to see that the customers can be viewed as data points in the clustering process, and the tables can be viewed as the clusters. Let $z_1, z_2, \ldots, z_n$ be the table label associated with each customer, let K_n be the number of tables in total, and let m_k be the number of customers sitting in the kth table, the probability of such an arrangement (a partition of n integers into K_n groups) is as follows:

$$p(z_1, z_2, \ldots, z_n) = p(z_1)p(z_2|z_1)\ldots p(z_n|z_{n-1},\ldots,z_1) = \frac{\alpha^{K_n} \prod_{k=1}^{K_n}(m_k - 1)!}{\alpha(\alpha+1)\ldots(\alpha+n-1)}$$

The expected number of tables K_n given n customers is:

$$E(K_n|\alpha) = \sum_{i=1}^{n} \frac{\alpha}{i-1+\alpha} \approx \alpha \log(1 + \frac{n}{\alpha}) = O(\alpha \log n)$$

In summary, CRP defines a distribution over partitions of the data points, that is, a distribution over all possible clustering structures with different number of clusters. Moreover, prior distributions can also be provided over cluster parameters, such as a Dirichlet prior over terms in LDA for each topic. A stochastic process called Dirichlet process combines the two types of priors, and thus is frequently used as the prior for mixture models, which is introduced in the following section.

3.2 Dirichlet Process

We now introduce Dirichlet process and Dirichlet process-based mixture model, the inference algorithms and applications are also briefly mentioned.

3.2.1 Overview of Dirichlet Process. Dirichlet process (DP) [33, 67, 68] is a stochastic process, which is a distribution defined on distributions. That is, if we draw a sample from a DP, it would be a distribution over values instead of a single value. In addition to CRP, which only considers the distribution over partitions of data points, DP also defines the data distribution for each cluster, with an analogy of the dishes served for each table in the Chinese restaurant metaphor. Formally, we say a stochastic process G is a Dirichlet process with base distribution H and concentration parameter α, written as $G \sim DP(\alpha, H)$, if for an arbitrary finite measurable partition $A_1, A_2, \ldots, A_r$ of the probability space of H, denoted as Θ, the following holds:

$$(G(A_1), G(A_2), \ldots, G(A_r)) \sim Dir(\alpha H(A_1), \alpha H(A_2), \ldots, \alpha H(A_r))$$

where $G(A_i)$ and $H(A_i)$ are the marginal probability of G and H over partition A_i. In other words, the marginal distribution of G must be Dirichlet distributed, and this is why it is called Dirichlet process. Intuitively, the base distribution H is the mean distribution of the DP, and the concentration parameter α can be understood as an inverse variance of the DP, namely, an larger α means a smaller variance and thus a more concentrated DP around the mean H. Notice that, although the base distribution H could be a continuous distribution, G will always be a discrete distribution, with point masses at most countably infinite. This can be understood by studying the random process of generating distribution samples ϕ_i's from G:

$$\phi_n | \phi_{n-1}, \ldots, \phi_1 = \begin{cases} \phi_k^*, & \text{with probability } \frac{m_k}{n-1+\alpha} \\ \text{new draw from } H, & \text{with probability } \frac{\alpha}{n-1+\alpha} \end{cases}$$

where ϕ_k^* represents the kth *unique* distribution sampled from H, indicating the distribution for kth cluster, and ϕ_i denotes the distribution for the ith sample, which could be a distribution from existing clusters or a new distribution.

In addition to the above definition, a DP can also be defined through a stick-breaking construction [62]. On one hand, the proportion of each cluster k among all the clusters, π_k, is determined by a stick-breaking process:

$$\beta_k \sim Beta(1, \alpha) \text{ and } \pi_k = \beta_k \prod_{l=1}^{k-1}(1 - \beta_l)$$

Metaphorically, assuming we have a stick with length 1, we first break it at β_1 that follows a Beta distribution with parameter α, and assign it to π_1; for the remaining stick with length $1 - \beta_1$, we repeat the process, break it at $\beta_2 \sim Beta(1, \alpha)$, and assign it $(\beta_2(1 - \beta_1))$ to π_2; we recursively break the remaining stick and get π_3, π_4 and so on. The stick-breaking distribution over π is sometimes written as $\pi \sim GEM(\alpha)$, where the letters GEM stand for the initials of the inventors. On the other hand, for each cluster k, its distribution ϕ_k^* is sampled from H. G is then a mixture model over these distributions, $G \sim \sum_k \pi_k \delta_{\phi_k^*}$, where $\delta_{\phi_k^*}$ denotes a point mass at ϕ_k^*.

Further, hierarchical Dirichlet process (HDP) [68] can be defined, where the base distribution H follows another DP. HDP can model topics across different collections of documents, which share some common topics across different corpora but may have some special topics within each corpus.

3.2.2 Dirichlet Process Mixture Model.

By using DP as priors for mixture models, we can get Dirichlet process mixture model (DPM) [48, 67, 68], which can model the number of components in a mixture model, and sometimes is also called infinite mixture model. For example, we can model infinite topics for topic modeling, infinite components in infinite Gaussian mixture model [57], and so on. In such mixture models, to sample a data value, it will first sample a distribution ϕ_i and then sample a value x_i according to the distribution ϕ_i. Formally, let $x_1, x_2, \ldots, x_n$ be n observed data points, and let $\theta_1, \theta_2, \ldots, \theta_n$ be the parameters for the distributions of latent clusters associated with each data point, where the distribution ϕ_i with parameter θ_i is drawn i.i.d from G, the generative model for x_i is then:

$$x_i | \theta_i \sim F(\theta_i)$$
$$\theta_i | G \sim G$$
$$G | \alpha, H \sim DP(\alpha, H)$$

where $F(\theta_i)$ is the distribution for x_i with the parameter θ_i. Notice that, since G is a discrete distribution, multiple θ_i's can share the same value.

From the generative process point of view, the observed data x_i's are generated by:

1. Sample π according to $\pi | \alpha \sim GEM(\alpha)$, namely, the stick-breaking distribution;

2. Sample the parameter θ_k^* for each distinctive cluster k according to $\theta_k | H \sim H$;

3. For each x_i,
   (a) first sample its hidden cluster label z_i by $z_i | \pi \sim \mathcal{M}(1; \pi)$,
   (b) then sample the value according to $x_i | z_i, \{\theta_k^*\} \sim F(\theta_{z_i}^*)$.

where $F(\theta_k^*)$ is the distribution of data in component k with parameter θ_k^*. That is, each x_i is generated from a mixture model with component parameters θ_k^*'s and the mixing proportion π.

3.2.3 The Learning Algorithms.

As DPM is a nonparametric model with infinite number of parameters in the model, EM algorithm cannot be directly used in the inference for DPM. Instead, MCMC approaches [48] and variational inference [10] are standard inference methods for DPM.

The general goal for learning DPM is to learn the hidden cluster labels z_i's and the parameters θ_i's for its associated cluster component for all the observed data points. It turns out that Gibbs sampling is very convenient to implement for such models especially when G is the conjugate prior for the data distribution F, as the conditional distribution of both θ_i and z_i can be easily computed and thus the posterior distribution of these parameters or hidden cluster labels can be easily simulated by the obtained Markov chain. For more details, please refer to [48], where several MCMC-based algorithms are provided and discussed.

The major disadvantages of MCMC-based algorithms are that the sampling process can be very slow and the convergence is difficult to diagnose. Therefore Blei et al. [10] proposed an alternative approach called variational inference for DPMs, which is a class of *deterministic* algorithms that convert inference problems into optimization problems. The basic idea of variational inference methods is to relax the original likelihood function or posterior probability function P into a simpler variational distribution function Q_μ, which is indexed by new free variables μ that are called variational parameters. The goal is to compute the variational parameters μ that minimizes the KL divergence between the variation distribution and the original distribution:

$$\mu^* = \arg\min D(Q_\mu || P)$$

where D refers to some distance or divergence function. And then Q_{μ^*} can be used to approximate the desired P. Please refer to [10] for more details of variational inference for DPM.

3.2.4 Applications in Text Mining.

There are many successful applications in text mining by using DPMs, and we select some of the most representative ones in the following.

In [9], a hierarchical LDA model (LDA) that based on nested Chinese restaurant process is proposed, which can detect hierarchical topic models instead of topic models in a flat structure from a collection of documents. In addition, hLDA can detect the number of topics automatically, which is the number of nodes in the hierarchical tree of topics. Compared with original LDA, hLDA can detect topics with higher interpretability and has higher predictive held-out likelihood in the testing set.

In [73], a time-sensitive Dirichlet process mixture model is proposed to detect clusters from a collection of documents with time information, for example, detecting subject threads for emails. Instead of considering each document equally important, the weights of history documents are discounted in the cluster. A time-sensitive DPM (tDPM) is then built based on the idea, which can not only output the number of clusters, but also introduce the temporal dependencies between documents, with less influence from older documents.

Evolution structure can also be detected using DPMs. A temporal Dirichlet process mixture model (TDPM) [2] is proposed as a framework to model the evolution of topics, such as retain, die out or emerge over time. In [1], an infinite dynamic topic model (iDTM) is further proposed to allow each document to be generated from multiple topics, by modeling documents in each time epoch using HDP instead of DP. An evolutional hierarchical Dirichlet process approach (EvoHDP) is proposed in [72] to detect evolutionary topics from multiple but correlated corpora, which can discover different evolving patterns of topics, including emergence, disappearance, evolution within a corpus and across different corpora.

3.3 Pitman-Yor Process

Pitman-Yor process [52, 66], also known as two-parameter Poisson-Dirichlet process, is a generalization over DP, which can successfully model data with power law [18] distributions. For example, if we want to model the distribution of all the words in a corpus, Pitman-Yor process is a better option than DP, where each word can be viewed as a table and the number of occurrences of the word can be viewed as the number of customers sitting in the table, in a restaurant metaphor.

Compared with CP, Pitman-Yor process has one more discount parameter $0 \leq d < 1$, in addition to the strength parameter $\alpha > -d$, which is written as $G \sim PY(d, \alpha, H)$, where H is the base distribution. This can be understood by studying the random process of generating distribution samples ϕ_i's from G:

$$\phi_n | \phi_{n-1}, \ldots, \phi_1 = \begin{cases} \phi_k^*, & \text{with probability } \frac{m_k - d}{n - 1 + \alpha} \\ \text{new draw from } H, & \text{with probability } \frac{\alpha + dK_n}{n - 1 + \alpha} \end{cases}$$

where ϕ_k^* is the distribution of table k, m_k is the number of customers sitting at table k, and K_n is the number of tables so far. Notice that when $d = 0$, Pitman-Yor process reduces to DP.

Two salient features of Pitman-Yor process compared with CP are: (1) given more occupied tables, the chance to have even more tables is higher; (2) tables with small occupancy number have a lower chance to get more customers. This implies that Pitman-Yor process has a power law (e.g., Zipf's law) behavior. The expected number of tables is $O(\alpha n^d)$, which has the power law form. Compared with the expected number of tables $O(\alpha \log n)$ for DP, Pitman-Yor process indicates a faster growing in the expected number of tables.

In [66], a hierarchial Pitman-Yor n-gram language model is proposed. It turns out that the proposed model has the best performance compared with the state-of-the-art methods, and has demonstrated that Bayesian approach can be competitive with the best smoothing techniques in languages modeling.

3.4 Others

There are many other stochastic processes that can be used in Bayesian nonparametric models, such as Indian buffet process [27], Beta process [69], Gaussian process [58] for infinite Gaussian mixture model, Gaussian process regression, and so on. We now briefly introduce them in the following, and the readers can refer to the references for more details.

- Indian buffet process. In mixture models, one data point can only belong to one cluster, with the probability determined by the mixing proportions. However, sometimes one data point can have multiple features. For example, a person can participate in a number of communities, all with a large strength. Indian buffet process is a stochastic process that can define the infinite-dimensional features for data points. It has a metaphor of people choosing (infinite) dishes arranged in a line in Indian buffet restaurant, which is where the name "Indian buffet process" is from.

- Beta process. As mentioned in [69], a beta process (BP) plays the role for the Indian buffet process that the Dirichlet process plays for the Chinese restaurant process. Also, a hierarchical beta process (hBP)-based method is proposed in [69] for the document classification task.

- Gaussian process. Intuitively, a Gaussian process (GP) extends a multivariate Gaussian distribution to the one with infinite dimensionality, similar to DP's role to Dirichlet distribution. Any finite subset of the random variables in a GP follows a multivariate Gaussian distribution. The applications for GP include Gaussian process regression, Gaussian process classification, and so on, which are discussed in [58].

4. Graphical Models

A Graphical model [32, 36] is a probabilistic model for which a graph denotes the conditional independence structure between random variables. Graphical model provides a simple way to visualize the structure of a probabilistic model and can be used to design and motivate new models. In a probabilistic graphical model, each node represents a random variable, and the links express probabilistic relationships between these variables. The graph then captures the way in which the joint distribution over all of the random variables can be decomposed into a product of factors each depending only on a subset of the variables. There are two branches of graphical representations of distributions that are commonly used: *directed* and *undirected*. In this chapter, we discuss the key aspects of graphical models and their applications in text mining.

4.1 Bayesian Networks

Bayesian networks (BNs), also known as *belief networks* (or Bayes nets for short), belong to the *directed graphical models*, in which the links of the graphs have a particular directionality indicated by arrows.

4.1.1 Overview. Formally, BNs are directed acyclic graphs (DAG) whose nodes represent random variables, and edges represent conditional dependencies. For example, a link from x to y can be informally interpreted as indicating that x "causes" y.

Conditional Independence. The simplest conditional independence relationship encoded in a BN can be stated as follows: a node is conditionally independent of its non-descendants given its parents, where the parent relationship is with respect to some fixed topological ordering of the nodes. This is also called *local Markov property*, denoted by $X_v \perp\!\!\!\perp X_{V \setminus de(v)} | X_{pa(v)}$ for all $v \in V$, where $de(v)$ is the set of descendants of v. For example, as shown in Figure 8.1(a), we obtain $x_1 \perp\!\!\!\perp x_3 | x_2$.

Factorization Definition. In a BN, the joint probability of all random variables can be factored into a product of density functions for all of the nodes in the graph, conditional on their parent variables. More precisely, for a graph with n nodes (denoted as $x_1, ..., x_n$), the joint distribution is given by:

$$p(x_1, ..., x_n) = \Pi_{i=1}^{n} p(x_i | pa_i), \qquad (8.1)$$

where pa_i is the set of parents of node x_i. By using the chain rule of probability, the above joint distribution can be written as a product of conditional distributions, given the topological order of these random variables:

$$p(x_1, ..., x_n) = p(x_1) p(x_2 | x_1) ... p(x_n | x_{n-1}, ..., x_1). \qquad (8.2)$$

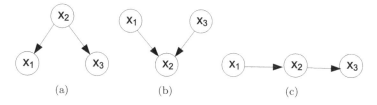

Figure 8.1. Examples of directed acyclic graphs describing the joint distributions.

The difference between the two expressions is the *conditional independence* of the variables encoded in a BN, that variables are conditionally independent of their non-descendants given the values of their parent variables.

Consider the graph shown in Figure 8.1, we can go from this graph to the corresponding representation of the joint distribution written in terms of the product of a set of conditional probability distributions, one for each node in the graph. The joint distributions for Figure 8.1(a)-(c) are therefore $p(x_1, x_2, x_3) = p(x_1|x_2)p(x_2)p(x_3|x_2)$, $p(x_1, x_2, x_3) = p(x_1)p(x_2|x_1, x_3)p(x_3)$, and $p(x_1, x_2, x_3) = p(x_1)p(x_2|x_1)p(x_3|x_2)$, respectively.

4.1.2 The Learning Algorithms. Because a BN is a complete model for the variables and their relationships, a complete joint probability distribution (JPD) over all the variables is specified for a model. Given the JPD, we can answer all possible inference queries by summing out (marginalizing) over irrelevant variables. However, the JPD has size $O(2^n)$, where n is the number of nodes, and we have assumed each node can have 2 states. Hence summing over the JPD takes exponential time. The most common *exact inference* method is **Variable Elimination** [19]. The general idea is to perform the summation to eliminate the non-observed non-query variables one by one by distributing the sum over the product. The reader can refer to [19] for more details. Instead of exact inference, a useful *approximate algorithm* called *Belief propagation* [46] is commonly used on general graphs including Bayesian network, which will be introduced in Section 4.3.

4.1.3 Applications in Text Mining. Bayesian networks have been widely used in many applications in text mining, such as spam filtering [61] and information retrieval [20]. In [61], a Bayesian approach is proposed to identify spam email by making use of a naive Bayes classifier. The intuition is that particular words have particular probabilities of occurring in spam emails and in legitimate emails. For

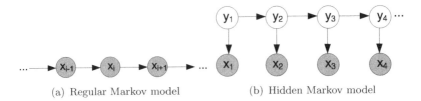

Figure 8.2. Graphical structures for the regular and hidden Markov model.

instance, the words "free" and "credit" will frequently appear in spam emails, but will seldom occur in other emails. To train the filter, the user must manually indicate whether an email is spam or not for a training set. With such a training dataset, Bayesian spam filters will learn a spam probability for each word, e.g., a high spam probability for the words "free" and "credit", and a relatively low spam probability for words such as the names of friends. Then, the email's spam probability is computed over all words in the email, and if the total exceeds a certain threshold, the filter will mark the email as a spam.

4.2 Hidden Markov Models

In a regular Markov model as Figure 8.2(a), the state x_i is directly visible to the observer, and therefore the state transition probabilities $p(x_i|x_{i-1})$ are the only parameters. Based on the Markov property, the joint distribution for a sequence of n observations under this model is given by

$$p(x_1, ..., x_n) = p(x_1) \prod_{i=2}^{n} p(x_i|x_{i-1}). \tag{8.3}$$

Thus if we use such a model to predict the next observation in a sequence, the distribution of predictions will depend on the value of the immediately preceding observation and will be independent of all earlier observations, conditional on the preceding observation.

4.2.1 Overview.
A hidden Markov model (HMM) can be considered as the simplest dynamic Bayesian network. In a hidden Markov model, the state y_i is not directly visible, and only the output x_i is visible, which is dependent on the state. The hidden state space is discrete, and is assumed to consist of one of N possible values, which is also called latent variable. The observations can be either discrete or continuous, which are typically generated from a categorical distribution or a Gaussian distribution. Generally, a HMM can be considered as a

Probabilistic Models for Text Mining

generalization of a *mixture model* where the hidden variables are related through a Markov process rather than independent of each other.

Suppose the latent variables form a first-order Markov chain as shown in Figure 8.2(b). The random variable y_t is the hidden state at time t, and the random variable x_t is the observation at time t. The arrows in the figure denote conditional dependencies. From the diagram, it is clear that y_{t-1} and y_{t+1} are independent given y_t, so that $y_{t+1} \perp\!\!\!\perp y_{t-1}|y_t$. This is the key conditional independence property, which is called the *Markov property*. Similarly, the value of the observed variable x_t only depends on the value of the hidden variable y_t. Then, the joint distribution for this model is given by

$$p(x_1, ..., x_n, y_1, ..., y_n) = p(y_1) \prod_{t=2}^{n} p(y_t|y_{t-1}) \prod_{t=1}^{n} p(x_t|y_t), \qquad (8.4)$$

where $p(y_t|y_{t-1})$ is the state transition probability, and $p(x_t|y_t)$ is the observation probability.

4.2.2 The Learning Algorithms. Given a set of possible states $\Omega_Y = \{q_1, ..., q_N\}$ and a set of possible observations $\Omega_X = \{o_1, ..., o_M\}$. The parameter learning task of HMM is to find the best set of state transition probabilities $A = \{a_{ij}\}$, $a_{ij} = p(y_{t+1} = q_j|y_t = q_i)$ and observation probabilities $B = \{b_i(k)\}$, $b_i(k) = p(x_t = o_k|y_t = q_i)$ as well as the initial state distribution $\Pi = \{\pi_i\}$, $\pi_i = p(y_0 = q_i)$ for a set of output sequences. Let $\Lambda = \{A, B, \Pi\}$ denote the parameters for a given HMM with fixed Ω_Y and Ω_X. The task is usually to derive the *maximum likelihood estimation* of the parameters of the HMM given the set of output sequences. Usually a local maximum likelihood can be derived efficiently using the *Baum-Welch algorithm* [5], which makes use of *forward-backward algorithm* [55], and is a special case of the generalized EM algorithm [22].

Given the parameters of the model Λ, there are several typical inference problems associated with HMMs, as outlined below. One common task is to compute the probability of a particular output sequence, which requires summation over all possible state sequences: The probability of observing a sequence $X_1^T = o_1, ..., o_T$ of length T is given by $P(X_1^T|\Lambda) = \sum_{Y_1^T} P(X_1^T|Y_1^T, \Lambda) P(Y_1^T|\Lambda)$, where the sum runs over all possible hidden-node sequences $Y_1^T = y_1, ..., y_T$.

This problem can be handled efficiently using the **forward-backward algorithm**. Before we describe the algorithm, let us define the forward (alpha) values and backward (beta) values as follows: $\alpha_t(i) = P(x_1 = o_1, ..., x_t = o_t, y_t = q_i|\Lambda)$ and $\beta_t(i) = P(x_{t+1} = o_{t+1}, ..., x_T = o_T|y_t =$

q_i, Λ). Note the forward values enable us to solve the problem through marginalizing, then we obtain

$$P(X_1^T|\Lambda) = \sum_{i=1}^{N} P(o_1, ..., o_T, y_T = q_i|\Lambda) = \sum_{i=1}^{N} \alpha_T(i).$$

The forward values can be computed efficiently with the principle of *dynamic programming*:

$$\alpha_1(i) = \pi_i b_i(o_1),$$
$$\alpha_{t+1}(j) = \left[\sum_{i=1}^{N} \alpha_t(i) a_{ij}\right] b_j(o_{t+1}).$$

Similarly, the backward values can be computed as

$$\beta_T(i) = 1,$$
$$\beta_t(i) = \sum_{j=1}^{N} a_{ij} b_j(o_{t+1}) \beta_{t+1}(j).$$

The backward values will be used in the Baum-Welch algorithm.

Given the parameters of HMM and a particular sequence of observations, another interesting task is to compute the most likely sequence of states that could have produced the observed sequence. We can find the most likely sequence by evaluating the joint probability of both the state sequence and the observations for each case. For example, in part-of-speech (POS) tagging [37], we observe a token (word) sequence $X_1^T = o_1, ..., o_T$, and the goal of POS tagging is to find a stochastic optimal tag sequence $Y_1^T = y_1 y_2 ... y_T$ that maximizes $P(Y_1^n, X_1^n)$. In general, finding the most likely explanation for an observation sequence can be solved efficiently using the **Viterbi algorithm** [24] by the recurrence relations:

$$V_1(i) = b_i(o_1) \pi_i,$$
$$V_t(j) = b_i(o_t) \max_i \left(V_{t-1}(i) a_{ij}\right).$$

Here $V_t(j)$ is the probability of the most probable state sequence responsible for the first t observations that has q_j as its final state. The Viterbi path can be retrieved by saving back pointers that remember which state $y_t = q_j$ was used in the second equation. Let $Ptr(y_t, q_i)$ be the function that returns the value of y_{t-1} used to compute $V_t(i)$ as follows:

$$y_T = \arg \max_{q_i \in \Omega_Y} V_T(i),$$
$$y_{t-1} = Ptr(y_t, q_i).$$

The complexity of this algorithm is $O(T \times N^2)$, where T is the length of observed sequence and N is the number of possible states.

Now we need a method of adjusting the parameters Λ to maximize the likelihood for a given training set. The **Baum-Welch algorithm** [5] is used to find the unknown parameters of HMMs, which is a particular case of a generalized EM algorithms [22]. We start by choosing arbitrary values for the parameters, then compute the expected frequencies given the model and the observations. The expected frequencies are obtained by weighting the observed transitions by the probabilities specified in the current model. The expected frequencies obtained are then substituted for the old parameters and we iterate until there is no improvement. On each iteration we improve the probability of being observed from the model until some limiting probability is reached. This iterative procedure is guaranteed to converge to a local maximum [56].

4.2.3 Applications in Text Mining. HMM models have been applied to a wide variety of problems in information extraction and natural language processing, which have been introduced in Chapter 2, including POS tagging [37] and named entity recognition [6]. Taking POS tagging [37] as an example, each word is labeled with a tag indicating its appropriate part of speech, resulting in annotated text, such as: "[VB heat] [NN water] [IN in] [DT a] [JJ large] [NN vessel]". Given a sequence of words X_1^n, e.g., "heat water in a large vessel", the task is to assign a sequence of labels Y_1^n, e.g., "VB NN IN DT JJ NN", for the words. Based on HMM models, we can determine the sequence of labels by maximizing a joint probability distribution $p(X_1^n, Y_1^n)$.

With the success of HMMs in POS tagging, it is natural to develop a variant of an HMM for the name entity recognition task [6]. Intuitively, the locality of phenomena may indicate names in the text, such as titles like "Mr." preceding a person's name. The HMM classifier models such kinds of dependencies, and performs sequence classification by assigning each word to one of the named entity types. The states in the HMM are organized into regions, one region for each type of named entity. Within each of the regions, a statistical bi-gram language model is used to compute the likelihood of words occurring within that region (named entity type). The transition probabilities are computed by deleted interpolation, and the decoding is done through the Viterbi algorithm.

4.3 Markov Random Fields

Now we turn to another major class of graphical models that are described by undirected graphs and that again specify both a factorization and a set of conditional independence relations.

4.3.1 Overview. A Markov random field (MRF), also known as an undirected graphical model [35], has a set of nodes each of which corresponds to a variable or group of variables, as well as a set of links each of which connects a pair of nodes. The links are undirected, that is they do not carry arrows.

Conditional Independence. Given three sets of nodes, denoted A, B, and C, in an undirected graph G, if A and B are separated in G after removing a set of nodes C from G, then A and B are conditionally independent given the random variables C, denoted as $A \perp\!\!\!\perp B|C$. The conditional independence is determined by simple graph separation. In other words, a variable is conditionally independent of all other variables given its neighbors, denoted as $X_v \perp\!\!\!\perp X_{V \setminus \{v \cup ne(v)\}} | X_{ne(v)}$, where $ne(v)$ is the set of neighbors of v. In general, an MRF is similar to a Bayesian network in its representation of dependencies, and there are some differences. On one hand, an MRF can represent certain dependencies that a Bayesian network cannot (such as cyclic dependencies); on the other hand, MRF cannot represent certain dependencies that a Bayesian network can (such as induced dependencies).

Clique Factorization. As the Markov properties of an arbitrary probability distribution can be difficult to establish, a commonly used class of MRFs are those that can be factorized according to the cliques of the graph. A *clique* is defined as a subset of the nodes in a graph such that there exists a link between all pairs of nodes in the subset. In other words, the set of nodes in a clique is fully connected.

We can therefore define the factors in the decomposition of the joint distribution to be functions of the variables in the cliques. Let us denote a clique by C and the set of variables in that cliques by x_C. Then the joint distribution is written as a product of *potential functions* $\psi_C(x_C)$ over the maximal cliques of the graph

$$p(x_1, x_2, ..., x_n) = \frac{1}{Z} \Pi_C \psi_C(x_C),$$

where the *partition function* Z is a normalization constant and is given by $Z = \sum_x \Pi_C \psi_C(x_C)$. In contrast to the factors in the joint distribution for a directed graph, the potentials in an undirected graph do

Probabilistic Models for Text Mining

not have a specific probabilistic interpretation. Therefore, how to motivate a choice of potential function for a particular application seems to be very important. One popular potential function is defined as $\psi_C(x_C) = \exp(-\epsilon(x_C))$, where $\epsilon(x_C) = -\ln\psi_C(x_C)$ is an *energy function* [45] derived from statistical physics. The underlying idea is that the probability of a physical state depends inversely on its energy. In the logarithmic representation, we have

$$p(x_1, x_2, ..., x_n) = \frac{1}{Z}\exp\left(-\sum_C \epsilon(x_C)\right).$$

The joint distribution above is defined as the product of potentials, and so the total energy is obtained by adding the energies of each of the maximal cliques.

A *log-linear model* is a Markov random field with feature functions f_k such that the joint distribution can be written as

$$p(x_1, x_2, ..., x_n) = \frac{1}{Z}\exp\left(\sum_{k=1}^{K}\lambda_k f_k(x_{C_k})\right),$$

where $f_k(x_{C_k})$ is the function of features for the clique C_k, and λ_k is the weight vector of features. The log-linear model provides a much more compact representation for many distributions, especially when variables have large domains such as text.

4.3.2   The Learning Algorithms.   In MRF, we may compute the conditional distribution of a set of nodes given values A to another set of nodes B by summing over all possible assignments to $v \notin A, B$, which is called *exact inference*. However, the exact inference is computationally intractable in the general case. Instead, approximation techniques such as MCMC approach [3] and loopy *belief propagation* [46, 8] are often more feasible in practice. In addition, there are some particular subclasses of MRFs that permit efficient maximum-a-posterior (MAP) estimation, or more likely assignment, inference, such as associate networks. Here we will briefly describe belief propagation algorithm.

Belief propagation is a message passing algorithm for performing inference on graphical models, including Bayesian networks and MRFs. It calculates the marginal distribution for each unobserved node, conditional on any observed nodes. Generally, belief propagation operates on a factor graph, which is a bipartite graph containing nodes corresponding to variables V and factors U, with edges between variables and the factors in which they appear. Any Bayesian network and MRF can be represented as a factor graph. The algorithm works by passing

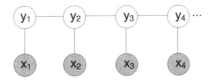

Figure 8.3. Graphical structure for the conditional random field model.

real valued function called *messages* along the edges between the nodes. Taking pairwise MRF as an example, let $m_{ij}(x_j)$ denote the message from node i to node j, and a high value of $m_{ij}(x_j)$ means that node i "believes" the marginal value $P(x_j)$ to be high. Usually the algorithm first initializes all messages to uniform or random positive values, and then updates message from i to j by considering all messages flowing into i (except for message from j) as follows:

$$m_{ij}(x_j) = \sum_{x_i} f_{ij}(x_i, x_j) \prod_{k \in ne(i) \setminus j} m_{ki}(x_i),$$

where $f_{ij}(x_i, x_j)$ is the potential function of the pairwise clique. After enough iterations, this process is likely to converge to a consensus. Once messages have converged, the marginal probabilities of all the variables can be determines by

$$p(x_i) \propto \prod_{k \in ne(i)} m_{ki}(x_i).$$

The reader can refer to [46] for more details. The main cost is the message update equation, which is $O(N^2)$ for each pair of variables (N is the number of possible states).

4.3.3 Applications in Text Mining.

Recently, MRF has been widely used in many text mining tasks, such as text categorization [16] and information retrieval [44]. In [44], MRF is used to model the term dependencies using the joint distribution over queries and documents. The model allows for arbitrary text features to be incorporated as evidence. In this model, an MRF is constructed from a graph G, which consists of query nodes q_i and a document node D. The authors explore full independence, sequential dependence, and full dependence variants of the model. Then, a novel approach is developed to train the model that directly maximizes the mean average precision. The results show significant improvements are possible by modeling dependencies, especially on the larger web collections.

4.4 Conditional Random Fields

So far, we have described the Markov network representation as a joint distribution. In this subsection, we introduce one notable variant of an MRF, i.e., conditional random field (CRF) [38, 65], which is yet another popular model for sequence labeling and has been widely used in information extraction as described in Chapter 2.

4.4.1 Overview. A CRF is an undirected graph whose nodes can be divided into exactly two disjoint sets, the observed variables X and the output variables Y, which can be parameterized as a set of factors in the same way as an ordinary Markov network. The underlying idea is that of defining a conditional probability distribution $p(Y|X)$ over label sequences Y given a particular observation sequence X, rather than a joint distribution over both label and observation sequences $p(Y, X)$. The primary advantage of CRFs over HMMs is their conditional nature, resulting in the relaxation of the independence assumptions required by HMMs in order to ensure tractable inference.

Considering a linear-chain CRF with $Y = \{y_1, y_2, ..., y_n\}$ and $X = \{x_1, x_2, ..., x_n\}$ as shown in Figure 8.3, an input sequence of observed variable X represents a sequence of observations and Y represents a sequence of hidden state variables that needs to be inferred given the observations. The y_i's are structured to form a chain, with an edge between each y_i and y_{i+1}. The distribution represented by this network has the form:

$$p(y_1, y_2, ..., y_n | x_1, x_2, ..., x_n) = \frac{1}{Z(X)} \exp\left(\sum_{k=1}^{K} \lambda_k f_k(y_i, y_{i-1}, x_i)\right),$$

where $Z(X) = \sum_{y_i} \exp\left(\sum_{k=1}^{K} \lambda_k f_k(y_i, y_{i-1}, x_i)\right)$.

4.4.2 The Learning Algorithms. For general graphs, the problem of exact inference in CRFs is intractable. Basically, the inference problem for a CRF is the same as for an MRF. If the graph is a chain or a tree, as shown in Figure 8.3, message passing algorithms yield exact solutions, which are similar to the forward-backward [5, 55] and Viterbi algorithms [24] for the case of HMMs. If exact inference is not possible, generally the inference problem for a CRF can be derived using approximation techniques such as MCMC [48, 3], loopy *belief propagation* [46, 8], and so on. Similar to HMMs, the parameters are typically learned by maximizing the likelihood of training data. It can be solved using an iterative technique such as iterative scaling [38] and gradient-descent methods [63].

4.4.3 Applications in Text Mining.
CRF has been applied to a wide variety of problems in natural language processing, including POS tagging [38], shallow parsing [63], and named entity recognition [40], being an alternative to the related HMMs. Based on HMM models, we can determine the sequence of labels by maximizing a joint probability distribution $p(\mathcal{X}, \mathcal{Y})$. In contrast, CRMs define a single log-linear distribution, i.e., $p(\mathcal{Y}|\mathcal{X})$, over label sequences given a particular observation sequence. The primary advantage of CRFs over HMMs is their conditional nature, resulting in the relaxation of the independence assumptions required by HMMs in order to ensure tractable inference. As expected, CRFs outperform HMMs on POS tagging and a number of real-word sequence labeling tasks [38, 40].

4.5 Other Models

Recently, there are many extensions of basic graphical models as mentioned above. Here we just briefly introduce the following two models, probabilistic relational model (PRM) [25] and Markov logic network (MLN) [59]. A Probabilistic relational model is the counterpart of a Bayesian network in statistical relational learning, which consists of relational schema, dependency structure, and local probability models. Compared with BN, PRM has some advantages and disadvantages. PRMs allow the properties of an object to depend probabilistically both on other properties of that object and on properties of related objects, while BN can only model relationships between at most one class of instances at a time. In PRM, all instances of the same class must use the same dependency mode, and it cannot distinguish two instances of the same class. In contrast, each instance in BN has its own dependency model, but cannot generalize over instances. Generally, PRMs are significantly more expressive than standard models, such as BNs. The well-known methods for learning BNs can be easily extended to learn these models.

A Markov logic network [59] is a probabilistic logic which combines first-order logic and probabilistic graphical models in a single representation. It is a first-order knowledge base with a weight attached to each formula, and can be viewed as a template for constructing Markov networks. Basically, probabilistic graphical models enable us to efficiently handle uncertainty. First-order logic enables us to compactly represent a wide variety of knowledge. From the point of view of probability, MLNs provide a compact language to specify very large Markov networks, and the ability to flexibly and modularly incorporate a wide range of domain knowledge into them. From the point of view of first-order logic, MLNs add the ability to handle uncertainty, tolerate imperfect and contradictory knowledge, and reduce brittleness. The inference in MLNs can be performed using standard Markov network inference techniques over the minimal subset of the relevant Markov network required for answering the query. These techniques include belief propagation [46] and Gibbs sampling [23, 3].

5. Probabilistic Models with Constraints

In probabilistic models, domain knowledge is encoded into the model implicitly for most of the time. In this section, we introduce several situations that domain knowledge can be modeled as explicit constraints to the original probabilistic models.

By merely using PLSA or LDA, we may derive different topic models when the algorithms converge to different local maximums. It will be very useful if users can explicitly state which topic model they favor. A simple way to handle this issue is to list the terms that are desired by the users in each topic. For example, if "sport", "football" must be contained in Topic 1, users can indicate a related term distribution as prior distribution for Topic 1. This prior can be integrated into PLSA. Another sort of guidance is to specify which terms should have similar probabilities in one topic (must-link) and which terms should not have similar probabilities in any topic (cannot-link). This kind of prior can be modeled as Dirichlet forest prior, which is discussed in [4].

In traditional topic models, documents are considered indepedent with each other. However, in reality there could be correlations among documents. For example, linked webpages tends to be similar with each other, a paper cites another paper indicates the two papers are somehow similar. NetPLSA [41] and iTopicModel [64] are two algorithms that improve the original PLSA by consider the network constraints among the documents. NetPLSA takes the network constraints as an additional graph regularization term that forces two linked documents much similar, while iTopicModel models the network constraints using a Markov random field and also considers the direction of links in the network. These algorithms can still be solved by EM algorithm. By looking at the E-step, we can see that the constraints can be integrated into E-step, with a nice interpretation.

In [26], it proposes a framework of posterior regularization for probabilistic models. Different from traditional priors that are directly applied onto parameters, posterior regularization framework allows users to specify the constraints which are dependent on data space. For example, in an unsupervised part-of-speech tagging task, users may require each sentence have at least one verb according to domain knowledge, which is difficult to encode as priors for model parameters only. In order to take the data-dependent constraints into consideration, a posterior regularization likelihood is proposed, which integrates both the model likelihood and the constraints. By studying several tasks with different constraint types, the new method has shown its flexibility and effectiveness.

Another line of systematical study of integrating probabilistic models and constraints are Constrained Conditional Models (CCMs) [54, 60, 13, 14]. CCM is a learning and inference framework that augments the learning of conditional models with declarative constraints. The objective function of a CCM includes two parts, one part involves the features of a task, and the other involves the penalties when constraints are violated. To keep the simplicity of the probabilistic model, complex global constraints are encoded as constraints instead of features. Usually, the inference problem given a trained model can be solved using integer linear programming. There are two strategies for the training stage, a local model that decouples learning and inference and a global model (joint learning) that optimizes the whole objective function. In practice, the local model for training is especially beneficial when joint learning for global model is computationally intractable or when training data is not available for joint learning. That is, it is more practical to train simple models using limited training data but inference with both the trained model and the global constraints at the decision stage.

6. Parallel Learning Algorithms

The efficiency of the learning algorithms is always an issue, especially for large scale of datasets, which is quite common for text data. In order to deal with such large datasets, algorithms with linear or even sub-linear time complexity are required, for which parallel learning algorithms provide a way to speed up original algorithms significantly. We now introduce several such algorithms among them.

The time complexity for original EM learning algorithm for PLSA is about linear to the total document-word occurrences in the corpus and the number of topics. By partitioning the document-word occurrence table into blocks, the calculation of the conditional probability for each term in each document can be parallelized for blocks with no conflicts. The tricky part is to partition the blocks such that the workload for each processing unit is balanced. Under this idea, [31] proposes a parallelized PLSA with 6 times' speedup on an eight-processor machine compared with the baseline. In [15], a Graphic Processing Unit

(GPU) instead of a multi-core machine is used to parallelize PLSA. GPU has a hundreds-of-core structure and high memory bandwith. It was designed to handle high-granularity graphics-related applications where many workloads can be simultaneously dispatched to processor elements, and now gradually becomes a general platform for parallel computing. In [15], both co-occurrence table-based partition and document-based partition are studied for the parallelization, which turn out to gain a significant speedup.

There are also some parallel learning algorithms for fast computing LDA. [47] proposes parallel version of algorithms based on variational EM algorithm for LDA. Two settings of implementations are considered, one is in a multiprocessor architecture and the other is in a distributed environment. In both settings, multiple threads or machines calculate E-step simultaneously for different partitions of the dataset, and a main program or a master machine will aggregate all the information and calculate M-step. In [49], parallel algorithms for LDA are based on Gibbs sampling algorithm. Two versions of algorithms, AD-LDA and HD-LDA, are proposed. AD-LDA is an approximate algorithm that applies local Gibbs sampling on each processor with periodic updates. HD-LDA is an algorithm with a theoretical guarantee to converge to Gibbs sampling using only one processor, which relies on a hierarchical Bayesian extension of the standard LDA model. Both algorithms have similar effectiveness performance as the single-processor learning, but with a significant time speedup. In PLDA [70], a further improvement is made by implementing AD-LDA on MPI (Message Passing Interface) and MapReduce, where MPI is a standardized and portable message-passing system for communicating between parallel computers, and MapReduce is a software framework introduced by Google to support distributed computing on clusters of computers.

In [17], instead of parallelizing one algorithm at a time, it proposes a broadly applicable paralleling programming method, which is easy to apply to many different learning algorithms. In the paper, it demonstrates the effectiveness of their methodology on a variety of learning algorithms by using MapReduce paradigm, which include locally weighted linear regression (LWLR), k-means, logistic regression (LR), naive Bayes (NB), SVM, ICA, PCA, gaussian discriminant analysis (GDA), EM, and backpropagation (NN).

7. Conclusions

In this chapter, we have introduced the most frequently used probabilistic models in text mining, which include mixture models with the applications of PLSA and LDA, the nonparametric models that use stochastic processes as priors and thus can model infinite-dimensional data, the well-known graphical models including Bayesian networks, HMM, Markov random field and conditional random field. In some scenarios, it will be helpful to model user guidance as constraints to the existing probabilistic models.

The goal of learning algorithms for these probabilistic models are to find MLE, MAP estimators for parameters in these models. Most of the time, no closed form solutions can be provided. Iterative algorithms such as EM algorithm is a powerful tool to learn mixture models. In other cases, exact solutions are difficult to obtain, and sampling methods based on MCMC, belief propagation or variational inference methods are the options. When dealing with large scale of text data, parallel algorithms could be the right way to go.

References

[1] A. Ahmed and E. Xing. Timeline: A dynamic hierarchical dirichlet process model for recovering birth/death and evolution of topics in text stream. *Uncertainty in Artificial Intelligence*, 2010.

[2] A. Ahmed and E. P. Xing. Dynamic non-parametric mixture models and the recurrent chinese restaurant process: with applications to evolutionary clustering. In *SDM*, pages 219–230, 2008.

[3] C. Andrieu, N. De Freitas, A. Doucet, and M. Jordan. An introduction to mcmc for machine learning. *Machine learning*, 50(1):5–43, 2003.

[4] D. Andrzejewski, X. Zhu, and M. Craven. Incorporating domain knowledge into topic modeling via dirichlet forest priors. In *Proceedings of the 26th Annual International Conference on Machine Learning*, ICML '09, pages 25–32, New York, NY, USA, 2009. ACM.

[5] L. Baum, T. Petrie, G. Soules, and N. Weiss. A maximization technique occurring in the statistical analysis of probabilistic functions of markov chains. *The annals of mathematical statistics*, 41(1):164–171, 1970.

[6] D. Bikel, R. Schwartz, and R. Weischedel. An algorithm that learns what's in a name. *Machine learning*, 34(1):211–231, 1999.

[7] J. Bilmes. A gentle tutorial of the EM algorithm and its application to parameter estimation for Gaussian mixture and hidden Markov models. Technical Report TR-97-021, ICSI, 1997.

[8] C. Bishop. Pattern recognition and machine learning. Springer, New York, 2006.

[9] D. M. Blei, T. L. Griffiths, and M. I. Jordan. The nested chinese restaurant process and bayesian nonparametric inference of topic hierarchies. *J. ACM*, Aug 2009.

[10] D. M. Blei and M. I. Jordan. Variational inference for dirichlet process mixtures. *Bayesian Analysis*, 1:121–144, 2005.

[11] D. M. Blei, A. Ng, and M. Jordan. Latent dirichlet allocation. *JMLR*, 3:993–1022, 2003.

[12] S. Borman. The expectation maximization algorithm: A short tutorial. *Unpublished Technical report*, 2004.
Available online at http://www.seanborman.com/publications.

[13] M. Chang, D. Goldwasser, D. Roth, and V. Srikumar. Discriminative learning over constrained latent representations. In *Proc. of the Annual Meeting of the North American Association of Computational Linguistics (NAACL)*, 6, 2010.

[14] M.-W. Chang, N. Rizzolo, and D. Roth. Integer linear programming in nlp – constrained conditional models. Tutorial, *NAACL*, 2010.

[15] H. Chen. Parallel implementations of probabilistic latent semantic analysis on graphic processing units. Computer science, University of Illinois at Urbana–Champaign, 2011.

[16] S. Chhabra, W. Yerazunis, and C. Siefkes. Spam filtering using a markov random field model with variable weighting schemas. In *ICDM Conference*, pages 347–350, 2004.

[17] C. T. Chu, S. K. Kim, Y. A. Lin, Y. Yu, G. R. Bradski, A. Y. Ng, and K. Olukotun. Map-Reduce for machine learning on multicore. In *NIPS*, pages 281–288, 2006.

[18] A. Clauset, C. R. Shalizi, and M. E. J. Newman. Power-law distributions in empirical data. *SIAM Rev.*, 51:661–703, November 2009.

[19] F. Cozman. Generalizing variable elimination in bayesian networks. In *Workshop on Probabilistic Reasoning in Artificial Intelligence*, pages 27–32, 2000.

[20] L. de Campos, J. Fernández-Luna, and J. Huete. Bayesian networks and information retrieval: an introduction to the special issue. *Information processing & management*, 40(5):727–733, 2004.

[21] F. Dellaert. The expectation maximization algorithm. Technical report, 2002.

[22] A. P. Dempster, N. M. Laird, and D. B. Rubin. Maximum likelihood from incomplete data via the em algorithm. *Journal of the Royal Statistical Society, Series B*, 39(1):1–38, 1977.

[23] J. R. Finkel, T. Grenager, and C. D. Manning. Incorporating non-local information into information extraction systems by gibbs sampling. In *ACL*, 2005.

[24] G. Forney Jr. The viterbi algorithm. *Proceedings of the IEEE*, 61(3):268–278, 1973.

[25] N. Friedman, L. Getoor, D. Koller, and A. Pfeffer. Learning probabilistic relational models. In *International Joint Conference on Artificial Intelligence*, volume 16, pages 1300–1309, 1999.

[26] K. Ganchev, J. A. Graça, J. Gillenwater, and B. Taskar. Posterior regularization for structured latent variable models. *Journal of Machine Learning Research*, 11:2001–2049, Aug. 2010.

[27] T. Griffiths and Z. Ghahramani. Infinite latent feature models and the indian buffet process. In *NIPS*, pages 475–482, 2005.

[28] T. L. Griffiths and M. Steyvers. Finding scientific topics. *PNAS*, 101(suppl. 1):5228–5235, 2004.

[29] T. Hofmann. Probabilistic latent semantic analysis. In *Proceedings of Uncertainty in Artificial Intelligence, UAI*, 1999.

[30] T. Hofmann. Probabilistic latent semantic indexing. In *ACM SIGIR Conference*, pages 50–57, 1999.

[31] C. Hong, W. Chen, W. Zheng, J. Shan, Y. Chen, and Y. Zhang. Parallelization and characterization of probabilistic latent semantic analysis. *International Conference on Parallel Processing*, 0:628–635, 2008.

[32] M. I. Jordan. Graphical models. *Statistical Science*, 19(1):140–155, 2004.

[33] M. I. Jordan. Dirichlet processes, chinese restaurant processes and all that. *Tutorial presentation at the NIPS Conference*, 2005.

[34] C. T. Kelley. Iterative methods for optimization. *Frontiers in Applied Mathematics*, SIAM, 1999.

[35] R. Kindermann, J. Snell, and A. M. Society. *Markov random fields and their applications*. American Mathematical Society Providence, RI, 1980.

[36] D. Koller and N. Friedman. *Probabilistic graphical models*. MIT press, 2009.

[37] J. Kupiec. Robust part-of-speech tagging using a hidden markov model. *Computer Speech & Language*, 6(3):225–242, 1992.

[38] J. D. Lafferty, A. McCallum, and F. C. N. Pereira. Conditional random fields: Probabilistic models for segmenting and labeling sequence data. In *ICML*, pages 282–289, 2001.

[39] J.-M. Marin, K. L. Mengersen, and C. Robert. Bayesian modelling and inference on mixtures of distributions. In D. Dey and C. Rao, editors, *Handbook of Statistics: Volume 25*. Elsevier, 2005.

[40] A. McCallum and W. Li. Early results for named entity recognition with conditional random fields, feature induction and web-enhanced lexicons. In *Proceedings of the seventh conference on Natural language learning at HLT-NAACL 2003-Volume 4*, pages 188–191. Association for Computational Linguistics, 2003.

[41] Q. Mei, D. Cai, D. Zhang, and C. Zhai. Topic modeling with network regularization. In *WWW Conference*, 2008.

[42] Q. Mei, X. Ling, M. Wondra, H. Su, and C. Zhai. Topic sentiment mixture: modeling facets and opinions in weblogs. In *WWW Conference*, pages 171–180, 2007.

[43] Q. Mei and C. Zhai. A mixture model for contextual text mining. In *ACM KDD Conference*, pages 649–655, 2006.

[44] D. Metzler and W. Croft. A markov random field model for term dependencies. In *ACM SIGIR Conference*, pages 472–479, 2005.

[45] T. Minka. Expectation propagation for approximate bayesian inference. In *Uncertainty in Artificial Intelligence*, volume 17, pages 362–369, 2001.

[46] K. Murphy, Y. Weiss, and M. Jordan. Loopy belief propagation for approximate inference: An empirical study. In *Proceedings of Uncertainty in AI*, volume 9, pages 467–475, 1999.

[47] R. Nallapati, W. Cohen, and J. Lafferty. Parallelized variational em for latent dirichlet allocation: An experimental evaluation of speed and scalability. In *Proceedings of the Seventh IEEE International Conference on Data Mining Workshops*, pages 349–354, 2007.

[48] R. M. Neal. Markov chain sampling methods for dirichlet process mixture models. *Journal of Computational and Graphical Statistics*, 9(2):249–265, 2000.

[49] D. Newman, A. Asuncion, P. Smyth, and M. Welling. Distributed inference for latent dirichlet allocation. In *NIPS Conference*, 2007.

[50] K. Nigam, A. K. McCallum, S. Thrun, and T. Mitchell. Text classification from labeled and unlabeled documents using em. *Machine Learning*, 39:103–134, May 2000.

[51] P. Orbanz and Y. W. Teh. Bayesian nonparametric models. In *Encyclopedia of Machine Learning*, pages 81–89. 2010.

[52] J. Pitman and M. Yor. The Two-Parameter Poisson-Dirichlet distribution derived from a stable subordinator. *The Annals of Probability*, 25(2):855–900, 1997.

[53] I. Porteous, D. Newman, A. Ihler, A. Asuncion, P. Smyth, and M. Welling. Fast collapsed gibbs sampling for latent dirichlet allocation. In *ACM KDD Conference*, pages 569–577, 2008.

[54] V. Punyakanok, D. Roth, W. Yih, and D. Zimak. Learning and inference over constrained output. In *Proc. of the International Joint Conference on Artificial Intelligence (IJCAI)*, pages 1124–1129, 2005.

[55] L. Rabiner. A tutorial on hidden markov models and selected applications in speech recognition. *Proceedings of the IEEE*, 77(2):257–286, 1989.

[56] L. R. Rabiner and B. H. Juang. An introduction to hidden Markov models. *IEEE ASSP Magazine*, pages 4–15, January 1986.

[57] C. E. Rasmussen. The infinite gaussian mixture model. In *In Advances in Neural Information Processing Systems 12*, volume 12, pages 554–560, 2000.

[58] C. E. Rasmussen and C. Williams. *Gaussian Processes for Machine Learning*. MIT Press, 2006.

[59] M. Richardson and P. Domingos. Markov logic networks. *Machine Learning*, 62(1):107–136, 2006.

[60] D. Roth and W. Yih. Integer linear programming inference for conditional random fields. In *International Conference on Machine Learning (ICML)*, pages 737–744, 2005.

[61] M. Sahami, S. Dumais, D. Heckerman, and E. Horvitz. A bayesian approach to filtering junk e-mail. In *AAAI Workshop on Learning for Text Categorization*, 1998.

[62] J. Sethuraman. A constructive definition of dirichlet priors. *Statistica Sinica*, 4:639–650, 1994.

[63] F. Sha and F. Pereira. Shallow parsing with conditional random fields. In *Proceedings of the 2003 Conference of the North American Chapter of the Association for Computational Linguistics on Human Language Technology-Volume 1*, pages 134–141, 2003.

[64] Y. Sun, J. Han, J. Gao, and Y. Yu. itopicmodel: Information network-integrated topic modeling. In *ICDM*, pages 493–502, 2009.

[65] C. Sutton and A. McCallum. An introduction to conditional random fields for relational learning. *Introduction to statistical relational learning*, pages 95–130, 2006.

[66] Y. W. Teh. A hierarchical bayesian language model based on pitman-yor processes. In *Proceedings of the 21st International Conference on Computational Linguistics and the 44th annual meeting*

of the Association for Computational Linguistics, ACL-44, pages 985–992, 2006.

[67] Y. W. Teh. Dirichlet processes. In *Encyclopedia of Machine Learning*. Springer, 2010.

[68] Y. W. Teh, M. I. Jordan, M. J. Beal, and D. M. Blei. Hierarchical Dirichlet processes. *Journal of the American Statistical Association*, 101(476):1566–1581, 2006.

[69] R. Thibaux and M. I. Jordan. Hierarchical beta processes and the indian buffet process. *Journal of Machine Learning Research – Proceedings Track*, 2:564–571, 2007.

[70] Y. Wang, H. Bai, M. Stanton, W.-Y. Chen, and E. Y. Chang. Plda: Parallel latent dirichlet allocation for large-scale applications. In *Proceedings of the 5th International Conference on Algorithmic Aspects in Information and Management*, pages 301–314, 2009.

[71] C. Zhai, A. Velivelli, and B. Yu. A cross-collection mixture model for comparative text mining. In *ACM KDD Conference*, pages 743–748, 2004.

[72] J. Zhang, Y. Song, C. Zhang, and S. Liu. Evolutionary hierarchical dirichlet processes for multiple correlated time-varying corpora. In *ACM KDD Conference*, pages 1079–1088, New York, NY, USA, 2010. ACM.

[73] X. Zhu, Z. Ghahramani, and J. Lafferty. Time-sensitive dirichlet process mixture models. Technical report, Carnegie Mellon University, 2005.

Chapter 9

MINING TEXT STREAMS

Charu C. Aggarwal
IBM T. J. Watson Research Center
Hawthorne, NY 10532, USA
charu@us.ibm.com

Abstract The large amount of text data which are continuously produced over time in a variety of large scale applications such as social networks results in massive streams of data. Typically massive text streams are created by very large scale interactions of individuals, or by structured creations of particular kinds of content by dedicated organizations. An example in the latter category would be the massive text streams created by news-wire services. Such text streams provide unprecedented challenges to data mining algorithms from an efficiency perspective. In this chapter, we review text stream mining algorithms for a wide variety of problems in data mining such as clustering, classification and topic modeling. We also discuss a number of future challenges in this area of research.

Keywords: Text Mining, Data Streams

1. Introduction

Text streams have become ubiquitous in recent years because of a wide variety of applications in social networks, news collection, and other forms of activity which result in the continuous creation of massive streams. Some specific examples of applications which create text streams are as follows:

- In social networks, users continuously communicate with one another with the use of text messages. This results in massive volumes of text streams which can be leveraged for a variety of mining and search purposes in the social network. This is because the text messages are reflective of user interests. A similar observation applies to chat and email networks.

- Many news aggregator services [1] may receive large volumes of news articles continuously over time. Such articles are often longer and more well structured than the kinds of messages which are seen in social, chat, or email networks.

- Many web crawlers may collect a large volume of documents from networks in a small time frame. In many cases, such documents are restricted to those which have been modified in a small time frame. This naturally results in a stream of modified documents.

Text streams create a huge challenge from the perspective of a wide variety of mining applications, because of the massive volume of the data which must be processed in online fashion. Data streams have been studied extensively in recent years not just in the text domain, but also in the context of a wide variety of multi-dimensional applications such as numerical and categorical data. A detailed discussion of mining algorithms for stream data may be found in [1]. While many of the techniques proposed for multi-dimensional data [1] can be generalized to text data at the high level, the details can be quite different because of the very different format and lack of structure of text data. Since stream mining techniques are generally dependent upon summarization, it follows that methods for online summarization need to be designed which work well for the unstructured nature of text data.

In the case of multi-dimensional and time-series data, such summarization often takes the form of methods such as histograms, wavelets, and sketches which can be used in order to create a structured summary of the underlying data [1]. However, the unstructured nature of text makes the use of such summaries quite challenging. While sketches have been used to some effect in the text domain [23], it has generally been difficult to generalize wavelet and histogram methods to the text domain. As we will see later in this section, the summarization methods designed for text streaming problems may vary a lot, and may often need to be tailored to the problem at hand. One of the goals of this chapter is to provide a broad spectrum of the different methods which are used for text mining, which can provide an overview of the tools which can be most effectively used for the text stream scenario. We will also present the future challenges and research directions associated with text stream mining.

---

[1] An example would be the *Google News* service.

This chapter is organized as follows. In section 2, we will present a variety of the well known algorithms for clustering text streams. This includes popular methods for topic detection and tracking in text stream. This is because the process of event detection is closely related to the clustering problem. The methods for classification of text streams are reviewed in section 3. Section 4 presents methods for evolution analysis of text stream. The conclusions and summary are presented in section 5.

2. Clustering Text Streams

The problem of clustering text streams has been widely studied in the context of numerical data [1, 4, 12]. Other popular methods which have been studied in the machine learning literature include the COBWEB and CLASSIT methods [20, 25]. The COBWEB algorithm assumes nominal attributes, whereas the CLASSIT algorithm assumes real-valued attributes. Many of these methods [4, 12] are extensions of the k-means method as extended to the stream scenario. This trend has also been applied to the case of text streams. One of the earliest methods for streaming text clustering is discussed in [54]. This technique is referred to as the *Online Spherical k-Means Algorithm (OSKM)*, which reflects the broad approach used by the methodology. This techniques divides up the incoming stream into small segments, each of which can be processed effectively in main memory. A set of k-means iterations are applied to each such data segment in order to cluster them. The advantage of using a segment-wise approach for clustering is that since each segment can be held in main memory, we can process each data point multiple times as long as it is held in main memory. In addition, the centroids from the previous segment are used in the next iteration for clustering purposes. A decay factor is introduced in order to age-out the old documents, so that the new documents are considered more important from a clustering perspective. This approach has been shown to be extremely effective in clustering massive text streams in [54].

The method in [54] is designed as a flat clustering algorithm, in which there is a single level to the clustering process. In many applications, it is useful to design hierarchical clustering algorithms in which different levels of the clustering can be explored. In the context of text data, this implies that different levels of topics and subtopics can be explored with the use of a hierarchical clustering process. A distributional modeling method for hierarchical clustering of streaming documents has been proposed in [37]. The method extends the COBWEB and CLASSIT algorithms [20, 25] to the case of text data. The work in [37] studies the different kinds of distributional assumptions of words in documents. We note that these distributional assumptions are required to adapt these algorithms to the case of text data. The approach essentially changes the distributional assumption so that the method can work effectively for text data.

A different method for clustering massive text and categorical data streams is discussed in [3]. The method discussed in [3] uses an approach which examines the relationship between outliers, emerging trends, and clusters in the underlying data. Old clusters may become inactive, and eventually get replaced by new clusters. Similarly, when newly arriving data points do not naturally fit in any particular cluster, these need to be initially classified as outliers. However, as time progresses, these new points may create a distinctive pattern of activity which can be recognized as a new cluster. The temporal locality of the data stream is manifested by these new clusters. For example, the first web page belonging to a particular category in a crawl may be recognized as an outlier, but may later form a cluster of documents of its own. On the other hand, the new outliers may not necessarily result in the formation of new clusters. Such outliers are true short-term abnormalities in the data since they do not result in the emergence of sustainable patterns. The approach discussed in [3] recognizes new clusters by first recognizing them as outliers.

This approach works with the use of a summarization methodology, which is motivated by the micro-clustering approach proposed in [4]. While the concept of micro-clustering was designed for numerical data, it can also be extended to the case of text and categorical data streams. This methodology essentially creates summaries from the data points which are used in order to estimate the assignment of incoming data points to clusters. The concept of micro-clusters is generalized to that of *condensed droplets* in [3].

In order to ensure greater importance of more recent data, a time-sensitive weightage is assigned to each data point. It is assumed that each data point has a time-dependent weight defined by the function $f(t)$. The function $f(t)$ is also referred to as the *fading function*. The fading function $f(t)$ is a non-monotonic decreasing function which decays uniformly with time t. In order to formalize this concept, we will define the *half-life* of a point in the data stream.

DEFINITION 9.1 *The half life t_0 of a point is defined as the time at which $f(t_0) = (1/2)f(0)$.*

Conceptually, the aim of defining a half life is to quantify the rate of decay of the importance of each data point in the stream clustering process. The *decay-rate* is defined as the inverse of the half life of the data stream. We denote the decay rate by $\lambda = 1/t_0$. We denote the weight function of each point in the data stream by $f(t) = 2^{-\lambda \cdot t}$. From the perspective of the clustering process, the weight of each data point is $f(t)$. It is easy to see that this decay function creates a half life of $1/\lambda$. It is also evident that by changing the value of λ, it is possible to change the rate at which the importance of the historical information in the data stream decays. The higher the value of λ, the lower the importance of the historical information compared to more recent data. For more stable data streams, it is desirable to pick a smaller value of λ, whereas for rapidly evolving data streams, it is desirable to pick a larger value of λ.

When a cluster is created during the streaming process by a newly arriving data point, it is allowed to remain as a trend-setting outlier for at least one half-life. During that period, if at least one more data point arrives, then the cluster becomes an active and mature cluster. On the other hand, if no new points arrive during a half-life, then the trend-setting outlier is recognized as a true anomaly in the data stream. At this point, this anomaly is removed from the list of current clusters. We refer to the process of removal as *cluster death*. Thus, a new cluster containing one data point dies when the (weighted) number of points in the cluster is 0.5. The same criterion is used to define the death of mature clusters. A necessary condition for this criterion to be met is that the inactivity period in the cluster has exceeded the half life $1/\lambda$. The greater the number of points in the cluster, the greater the level by which the inactivity period would need to exceed its half life in order to meet the criterion. This is a natural solution, since it is intuitively desirable to have stronger requirements (a longer inactivity period) for the death of a cluster containing a larger number of points.

Next, we describe the process of creation of a condensed-droplet from the underlying text stream. An important point to remember is that a text data set can be treated as a *sparse numeric* data set. This is because most documents contain only a small fraction of the vocabulary with non-zero frequency. For a cluster of documents $\mathcal{C}$ at time t, we denote the corresponding condensed droplet by $\mathcal{D}(t, \mathcal{C})$.

DEFINITION 9.2 *A cluster droplet $\mathcal{D}(t, \mathcal{C})$ for a set of text data points $\mathcal{C}$ at time t is defined to as a tuple $(\overline{DF2}, \overline{DF1}, n, w(t), l)$. Each tuple component is defined as follows:*

- *The vector $\overline{DF2}$ contains $3 \cdot wb \cdot (wb-1)/2$ entries. Here wb is the number of distinct words in the cluster $\mathcal{C}$. For each pair of dimensions, we maintain a list of the pairs of word ids with non-*

zero counts. We also maintained the sum of the weighted counts for such word pairs.

- The vector $\overline{DF1}$ contains $2 \cdot wb$ entries. We maintain the identities of the words with non-zero counts. In addition, we maintain the sum of the weighted counts for each word occurring in the cluster.

- The entry n contains the number of data points in the cluster.

- The entry $w(t)$ contains the sum of the weights of the data points at time t. We note that the value $w(t)$ is a function of the time t and decays with time unless new data points are added to the droplet $\mathcal{D}(t)$.

- The entry l contains the time stamp of the last time that a data point was added to the cluster.

The concept of cluster droplet has some interesting properties that will be useful during the maintenance process. These properties relate to the additivity and decay behavior of the cluster droplet.

OBSERVATION 2.1 *Consider the cluster droplets* $\mathcal{D}(t, \mathcal{C}_1) = (\overline{DF2_1}, \overline{DF1_1}, n_1, w(t)_1, l_1)$ *and* $\mathcal{D}(t, \mathcal{C}_2) = (\overline{DF2_2}, \overline{DF1_2}, n_2, w(t)_2, l_2)$. *Then the cluster droplet* $\mathcal{D}(t, \mathcal{C}_1 \cup \mathcal{C}_2)$ *is defined by the tuple* $(\overline{DF2_1} + \overline{DF2_2}, \overline{DF1_1} + \overline{DF1_2}, n_1 + n_2, w(t)_1 + w(t)_2, max\{l_1, l_2\})$.

The cluster droplet for the union of two clusters is the sum of individual entries. The only exception is the last entry which is the maxima of the two last-update times. We note that the additivity property provides considerable convenience for data stream processing since the entries can be updated efficiently using simple additive operations.

The second observation relates to the rate of decay of the condensed droplets. Since the weights of each data point decay with the passage of time, the corresponding entries also decay at the same rate. Correspondingly, we make the following observation:

OBSERVATION 2.2 *Consider the cluster droplet* $\mathcal{D}(t, \mathcal{C}) = (\overline{DF2}, \overline{DF1}, n, w(t), l)$. *Then the entries of of the same cluster droplet* $\mathcal{C}$ *at a time* $t' > t$ *are given by* $\mathcal{D}(t', \mathcal{C}) = (\overline{DF2} \cdot 2^{-\lambda \cdot (t'-t)}, \overline{DF1} \cdot 2^{-\lambda \cdot (t'-t)}, n, w(t) \cdot 2^{-\lambda \cdot (t'-t)}, l)$.

The second observation is critical in regulating the rate at which the cluster droplets are updated during the clustering process. Since all cluster droplets decay at essentially the same rate (unless new data points are added), it follows that it is not necessary to update the decay statistics

Mining Text Streams 303

at each time stamp. Rather, each cluster droplet can be updated lazily, whenever new data points are added to it.

The overall algorithm proceeds as follows. At the beginning of algorithmic execution, we start with an empty set of clusters. As new data points arrive, unit clusters containing individual data points are created. Once a maximum number k of such clusters have been created, we can begin the process of online cluster maintenance. Thus, we initially start off with a trivial set of k clusters. These clusters are updated over time with the arrival of new data points.

When a new data point $\overline{X}$ arrives, its similarity to each cluster droplet is computed. In the case of text data sets, the cosine similarity measure [17, 40] between $\overline{DF1}$ and $\overline{X}$ is used. The similarity value $S(\overline{X}, C_j)$ is computed from the incoming document $\overline{X}$ to every cluster C_j. The cluster with the maximum value of $S(\overline{X}, C_j)$ is chosen as the relevant cluster for data insertion. Let us assume that this cluster is C_{mindex}. We use a threshold denoted by *thresh* in order to determine whether the incoming data point is an outlier. If the value of $S(\overline{X}, C_{mindex})$ is larger than the threshold *thresh*, then the point $\overline{X}$ is assigned to the cluster C_{mindex}. Otherwise, we check if some inactive cluster exists in the current set of cluster droplets. If no such inactive cluster exists, then the data point $\overline{X}$ is added to C_{mindex}. On the other hand, when an inactive cluster does exist, a new cluster is created containing the solitary data point $\overline{X}$. This newly created cluster replaces the inactive cluster. We note that this new cluster is a potential true outlier or the beginning of a new trend of data points. Further understanding of this new cluster may only be obtained with the progress of the data stream.

In the event that $\overline{X}$ is inserted into the cluster C_{mindex}, we need to perform two steps:

- We update the statistics to reflect the decay of the data points at the current moment in time. This updating is performed using the computation discussed in Observation 2.2. Thus, the relevant updates are performed in a "lazy" fashion. In other words, the statistics for a cluster do not decay, until a new point is added to it. Let the corresponding time stamp of the moment of addition be t. The last update time l is available from the cluster droplet statistics. We multiply the entries in the vectors $\overline{DC2}$, $\overline{DC1}$ and $w(t)$ by the factor $2^{-\lambda \cdot (t-l)}$ in order to update the corresponding statistics. We note that the lazy update mechanism results in stale decay characteristics for most of the clusters. This does not however affect the afore-discussed computation of the similarity measures.

- In the second step, we add the statistics for each newly arriving data point to the statistics for $\mathcal{C}_{mindex}$ by using the computation discussed in Observation 2.2.

In the event that the newly arriving data point does not naturally fit in any of the cluster droplets and an inactive cluster does exist, then we replace the most inactive cluster by a new cluster containing the solitary data point $\overline{X}$. In particular, the replaced cluster is the least recently updated cluster among all inactive clusters. This process is continuously performed over the life of the data stream, as new documents arrive over time. The work in [3] also presents a variety of other applications of the stream clustering technique such as evolution and correlation analysis.

A different way of utilizing the temporal evolution of text documents in the clustering process is described in [26]. Specifically, the work in [26] uses *bursty features* as markers of new topic occurrences in the data stream. This is because the semantics of an up-and-coming topic are often reflected in the frequent presence of a few distinctive words in the text stream. A specific example illustrates the bursty features in two topics corresponding to the two topics of *"US Mid-Term Elections"* and *"Newt Gingrich resigns from House"* respectively. The corresponding bursty features [26] which occurred frequently in the newsstream during the period for these topics were as follows:

US Mid-term Elections (Nov. 3, 1998 burst):
election, voters, Gingrich, president, Newt, ...

Newt Gingrich resigns from house (Nov 6, 1998 burst):
House, Gingrich, Newt, president, Washington ...

It is evident that at a given period in time, the nature of relevant topics could lead to bursts in specific features of the data stream. Clearly, such features are extremely important from a clustering perspective. Therefore, the method discussed in [26] uses a new representation, which is referred to as the *bursty feature representation* for mining text streams. In this representation, a time-varying weight is associated with the features depending upon its burstiness. This also reflects the varying importance of the feature to the clustering process. Thus, it is important to remember that a particular document representation is dependent upon the particular instant in time at which it is constructed.

Another issue which is handled effectively in this approach is an implicit reduction in dimensionality of the underlying collection. Text is inherently a high dimensional data domain, and the pre-selection of some of the features on the basis of their burstiness can be a natural way to reduce the dimensionality of document representation. This can help in both the effectiveness and efficiency of the underlying algorithm.

The first step in the process is to identify the bursty features in the data stream. In order to achieve this goal, the approach uses Kleinberg's 2-state finite automaton model [27]. Once these features have been identified, the bursty features are associated with weights which depend upon their level of burstiness. Subsequently, a bursty feature representation is defined in order to reflect the underlying weight of the feature. Both the identification and the weight of the bursty feature are dependent upon its underlying frequency. A standard k-means approach is applied to the new representation in order to construct the clustering. It was shown in [26] that the approach of using burstiness improves the cluster quality. Once criticism of the work in [26] is that it is mostly focussed on the issue of improving effectiveness with the use of temporal characteristics of the data stream, and does not address the issue of efficient clustering of the underlying data stream.

In general, it is evident that feature extraction is important for all clustering algorithms. While the work in [26] focusses on using temporal characteristics of the stream for feature extraction, the work in [32] focusses on using *phrase extraction* for effective feature selection. This work is also related to the concept of topic-modeling, which will be discussed somewhat later. This is because the different topics in a collection can be related to the clusters in a collection. The work in [32] uses topic-modeling techniques for clustering. The core idea in the work of [32] is that individual words are not very effective for a clustering algorithm because they miss the context in which the word is used. For example, the word "star" may either refer to a celestial body or to an entertainer. On the other hand, when the phrase "fixed star" is used, it becomes evident that the word "star" refers to a celestial body. The phrases which are extracted from the collection are also referred to as *topic signatures*.

The use of such phrasal clarification for improving the quality of the clustering is referred to as *semantic smoothing* because it reduces the noise which is associated with semantic ambiguity. Therefore, a key part of the approach is to extract phrases from the underlying data stream. After phrase extraction, the training process determines a translation probability of the phrase to terms in the vocabulary. For example, the word "planet" may have high probability of association with the phrase "fixed star", because both refer to celestial bodies. Therefore, for a given document, a rational probability count may also be assigned to all terms. For each document, it is assumed that all terms in it are generated either by a topic-signature model, or a background collection model.

The approach in [32] works by modeling the soft probability $p(w|C_j)$ for word w and cluster C_j. The probability $p(w|C_j)$ is modeled as a linear combination of two factors; (a) A maximum likelihood model which computes the probabilities of generating specific words for each cluster (b) An indirect (translated) word-membership probability which first determines the maximum likelihood probability for each topic-signature, and then multiplying with the conditional probability of each word, given the topic-signature. We note that we can use $p(w|C_j)$ in order to estimate $p(d|C_j)$ by using the product of the constituent words in the document. For this purpose, we use the frequency $f(w,d)$ of word w in document d.

$$p(d|C_j) = \prod_{w \in d} p(w|C_j)^{f(w,d)} \quad (9.1)$$

We note that in the static case, it is also possible to add a background model in order to improve the robustness of the estimation process. This is however not possible in a data stream because of the fact that the background collection model may require multiple passes in order to build effectively. The work in [32] maintains these probabilities in online fashion with the use of a *cluster profile*, that weights the probabilities with the use of a fading function. We note that the concept of cluster profile is analogous to the concept of condensed droplet introduced in [3]. The key algorithm (denoted by OCTS) is to maintain a dynamic set of clusters into which documents are progressively assigned with the use of similarity computations. It has been shown in [32] how the cluster profile can be used in order to efficiently compute $p(d|C_j)$ for each incoming document. This value is then used in order to determine the similarity of the documents to the different clusters. This is used in order to assign the documents to their closest cluster. We note that the methods in [3, 32] share a number of similarities in terms of (a) maintenance of cluster profiles, and (b) use of cluster profiles (or condensed droplets) to compute similarity and assignment of documents to most similar clusters. (c) The rules used to decide when a new singleton cluster should be created, or one of the older clusters should be replaced.

The main difference between the two algorithms is the technique which is used in order to compute cluster similarity. The OCTS algorithm uses the probabilistic computation $p(d|C_j)$ to compute cluster similarity, which takes the phrasal information into account during the computation process. One observation about OCTS is that it may allow for very similar clusters to co-exist in the current set. This reduces the space available for distinct cluster profiles. A second algorithm called OCTSM is also proposed in [32], which allows for merging of very similar clusters. Before each assignment, it checks whether pairs of similar clusters can be merged on the basis of similarity. If this is the case, then we allow the merging of the similar clusters and their corresponding cluster profiles. Detailed experimental results on the different clustering algorithms and their effectiveness are presented in [32].

Another method [42] uses a combination of a spectral partitioning and probabilistic modeling method for novelty detection and topic tracking. This approach uses a HITS-like spectral technique within a probabilistic framework. The probabilistic part is an unsupervised boosting method, which is closer to semi-parametric maximum likelihood methods.

A closely related area to clustering is that of topic modeling, which is a problem closely related to that of clustering. In the problem of topic modeling, we perform a *soft* clustering of the data in which each document has a membership probability to one of a universe of topics rather than a deterministic membership. A variety of mixture modeling techniques can be used in order to determine the topics from the underlying data. Recently, the method has also been extended to the *dynamic* case which is helpful for topic modeling of text streams [10]. A closely related topic is that of topic *detection* and *tracking*, which is discussed below.

Recently, a variety of methods for maintaining topic models in a streaming scenario have been proposed in [49]. The work evaluates a number of different methods for adapting topic models to the streaming scenario. These include methods such as Gibbs sampling and variational inference. In addition, a method is also proposed, which is based on text classification. The latter has the advantage of requiring only a single matrix multiplication, and is therefore much more efficient. A method called *SparseLDA* is proposed, which is an effective method for evaluating Gibbs sampling distributions. The results in [49] show that this method is 20 times faster than traditional LDA.

In some applications, it is desirable to have clusters of approximately balanced size, in which a particular cluster is not significantly larger than the others. A competitive online learning for determining such balanced clusters have been proposed in [9]. The essential approach in [9] is to design a model in which it is harder for new data points to join larger clusters. This is achieved by penalizing the imbalance into the objective function criterion. It was shown in [9] that such an approach can determine well balanced clusters.

2.1 Topic Detection and Tracking in Text Streams

A closely related problem to clustering is that of *topic detection and tracking* [5, 11, 24, 46, 47, 51]. In this problem, we create unsupervised clusters from a text stream, and then determine the sets of clusters which match real events. These real events may correspond to documents which are identified by a human. Since the problem of online topic detection is closely related to that of clustering, we will discuss this problem as a subsection of our broader discussion on clustering, though not all methods for topic detection use clustering techniques. In this subsection, we will discuss all the methods for topic detection, whether they use clustering or not.

The earliest work on topic detection and tracking was performed in [5, 47]. The work arose out of a DARPA initiative [55] which formally defined this problem and proposed the initial set of algorithms for the task. An interesting technique in [47] designs methods which can be used for both retrospective and online topic tracking and detection. In retrospective event detection, we create groups from a corpus of documents (or stories), and each group corresponds to an event. The online version is applicable to the case of data streams, and in this case, we process documents sequentially in order to determine whether an incoming document corresponds to a new event. An online clustering algorithm can also be used in order to track the different events in the data in the form of clusters. For each incoming document $\overline{X}$ we compute its similarity to the last m documents $\overline{Y_{t-1}} \ldots \overline{Y_{t-m}}$. The score for the incoming document $\overline{X}$ is computed as follows:

$$score(\overline{X}) = \max_{i \in \{t-m \ldots t-1\}} (1 - sim(\overline{X}, \overline{Y_i})) \quad (9.2)$$

A document is considered novel, when its score is above a pre-defined threshold. In addition, a decay-weighted version is designed in which the weight of documents depends upon its recency. The idea here is that a document is considered to be a new event when the last occurrence of a similar document did not occur recently "enough". In this case, the corresponding score is designed as follows:

$$score(\overline{X}) = \max_{i \in \{t-m \ldots t-1\}} (1 - \frac{(m+i-t+1)}{m} \cdot sim(\overline{X}, \overline{Y_i})) \quad (9.3)$$

We note that each $\overline{Y_i}$ need not necessarily represent a single document, but may also represent a cluster or a larger grouping. We also note that the method for detecting a new event can be combined with a cluster tracking task. This is because the determination of a new event is indicative of a new event (or singleton cluster) in the data.

The work in [5] addressed the problem of new event detection by examining the relationship of a current document to the previous documents in the data. The key idea is to use feature extraction and selection techniques in order to design a query representation of the document content. We determine the query's initial threshold by comparing the document content with the query, and set it as the triggering threshold. Then, we determine if the document triggers any queries from the previous documents in the collection. If no queries are triggered, then the document is deemed to be a new event. Otherwise, this document is not a new event.

One general observation about the online topic detection and tracking problem [6, 48] is that this problem is quite hard in general, and the performance of first event detection can degrade under a variety of scenarios. In order to improve the effectiveness of first event detection, the work in [48] proposes to use the training data of old events in order to learn useful statistics for the prediction of new events. The broad approach in [48] contains the following components:

- The documents are classified into broad topics, each of which consists of multiple events.

- Named entities are identified, optimizing their weight relative to normal words for each topic, and computing a stopword list for each topic.

- Measuring the novelty of a new document conditioned on the system-predicted topic for that document.

Clearly such an approach has the tradeoff that it requires prior knowledge about the collection in the form of training data, but provides better accuracy. More details on the approach may be found in [48].

A method proposed in [11] is quite similar to that proposed in [47], except that it proposes a number of improvements in how the tf–idf model is incrementally maintained for computation of similarity. For example, a *source-specific* tf–idf model is maintained in which the statistics are specific to each news source. Similarly, the approach normalizes for the fact that documents which come from the same source tend to have a higher similarity because of the vocabulary conventions which are often used in the similarity computation. In order to achieve this goal, the approach computes the average similarity between the documents from a particular pair of sources and subtracts this average value while computing the similarity between a pair of documents for the purpose of new event detection. A number of other techniques for improving the quality of event detection (such as using inverse event frequencies) are discussed in [11]. A method for online topic detection and tracking is presented in [51] as an application of stream clustering. This work uses a probabilistic LDA model in order to create an online model for estimating the growing number of clusters. The general approach in this work is similar to the concepts already proposed in [47]. The main novelty of the work is the design of an online approach for probabilistic clustering.

Another method for fast and parameter-free bursty event detection is proposed in [24]. This approach focusses on finding bursty features which characterize the presence of an event. In order to achieve this goal, the technique in [24] proposes a feature-pivot clustering, which groups features on the basis of bursty co-occurrence. The approach is designed to be parameter-free, which gives it an advantage in a number of scenarios.

The problem of event detection has also been studied with the use of *keyword graphs* in [39]. The work in [39] builds a keywords graph from a text stream in which a node corresponds to a keyword, and an edge is added to the keyword graphs when a pair of words occur together in the document. Events are characterized as communities of keywords in this network. This broad approach is used in the context of a *window-based technique* for the case of social streams.

In the context of social network streams, a natural question arises, as to whether one can use any of the *social dimensions* of the underlying stream in order to improve the underlying event detection process. Such an approach has been proposed in [53], in which events are determined by combining text clustering, social network structure clustering, and temporal segmentation. In addition, a multi-dimensional visualization tool is provided, which discusses ways of visualizing the relationships between the events along the three dimensions. In this case, an event is defined as a set of text documents which are semantically coherent, and are structurally well connected both in terms of social network structure and time. These three different characteristics are used in the following ways:

- First, the content is used in order to create a hierarchy of topics from the social text stream.

- While the topical patterns are useful for distinguishing content, the temporal segmentation is used in order to distinguish different events. The assumption is that the communication between different parties happen during a short contiguous time period.

- Since it is assumed that the events occur between connected entities, we use the connectivity between the different events in order to further segment the events. An edge between a pair of nodes corresponds to a communication between the social entities. Multiple edges are allowed between pairs of nodes. This structure is used in order to determine the event-dependent communities.

Another method [36] has been proposed in the context of the *Twitter* social network data set. In this paper, a locality sensitive hashing method (LSH) is used in order to keep track of the different documents. The idea in LSH [14] is to use a hashing scheme in which the probability of hashing two documents into the same bucket is proportional to the cosine of the angle between the two documents. For a given query document, we check all the documents in the same bucket, and then perform the similarity calculation explicitly with all documents in the same bucket. The closest document is returned as the nearest neighbor. We note that the problem of first-story detection can be considered an application of repeated nearest neighbor querying in which an incoming document is compared to the previously seen documents from the data stream. While LSH can be used directly in conjunction with a nearest neighbor search for first story detection, such an approach typically leads to poor results. This is because LSH works effectively only if the true nearest neighbor is close to the query point. Otherwise, such an approach is unable to find the true nearest neighbor. Therefore, the approach in [36] uses the strategy of using LSH only for declaring a document to be sufficiently new on an aggregate basis. For such cases, the document is compared against a fixed number of the most recent documents. In the event that the corresponding distance is above a given threshold, we can declare the underlying story as novel. The main advantage of this technique over many of the aforementioned techniques is that of *speed*, which is especially important, when dealing with social streams of very high volume. This speed is achieved because of the use of the LSH technique, though there is some loss in accuracy because of the approximation process. This technique was compared [36] against the *UMass system* [7], and it was found that more than an order of magnitude improvement in speed was obtained with only a minor loss in accuracy.

Most of the work in text stream mining and topic detection is performed in the context of a *single news stream*. In many cases, we have multiple text streams [44], which are indexed by the same set of time points (called coordinated text streams). For example, when a major event happens, all the news articles published by different agencies in different languages cover the same event in that period. This is referred to as a *correlated bursty topic pattern* in the different news article streams. In some cases, when the correlated streams are multi-lingual, they may even have completely different vocabulary. The discovery of bursty topic patterns can determine the interesting events and associations between different streams. Such an approach can also determine interesting local and stream-specific patterns by factoring out the global noise which is common to the different streams.

In order to achieve this goal, the technique in [44] aligns the text samples from different streams on the basis of the shared time-stamps. This is used in order to discover topics from multiple streams simultaneously with a single probabilistic mixture model. The approach of constructing independent topic models from different streams is that the topic models from the different streams may not match each other very well, or at least, create a challenge in matching the different topic models, if it is done at the end of the process of model construction. Therefore, it is important to make the mixture models for the different streams communicate with one another during the modeling process, so that a single mixture model is constructed across the different streams. In order to achieve this goal, the stream samples at a given point are merged together, while keeping the stream identities. In order to achieve this, we align the topic models from different streams while keeping their identities. While the topic models from different streams are separate, the global mixture model is designed for the overall text stream sample. Such a mixture model is coordinated, because it would emphasize topics which tend to occur across multiple streams. Once the coordinated mixture model is obtained, the topical models for the different streams can be extracted by fitting the mixture model to the different streams. As the topic models for the different streams ate aligned with one another, we can obtained a correlated bursty topic pattern, when the corresponding topic is bursty during the same period. A key aspect of this approach is that it does not require the different streams to share vocabulary. Rather it is assumed the topics involved in a correlated bursty topic pattern tend to have the same distribution across streams, and this can be used in order to match topics from different streams. More details on the approach may be found in [44].

3. Classification of Text Streams

The problem of classification of data streams has been widely studied by the data mining community in the context of different kinds of data [1, 2, 43, 50]. Many of the methods for classifying numerical data can be easily extended to the case of text data, just as the numerical clustering method in [4] has been extended to a text clustering method in [3]. In particular, the classification method of [2] can be extended to the text domain, by adapting the concept of numerical micro-clusters to condensed droplets as discussed in [3]. With the use of the condensed droplet data structure, it is possible to extend all the methods discussed in [2] to the text stream scenario. Similarly, the core idea in [43] uses an ensemble based approach on chunks of the data stream. This broad approach is essentially a *meta-algorithm* which is not too dependent upon the specifics of the underlying data format. Therefore, the broad method can also be easily extended to the case of text streams.

The problem of text stream classification arises in the context of a number of different IR tasks. The most important of these is *news filtering* [30], in which it is desirable to automatically assign incoming documents to pre-defined categories. A second application is that of email spam filtering [8], in which it is desirable to determine whether incoming email messages are spam or not. We note that the problem of text stream classification can arise in two different contexts, depending upon whether the training or the test data arrives in the form of a stream:

- In the first case, the training data may be available for batch learning, but the test data may arrive in the form of a stream.

- In the second case, both the training and the test data may arrive in the form of a stream. The patterns in the training data may continuously change over time, as a result of which the models need to be updated dynamically.

The first scenario is usually easy to handle, because most classifier models are compact and classify individual test instances efficiently. On the other hand, in the second scenario, the training model needs to be constantly updated in order to account for changes in the patterns of the underlying training data. The easiest approach to such a problem is to incorporate temporal decay factors into model construction algorithms, so as to age out the old data. This ensures that the new (and more timely data) is weighted more significantly in the classification model. An interesting technique along this direction has been proposed in [38], in which a temporal weighting factor is introduced in order to modify the classification algorithms. Specifically, the approach has been applied to the Naive Bayes, Rocchio, and k-nearest neighbor classification algorithms. It has been shown that the incorporation of temporal weighting factors is useful in improving the classification accuracy, when the underlying data is evolving over time.

A number of methods have also been designed specifically for the case of text streams. In particular, the method discussed in [23] studies methods for classifying text streams in which the classification model may evolve over time. This problem has been studied extensively in the literature in the context of multi-dimensional data streams [2, 43]. For example, in a spam filtering application, a user may generally delete the spam emails for a particular topic, such as those corresponding to political advertisements. However, in a particular period such as the presidential elections, the user may be interested in the emails for that topic, it may not be appropriate to continue to classify that email as spam.

The work in [23] looks at the particular problem of classification in the context of user-interests. In this problem, the label of a document is considered either *interesting* or *non-interesting*. In order to achieve this goal, the work in [23] maintains the interesting and non-interesting topics in a text stream together with the evolution of the theme of the interesting topics. A document collection is classified into multiple topics, each of which is labeled either interesting or non-interesting at a given point. A concept refers to the main theme of interesting topics. A concept drift refers to the fact that the main theme of the interesting topic has changed.

The main goals of the work are to maximize the accuracy of classification and minimize the cost of re-classification. In order to achieve this goal, the method in [23] designs methods for detecting both *concept drift* as well as *model adaptation*. The former refers to the change in the theme of user-interests, whereas the latter refers to the detection of brand new concepts. In order to detect concept drifts, the method in [23] measures the classification error-rates in the data stream in terms of true and false positives. When the stream evolves, these error rates will increase, if the change in the concepts are not detection. In order to determine the change in concepts, techniques from statistical quality control are used, in which we determine the mean μ and standard deviation σ of the error rates, and determine whether this error rate remains within a particular tolerance which is $[\mu - k \cdot \sigma, \mu + k \cdot \sigma]$. Here the tolerance is regulated by the parameter k. When the error rate changes, we determine when the concepts should be dropped or included. In addition, the drift rate is measured in order to determine the rate at which the concepts should be changed for classification purposes. In addition, methods for dynamic construction and removal of sketches are discussed in [23].

Another related work is that of one-class classification of text streams [52], in which only training data for the positive class is available, but there is no training data available for the negative class. This is quite common in many real applications in which it easy to find representative documents for a particular topic, but it is hard to find the representative documents in order to model the background collection. The method works by designing an ensemble of classifiers in which some of the classifiers corresponds to a recent model, whereas others correspond to a long-term model. This is done in order to incorporate the fact that the classification should be performed with a combination of short-term and long-term trends.

A rule-based technique, which can learn classifiers incrementally from data streams is the *sleeping-experts systems* [15, 21]. One characteristic of this rule-based system is that it uses the position of the words in generating the classification rules. Typically, the rules correspond to sets of words which are placed close together in a given text document. These sets of words are related to a class label. For a given test document, it is determined whether these sets of words occur in the document, and are used for classification. This system essentially learns a set of rules (or sleeping experts), which can be updated incrementally by the system. While the technique was proposed prior to the advent of data stream technology, its online nature ensures that it can be effectively used for the stream scenario.

One of the classes of methods which can be easily adapted to stream classification is the broad category of *neural networks* [41, 45]. This is because neural networks are essentially designed as a classification model with a network of perceptrons and corresponding weights associated with the term-class pairs. Such an incremental update process can be naturally adapted to the streaming context. These weights are incrementally learned as new examples arrive. The first neural network methods for online learning were proposed in [41, 45]. In these methods, the classifier starts off by setting all the weights in the neural network to the same value. The incoming training example is classified with the neural network. In the event that the result of the classification process is correct, then the weights are not modified. On the other hand, if the classification is incorrect, then the weights for the terms are either increased or decreased depending upon which class the training example belongs to. Specifically, if class to which the training example belongs is a positive instance, the weights of the corresponding terms (in the training document) are increased by α. Otherwise, the weights of these terms are reduced by α. The value of α is also known as the *learning rate*. Many other variations are possible in terms of how the weights may be modified. For example, the method in [18] uses a multiplicative update rule, in which two multiplicative constants $\alpha_1 > 1$ and $\alpha_2 < 1$ are used for the classification process. The weights are multiplied by α_1, when the example belongs to the positive class, and is multiplied by α_2 otherwise. Another variation [31] also allows the modification of weights, when the classification process is correct. A number of other online neural network methods for text classification (together with background on the topic) may be found in [16, 22, 34, 35]. A Bayesian method for classification of text streams is discussed in [13]. The method in [13] constructs a Bayesian model of the text which can be used for online classification. The key components of this approach are the design of a Bayesian online perceptron and a Bayesian online Gaussian process, which can be used effectively for online learning.

4. Evolution Analysis in Text Streams

A key problem in the case of text is to determine evolutionary patterns in temporal text streams. An early survey on the topic of evolution analysis in text streams may be found in [28]. Such evolutionary patterns can be very useful in a wide variety of applications, such as summarizing events in news articles and revealing research trends in the scientific literature. For example, an event may have a life cycle in the underlying theme patterns such as the beginning, duration, and end of a particular event. Similarly, the evolution of a particular topic in the research literature may have a life-cycle in terms of how the different topics affect one another. This problem was defined in [33], and contains three main parts: (a) Discovering the themes in text; (b) creating an evolution graph of themes; and (c) studying the life cycle of themes.

A *theme* is defined as a semantically related set of words, with a corresponding probability distribution, which coherently represents a particular topic or sub-topic. This corresponds to a model, which is represented by θ. The *span* of such a theme represents the starting and end point of the corresponding theme in terms of the time-interval (s, t). Thus, the theme span is denoted by the triple $\gamma = (\theta, s, t)$. As time passes, a particular theme γ_1 may perform a transition into another theme γ_2. A theme γ_1 is said to have evolved into another theme γ_2, if there is sufficient similarity between their corresponding spans. It is possible to create a *theme evolution graph* $G = (N, A)$, in which each node in N corresponds to a theme, and each edge in A corresponds to a transition between two themes. The overall approach requires three steps:

- In the first step, we segment the text stream into a number of possibly overlapping sub-collections with fixed or variable time spans. This is done in an application-specific way.

- The salient themes are determined from each collection with the use of a probabilistic mixture model. A standard mixture model technique [19] was used for this purpose.

- Finally, all the evolution transitions are determined from these theme patterns. This is done with the use of the KL-divergence measure in order to compute the evolution distance between two themes. In the event that the similarity is above a given threshold, the transition is considered valid.

In addition, a method is proposed in [33] for analyzing the entire theme life cycle by measuring the strength of the theme over different periods. A method based on HMM is proposed to measure the theme-shifts over the entire period as well.

The problem of tracking new topics, ideas, and memes across the Web has been studied in [29]. This problem is related to that of the topic detection and tracking discussed earlier. However, the rate of evolution in the web and blog scenario is significantly greater than the models which have been discussed in earlier work. In the context of the web and social networks, the content spreads widely and then fades over time scales on the order of days. The work is [29] develops a framework for tracking short, distinctive phrases that persistently appear in online text over time. In addition, scalable algorithms were proposed for clustering textual variants of such phrases. In addition, the approach is able to perform local analysis for specific threads. This includes the determination of peak intensity and the rise and decay in the intensity of specific threads. The relationship between the news cycle and blogs is examined in terms of how events propagate from one to the other, and the corresponding time-lag for the propagation process.

5. Conclusions

This chapter studies the problem of mining text streams. The challenge in the case of text stream arises because of its temporal and dynamic nature in which the patterns and trends of the stream may vary rapidly over time. The determination of the changes in the underlying patterns and trends is very useful in the context of a wide variety of applications. A variety of problems in text stream mining are examined such as clustering, classification, evolution analysis, and event detection. In addition, we studied the applications of some of these techniques in the context of new applications such as social networks. A lot of interesting avenues for research remain in the context of social media analytics, and the use of social dimensions in order to enhance text stream mining. In particular, the incorporation of network structure into the mining of social streams such as *Twitter* remains a relatively unexplored area, which can be a fruitful avenue for future research.

References

[1] C. C. Aggarwal. Data Streams: Models and Algorithms, *Springer*, 2007.

[2] C. C. Aggarwal, J. Han, J. Wang, P. Yu. On Demand Classification of Data Streams, *KDD Conference*, 2004.

[3] C. C. Aggarwal, P. S. Yu. A Framework for Clustering Massive Text and Categorical Data Streams, *SIAM Conference on Data Mining*, 2006.

[4] C. C. Aggarwal, J. Han. J. Wang, P. Yu. A Framework for Clustering Evolving Data Streams, *VLDB Conference*, 2003.

[5] J. Allan, R. Papka, V. Lavrenko. On-line new event detection and tracking. *ACM SIGIR Conference*, 1998.

[6] J. Allan, V. Lavrenko, H. Jin. First story detection in tdt is hard. *ACM CIKM Conference*, 2000.

[7] J. Allan, V. Lavrenko, D. Malin, R. Swan. Detections, bounds and timelines: Umass and tdt3, *Proceedings of the Topic Detection and Tracking Workshop*, 2000.

[8] I. Androutsopoulos, J. Koutsias, K. V. Chandrinos, C. D. Spyropoulos. An experimental comparison of naive Bayesian and keyword-based anti-spam filtering with personal e-mail messages. *Proceedings of the ACM SIGIR Conference*, 2000.

[9] A. Banerjee, J. Ghosh. Competitive learning mechanisms for scalable, balanced and incremental clustering of streaming texts, *NIPS Conference*, 2003.

[10] D. Blei, J. Lafferty. Dynamic topic models. *ICML Conference*, 2006.

[11] T. Brants, F. Chen, A. Farahat. A system for new event detection. *ACM SIGIR Conference*, 2003.

[12] L. O'Callaghan, A. Meyerson, R. Motwani, N. Mishra, S. Guha. Streaming-Data Algorithms for High-Quality Clustering. *ICDE Conference*, 2002.

[13] K. Chai, H. Ng, H. Chiu. Bayesian Online Classifiers for Text Classification and Filtering, *ACM SIGIR Conference*, 2002.

[14] M. Charikar. Similarity Estimation Techniques from Rounding Algorithms, *STOC Conference*, 2002.

[15] W. Cohen, Y. Singer. Context-sensitive learning methods for text categorization. *ACM Transactions on Information Systems*, 17(2), pp. 141–173, 1999.

[16] K. Crammer, Y. Singer. A New Family of Online Algorithms for category ranking, *ACM SIGIR Conference*, 2002.

[17] D. Cutting, D. Karger, J. Pedersen, J. Tukey. Scatter/Gather: A Cluster-based Approach to Browsing Large Document Collections. *Proceedings of the SIGIR*, 1992.

[18] I. Dagan, Y. Karov, D. Roth. Mistake-driven learning in text categorization. *Conference Empirical Methods in Natural Language Processing*, 1997.

[19] A. P. Dempster, N. M. Laird, D. B. Rubin. Maximum likelihood from incomplete data via the EM algorithm. *Journal of Royal Statistical Society* 39: pp. 1–38, 1977.

[20] D. Fisher. Knowledge Acquisition via incremental conceptual clustering. *Machine Learning*, 2: pp. 139–172, 1987.

[21] Y. Freund, R. Schapire, Y. Singer, M. Warmuth. Using and combining predictors that specialize. *Proceedings of the 29th Annual ACM Symposium on Theory of Computing*, pp. 334–343, 1997.

[22] Y. Freund, R. Schapire. Large Margin Classification using the perceptron Algorithm, *COLT*, 1998.

[23] G. P. C. Fung, J. X. Yu, H. Lu. Classifying text streams in the presence of concept drifts. *PAKDD Conference*, 2004.

[24] G. P. C. Fung, J. X. Yu, P. Yu, H. Lu. Parameter Free Bursty Events Detection in Text Streams, *VLDB Conference*, 2005.

[25] J. H. Gennari, P. Langley, D. Fisher. Models of incremental concept formation. *Journal of Artificial Intelligence*, 40: pp. 11–61, 1989.

[26] Q. He, K. Chang, E.-P. Lim, J. Zhang. Bursty feature representation for clustering text streams. *SDM Conference*, 2007.

[27] J. Kleinberg, Bursty and hierarchical structure in streams, *ACM KDD Conference*, pp. 91–101, 2002.

[28] A. Kontostathis, L. Galitsky, W. M. Pottenger, S. Roy, D. J. Phelps. A survey of emerging trend detection in textual data mining. *Survey of Text Mining*, pp. 185–224, 2003.

[29] J. Leskovec, L. Backstrom, J. Kleinberg. Meme Tracking and the Dynamics of the News Cycle, *KDD Conference*, 2009.

[30] D. Lewis. The TREC-4 filtering track: description and analysis. *Proceedings of TREC-4, 4th Text Retrieval Conference*, pp. 165–180, 1995.

[31] D. Lewis, R. E. Schapire, J. P. Callan, R. Papka. Training algorithms for linear text classifiers. *ACM SIGIR Conference*, 1996.

[32] Y.-B. Liu, J.-R. Cai, J. Yin, A. W.-C. Fu. Clustering Text Data Streams, *Journal of Computer Science and Technology*, Vol. 23(1), pp. 112–128, 2008.

[33] Q. Mei, C.-X. Zhai. Discovering Evolutionary Theme Patterns from Text- An Exploration of Temporal Text Mining, *ACM KDD Conference*, 2005.

[34] H. T. Ng, W. B. Goh, K. L. Low. Feature selection, perceptron learning, and a usability case study for text categorization. *SIGIR Conference*, 1997.

[35] F. Rosenblatt. The perceptron: A probabilistic model for information and storage organization in the brain, *Psychological Review*, 65: pp. 386–407, 1958.

[36] S. Petrovic, M. Osborne, V. Lavrenko. Streaming First Story Detection with Application to Twitter. *Proceedings of the ACL Conference*, pp. 181–189, 2010.

[37] N. Sahoo, J. Callan, R. Krishnan, G. Duncan, R. Padman. Incremental Hierarchical Clustering of Text Documents, *ACM CIKM Conference*, 2006.

[38] T. Salles, L. Rocha, G. Pappa, G. Mourao, W. Meira Jr., M. Goncalves. Temporally-aware algorithms for document classification. *ACM SIGIR Conference*, 2010.

[39] H. Sayyadi, M. Hurst, A. Maykov. Event Detection in Social Streams, *AAAI*, 2009.

[40] H. Schutze, C. Silverstein. Projections for Efficient Document Clustering, *ACM SIGIR Conference*, 1997.

[41] H. Schutze, D. Hull, J. Pedersen. A comparison of classifiers and document representations for the routing problem. *ACM SIGIR Conference*, 1995.

[42] A. Surendran, S. Sra. Incremental Aspect Models for Mining Document Streams. *PKDD Conference*, 2006.

[43] H. Wang, W. Fan, P. Yu, J. Han, Mining Concept-Drifting Data Streams with Ensemble Classifiers, *KDD Conference*, 2003.

[44] X. Wang, C.-X. Zhai, X. Hu, R. Sproat. Mining Correlated Bursty Topic Patterns from Correlated Text Streams, *ACM KDD Conference*, 2007.

[45] E. Wiener, J. O. Pedersen, A. S. Weigend. A Neural Network Approach to Topic Spotting. *SDAIR*, pp. 317–332, 1995.

[46] Y. Yang, J. Carbonell, R. Brown, T. Pierce, B. T. Archibald, X. Liu. Learning approaches for detecting and tracking news events. *IEEE Intelligent Systems*, 14(4):32–43, 1999.

[47] Y. Yang, T. Pierce, J. Carbonell. A study on retrospective and on-line event detection. *ACM SIGIR Conference*, 1998.

[48] Y.Yang, J. Carbonell, C. Jin. Topic-conditioned Novelty Detection. *ACM KDD Conference*, 2002.

[49] L. Yao, D. Mimno, A. McCallum. Efficient methods for topic model inference on streaming document collections, *ACM KDD Conference*, 2009.

[50] K. L. Yu, W. Lam. A new on-line learning algorithm for adaptive text filtering. *ACM CIKM Conference*, 1998.

[51] J. Zhang, Z. Ghahramani, Y. Yang. A probabilistic model for on-line document clustering with application to novelty detection. In *Saul L., Weiss Y., Bottou L. (eds) Advances in Neural Information Processing Letters*, 17, 2005.

[52] Y. Zhang, X. Li, M. Orlowska. One Class Classification of Text Streams with Concept Drift, *ICDMW Workshop*, 2008.

[53] Q. Zhao, P. Mitra. Event Detection and Visualization for Social Text Streams, *ICWSM*, 2007.

[54] S. Zhong. Efficient Streaming Text Clustering. *Neural Networks*, Volume 18, Issue 5-6, 2005.

[55] http://projects.ldc.upenn.edu/TDT/

Chapter 10

TRANSLINGUAL MINING FROM TEXT DATA

Jian-Yun Nie
University of Montreal
Montreal, H3C 3J7, Quebec, Canada
nie@iro.umontreal.ca

Jianfeng Gao
Microsoft Corporation
Redmond, WA, USA
jfgao@microsoft.com

Guihong Cao
Microsoft Corporation
Redmond, WA, USA
gucao@microsoft.com

Abstract Like full-text translation, cross-language information retrieval (CLIR) is a task that requires some form of knowledge transfer across languages. Although robust translation resources are critical for constructing high quality translation tools, manually constructed resources are limited both in their coverage and in their adaptability to a wide range of applications. Automatic mining of translingual knowledge makes it possible to complement hand-curated resources. This chapter describes a growing body of work that seeks to mine translingual knowledge from text data, in particular, data found on the Web. We review a number of mining and filtering strategies, and consider them in the context of statistical machine translation, showing that these techniques can be effective in collecting large quantities of translingual knowledge necessary for CLIR.

Keywords: cross-lingual mining, translingual mining, cross-lingual information retrieval

1. Introduction

The principle goal of text mining is to discover knowledge from text data. Various forms of knowledge may be involved, including possibly concepts and relations among them. While the bulk of work on text mining has been conducted on monolingual texts, relating to identifying concepts and relations among them in a single language, a by-no-means negligible class of applications involves more than one language. The prototypical member of this class is Machine translation (MT), which seeks to transfer a sentence or a text from a language into another. To do this, one has to create or extract various types of *translingual* knowledge such as word translation (usually in the form of a bilingual dictionary or a statistical translation model) and methods of syntactic transfer. Whereas classical MT systems were once constructed using manually defined rules and dictionaries, modern MT systems exploit large bilingual text data from which to obtain translational knowledge automatically. The extraction of this translational knowledge is, in its essence, a form of translingual text mining. Another important application that calls for translingual text mining is cross-language information retrieval (CLIR), in which one tries to retrieve documents in a language different from the language of the original query. A person may wish, for example, to retrieve documents in English using a query in Chinese. Although additional translational knowledge may need to be brought to bear in order to compare the returned documents and the query in two languages, the informational goal of CLIR is distinct from that of full text MT, and the process of extracting translingual knowledge differs accordingly.

In this chapter, we survey some of the approaches used to extract translingual knowledge from texts for different purposes, in particular, MT and CLIR. We will begin with a description of the classical approaches to statistical machine translation, and describing how statistical translation models can be constructed from parallel texts, and examining extensions to the classical approaches that attempt to go beyond word-based translation. In the remaining sections, we consider a variety of methods for translingual text mining for CLIR applications.

2. Traditional Translingual Text Mining – Machine Translation

The goal of machine translation (MT) is to use a computer system to translate a text written in a source language (e.g., Chinese) into a target language (e.g., English). In this section, we provide an overview of translation models that are widely used in state-of-the-art statistical machine translation (SMT) systems. A comprehensive review is provided in a very readable form in Koehn (2009). Although these models are designed for translating regular natural language sentences, they can also be adapted to the task of search query translation for cross-lingual information retrieval (CLIR), as will be discussed in Section 4. The query translation task differs from conventional text translation mainly in its treatment of word order. In text translation word order is crucial to the readability of the translated sentences which are presented directly to the end users. In CLIR, query translation is an intermediate step that provides a set of translated query terms so that the search engine can retrieve documents in the target language. Word order thus has little impact on the search quality as long as the translation preservers the underlying search intent, rather than the form, of the original query. This section focuses only on statistical translation models for regular text. Readers who are interested in statistical models for query translation may refer to Gao and Nie (2006) and Gao et al. (2001, 2002).

2.1 SMT and Generative Translation Models

SMT is typically formulated within the framework of the noisy channel model. Given a source sentence (in Chinese) $C = c_1 \ldots c_J$, we want to find the best English translation $E = e_1 \ldots e_I$ among all possible translations:

$$E^* = \underset{E}{\operatorname{argmax}} P(E|C) \qquad (10.1)$$

where the argmax operation denotes the decoder, i.e., the search algorithm used to find the target sentence with the highest probability among all possible targets.

Applying Bayes' decision rule and dropping the constant denominator, we have

$$E^* = \underset{E}{\operatorname{argmax}} P(C|E)P(E) \qquad (10.2)$$

where $P(E)$ is the language model, assessing the overall well-formedness of the target sentence, and $P(C|E)$ is the translation model, modeling the transformation probability from E to C. In this section, we focus our discussion on the translation model only. Notice that, mathematically, the translation direction changes from $P(E|C)$ in Equation (2.1) to

$P(C|E)$ in Equation (2.2) when Bayes rule is applied. Following Koehn (2009), we will seek to avoid potential confusion that might arise from this alternation by adhering to the notation $P(C|E)$.

In a significant generalization of the noisy channel model, Och and Ney (2002) introduced a log-linear model that models $P(E|C)$ directly. This log-linear model is currently adopted by most of state-of-the-art SMT systems and is of the form

$$P(E|C) = \frac{1}{Z(C,E)} \exp \sum_i \lambda_i h_i(C,E) \qquad (10.3)$$

where Z is the normalization constant, $h(\cdot)$ are a set of features computed over the translation and λ's are feature weights optimized on development data using e.g., minimum error rate training (Och 2003). The features used in the log-linear model can be binary features or real-value features derived from probabilistic models. For example, we can define the logarithm of language model and translation model probabilities in Equation (2.2) as two features, thereby subsuming the noisy channel model as a special case. The log-linear model thus provides a flexible mathematical framework with which to incorporate a wide variety of features useful for MT.

Conceptually, a translation model tries to "remember" to the extent possible how likely it is that a source sentence translates into a target sentence in training data. Figure 1 shows a Chinese sentence paired with its English translation. Ideally, if the translation model could remember such translation pairs for all possible Chinese sentences, we would have a perfect translation system. Unfortunately, a training corpus, no matter how large, can cover only a tiny fraction of all possible sentences. Given limited training data, it is usual to break the sentences in the training corpus into smaller *translation units* (e.g., words) whose distribution (i.e., translation probabilities) can be more easily modeled. In Figure 1, although the translation of the full sentence is unlikely to occur in training data, individual word translation pairs such as (rescue, 救援) will be found. Given an input sentence that is unseen in training data, an SMT system can be expected to perform a translation process that runs broadly as follows: first the input source sentence is broken into smaller translation units, then each unit is translated into a target language, and finally the translated units are glued together to form a target sentence. The translation models that we detail in the sections below differ in how the translation units are defined, translated and reassembled. The method we use to formulate a translation model is called **generative modeling**, and consists of three steps:

- Story making: formulating a generative story about how a target sentence is generated step by step from a source sentence.

- Mathematical formulation: modeling each generation step in the generative story using a probability distribution.

- Parameter estimation: implementing an effective way of estimating the probability distributions from training data.

These three modeling tasks are closely interrelated. The way in which we break the generation process into smaller steps in our story determines the complexity of the probabilistic models, which in turn determines the set of the model parameters that need to be estimated. We can view the three tasks as straddling the artistic (story making), the scientific (mathematical formulation), and the engineering (parameter estimation). The overall challenge of generative modeling is to find a harmonic combination of the three, an intellectual endeavor that attracts the talent of some of the best computer scientists all over the world.

State-of-the-art translation models used for conventional text translation broadly fall into three categories: word-based models, phrase-based models, and syntax-based models. In what follows, we will describe them in turn starting with the generative story, then describing the mathematical formulation and the way in which the model parameters are estimated on the training data.

```
救援(rescue) 人员(staff) 在(in) 倒塌的(collapsed) 房屋(house)
里(in) 寻找(search) 生还者(survivors)。
Rescue workers search for survivors in collapsed houses.
```

Figure 10.1. A Chinese sentence and its English translation

2.2 Word-Based Models

Word-based models use words as translation units. The models stem from pioneering work on statistical machine translation conducted by an IBM group in the early 1990s. In what has become classical paper Brown et al. (1993) proposed a series of word-based translation models of increasing complexity that come to be known as the IBM Models.

IBM Model 1, one of the simplest and most widely used word-based models, is what is termed a *lexical translation model*, in which the order of the words in the source and target sentence is ignored. The generative story about how the target sentence E is generated from the source sentence C, runs as follows:

1. First choose the length for the target sentence I, according to the distribution $P(I|C)$.

2. Then, for each position $i(i = 1 \ldots I)$ in the target sentence, we choose a position j in the source sentence from which to generate the i-th target word e_i according to the distribution $P(j|C)$, and generate the target word by translating c_j according to the distribution $P(e_i|c_j)$. We include in position zero of the source sentence an artificial "null word", denoted by <null>the purpose of which is to allow the insertion of additional target words.

Now, let us formulate the above story mathematically. In Step 1, we assume that the choice of the length is independent of C and I, thus we have $P(I|C) = \epsilon$, where ϵ is a small constant. In Step 2, we assume that all positions in the source sentence, including position zero for the null word, are equally likely to be chosen. Thus we have $P(j|C) = \frac{1}{J+1}$. Then the probability of generating e_i given C is the sum over all possible positions, weighted by $P(j|C)$: $P(e_i|C) = \sum_j P(j|C)P(e_i|c_j) = \frac{1}{J+1}\sum_j P(e_i|c_j)$. Assuming that each target word is generated independently from C, we end up with the final form of IBM Model 1.

$$P(E|C) = P(I|C) \prod_{i=1}^{I} P(e_i|C) \qquad (10.4)$$

$$= \frac{\epsilon}{(J+1)^I} \prod_{i=1}^{I} \sum_{j=0}^{J} P(e_i|c_j) \qquad (10.5)$$

We can see that IBM Model 1 has only one type of parameter to estimate, the lexical translation probabilities $P(e|c)$. If the training data consists of sentence pairs that are word-aligned as shown in Figure 2, $P(e|c)$ can be computed via Maximum Likelihood Estimation (MLE) as follows:

$$P(e|c) = \frac{N(c,e)}{\sum_{e'} N(c,e')} \qquad (10.6)$$

where $N(c,e)$ is the number of times that the word pair (c,e) is aligned in training data. In practice, it is more realistic to assume that training data is aligned at the sentence level but not at the word level. Accordingly, we apply the Expectation Maximization (EM) algorithm to compute the values of $P(e|c)$ and the word alignment iteratively. This process will determine the best $P(e|c)$ that maximizes the probability of the given alignment between sentences. The algorithm works as follows:

1. Initialize the model with a uniform translation probability distribution.

2. Apply the model to the data, computing the probabilities of all possible word alignments.

3. (Re-)estimate the model by collecting counts for word translation over all possible alignments, weighted by their probabilities computed in the Step 2.

4. Iterate through Steps 2 and 3 until convergence.

Since at every EM iteration the likelihood of the model given the training data is guaranteed not to decrease, the EM algorithm is guaranteed to converge. In the case of IBM Model 1, it is guaranteed to reach a global maximum.

Brown et al. (1993) presents five word-based translation models of increasing complexity, namely IBM Model 1 through 5. In IBM Model 1 the order of the words in the source and target sentences is ignored, and the model assumes that all word alignments are equally likely. Model 2 improves on Model 1 by adding an absolute alignment model in which words that follow each other in the source language have translations that follow each other in the target language. Models 3, 4, and 5 model the "fertility" of the generation process with increasing complexity. Fertility is a notion reflecting the observation that an input word in a source language tends to produce a specific number of output words in a target language. The fertility model captures the information that some Chinese words are more likely than others to generate multiple English words. All these models have their individual generative stories and corresponding mathematical formulations, and their model parameters are estimated using the EM algorithm. Readers may refer to Brown et al. (1993) for details.

2.3 Phrase-Based Models

Phrase-based models are the basis for most state-of-the-art SMT systems. Like the word-based models, these are generative models that translate an input sentence in a source language C into a sentence in a target language E. Instead of translating single words in isolation, however, phrase-based models translate sequences of words (i.e., phrases) in C into sequences of words in E. The use of phrases as translation units is motivated by the observation that one word in a source language frequently translates into multiple words in a target language, or vice versa. Word-based models cannot handle these cases adequately: the

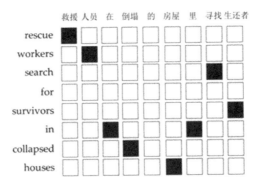

Figure 10.2. Word alignment: words in the English sentence (rows) are aligned to words in the Chinese sentence (columns) as indicated by the filled boxes in the matrix

English phrase "stuffy nose", for example, translates the Chinese word "鼻塞" with relatively high probability, but neither of the individual English words "stuffy" and "nose" has a high word translation probability to "鼻塞".

The generative story behind the phrase-based models can be stated as follows. First, the input source sentence C is segmented into K non-empty word sequences $\mathbf{c}_1, \ldots, \mathbf{c_K}$. Then each is translated to a new non-empty word sequence $\mathbf{e}_1, \ldots, \mathbf{e_K}$. Finally these phrases are permuted and concatenated to form the target sentence E. Here $\mathbf{c}$ and $\mathbf{e}$ denote consecutive sequences of words.

To formalize this generative process, let S denote the segmentation of C into K phrases $\mathbf{c}_1 \ldots \mathbf{c}_K$, and let T denote the K translation phrases $\mathbf{e}_1 \ldots \mathbf{e}_K$ We refer to these $(\mathbf{c}_i, \mathbf{e}_i)$ pairs as *bilingual phrases*. Finally, let M denote a permutation of K elements representing the final reordering step. Figure 3 demonstrates the generative procedure.

Next let us place a probability distribution over translation pairs. Let $B(C, E)$ denote the set of S, T, M triples that translate C into E. If we assume a uniform probability over segmentations, then the phrase-based translation model can be defined as:

$$P(E|C) \propto \sum_{(S,T,M) \in B(C,E)} P(T|C,S) \cdot P(M|C,S,T) \quad (10.7)$$

It is common practice in SMT to use the maximum approximation to the sum: the maximum probability assignment can be found efficiently by using a dynamic programming approach:

$$P(E|C) \approx \max_{(S,T,M) \in B(C,E)} P(T|C,S) \cdot P(M|C,S,T) \quad (10.8)$$

Figure 10.3. Example demonstrating the generative procedure behind the phrase-based model.

Reordering is handled by a distance-based reordering model (Koehn et al. 2003) relative to the previous phrase. We define $start_i$ as the position of the first word of the Chinese input phrase that translates to the i-th English phrase, and end_i as the position of the last word of that Chinese phrase. The reordering distance is computed as $start_i - end_i - 1$, i.e., the number of words skipped when taking foreign words out of sequence. We also assume that a phrase-segmented English sentence $T = \mathbf{e}_1 \ldots \mathbf{e}_K$ is generated from left to right by translating each phrase $\mathbf{c}_1 \ldots \mathbf{c}_K$ independently. This yields one of the best-known forms of phrase-based model:

$$P(E|C) \propto \max_{(S,T,M) \in B(C,Q)} \prod_{k=1}^{K} P(\mathbf{e_k}|\mathbf{c_k}) d(start_i - end_{i-1} - 1) \quad (10.9)$$

In Equation (10.9) the only parameter to be estimated is the translation probabilities on the bilingual phrases $P(\mathbf{e}|\mathbf{c})$. In what follows, we rely mainly on work by Och and Ney (2002) and Koehn et al. (2003) to describe how bilingual phrases are extracted from the parallel data and $P(\mathbf{e}|\mathbf{c})$ is estimated.

First, we learn two word translation models via EM training of a word-based model (i.e., IBM Model 1 or 4) on sentence pairs in two directions: from source to target and from target to source. We then perform Viterbi word alignment in each direction according to the corresponding model for that direction. The two alignments are combined, starting with the intersection of the two alignments, and gradually including more alignment links according to heuristic rules detailed in Och and Ney (2002). Finally, bilingual phrases that are consistent with the word alignment are extracted. Consistency here implies two things. First, there must be at least one aligned word pair in the bilingual phrase. Second, there must be no word alignments from words inside the bilingual phrase to words outside the bilingual phrase. That is, we do not extract a phrase pair if there is an alignment from within the phrase pair to an element outside the phrase pair. Figure 4 illustrates the bilingual phrases we can generate from the word-aligned sentence pair by this process.

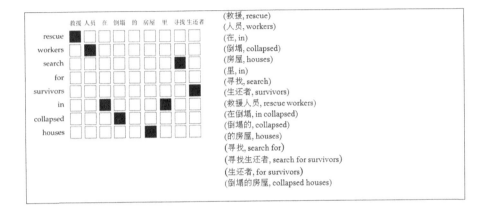

Figure 10.4. An example of a word alignment and the bilingual phrases containing up to 3 words that are consistent with the word alignment.

After gathering all such bilingual phrases from the training data, we can estimate conditional relative frequency estimates without smoothing. For example, the phrase transformation probability $P(\mathbf{e}|\mathbf{c})$ in Equation (2.7) can be estimated approximately as:

$$P(\mathbf{e}|\mathbf{c}) = \frac{N(\mathbf{c}, \mathbf{e})}{\sum_{\mathbf{e}'} N(\mathbf{c}, \mathbf{e}')} \qquad (10.10)$$

where $N(\mathbf{c}, \mathbf{e})$ is the number of times that $\mathbf{c}$ is aligned to $\mathbf{e}$ in training data. These estimates are useful for contextual lexical selection when there is sufficient training data, otherwise can be subject to data sparsity issues.

An alternate means of estimating translation probabilities that is less susceptible to data sparsity is the so-called *lexical weight* estimate. Assume we have a word translation distribution $t(e|c)$ (defined over individual words, not phrases), and a word alignment A between $\mathbf{e}$ and $\mathbf{c}$; here, the word alignment contains (i, j) pairs, where $i \in 1 \ldots |\mathbf{e}|$ and $j \in 0 \ldots |\mathbf{c}|$, with 0 indicating an inserted word. Then we can use the following estimate:

$$P_w(\mathbf{e}|\mathbf{c}, A) = \prod_{i=1}^{|e|} \frac{1}{|\{j|(j,i) \in A\}|} \sum_{\forall (i,j) \in A} t(e_i|c_j) \qquad (10.11)$$

We assume that for every position in **e**, there is either a single alignment to 0, or multiple alignments to non-zero positions in **c**. In effect, this computes a product of per-word translation scores; the per-word scores are averages of all the translations for the alignment links of that word. We estimate the word translation probabilities using counts from the word aligned corpus: $t(e|c) = \frac{N(c,e)}{\sum_{e'} N(c,e')}$. Here $N(c,e)$ is the number of times that the words (not phrases as in Equation (2.8)) c and e are aligned in the training data. These word-based scores of bilingual phrases, though not as effective in contextual selection as previous ones, are more robust to noise and sparsity. Both model forms of Equation (2.8) and (2.9) are used as features in the log-linear model for SMT as Equation (2.3).

2.4 Syntax-Based Models

The possibility of incorporating syntax information in SMT has been a long-standing topic of research. Syntax-based translation models have begun to perform as well as state-of-the-art phrase-based models, and in the case of some language pairs may even outperform their phrase-based counterpart. Research on syntax-based models is a fast-moving area, with numerous open questions. Our description in this section focuses on some basic underlying principles, illustrated by examples from the most successful models proposed so far (e.g., Chiang 2005; Galley et al. 2004).

Syntax-based models rely on parsing the sentence in either the source or the target language, or in some cases in both. Figure 5 depicts the sentence pair from Figure 1, but with constituent parses added. These parses are generated from a statistical parser trained on Penn Treebank. Each parse is a rooted tree where the leaves are original words of the sentence and the internal nodes cover a contiguous sequence of the words in the sentence, called a constituent Each constituent is associated with a phrase label describing the syntactic role of the words under its node.

The tree-structured parse plays similar roles in syntax-based models to those of a phrase in phrase-based models. The first role is to identify translation units in an input sentence. While in phrase-based models the units are phrases, in syntax-based models they are constituents of the kind seen in Figure 5. The second is to guide how best to glue those translated constituents into a well-formed target sentence. Again, we assume a generative story, similar to that for phrase-based models:

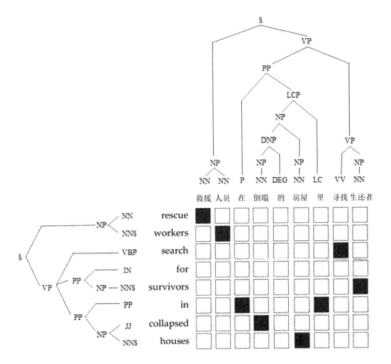

Figure 10.5. A pair of word-aligned Chinese and English sentences and their parse trees.

1. Parse an input Chinese sentence into a parse tree

2. Translate each Chinese constituent into English

3. Glue the English constituents into a well-formed English sentence.

This generative process is typically formulated under the framework of weighted synchronous Context Free Grammar (CFG) (Chiang, 2005), which consists of a set of rewriting rules r of the form:

$$X \to (\gamma, \alpha, \sim) \tag{10.12}$$

where X is a nonterminal, γ and α are both strings of terminals and non-terminals corresponding respectively to source and target strings, and $\sim$ indicates that any non-terminals in the source and target strings are aligned. For example, a rule extracted from the example in Figure 5 is:

$$\text{VP} \to (\text{PP寻找NP, search for NP PP}, \sim) \tag{10.13}$$

where ~ indicates that PP and NP in the source and target languages are aligned. We can see that these non-terminals generalize the phrases used in the phrase-based models described in Section 2.2.

We now define a derivation D as a sequence of K rewrites $r_1, \ldots, r_K$, each of which picks a rewriting rule from the grammar, and rewrites a constituent in Chinese into English, until an English sentence is generated. Let $E(D)$ be the English strings generated by D, and $C(D)$ be the Chinese strings generated by D. Assuming that the parse tree of the input Chinese sentence is $Tree(C)$, the translation model can be formulated as

$$P(E|C, Tree(C)) = \sum_{\substack{D:E(D)=E \\ \text{and } C(D)=C}} P(D) \qquad (10.14)$$

As when formulating the phrase-based models, we use the maximum approximation to the sum:

$$P(E|C, Tree(C)) \propto \max_{\substack{D:E(D)=E \\ \text{and } C(D)=C}} P(D) \qquad (10.15)$$

A synchronous CFG assumes that each rewriting rule application depends only on a non-terminal, and not on any surrounding context. Thus we have:

$$P(D) = \prod_{k=1}^{K} P(r_k) \qquad (10.16)$$

Rewriting rule (2.11) not only specifies lexical translations but also encapsulates nicely the kind of reordering involved when translating Chinese verb complexes into English. As a result, searching for the derivation that has the maximum probability assignment, as in Equation (2.13), simultaneously accomplishes the two tasks of constituent translation and sentence reordering (as in Steps 2 and 3 in our generative story). The search can be achieved by chart parsing.

The synchronous grammar proposed in Chiang (2005) illustrates how these rewriting rules may be extracted from data and how their probabilities are estimated. The grammar has not underlying linguistic interpretation and uses only one non-terminal X. Assume that we have the word-alignment sentence pair, as shown in figure 4. First, we extract initial bi-phrases that are consistent with the word-alignment, as described in Section 2.2. We write these bilingual phrases in the form of synchronous grammar:

$X \rightarrow$ (在倒塌的房屋里寻找生还者, srch. for surviv. in collap. home, ~)

We then generalize these rules by replacing some substrings with the nonterminal symbol X:

$$X \to (在X_1里寻找X_2, \text{search for } X_2 \text{ in } X_1, \sim)$$

using subscript indices to indicate which occurrences of X are linked by $\sim$. This rule captures information about both lexical translation and word reordering, with the result that the learned grammar can be viewed as a significant generalization of phrase-based models capable of handle longer range word reordering.

To limit the number of rules generated in this fashion, the rewrite rules are constrained: (a) to contain at least one and at most 5 lexical items per language, (b) to have no sequences of non-terminals, (c) to have at most two non-terminals, and (d) to span at most 15 words. Once the rewrite rules are extracted, their probabilities are estimated on the word-aligned sentence pairs using a method analogous with that for the phrase-based models. Readers may refer to Chiang (2005) for a detailed description.

3. Automatic Mining of Parallel texts

The previous section provides an overview of the state of the art in SMT. It also describes the most traditional way to exploit a parallel corpus to extract translational knowledge in form of translation models. These models are the basis for many applications in which translation is required.

SMT requires a large number of parallel texts for model training. Traditionally, one assumed that such parallel texts are available. Indeed, there have been several manually compiled large parallel corpora available. The Canadian Hansard[1] is probably the most widely used and best known. This corpus contains all the debates in the Canadian parliament in both English and French. Translation is made by professionals and it is of very high quality. The first research work on statistical MT has been carried out using this corpus. Later on, several other parallel corpora became available, in particular, the Hong Kong law documents in English and Chinese[2] and the documents of the European Parliament in several European languages[3]. These manually compiled parallel texts can be used in the methods presented in the previous section, often after a step of sentence alignment (Gale and Church 1993).

---

[1] http://www.parl.gc.ca/ParlBusiness.aspx?Language=E
[2] http://www.legco.gov.hk/english/index.htm
[3] http://www.europarl.europa.eu/

However, despite the high quality of translation in these parallel corpora, we do encounter several problems when they are used for the translation purposes. Indeed, although the size of the corpora are large, it is still limited for the purpose of model training, leaving a considerable proportion of the translation phenomena either uncovered or insufficiently covered for general translation applications In particular:

- **Vocabulary** The documents in these manually compiled parallel corpora are formal in style and vocabulary. They do not provide good coverage of terms or words used in less formal discussions and communications on the Web. Many terms and words in the latter will be "unknown" by the models trained on these data

- **Structure** High quality documents and their translations are written in correct syntax. This is not the case for Web documents and search queries. A syntax-based SMT system trained on these data will be inadequate to cope with the flexible structure of texts and queries on the Web.

- **Adaptability** Because the statistical translation models are trained on the parallel texts, they tend to fit the latter, including the frequency of word usage and word translation. Even if a word or a term is well covered by translation model, the suggested translations may not be suitable for the intended application.

One possible solution to the above problems is to develop automatic tools to collect appropriate parallel documents according to one's requirement. The Web is an excellent resource for this purpose, and indeed it is a truly multilingual resource, one on which documents in many different languages are published. A certain proportion of the documents are parallel, i.e. the same documents are published in several languages. These documents virtually constitute a large parallel corpus. The key problem is to collect those parallel texts without including (too many) non-parallel ones.

Attempts to collect parallel texts from the Web date to the late 1990s, with Resnik (1998) and Nie et al. (1999). Both studies exploit two factors to determine whether two texts are parallel: the Web structure in which the texts are stored and published, the text structure of the documents themselves.

3.1 Using Web structure

Resnik (1998) observed that in many cases, parallel Web pages are linked from an entry page (home page) on a website, each with a language identifier such as "English" and "Français" as anchor text. For example, the following website (Natural Resources of Canada) is organized in the manner shown in Figure 6.

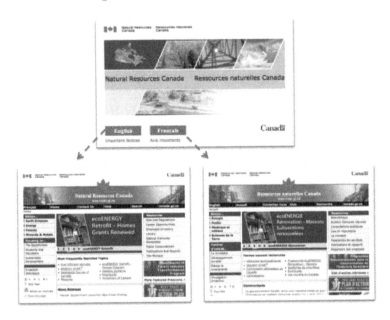

Figure 10.6. An example of parallel pages linked from a home page

The mining system STRAND (Resnik, 1998) identifies the referred pages as candidate parallel web pages. In the query language of Alta Vista used by STRAND, the following query will retrieve parent pages referring to two child pages in the relevant languages:

anchor:"english" AND anchor:"français"

However, the above criterion can only detect a limited number of parallel Web pages. More commonly sites are organized so that each of the pages contains a link to the corresponding parallel page, as shown in Figure 7. Again, the link usually contains anchor text that identifies the language.

To retrieve those pages, the following Alta Vista query can be used to retrieve the French documents containing an anchor text to an English page:

anchor: "english" OR anchor: "anglais"

while setting the language of the documents to French. Analogously, one can retrieve English documents containing anchor text linking to a French page.

Translingual Mining from Text Data 339

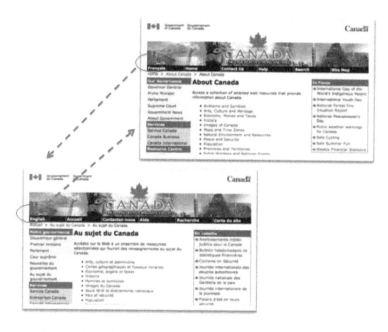

Figure 10.7. An example of mutually linked parallel pages.

This second criterion is the main approach taken by PTMiner (Nie et al. 1999, Chen and Nie 2000) to identify candidate pages. PTMiner additionally used a site crawler to download all the pages from candidate sites (the sites that contain some candidate parallel pages) in order to find more Web pages on those sites that are not indexed by the search engine.

3.2 Matching parallel pages

Once two sets of candidate pages are determined, the next task is to pair the pages up. The contents of the pages will eventually be used, but first heuristics are applied to quickly identify candidate parallel pages Since parallel Web pages are usually assigned similar file names two Web pages with the names "description_en.htm" and "description_fr.htm" are likely parallel. Similarly, Web sites may use two separate directories to store pages in two languages, in which case, the names of the directories may be slightly different, e.g. "www.website.com/English/file1.html" vs. "www.website.com/French/file1.html". In both cases the difference between file and directory names is often related to the language, and this can be recognized using simple heuristics. Such heuristics are used in PTMiner to pair up mined candidate Web pages efficiently.

To further filter out non-parallel pages, additional checks on the pages' contents can then be applied:

- Are the HTML structures of the two pages similar? The assumption is that parallel pages are usually created with the same or similar HTML structures. In both STRAND and PTMiner, the HTML markup sequence of each page is extracted, and the pages are considered to be parallel if their HTML markup sequences resemble each other However, more sophisticated comparison of document structure can be performed. For example, one can use the DOM tree of the Web page (Shi et al. 2006).

- Are the two pages of similar lengths? It is generally observed that the lengths of parallel texts are similar (or proportional to the length ratio of texts in the two languages). This is an easy way to filter out candidate pairs whose content cannot be parallel.

- Finally, what is the content translation probability? If the texts in the two pages have a high mutual translation probability, then the pages are likely to be parallel. Although an effective means of confirming textual parallelism, this ultimate step is costly to implement and has not been widely used.

The precision of the Web pages identification by STRAND and PTMiner is impressive: it is estimated that more than 90% of the identified pairs of Web pages are indeed parallel. Evaluation of recall, on the other hand, presents greater difficulty: Resnik calculated recall of STRAND at 62.5%, while Nie et al. estimated the lower bound of recall of PTMiner at a little over 50% on the assumption that every Web page from a candidate website has a parallel page in another language, an assumption that obviously overestimates the case Nevertheless, lower recall ratios can be tolerated because the number of potential parallel pages on the Web is very large, and it is more important to have a mining process with high precision than high recall.

In term of volume, STRAND has mined a relatively small number of parallel pages while PTMiner has successfully collected large amounts of parallel page data in English-French, English-Chinese, English-German, etc., chiefly by exploiting criteria that correspond to more commonly employed techniques of organizing parallel pages on the Web, as well as the site crawling process.

The above mining strategy has been used in a number of studies pertaining to different language pairs: Ma and Liberman (1999) used a similar approach to PTMiner to mine parallel pages in German and English, with some slight differences in the process: the similarity between the file names of candidate pages is measured by edit distance, and the known translations are mapped with some position constraint within the texts. Similar approaches have been used by Nagata et al. (2001) and Yang and Li (2003) to mine English-Japanese and English-Chinese parallel pages. Resnik et al. (2003) have further explored the mining of parallel Web pages from Web archives.

The above processes are designed for mining on general websites. It is possible to incorporate additional criteria according to the specific organization of a website. For example, parallel texts on the same website (e.g. Wikipedia, newswire websites) can share common resources such as pictures. Metadata can also be incorporated in documents. The use of such indications can further improve the mining process.

Bilingual and multilingual newswire websites are a common source from which parallel texts are mined. Many newswire publishers publish articles in several languages, and in many cases, the articles in different languages are translations. For example, China Daily publishes certain bilingual news articles that are aligned in paragraphs. Some of the news articles are translated and published in several languages such as Chinese, English and French. Several European newspapers also publish simultaneously articles in several languages. This provides an easy way to collect parallel news articles. However, the collection of parallel news articles depends on the specific organization of each newspaper. In some cases, there is a systematic schema of correspondence, while in other cases no clear structural information is available to determine whether two articles are parallel. In the latter case, the mined result is often comparable texts rather than parallel texts. We will describe some attempts of this kind in Section 5.

4. Using Translation Models in CLIR

It is safe to assume that not all automatically mined Web pages are strictly parallel. Indeed, during manual evaluation, it turns out that some pages, which presumably should contain the same information, are not parallel in content: one of the pages can be outdated, contain only part of the information, or even consist of an "under construction" message. The precision and recall numbers mentioned earlier are subject to human judgment: If the contents are parallel above some threshold, we consider the pages to be parallel a situation that is less ideal than the Hansard corpus, especially for tasks such as full-text machine translation that call for high quality parallel texts for training.

For other less demanding tasks such as CLIR, however, translation models trained from automatically collected Web pages can perform very well. A translation model trained on a parallel corpus can be naturally integrated into the CLIR process. General IR can be processed using a language modeling approach as follows:

$$Score(Q, D) = \sum_{t \in V} P(t|M_Q) \log P(t|M_D) \qquad (10.17)$$

where M_Q and M_D are respectively a statistical language model estimated for the query and the document, and V is the vocabulary. Notice that both M_Q and M_D are generation models, i.e. no word order or relationship is taken into account. Such an approach is often called "bag of words" approach. Using such an approach, translation in CLIR is also performed at the word level: Each word is translated independently. Therefore, the simple IBM Model 1 is widely used.

For CLIR, either the document or the query should be translated. One can of course use an MT system to translate them. Because the Web search engine only uses words and ignores word order, MT offers more than what is needed. One may argue that this is not necessarily a bad thing to have a tool offering more than required. Indeed, in the CLIR experiments, it is usually found that a high-quality MT system leads to a good CLIR result when it is used to translate queries or documents.

However, off-the-shelf MT systems also have weaknesses:

- An MT system chooses only one translation word (or expression) for each source word. In reality, there may be multiple translations. For example, "drug" (illegal substance) can be translated into "drogue" or "stupéfiant" in French. By limiting to one translation, documents in French using the other term cannot be found. For CLIR, keeping multiple translations for a word is often preferred.

- The translation by an MT system is limited to the true "translations" of the words in a query or a document. In IR, on the other hand, it is usually preferred to add related terms in the query (or document) to expand it. Query (or document) expansion is a common method in IR to increase retrieval effectiveness. By including only true translation words in a query translation, CLIR does not benefit from query expansion. It is preferred to include also related terms in the target language when doing query translation in CLIR.

- The final translation result by an MT system does not distinguish the words in their importance, i.e. all the words are un-weighted.

In IR, the weighting of terms in the query is crucial, and the translation probability or weight can greatly help distinguishing important terms vs. unimportant ones.

The above reasons have motivated a number of attempts to design CLIR approaches without, or in addition to, the use of MT systems. The principle of CLIR can be well described within the language-modeling framework for IR. It includes a translation of the query or document model as follows:

$$Score(Q, D) = \sum_{t \in V_t} [\sum_{s \in V_s} P(t|s)P(s|M_Q)] \log P(t|M_D) \qquad (10.18)$$

$$Score(Q, D) = \sum_{t \in V_t} P(s|M_Q) \log [\sum_{t \in V_t} P(s|t)P(t|M_D)] \qquad (10.19)$$

in which V_s and V_t are respectively the vocabulary in source and target languages, and $P(s|t)$ and $P(t|s)$ are translation probability (in IBM model 1) of a target language term (t) to a source language term (s) and vice versa. In practice, rather than using the whole vocabulary in $\sum_{s \in V_s} P(t|s)P(s|M_Q)$ and $\sum_{t \in V_t} P(s|t)P(t|M_D)$, one can select a subset of the translation terms, for example, the translation terms whose translation probability is higher than a threshold, or the N best translation terms for the query. Different from general MT, query or document translation in CLIR usually selects multiple translation words (rather than the best one), thereby producing a desired expansion effect. In addition, the translation probability is used explicitly to determine term weighting for the retrieval process.

The use of automatically mined parallel corpora in CLIR has been successful. In an early experiment on CLIR, Nie et al. (1999) reported that using the Web corpus, the CLIR effectiveness is very similar to using the Hansard corpus. Further experiments (Kraaij et al. 2003) have shown that CLIR using the Web parallel corpus outperforms methods that use an existing dictionary-based MT system - Systran. These results indicate that CLIR does not require as high quality corpora for training translation models. A noisy corpus can be as effective as a manually compiled high-quality corpus. In addition, a query or document translation properly incorporated into the retrieval model (as Equations 2.16 and 2.17) is a better solution than using an MT system as an individual tool, separated from the retrieval model.

5. Collecting and Exploiting Comparable Texts

The success of using a noisy parallel corpus in CLIR indicates that one can tolerate certain noise in the text data used for model training. To what extent is the process tolerant to noise? There is no clear answer to this question, but there is a series of experiments using comparable texts for CLIR, which have shown encouraging results: comparable texts are good complements to other translation resources.

In general, comparable texts are defined as texts that are not necessarily parallel, but describe the same event. Other terminologies are also used. Fung and Cheung (2004) defined quasi-comparable and comparable documents because they were written independently but on more or less the same topic. Noisy-parallel documents refer to a pair of source and translated documents that were either adapted or evolved in different ways such as Wikipedia articles. There are indeed a variety of comparable texts with different degrees of relatedness. Fung (1995) considers a continuum from parallel, comparable to unrelated texts. Brashchler and Schaüble (1998) defined the following levels of relatedness:

1. Same story: The two documents deal with the same event.

2. Related story: The two documents deal with the same event or topic from slightly different viewpoints. Or one of them deals with the topic from a broader story.

3. Shared aspect: The documents deal with related events. They may share locations or persons.

4. Common terminology: The events or topics are not directly related, but the documents share a considerable amount of terminology.

5. Unrelated: The similarities between the documents are slight or nonexistent.

Depending on the process used, different types of comparable texts can be collected. In general, the following indicators can be used to determine comparable texts from a website, especially from a newswire:

- The publication dates of two comparable texts should be the same or close;

- Some articles incorporate metadata to describe the content categories, in which case, the category of the comparable texts should be the same;

- The fact that two texts contain links to the same objects (e.g. pictures) increases the chance that the texts are about the same event;

- Although one cannot expect exact mutual translation at sentence level between comparable texts, the main vocabulary should be translatable and this can be verified using a simple resource such as a bilingual dictionary;

- The texts may contain similar special elements: named entities – they talk about the same persons and describe events of the same dates, or domain-specific words and their translations;

- Using a CLIR method, one can form a query with a source language text, and retrieve a set of potential comparable texts in the target language.

Sheridan and Ballerini (1996) are among the first to exploit comparable texts for CLIR. They mined newspaper articles in German and Italian from the website of *Schweizerische Depeschenagentur* (SDA) using content descriptor metadata and publication dates.

In their study, Brashchler and Schaüble (1998) "translated" the named entities as well as words from the source language text, and used the translation to retrieve comparable texts. Their evaluation revealed that about 60% of the texts mined are documents that share one or more events, and 75% of them share a common terminology. The mined texts have been used in a CLIR task, leading to a retrieval result only slightly worse than the best participants in TREC-7. Similar approaches have been used in other studies (e.g. Talvensaari et al, 2006, Talvensaari 2007, Huang et al. 2010). Huang et al (2010) investigated the translation of key terms in the above process: Not only single-word terms but also multi-word terms are extracted from the source-language document and translated. By doing so, they reduced the translation ambiguity, and produced more precise description in the target language.

Instead of using the translated terms in a CLIR process to mine comparable texts, in several studies, the frequencies and ranks of the source terms and their translations have also been used. Fung and Lo (1998), Fung and Cheng (2004) and Carpuat et al. (2006) used a different approach to align comparable texts. They use a set of seed words, for which the translations are known. Seed words in source- and target-language texts are extracted and their frequencies are compared. It is assumed that the seed words should be comparable in their frequency ranks. Tao et al. (2005) used a more elaborated method based on Pearson correlation: words and their known translations in a pair of comparable texts should have a strong correlation in their ranks.

The mined comparable texts can be used to derive a general bilingual lexicon (Rapp, 1995) or for translations of specific named entities (Fung, 1995, Ji, 2009). In general, it is more difficult to train a translation model using comparable texts than using parallel texts. A less strict bilingual term similarity is determined instead. The principle is analogous to word co-occurrence analysis in monolingual texts: two terms in different languages have a strong translingual relationship if they co-occur often in comparable texts in respective languages. The following formula (or some variants) can be used:

$$sim(w_s, w_t) = \frac{coocc(w_s, w_t)}{Z} \quad (10.20)$$

where w_s and w_t are source and target words, $coocc(w_s, w_t)$ is a measure of their co-occurrence and Z a normalization factor. $coocc(w_s, w_t)$ can take different forms: the number of pairs of comparable documents which contain the two words respectively, the minimal frequency of the two terms in the respective document, or some transformed measure based on these. As not all the words in the source document have their translations in the target document, the translingual relationships can be built up only for the most frequent words, or for named entities (Fung, 1995, Ji, 2009). Needless to say, the translingual relationships are much less precise than those extracted from truly parallel texts. There are two main reasons:

- The comparable texts are noisier by nature. A pair of comparable documents is not mutual translation, and the relationships between terms extracted from them are more translingual related than translation relations.

- As no process similar to sentence alignment on parallel texts can be performed, it is usually assumed that a word in a document corresponds to any word in the document in another language. In other words, the correspondence is not bound within a smaller portion of text than the entire document. The translingual relationships extracted are very noisy.

The translingual relationships can be hardly used alone for MT. At best, it can be used to complement other translation resources. For CLIR, the noisy translingual relationships extracted from comparable corpora have been found to perform quite well (Braschler and Shaüble, 1998) indicating that the utility of comparable texts, when exploited in a simple manner, is limited to less demanding tasks such as CLIR.

An alternative approach to exploiting a parallel/comparable corpus is pseudo-relevance feedback (Carbonell et al. 1997): Use a query in the source language to retrieve a set of texts in the parallel/comparable corpus. One can then select the set of corresponding texts in the target language, from which a set of terms can be extracted. These latter constitute a "translation" of the original query. As one may notice, this approach is similar to those on translingual term similarity. However, the difference is that, rather than determining the translingual relations between individual terms, this approach determines a translingual relation between sets of terms. There is potentially a larger effect of local context (Xu and Croft 1996).

Another approach is to construct a new representation space to which terms in both languages can be mapped. CLIR using Latent Semantic Indexing (Dumais et al. 1997) exploits this principle: parallel (comparable) texts are concatenated for form a composed document; A latent representation space is created and implicit translation is generated by mapping a term, a document or a query into the new space. One can also use a generative topic model instead of LSI.

In addition to the above methods, comparable texts can be exploited in a more refined manner by extracting a subset of strongly comparable or parallel parts (sentences) from them. We will describe these approaches in the next section.

6. Selecting Parallel Sentences, Phrases and Translation Words

The mining approaches described in the previous sections all rely on heuristics relating to the organization and other characteristics of parallel Web pages. Since some of the mining results are likely to non-parallel, or only partially parallel it is pertinent to ask whether it is possible and beneficial to clean the mined results in order to minimize noise.

There have been a number of attempts to extract a subset of high quality parallel texts or sentences from a corpus that has been initially mined by some other means. An original corpus can be extracted by an application such as PTMiner or STRAND. Or it might take the form a set of comparable texts minded from a newswire Web site. Even with the truly parallel corpora, a certain filtering is made. In fact, before translation models are trained on a set of parallel texts, the sentences in the texts are aligned (Gale and Church 1993) Different patterns of sentences alignment can be recognized: 0-1 or 1-0 (i.e. a sentence is aligned with no sentence), 1-1 (one sentence is aligned with one sentence), 1-2 or 2-1, and so forth. It has been observed that errors (i.e. non-parallel sentences) most often appear in alignments other than 1-1. For example, 1-0 or 0-1 alignments may be due to insertion and deletion during the manual translation. Therefore, a simple filtering process is to use only 1-1 aligned sentence pairs for model training.

It is also possible to clean up an initial parallel corpus using other heuristics. In Nie and Cai (2001), the following criteria are used to filter the data extracted by PTMiner:

- The length ratio of the text pair should be close to the standard length ratio of the two languages;
- The proportion of the 1-1 alignments of a text pair should be high;
- A relatively large percentage of the terms should be translatable into terms of another text using a dictionary.

Any text pair that does not comply with these conditions is removed from the corpus. The experiments of Nie and Cai show that a combination of the above criteria can effectively remove some non-parallel texts and retrain the parallel ones. They also observe that translation models trained on the resulting cleaned corpus mined by PTMiner are of higher quality, and are more effective when used in CLIR.

While Nie and Cai's study sought to filter out non-parallel documents from the corpus, other researchers have attempted to extract parallel sentences more directly from comparable corpora. Munteanu and Marcu (2005) use the following process to extract parallel sentences in Chinese, Arabic, and English: 1). Candidate document pairs are first selected using their publication dates (within a date window of 5 days). 2). Candidate sentence pairs from the paired documents are selected using criteria similar to those used by Nie and Cai (2001), i.e. sentence length ratio and percentage of terms that can be translated in another sentence using a dictionary. 3). Finally, a maximum entropy classifier is used to determine if the candidate sentence pair is likely to be parallel. Similar methods have also been taken by Zhao and Vogel (2002), Utiyama and Isahara (2003) and Hong et al. (2010), who estimate sentence similarity variously on the basis of sentence length ratio, sentence alignment, IBM-1 translation model and percentage of known translations using a dictionary. In manual evaluation, it has been found that the selected sentence pairs can have a precision of 90% (Utiyama and Isahara 2003). These studies demonstrate that selecting a set of parallel sentences from a comparable corpus is possible. The experiments also showed that the extracted parallel sentences are useful for MT in some context: SMT systems that use the selected sentence pairs in combination with an initial set of parallel texts generally produce a higher BLEU score in SMT experiments. However, when these parallel sentences are used alone, the performance is usually lower than that of using truly parallel texts.

Several studies use an iterative process to gradually select parallel sentences from a noisy corpus for model training. Fung and Leung (2004) first use a bilingual lexicon to select comparable texts and parallel sentences from the original set of documents. The selected parallel sentences are used to train a translation model, which is then used to complement the bilingual lexicon in a second round of document and sentence selection. Fung and Leung reported a precision of 67% in the extraction of parallel sentences using the adaptive method, 24% higher than a baseline method that only used a bilingual lexicon.

The common observation that word-based translation is too ambiguous for precise translation led researchers to propose phrase-based models (Ballesteros and Croft, 1997; Gao et al., 2001; Gao et al., 2006; Koehn et al. 2003). In the translingual relation mining task, likewise, one can go a step further Munteanu and Marcu (2006) word-align pairs of candidate sentences using IBM Model 1 in conjunction with additional heuristics, and treat a sequence of source language words (phrase) as parallel to a sequence of target language words if they have a strong mutual alignment score This principle is analogous to the case of phrase-based SMT (Koehn et al. 2003), where a sequence of words is considered to form a phrase if the constituent words are translated into a sequence of consecutive words in another language (see Section 2.2).

Again, the resulting translation model can be filtered so as to remove noise. One may choose to use only those translations whose probability is higher than a threshold (e.g. 0.01), or the N best translations for each word. One can also select translation terms according to the context, i.e. the query to be translated. One criterion that has been used in Gao et al. (2001; 2002; 2006) and Liu et al. (2005) is to assume that the resulting set of translation terms for a query should be consistent, i.e. they should co-occur often in the target language. Application of this criterion can remove unlikely translation terms that are inconsistent with the other words (or their translations) in the query. Interested readers can refer to these papers for details.

7. Mining Translingual Relations From Monolingual Texts

Translingual knowledge is by no means confined in texts in two different languages. It is by no means rare that one can find rich translingual knowledge within a "monolingual text", or more precisely, a mostly monolingual text that contains glosses (translations or transliterations) inlined in the text. For example, the following is a short text in Chinese, with personal names glossed in English.

斯科特·霍夫曼(Scott Huffman)和史蒂夫·常(Steve Chang)继续解读移动搜索。

Even if one does not understand the whole Chinese sentence, it is possible to guess that 斯科特·霍夫曼 is the Chinese transliteration of "Scott Huffman" and 史蒂夫·常 the transliteration of "Steve Chang". This phenomenon frequently appears in many (especially Asian) languages, in particular, when a personal name or technical term calls for a transliteration or translation gloss. Between languages written in the same script, glossing of named entities may not be necessary. Indeed, it is seldom necessary for a personal name to be transliterated from one European language into another. However, when languages are written in completely different scripts, transliteration or translation is usually necessary.

Since our present focus is on mining Web data, we will not discuss the mechanics of transliterating personal names. In general, rules or statistical translation models trained on a set of name translations are used to determine possible correspondences between phonemes in two languages and between characters/syllables and phonemes. Interested readers may refer to (Chen et al. 2006; Jeong et al. 1999; Kuo et al. 2006; Lam et al. 2007; Qu et al. 2003; Sproat et al. 2005) for details.

The huge volume of documents on the web containing glosses of the kind seen provides us with a rich resource for mining translingual knowledge for personal and organizational names and technical terms. One common approach is to manually define a set of common patterns of glossing. Zhang and Vines (2004) identified the following patterns in a monolingual text (identified here as the target language):

...translation (source_term) ...e.g. 斯科特· 霍夫曼(Scott Huffman)
...translation, source_term ...e.g. 美国花旗银行, Citibank, ...
...translation, or source_term ...e.g. 潜在语义信息检索模型, 或LSI...

These patterns reflect the common ways of specifying the corresponding terms (or their glosses) in their original language, especially when for names of persons and organizations and for technical terms.

A typical mining process based on manually defined patterns runs as follows (Zhang and Vine, 2004): First, given a source language term (English) for which translations are sought, the term is used as a search query to retrieve Chinese (target language) documents. Then the patterns are applied to the snippets of the returned results to identify the candidate translations. Further analysis of the candidates allows selection of the most frequent candidates. A number of studies have used the strategy (Cheng et al., 2004; Cao et al. 2007) to mine large numbers of translation relations from monolingual texts on the Web.

Additional mining criteria can be added to retrieve more relevant candidate snippets. For example, Zhang et al. (2005) and Huang et al. (2005) add related target language terms to the search query for snippets: To find a transliteration of "Leo Tolstoy" in Chinese (列夫.托尔斯泰), if one knows that the work "War and Peace" is closely connected to the author's name, then the Chinese terms "战争" (war) and "和平" (peace) can be added into the search query to locate highly related snippets.

Rather than exploiting a set of patterns to mine translingual relationships, Cheng et al. (2004) tries to mine related terms directly from the snippets returned by the search engine. Once a set of snippets is collected, a similarity measure is used to select terms that are related to the original term. Figure 10.8 shows an example using the query "yahoo" to retrieve documents in Chinese:

```
Yahoo!奇摩
... Yahoo!奇摩會員, 立即登入. ... 服務總覽 · 網站登錄 · 會員中心 · 服務條款
權政策 · 加值付費服務 · 關於Yahoo!奇摩 廣告刊登 · 影音服務 · 合作提案 ·
網路行銷 - Yahoo!奇摩徵才 · 連結Yahoo!奇摩. 雅虎國際資訊版權所有© 20
tw.yahoo.com/ - 29k - 2004年5月14日 - 頁庫存檔 - 類似網頁
Yahoo!奇摩新聞
Yahoo!奇摩新聞, 搜尋網站. 會員中心 | 服務說明 | Yahoo!奇摩. ... , 統一發票
摸彩號碼. 更多. 新聞電子報. ·, Yahoo!奇摩新聞電子報. 政治 社會 影視 國際
```

Figure 10.8. Results of search for Chinese documents using "Yahoo" as query. (from (Cheng et al., 2004))

The snippet results contain Chinese terms strongly correlated with "yahoo" such as 奇摩(Yahoo!'s name in Taiwan) and 搜索(search). Unsurprisingly, the extracted terms are more often related terms than translations, so they may not be appropriate for use in full-text translation, but appropriate for less demanding applications such as CLIR (Cheng et al 2004). The experiments on CLIR show that these glosses supplement existing dictionaries, and can reduce the number of unknown words in query translation. This mining approach can also find additional good translations for terms that are already covered by an existing resource.

8. Mining using hyperlinks

Modern search engines often view an anchor text linking to a Web page as an alternative description of the page. When different anchor texts link to the same Web page, those anchor texts can be considered strongly related. If the anchor texts are in different languages, moreover, then this relationship constitutes a kind of translingual/translational relationship. Figure 10.9, below, shows anchor texts in different languages pointing to the same Web page (www.yahoo.com).

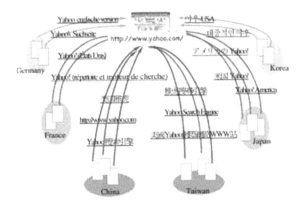

Figure 10.9. Possible hyperlinks and anchor texts to the web page www.yahoo.com. (from (Lu et al., 2004))

This is the principle used in (Lu et al. 2004) to extract translations using anchor texts. The terms "雅虎搜索引擎", "美国雅虎", "yahoo!", "Yahoo!モバイル", "Yahooの検索エンジン", etc. correspond to different names for "Yahoo!" in different languages. Lu et al. (2004) proposed a translingual similarity measure to determine relationships between terms in different languages. This approach is particularly suited to mining translations or transliterations of proper names (names of organizations and companies). It will find, for example, different transliterations of "Sony" in simplified Chinese "索尼" and in traditional Chinese "新力"; and translations and transliterations of "General Electric" or "GE" in simplified Chinese "通用电气" and in Traditional Chinese "奇异" (transliteration of "GE").

Mining on Wikipedia is a special case of hyperlink mining. Wikipedia is increasingly used in CLIR experiments to find equivalent expressions across languages, in particular proper names and technical terms. The encyclopedia contains numerous explicit links between different entries of the same entity in different languages that can be assumed to be mutual translations (Gamallo et al. 2010) For example, "Chang Kai-Shek", "Jiang Jieshi", 蒋中正and蒋介石are the different names of the same person, and they refer to the same page on Wikipedia. While the coverage provided by this resource is limited, one can further extend the mining process by also assuming that articles on the same topic in different languages are either "parallel" or comparable. These characteristics of

Wikipedia have been successfully exploited to extract translingual relations between elements in the two texts and used for CLIR (Potthast et al. 2008; Schönhofen et al. 2007; Smith et al. 2010).

9. Conclusions and Discussions

Translation is an essential component of MT and CLIR. Since manually constructed resources are limited in coverage there is an acute need to acquire translingual knowledge automatically. In this chapter, we have presented a broad overview of a growing body of work on mining parallel texts, parallel sentences and phrases on the Web. These studies show that the mining processes that employ heuristics based on the organization of parallel texts and the characteristics of parallel sentences, or translation knowledge already available (e.g. a bilingual dictionary), make it possible to harvest a large amount of parallel and comparable texts on the Web. The mined texts, without cleaning, can be too noisy for tasks such as MT. However, for tasks such as CLIR, which does not always demand high-quality text translations, parallel/comparable corpora mined using these mining approaches can be directly used to train models or learn term similarity measures for query translation. Experimental results show that one can obtain improved CLIR effectiveness compared with other resources such as MT and bilingual dictionaries.

For more demanding tasks like conventional text MT, refinements can be implemented to acquire more precise translation knowledge, including filtering of the mined corpus itself, and selection of parallel sentences or parallel phrases from the corpus. Experiments with SMT models indicate that the smaller and cleaner corpora obtained by filtering do in fact help improve the translation quality in terms of BLEU score and other metrics.

Although feasibility and utility of mining translingual knowledge on the Web is now well established, much room remains for methodological improvement. Despite application of filtering techniques, a significant percentage of the mined corpora still contain non-parallel data. Such corpora may be unreliable when used to train sophisticated translation models beyond the IBM-1 models employed in most CLIR studies. For MT purposes, moreover, it may be necessary to further refine the mining process itself in order to locate strictly parallel texts and sentences. On the other end of the spectrum, although a comparable corpus is considered too noisy to be suited to translation model training approaches to smoothing the models trained using strictly parallel texts and the ones using translingual term similarity with less strictly matched texts might be applicable to produce useful models.

While it is preferable to extract well-formed phrases for general MT tasks, the requirements for other tasks such as CLIR may be less stringent. A more flexible phrase-based query translation model may well be applicable, in which, for example, context is provided by pairs of query terms, with one word defining a context for the translation of the other even though the two words themselves may not form a single phrase.

Parallel texts are essential to translation, and identifying translingual resources remains a primary goal of mining parallel texts on the Web. But parallelism need not be viewed as limited to cases involving different languages. Other kinds of data can also potentially be regarded as parallel. For example, two sets of texts in the same language can be treated as parallel and used to train a "translation" model to capture the relationships between elements in that language, an approach that has been successfully used in monolingual IR (Burger and Lafferty, 1999; Gao et al. 2010). This notion can be further extended to mining trans-media knowledge: correspondences between images and textual annotations can be exploited to generate trans-media relations between visual features and words (Jeon et al. 2003; Oumohmed et al. 2005). These studies demonstrate that the SMT paradigm is applicable in tasks other than translation and hint at the possibility of interesting new approaches in other areas.

References

[1] Adafre, S.F. and de Rijke, M. (2006). Finding similar sentences acorss multiple languages in Wikipedia. *11th Conference of the European Chapter of the Association for Computational Linguistics*, pp. 62–69.

[2] Ballesteros, L. and Croft, W. (1997). Phrasal translation and query expansion techniques for cross-language information retrieval. In *Proceedings of SIGIR Conf.* pp. 84-91.

[3] Berger, A. and Lafferty, J. (1999). Information retrieval as statistical translation. In *Proceedings of SIGIR Conf.*, pp. 222-229.

[4] Braschler, M., and Schäuble, P. (1998). Multilingual information retrieval based on document alignment techniques. *ECDL '98: Proceedings of the Second European Conference on Research and Advanced Technology for Digital Libraries*, pp. 183–197.

[5] Braschler, M., and Schäuble, P. (2001). Experiments with the Eurospider Retrieval System for CLEF 2000, in *Proceedings of CLEF Conference.* pp. 140-148.

[6] Brown, P., Della Pietra, S., Della Pietra, V., and Mercer, R. (1993). The mathematics of statistical machine translation: Parameter estimation. *Computational Linguistics*, 19(2), pp. 263-311.

[7] Cao, G., Gao, J., Nie, J.Y. (2007) A system to mine large-scale bilingual dictionaries from monolingual Web pages, *MT Summit*, pp. 57-64.

[8] Carbonell, J.G, Yang, Y, Frederking, R.E., Brown, R., Geng, Y. and Lee, D. (1997) Translingual information retrieval: A comparative evaluation. In: Proceedings of the International Joint Conference on Arti?cial Intelligence (IJCAI '97).

[9] Chiang, D., (2005) A Hierarchical Phrase-Based Model for Statistical Machine Translation. *ACL*.

[10] Chen, J., Nie, J.Y., (2000) Automatic construction of parallel English-Chinese corpus for cross-language information retrieval. *ANLP pp.* 21-28

[11] Chen, H.H., Lin, W.C. and Yang, C.H. (2006). Translation-Transliterating Named Entities for Multilingual Information Access. *Journal of the American Society for Information Science and Technology*, 57(5):645-659

[12] Cheng, P., Teng, J., Chen, R., Wang, J., Lu, W., and Chien, L. (2004). Translating Unknown Queries with Web Corpora for Cross-Language Information Retrieval. In *Proceedings of SIGIR Conf.*, pp.162-169.

[13] Dumais, S. T., Letsche, T. A., Littman, M. L. and Landauer, T. K. (1997) Automatic cross-language retrieval using Latent Semantic Indexing. *AAAI Spring Symposuim on Cross-Language Text and Speech Retrieval*, March 1997.

[14] Franz, M., McCarley, J.S. and Koukos, S. (1999) Ad hoc and multilingual information retrieval at IBM. Proceedings of the Seventh Text Retrieval Conference (TREC-7), pp. 157–168.

[15] Fung, P. (1995). A Pattern Matching Method for Finding Noun and Proper Noun Translations from Noisy Parallel Corpora. *Proceedings of the Association for Computational Linguistics*, pp. 236-243.

[16] Pascale Fung and Yuen Yee Lo. 1998. An IR approach for translating new words from nonparallel, comparable texts. *Proceedings of COLING-ACL98*, pp. 414– 420.

[17] Fung, P. and McKeown, K. (1997) Finding terminology translations from non-parallel corpora. In: The 5th Annual Workshop on Very Large Corpora.

[18] Fung, P. and Cheung, P. (2004) Multilevel boot-strapping for extracting parallel sentences from a quasi parallel corpus. *Conference on Empirical Methods in Natural Language Processing (EMNLP 04)*, pp. 1051–1057.

[19] Gale, W. A., Church K. W. 1993. *A Program for Aligning Sentences in Bilingual Corpora*. Computational Linguistics, 19(3): 75-102.

[20] Galley, M., Hopkins, M., Knight, K., Marcu, D., (2004) What's in a translation rule? *HLT-NAACL*, pp. 273-280

[21] Pablo Gamallo Otero, Isaac Gonzalez Lopez, (2009) Wikipedia as Multilingual Source of Comparable Corpora, *Proceedings of the 3rd Workshop on Building and Using Comparable Corpora*, LREC 2010, pp. 21–25

[22] Gao, J., Nie, J.Y., Xun, E., Zhang, J., Zhou, M., and Huang, C. (2001). Improving query translation for cross-language information retrieval using statistical models. In *Proceedings of SIGIR Conf.*, pp. 96-104.

[23] Gao, J., Zhou, M., Nie, J.Y., He, H., Chen, W. (2002) Resolving query translation ambiguity using a decaying co-occurrence model and syntactic dependence relations. *SIGIR*, pp. 183-190

[24] Gao, J., Nie, J.Y. (2006) Study of Statistical Models for Query Translation: Finding a Good Unit of Translation. *SIGIR*, pp 194-201, 2006.

[25] Gao, J., He, X., Nie. J.Y. (2010) Clickthrough-based translation models for web search: from word models to phrase models. *CIKM*, pp 1139-1148, 2010.

[26] Hong, Gumwon, Li, Chi-Ho, Zhou, Ming and Rim, Hae-Chang (2010) An Empirical Study on Web Mining of Parallel Data, *COLING*, pp. 474–482.

[27] Huang, Degen, Zhao, Lian, Li, Lishuang Yu, Haitao (2010) Mining Large-scale Comparable Corpora from Chinese-English News Collections, *COLING*, pp. 472-480.

[28] Huang, F., Zhang, Y., and Vogel, S. (2005). Mining Key Phrase Translations from Web Corpora. In *Proceedings of HLT-EMNLP Conf.*, pp. 483-490.

[29] Jeon, J. Lavrenko, V. and Manmatha, R. (2003) Automatic Image Annotation and Retrieval using Cross-Media Relevance Models, *SIGIR*, pp. 119-126.

[30] Jeong, K.S., Myaeng, S.H., Lee, J.S, and Choi, K.S., (1999) Automatic identification and back-transliteration of foreign words for

information retrieval, *Information Processing and Management*, 35(4), pp. 523-540.

[31] Ji, Heng (2009) Mining Name Translations from Comparable Corpora by Creating Bilingual Information Networks, *Proceedings of the 2nd Workshop on Building and Using Comparable Corpora, ACL-IJCNLP 2009*, pages34–37.

[32] Koehn, P., Och, F.J., Marcus, D., (2003) Statistical phrase-based translation, In *Proceedings of HLT-NAACL*, pp. 48-54.

[33] Koehn, P. (2009) Statistical Machine Translation. Cambridge University Press.

[34] Kraaij, W., Nie, J.Y., and Simard, M. (2003). Embedding Web-Based Statistical Translation Models in Cross-Language Information Retrieval. *Computational Linguistics*, 29(3): 381-420.

[35] Kumano, T. and Tanaka, H., Tokunaga, T. (2007) Extracting phrasal alignments from comparable corpora by using joint probability SMT model. 11th International Conference on Theoretical and Methodological Issues in Machine Translation (TMI'07).

[36] Kuo, J.S., Li, H., and Yang Y.K (2006). Learning Transliteration Lexicon from the Web. In *the Proceedings of COLING/ACL*, pp.1129-1136

[37] Lam, W., Chan, S.K., and Huang, R. (2007). Named Entity Translation Matching and Learning: With Application for Mining Unseen Translations. *ACM Transactions on Information Systems*, 25(1), pp.

[38] Liu, Y., Jin R. and Chai, Joyce Y. (2005). A maximum coherence model for dictionary-based cross-language information retrieval, In *Proceedings of SIGIR conf.*, pp. 536-543.

[39] Lu, W. Chien, L.F. and Lee, H. (2004). Anchor Text Mining for Translation of Web Queries: A Transitive Translation Approach. *ACM Transactions on Information Systems*, Vol.22, pp. 242-269.

[40] Ma, X. and Liberman, M., (1999). Bits: A Method for Bilingual Text Search over the Web. *Proceedings of Machine Translation Summit VII*.

[41] Munteanu, D. S., Marcu, D. (2005) Improving Machine Translation Performance by Exploiting Non-Parallel Corpora. 2005. *Computational Linguistics*. 31(4). pp: 477-504.

[42] Munteanu, D. S. and Marcu D. (2006). Extracting parallel subsentential fragments from non-parallel corpora. *ACL*, pp. 81–88.

[43] Nagata, M., Saito, T., and Suzuki, K. (2001). Using the web as a bilingual dictionary. In *Proceedings of the Workshop on Data-Driven Methods in Machine Translation* (with ACL Conf.), pp. 1-8.

[44] Nie, J.Y., Cai, J. (2001) Filtering parallel corpora of web pages, IEEE symposium on NLP and Knowledge Engineering, pp. 453-458.

[45] Nie, J.Y., Simard, M., Isabelle, P., Durand, R. (1999) Cross-Language Information Retrieval based on Parallel Texts and Automatic Mining of Parallel Texts in the Web, In *Proceedings of SIGIR Conf.*, pp. 74-81

[46] Och, F., and Ney, H. (2002) Discriminative Training and Maximum Entropy Models for Statistical Machine Translation. *ACL*, pp. 295-302

[47] Och, F. (2003). Minimum error rate training in statistical machine translation. In *Proceedings of ACL.* pp. 160-67

[48] Oumohmed, A.I., Mignotte, M., Nie, J.Y. (2005) Semantic-Based Cross-Media Image Retrieval, *Pattern Recognition and Image Analysis: Third International Conference on Advances in Pattern Recognition (ICAPR)*, LNCS 3687, pp. 414-423.

[49] Potthast, M., Stein, B., Anderka, M. (2008) A Wikipedia-based Multilingual Retrieval Model. *ECIR*, LNCS 4956, pp. 522-530.

[50] Qu, Y., Grefenstette, G., and Evans, D. A. (2003). Automatic transliteration for Japanese-to-English text retrieval. In *Proceedings of SIGIR Conference*, pp. 353-360.

[51] Rapp, R. (1995). Identifying Word Translations in Non-Parallel Texts. *Proceedings of the 33rd Annual Meeting of the Association for Computational Linguistics*, pp. 320-322.

[52] Resnik, P., (1999) Mining the Web for Bilingual Text, 37th Annual Meeting of the Association for Computational Linguistics (ACL'99).

[53] Resnik P. and Smith. N.A. (2003) The Web as a Parallel Corpus, *Computational Linguistics*, 29(3), pp. 349-380, September 2003.

[54] Sheridan, P. and Ballerini, J. P. (1996). Experiments in multilingual information retrieval using the SPIDER system. In *Proceedings of SIGIR Conf.*, pp. 58-65.

[55] Schönhofen, P., Benczúr, A., Bíró, I., Csalogány, K. (2007) Performing cross-language retrieval with Wikipedia, CLEF-2007 (http://www.clef-campaign.org/2007/working_notes/schonhofenCLEF2007.pdf)

[56] Shi, L., Niu, C., Zhou, M., and Gao, J. (2006) A DOM Tree Alignment Model for Mining Parallel Data from the Web, *ACL*, pp. 489-496.

[57] Smith, J. R., Quirk, C., and Toutanova, K. (2010) Extracting parallel sentences from comparable corpora using document level alignment. *HLT*, pp. 403–411

[58] Sproat, R., Tao, T., Zhai, C. (2006) Named Entity Transliteration with Comparable Corpora. In *Proceedings of ACL*.

[59] Tuomas Talvensaari, Jorma Laurikkala, Kalervo Järvelin, Martti Juhola (2006) A study on automatic creation of a comparable document collection in cross-language information retrieval, Journal of Documentation, Vol. 62 No. 3, pp. 372-387

[60] Tuomas Talvensaari, Jorma Laurikkala, Kalervo Järvelin, Martti Juhola, and Heikki Keskustalo (2007). Creating and exploiting a comparable corpus in cross-language information retrieval. *ACM Trans. Inf. Syst.* 25, 1, Article 4.

[61] Utiyama M. and Isahara, H. (2003) Reliable Measures for Aligning Japanese-English News Articles and Sentences. *ACL*, pp. 72–79.

[62] Jinxi Xu, W. Bruce Croft (1996) Query Expansion Using Local and Global Document Analysis. *SIGIR*, pp. 4-11

[63] Yang, Christopher C., and Kar Wing Li. 2003. Automatic construction of English/Chinese parallel corpora. *Journal of the American Society for Information Science and Technology*, 54(8), pp. 730–742.

[64] Zhang, Y. and Vines, P. (2004). Using the Web for Automated Translation Extraction in Cross-Language Information Retrieval. In *Proceedings of SIGIR Conf.*, pp.162-169.

[65] Zhang, Y., Huang, F., Vogel, S. (2005) Mining Translations of OOV Terms from the Web through Cross-lingual Query Expansion, *SIGIR*, pp. 669-670.

[66] Zhao, B., and Vogel, S. (2002). Adaptive Parallel Sentences Mining from Web Bilingual News Collection. In Proceedings of IEEE international conference on data mining, pages 745-750.

Chapter 11

TEXT MINING IN MULTIMEDIA

Zheng-Jun Zha
School of Computing, National University of Singapore
zhazj@comp.nus.edu.sg

Meng Wang
School of Computing, National University of Singapore
wangm@comp.nus.edu.sg

Jialie Shen
Singapore Management University
jlshen@smu.edu.sg

Tat-Seng Chua
School of Computing, National University of Singapore
chuats@comp.nus.edu.sg

Abstract A large amount of multimedia data (e.g., image and video) is now available on the Web. A multimedia entity does not appear in isolation, but is accompanied by various forms of metadata, such as surrounding text, user tags, ratings, and comments etc. Mining these textual metadata has been found to be effective in facilitating multimedia information processing and management. A wealth of research efforts has been dedicated to text mining in multimedia. This chapter provides a comprehensive survey of recent research efforts. Specifically, the survey focuses on four aspects: (a) surrounding text mining; (b) tag mining; (c) joint text and visual content mining; and (d) cross text and visual content mining. Furthermore, open research issues are identified based on the current research efforts.

Keywords: Text Mining, Multimedia, Surrounding Text, Tagging, Social Network

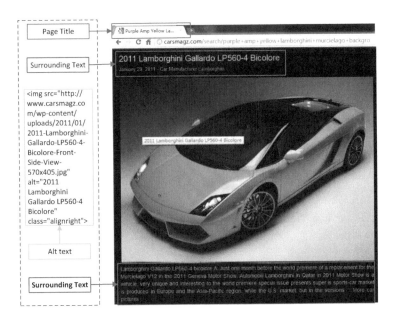

Figure 11.1. Illustration of textual metadata of an embedded image in a Web page.

1. Introduction

Lower cost hardware and growing communications infrastructure (e.g. Web, cell Phones, etc.) have led to an explosion in the availability of ubiquitous devices to produce, store, view and exchange multimedia entities (images, videos). A large amount of image and video data are now available. Take one of the most popular photo sharing services Flickr [1] as example, it has accumulated several billions of images. Another example is Youtube [2], which is a video sharing Web site that is hosting billions of videos. As the largest photo sharing site, Facebook [3] currently stores hundreds of hundreds of billions of photos.

On the other hand, a multimedia entity does not appear in isolation but is accompanied by various forms of textual metadata. One of the most typical examples is the surrounding text appearing around the embedded images or videos in the Web page (See Figure 11.1). With recent proliferation of social media sharing services, the newly emerging textual meatadata include user tags, ratings, comments, as well as

---

[1] http://www.flickr.com/
[2] http://www.youtube.com/
[3] http://www.facebook.com/

Figure 11.2. Illustration of textual metadata of an image on a photo sharing Web site.

the information about the uploaders and their social network (See Figure 11.2). These metadata, in particular the tags, have been found to be an important resource for facilitating multimedia information processing and management. Given the wealth of research efforts that has been done, there have been various studies in multimedia community on the mining of textual metadata. In this chapter, a multimedia entity refers to an image or a video. For the sake of simplicity and without lost of generality, we use the term image to refer to multimedia entity for the rest of this chapter.

In this chapter, we first review the related works on mining surrounding text for image retrieval as well as the recent research efforts that explore surrounding text for image annotation and clustering in Section 2. In Section 3, we provide a literature review on tag mining and show that the main focus of existing tag mining works includes three aspects: tag ranking, tag refinement, and tag information enrichment. In

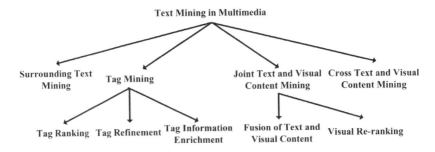

Figure 11.3. A taxonomy consisting of the research works reviewed in this chapter.

Section 4, we survey the recent progress in integrating textual metadata and visual content. We categorize the exiting works into two categories: the fusion of text and visual content as well as visual re-ranking. In Section 5, we provide a detailed discussion on recent research on cross text and visual content mining. We organize all the works reviewed in this chapter into a taxonomy as shown in Figure 11.3. The taxonomy provides an overview of state-of-the-art research and helps us to identify open research issues to be presented in Section 6.

2. Surrounding Text Mining

In order to enhance the content quality and improve user experience, many hosting Web pages include different kinds of multimedia entities, like image or video. These multimedia entities are frequently embedded as part of the text descriptions which we called the surrounding text. While there is no standard definition, surrounding text generally refers to the text consisting of words, phrases or sentences that surrounds or close to the embedded images, such as those that appear at the top, below, left or right region of images or connected via Web links. The effective use of surrounding texts is becoming increasingly important for multimedia retrieval. However, developing effective extraction algorithm for the comprehensive analysis of surrounding text has been a very challenging task. In many cases, automatically determining which page region is more relevant to the image than the others could be difficult. Moreover, how large the region nearby should be considered is still an open question. Further, the quality of surrounding texts could be low and inconsistent. These problems make it very hard to directly apply the surrounding text information to facilitate accurate retrieval. Thus, refinement process or combining it with other cues is essential.

The earliest efforts on modeling and analyzing surrounding texts to facilitate multimedia retrieval occurred in the 1990s. AltaVista's A/V

Photo Finder applies textual and visual cues to index image collections [1]. The indexing terms are precomputed based on the HTML documents containing the Web images. With a similar approach, the WebSeer system harvests the information for indexing Web images from two different sources: the related HTML text and the embedded image itself [12]. It extracts keywords from page title, file name, caption, alternative text, image hyperlinks, and body text titles. A weight is calculated for each keyword based on its location inside a page. In PICITION system [40], an interesting approach is developed to exploit both textual and visual information to index a pictorial database. Image captions are used as an important cue to identify faces appearing in a related newspaper photograph. The empirical study based on a data set containing 50 pictures and captions obtained from the *Buffalo News* and the *New York Times* is used to demonstrate the effectiveness of the PICITION system. While the system can be successfully adopted for accessing photographs in newspaper or magazine, it is not straightforward to apply it for Web image retrieval.

In [39], Smith and Chang proposed the WebSeek framework designed to search images from the Web. The key idea is to analyze and classify the Web multimedia objects into a predefined taxonomy of categories. Thus, an initial search can be performed to explore a catalog associated with the query terms. The image attribute (e.g., color histogram for images) is then computed for similarity matching within the category.

Besides its efficacy in image retrieval, surrounding text has been explored for image annotation recently. Feng et al. presented a bootstrapping framework to label and search Web images based on a set of predefined semantic concepts [9]. To achieve better annotation effectiveness, a co-training scheme is designed to explore the association between the text features computed using corresponding HTML documents and visual features extracted from image content. Observing that the links between the visual content and the surrounding texts can be modeled via Web page analysis, a novel method called Iterative Similarity Propagation is proposed to refine the closeness between the Web images and their annotations [50]. On the other hand, it is not hard to find that images from the same cluster may share many similar characteristics or patterns with respect to relevance to information needs. Consequently, accurate clustering is a very crucial technique to facilitate Web multimedia search and many algorithms have recently been proposed based on the analysis of surrounding texts and low level visual features [3][13][34]. For example, Cai et al. [3] proposed a hierarchical clustering method that exploits visual, textual, and link analysis. A webpage is partitioned into blocks, and the textual and link information of an image are extracted from the block containing that image. By using block-level link analysis techniques, an image graph is constructed. They then applied spectral techniques to find a Euclidean embedding of the images. As a result, each image has three types of representations: visual feature, textual feature, and graph-based representation. Spectral clustering techniques are employed to cluster search results into various clusters. Gao et al. [13] and Rege et al. [34] used a tripartite graph to model the relations among visual features, images and their surrounding text. The clustering is performed by partitioning this tripartite graph.

3. Tag Mining

In newly emerging social media sharing services, such as the Flickr and Youtube, users are encouraged to share multimedia data on the Web and annotate content with tags. Here a tag is referred to as a descriptive keyword that describes the multimedia content at semantic or syntactic level. These tags have been found to be an important resource for multimedia management and have triggered many innovative research topics [61][51][38][36]. For example, with accurate tags, the retrieval of multimedia content can be easily accomplished. The tags can be used to index multimedia data and support efficient tag-based search. Nowadays, many online media repositories, such as Flickr and Youtube, support tag-based multimedia search. However, since the tags are provided by grassroots Internet users, they are often noisy and incomplete and there is still a gap between these tags and the actual content of the images[20][26][48]. This deficiency has limited the effectiveness of tag-based applications.

Recently, a wealth of research has been proposed to enhance the quality of human-provided tags. The existing works mainly focus on the following three aspects: (a) tag ranking, which aims to differentiate the tags associated with the images with various levels of relevance; (b) tag refinement with the purpose to refine the unreliable human-provided tags; and (c) tag information enrichment, which aims to supplement tags with additional information [26]. In this section, we present a comprehensive review of existing tag ranking, tag refinement, and tag information enrichment methods.

3.1 Tag Ranking

As shown in [25], the relevance level of the tags cannot be distinguished from the tag list of an image. The lack of relevance information in the tag list has limited the application of tags. Recently, tag ranking has been studied to infer the relevance levels of tags associated with an

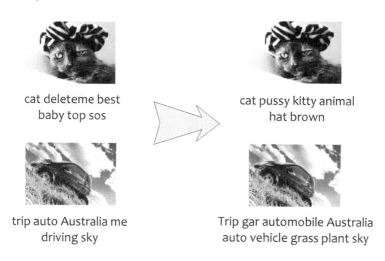

Figure 11.4. Examples of of tag refinement. The left side of the figure shows the original tags while the right side shows the refined tags. The technique is able to remove irrelevant tags and add relevant tags to obtain better description of multimedia contents.

image. As a pioneering work, Liu et al. [25] proposed to estimate tag relevance scores using kernel density estimation, and then employ random walk to boost this primary estimation. Li et al. [22] proposed a data driven method for tag ranking. They learned the relevance scores of tags by a neighborhood voting approach. Given an image and one of its associated tag, the relevance score is learned by accumulating the votes from the visual neighbors of the image. They then extended the work to multiple visual spaces [23]. They learned the relevance scores of tags and ranked them by neighborhood voting in different feature spaces, and the results are aggregated with a score fusion or rank fusion method. Different aggregation methods have been investigated, such as the average score fusion, Borda count and RankBoost. The results show that a simple average fusion of scores is already able to perform closed to supervised fusion methods like RankBoost.

3.2 Tag Refinement

User-provided tags are often noisy and incomplete. The study in [20] shows that when a tag appears in a Flickr image, there is only about a 50% chance that the tag is really relevant, and the study in [38] shows that more than half of Flickr images are associated with less than three tags. Tag refinement technologies are proposed aiming at obtaining more accurate and complete tags for multimedia description, as shown in Figure 11.4.

A lot of tag refinement approaches have been developed based on various statistical learning techniques. Most of them are based on the following three assumptions.

- The refined tags should not change too much from those provided by the users. This assumption is usually used to regularize the tag refinement.

- The tags of visually similar images should be closely related. This is a natural assumption that most automatic tagging methods are also built upon.

- Semantically close or correlative tags should appear with high correlation. For example, when a tag "sea" exists for an image, the tags "beach" and "water" should be assigned with higher confidence while the tag "street" should have low confidence.

For example, Chen et al. [6] first trained a SVM classifier for each tag with the loosely labeled positive and negative samples. The classifiers are used to estimate the initial relevance scores of tags. They then refined the scores with a graph-based method that simultaneously considers the similarity between images and semantic correlation among tags. Xu et al. [52] proposed a tag refinement algorithm from topic modeling point of view. A new graphical model named regularized latent Dirichlet allocation (rLDA) is presented to jointly model the tag similarity and tag relevance. Zhu et al. [64] proposed a matrix decomposition method. They used a matrix to represent the image-tag relationship: the (i, j)-th element is 1 if the i-th image is associated with the j-th tag, and 0 otherwise. The matrix is then decomposed into a refined matrix plus an error matrix. They enforced the error matrix to be sparse and the refined matrix to follow three principles: (a) let the matrix be low-rank; (b) if two images are visually similar, the corresponding rows are with high correlation; and (c) if two tags are semantically close, the corresponding vectors are with high correlation. Fan et al. [8] grouped images with a target tag into clusters. Each cluster is regarded as a unit. The initial relevance scores of the clusters are estimated and then refined by a random walk process. Liu et al. [24] adopted a three-step approach. The first step filters out tags that are intrinsically content-unrelated based on the ontology in WordNet. The second step refines the tags based on the consistency of visual similarity and semantic similarity of images. The last step performs tag enrichment, which expands the tags with their appropriate synonyms and hypericum.

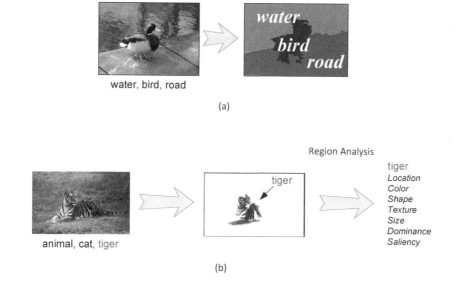

Figure 11.5. (a) An example of tag localization, which finds the regions that the tags describe. (b) An illustration of tag information enrichment. It first finds the corresponding region of the target tag and then analyze the properties of the region.

3.3 Tag Information Enrichment

In the manual tagging process, generally human labelers will only assign appropriate tags to multimedia entities without any additional information, such as the image regions depicted by the corresponding tags. But by employing computer vision and machine learning technologies, certain information of the tags, such as the descriptive regions and saliency, can be automatically obtained. We refer to these as tag information enrichment.

Most existing works employ the following two steps for tag information enrichment. First, tags are localized into regions of images or sub-clips of videos. Second, the characteristics of the regions or sub-clips are analyzed, and the information about the tags is enriched accordingly. Figure 11.5 (a) illustrates the examples of tag localization for image and video data. Liu et al. [28] proposed a method to locate image tags to corresponding regions. They first performed over-segmentation to decompose each image into patches and then discovered the relationship between patches and tags via sparse coding. The over-segmented regions are then merged to accomplish the tag-to-region process. Liu et al. extended the approach based on image search [29]. For a tag of the target image, they collected a set of images by using the tag as query with an image search engine. They then learned the relationship between the tag and the patches in this image set. The selected patches are used to reconstruct each candidate region, and the candidate regions are ranked based on the reconstruction error. Liu et al. [27] accomplished the tag-to-region task by regarding an image as a bag of regions and then performed tag propagation on a graph, in which vertices are images and edges are constructed based on the visual link of regions. Feng et al. [10] proposed a tag saliency learning scheme, which is able to rank tags according to their saliency levels to an image's content. They first located tags to images' regions with a multi-instance learning approach. In multi-instance learning, an image is regarded as a bag of multiple instances, i.e., regions [58]. They then analyzed the saliency values of these regions. It can provide more comprehensive information when an image is relevant to multiple tags, such as those describing different objects in the image. Yang et al. [55] proposed a method to associate a tag with a set of properties, including location, color, texture, shape, size and dominance. They employed a multi-instance learning method to establish the region that each tag is corresponding to, and the region is then analyzed to establish the properties, as shown in Figure 11.5 (b). Sun and Bhowmick [41] defined a tag's visual representativeness based on a large image set and the subset that is associated with the tag. They employed two distance metrics, cohesion and separation, to estimate the visual representativeness measure.

Ulges et al. [43] proposed an approach to localize video-level tags to keyframes. Given a tag, it regards whether a keyframe is relevant as a latent random variable. An EM-style process is then adopted to estimate the variables. Li et al. [21] employed a multi-instance learning approach to accomplish the video tag localization, in which video and shot are regarded as bag and shot, respectively.

By supplementing tags with additional information, a lot of tag-based applications can be facilitated, such as tag-based image/video retrieval and intelligent video browsing etc.

4. Joint Text and Visual Content Mining

Beyond mining pure textual metadata, researchers in multimedia community have started making progress in integrating text and content for multimedia retrieval via joint text and content mining. The integration of text and visual content has been found to be more effective than exploiting purely text or visual content separately. The joint text and content mining in multimedia retrieval often comes down to finding effective mechanisms for fusing multi-modality information from textual metadata and visual content. Existing research efforts can generally be categorized into four paradigms: (a) linear fusion; (b) latent-space-based fusion; (c) graph-based fusion; and (d) visual re-ranking that exploits visual information to refine text-based retrieval results. In this section, we first briefly review linear, latent space based, and graph based fusion methods and then provide comprehensive literature review on visual re-ranking technology.

Linear fusion combines the retrieval results from various modalities linearly [18][4][31]. In [18], visual content and text are combined in both online learning stage with relevance feedback and offline keyword propagation. In [31], linear, max, and average fusion strategies are employed to aggregate the search results from visual and textual modalities. Chang et al. [4] adopted a query-class-dependent fusion approach. The critical task in linear fusion is the estimation of fusion weights of different modalities. A certain amount of training data is usually required for estimating these weights. The latent space based fusion assumes that there is a latent space shared by different modalities and thus unify different modalities by transferring the features of these modalities into the shared latent space [63][62]. For example, Zhao et al. [63] adopted the Latent Semantic Indexing (LSI) method to fuse text and visual content. Zhang et al. [62] proposed a probabilistic context model to explicitly exploit the synergy between text and visual content. The synergy is represented as a hidden layer between the image and text modalities. This hidden layer constitutes the semantic concepts to be annotated through a probabilistic framework. An Expectation-Maximization (EM) based iterative learning procedure is developed to determine the conditional probabilities of the visual features and the words given a hidden concept class. Latent space based methods usually require a large amount of training samples for learning the feature mapping from each modality into the unified latent space. Graph based approach [49] first builds the relations between different modalities, such as relations between images and text using the Web page structure. The relations are then utilized to iteratively update the similarity graphs computed from different modalities. The difficulty of creating similarity graphs for billions of images on the Web makes this approach insufficiently scalable.

4.1   Visual Re-ranking

Visual re-ranking is emerging as one of the promising technique for automated boosting of retrieval precision [42] [30] [55]. The basic functionality is to reorder the retrieved multimedia entities to achieve the optimal rank list by exploiting visual content in a second step. In particular, given a textual query, an initial list of multimedia entities is returned using the text-based retrieval scheme. Subsequently, the most relevant results are moved to the top of the result list while the less relevant ones are reordered to the lower ranks. As such, the overall search precision at the top ranks can be enhanced dramatically. According to the statistical analysis model used, the existing re-ranking approaches can roughly be categorized into three categories including the clustering based, classification based and graph based methods.

Cluster analysis is very useful to estimate the inter-entity similarity. The clustering based re-ranking methods stem from the key observation that a lot of visual characteristics can be shared by relevant images or video clips. With intelligent clustering algorithms (e.g., mean-shift, K-means, and K-medoids), initial search results from text-based retrieval can be grouped by visual closeness. One good example of clustering based re-ranking algorithms is an Information Bottle based scheme developed by Hsu et al. [16]. Its main objective is to identify optimal clusters of images that can minimize the loss of mutual information. The cluster number is manually configured to ensure the each cluster contains the same number of multimedia entities (about 25). This method was evaluated using the TRECVID 2003-2005 data and significant improvements were observed in terms of MAP measures. In [19], a fast and accurate scheme is proposed for grouping Web image search results into semantic clusters. For a given query, a few related semantic clusters are identified in the first step. Then, the cluster names relating to query are derived and used as text keywords for querying image search engine. The empirical results from a set of user studies demonstrate an improvement in performance over Google image search results. It is not hard to show that the clustering based re-ranking methods can work well when the initial search results contain many near-duplicate media documents. However, for queries that return highly diverse results or without clear visual patterns, the performance of the clustering-based methods is not guaranteed. Furthermore, the number of clusters has large impact on the final effectiveness of the algorithms. However, determining the optimal cluster number automatically is still an open research problem.

In the classification based methods, visual re-ranking is formulated as a binary classification problem aiming to identify whether each search result is relevant or not. The major process for result list reordering consists of three major steps: (a) the selection of pseudo-positive and pseudo-negative samples; (b) use the samples obtained in step (a) to train a classification scheme; and (c) reorder the samples according to their relevance scores given by the trained classifier. For existing classification methods, pseudo relevance feedback (PRF) is applied to select the training examples. It assumes that: (a) a limited number of top-ranked entities in the initial retrieval results are highly relevant to the search queries; and (b) automatic local analysis over the entities can be very helpful to refine query representation. In [54], the query images or video clip examples are used as the pseudo-positive samples. The pseudo-negative samples are selected from either the least relevant samples in the initial result list or the databases that contain less samples related to the query. The second step of the classification based methods aim to train classifiers and a wide range of statistical classifiers can be adopted. They include the Support Vector Machine (SVM) [54], Boosting [53] and ListNet [57]. The main weakness for the classification based methods is that the number and quality of training data required play a very important role in constructing effective classifiers. However, in many real scenarios, the training examples obtained via PRF are very noisy and might not be adequate for training effective classifier. To address this issue, Fergus et al. [11] used RANSAC to sample a training subset with a high percentage of relevant images. A generative constellation model is learned for the query category while a background model is learned from the query "things". Images are re-ranked based on their likelihood ratio. Observing that discriminative learning can lead to superior results, Schroff et al. [35] first learned a query independent text based re-ranker. The top ranked results from the text based re-ranking are then selected as positive training examples. Negative training examples are picked randomly from the other queries. A binary SVM classifier is then used to re-rank the results on the basis of visual features. This classifier is found to be robust to label noise in the positive training set as long as the non-relevant images are not visually consistent. Better training data can be obtained from online knowledge resources if the set of queries restricted. For instance, Wang et al. [44] learned a generative text model from the query's Wikipedia [4] page and a discriminative image model from the Caltech [15] and Flickr data sets. Search results are then re-ranked on the basis of these learned probability models. Some user interactions are required to disambiguate the query.

Graphs provide a natural and comprehensive way to explore complex relations between data at different levels and have been applied to a wide range of applications [59][46][47][60]. With the graph based re-ranking methods, the multimedia entities in top ranks and their associations/dependencies can be represented as a collection of nodes (vertices) and edges. The local patterns or salient features discover using graph

---

[4] http://www.wikipedia.org/

analysis are very helpful to improve effectiveness of rank lists. In [16], Hsu et al. modeled the re-ranking process as a random walk over the context graph. In order to effectively leverage the retrieved results from text search, each sample corresponds to a "dongle" node containing ranking score based on text. For the framework, edges between "dongle" nodes are weighted with multi-modal similarities. In many cases, the structure of large scale graphs can be very complex and this easily makes related analysis process very expensive in terms of computational cost. Thus, Jing and Baluja proposed a VisualRank framework to efficiently model similarity of Google image search results with graph [17]. The framework casts the re-ranking problem as random walk on an affinity graph and reorders images according to the visual similarities. The final result list is generated via sorting the images based on graph nodes' weights. In [42], Tian et al., presented a Bayesian video search re-ranking framework formulating the re-ranking process as an energy minimization problem. The main design goal is to optimize the consistency of ranking scores over visually similar videos and minimize the disagreement between the optimal list and the initial list. The method achieves a consistently better performance over several earlier proposed schemes on the TRECVID 2006 and 2007 data sets. The graph based re-ranking algorithms mentioned above generally do not consider any initial supervision information. Thus, the performance is significantly dependent on the statistical properties of top ranked search results. Motivated by this observation, Wang et al, proposed a semi-supervised framework to refine the text based image retrieval results via leveraging the data distribution and the partial supervision information obtained from the top ranked images [45]. Indeed, graph analysis has been shown to be a very powerful tool for analyzing and identifying salient structure and useful patterns inside the visual search results. With recent progresses in graph mining, this research stream is expected to continue to make important contributions to improve visual re-ranking from different perspectives.

5. Cross Text and Visual Content Mining

Although the joint text and visual content mining approaches described above facilitate image retrieval, they require that the test images have associated text modality. However, in some real world applications, images may not always have associated text. For example, most surveillance images/videos in in-house repository are not accompanied with any text. Even on social media Website such as the Flickr, there exist a substantial number of images without any tags. In such cases, joint

Text Mining in Multimedia 375

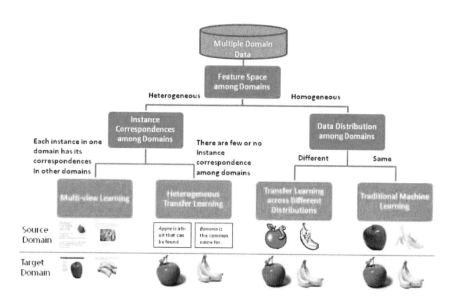

Figure 11.6. An illustration of different types of learning paradigms using image classification/clustering in the domains of apple and banana. Adapted from [56].

text and visual content mining cannot be applied due to missing text modality.

Recently, cross text and visual content mining has been studied in the context of transfer learning techniques. This class of techniques emphasizes the transferring of knowledge across different domains or tasks [32]. Cross text and visual content mining does not require that a test image has an associated text modality, and is thus beneficial to dealing with the images without any text by propagating the semantic knowledge from text to images [5]. It is also motivated by two observations. First, visual content of images is much more complicated than the text feature. While the textual words are easier to interpret, there exist a tremendous semantic gap between visual content and high-level semantics. Second, image understanding becomes particularly challenging when only a few labeled images are available for training. This is a common challenge, since it is expensive and time-consuming to obtain labeled images. On the contrary, labeled/unlabeled text data are relatively easier to collect. For example, millions of categorized text articles are freely available in Web

---

[5] Cross text and visual content can also facilitate text understanding in special cases by propagating knowledge from images to text.

text collections, such as Wikipedia, covering a wide range of topics from culture and arts, geography and places, history and events, to natural and physical science. A large number of Wikipedia articles are indexed by thousands of categories in these topics [33]. This provides abundant labeled text data. Thus, it is desirable to propagate semantic knowledge from text to images to facilitate image understanding. However, it is not trivial to transfer knowledge between various domains/tasks due to the following challenges:

- The target data may be drawn from a distribution different from the source data.

- The target and source data may be in different feature spaces (e.g., image and text) and there may be no correspondence between instances in these spaces.

- The target and source tasks may have different output spaces.

While the traditional transfer learning techniques focus on the distribution variance problem, the recent proposed heterogenous transfer learning approaches aim to tackle both the distribution variance and heterogenous feature space problems [56][7][65][33], or all the three challenges listed above [37]. Figure 11.6 from [56] presents an intuitive illustration of four learning paradigms, including traditional machine learning, transfer learning across different distributions, multi-view learning and heterogenous transfer learning. As we can see, heterogenous transfer learning is usually much more challenging due to the unknown correspondence across the distinct feature spaces. In order to learn the underlying correspondence for knowledge transformation, a "semantic bridge" is required. The "semantic bridge" can be obtained from the co-occurrence information between text and images or the linkage information in social media networks. For example, while the traditional webpages provide the co-occurrence information between text and images, the social media sites contain a large number of linked information between different types of entities, such as the text articles, tags, posts, images and videos. This linkage information provide a "semantic bridge" to learn the underlying correspondence [2].

Most existing works exploit the tag information that provide text-to-image linking information. As a pioneering work, Dai et al. [7] showed that such information can be effectively leveraged for transferring knowledge between text and images. The key idea of [7] is to construct a correspondence between the images and the auxiliary text data with the use of tags. Probabilistic latent semantic analysis (PLSA) model is employed to construct a latent semantic space which can be used for transferring knowledge. Chen et al. [56] proposed the concept of heterogeneous transfer learning and applied it to improve image clustering by leveraging auxiliary text data. They collected annotated images from the social web, and used them to construct a text to image mapping. The algorithm is referred to as aPLSA (Annotated Probabilistic Latent Semantic Analysis). The key idea is to unify two different kinds of latent semantic analysis in order to create a bridge between the text and images. The first kind of technique performs PLSA analysis on the target images, which are converted to an image instance-to-feature cooccurrence matrix. The second kind of PLSA is applied to the annotated image data from social Web, which is converted into a text-to-image feature co-occurrence matrix. In order to unify those two separate PLSA models, these two steps are done simultaneously with common latent variables used as a bridge linking them. It has been shown in [5] that such a bridging approach leads to much better clustering results. Zhu et al. [65] discussed how to create the connections between images and text with the use of tag data. They showed how such links can be used more effectively for image classification. An advantage of [65] is that it exploits unlabeled text data instead of labeled text as in [7].

In contrast to these methods that exploit tag information to link images and auxiliary text articles, Qi et al. [33] proposed to learn a "translator" which can directly establish the semantic correspondence between text and images even if they are new instances of the image data with unknown correspondence to the text articles. This capability increase the flexibility of the approach and makes it more widely applicable. Specifically, they created a new topic space into which both the text and images are mapped. A translator is then learned to link the instances across heterogeneous text and image spaces. With the resultant translator, the semantic labels can be propagated from any labeled text corpus to any new image by a process of cross-domain label propagation. They showed that the learned translator can effectively convert the semantics from text to images.

6. Summary and Open Issues

In this chapter, we have reviewed the active research on text mining in multimedia community, including surrounding text mining, tag mining, joint text and visual content mining, and cross text and visual content mining. Although research efforts in this filed have made great progress in various aspects, there are still many open research issues that need to be explored. Some examples are listed and discussed as follows.

Joint text and visual content multimedia ranking

Despite the success of visual re-ranking in multimedia retrieval, visual re-ranking only employs the visual content to refine text-based retrieval results; visual content has not been used to assist in learning the ranking model of search engine, and sometimes it is only able to bring in limited performance improvements. In particular, if text-based ranking model is biased or over-fitted, re-ranking step will suffer from the error that is propagated from the initial results, and thus the performance improvement will be negatively impacted. Therefore, it is worthwhile to simultaneously exploit textual metadata and visual content to learn a unified ranking model. A preliminary work has been done in [14], where a content-aware ranking model is developed to incorporate visual content into text-based ranking model learning. It shows that the incorporation of visual content into ranking model learning can result in a more robust and accurate ranking model since noise in textual features can be suppressed by visual information.

Scalable text mining for large-scale multimedia management

Despite of the success of existing text mining in multimedia, most existing techniques suffer from difficulties in handling large-scale multimedia data. Huge amount of training data or high computation powers are usually required by existing methods to achieve acceptable performance. However, it is too difficult, or even impossible, to meet this requirement in real-world applications. Thus there is a compelling need to develop scalable text mining techniques to facilitate large-scale multimedia management.

Multimedia social network mining

In recent years, we have witnessed the emergence of multimedia social network communities like Napster [6], Facebook [7], and Youtube, where millions of users and billions of multimedia entities form a large-scale multimedia social network. Multimedia social networking is becoming an important part of media consumption for Internet users. It brings in new and rich metadata, such as user preferences, interests, behaviors, social relationships, and social network structure etc. These information present new potential for advancing current multimedia analysis

---

[6] http://music.napster.com/
[7] http://www.facebook.com/

techniques and also trigger diverse multimedia applications. Numerous research topics can be explored, including (a) the combination of conventional techniques with information derived from social network communities; (b) fusion analysis of content, text, and social network data; and (c) personalized multimedia analysis in social networking environments.

Acknowledgements

This work was in part supported by A*Star Research Grant R-252-000-437-305 and NRF (National Research Foundation of Singapore) Research Grant R-252-300-001-490 under the NExT Center.

References

[1] Altavista's a/v photo finder. http://www.altavista.com/sites/search/simage.

[2] C. C. Aggarwal, H. Wang. *Text Mining in Social Networks.* Social Network Data Analytics, Springer, 2011.

[3] D. Cai, X. He, Z. Li, W.-Y. Ma, and J.-R. Wen. Hierarchical clustering of www image search results using visual, textual and link information. In *Proceedings of the ACM Conference on Multimedia*, 2004.

[4] S.-F. Chang, W. Hsu, W. Jiang, L. Kennedy, D. Xu, A. Yanagawa, and E. Zavesky. Columbia university trecvid-2006 video search and high-level feature extraction. In *Proceedings of NIST TRECVID workshop*, 2006.

[5] L. Chen and A. Roy. Event detection from Flickr data through wavelet-based spatial analysis. In *Proceedings of the ACM conference on Information and knowledge management*, pages 523–532. ACM, 2009.

[6] L. Chen, D. Xu, I. W. Tsang, and J. Luo. Tag-based web photo retrieval improved by batch mode re-tagging. In *Proceedings of the IEEE International Conference on Computer Vision and Pattern Recognition*, 2010.

[7] W. Dai, Y. Chen, G.-R. Xue, Q. Yang, and Y. Yu. Translated learning: Transfer learning across difference feature spaces. In *NIPS*, pages 353–360, 2008.

[8] J. Fan, Y. Shen, N. Zhou, and Y. Gao. Harvesting large-scale weakly-tagged image databases from the web. In *Proceedings of the IEEE International Conference on Computer Vision and Pattern Recognition*, 2010.

[9] H. Feng, R. Shi, and T.-S. Chua. A bootstrapping framework for annotating and retrieving www images. In *Proceedings of the ACM Conference on Multimedia*, 2004.

[10] S. Feng, C. Lang, and D. Xu. Beyond tag relevance: integrating visual attention model and multi-instance learning for tag saliency ranking. In *Proceedings of International Conference on Image and Video Retrieval*, 2010.

[11] R. Fergus, P. Perona, and A. Zisserman. A visual category filter for google images. In *Proceedings of the European Conference on Computer Vision*, 2004.

[12] C. Frankel, M. J. Swain, and V. Athitsos. Webseer: An image search engine for the world wide web. Technical report, University of Chicago, Computer Science Department, 1996.

[13] B. Gao, T.-Y. Liu, Q. Tao, X. Zheng, Q. Cheng, and W.-Y. Ma. Web image clustering by consistent utilization of visual features and surrounding texts. In *Proceedings of the ACM Conference on Multimedia*, 2005.

[14] B. Geng, L. Yang, C. Xu, and X.-S. Hua. Content-aware ranking for visual search. In *Proceedings of the International Conference on Computer Vision and Pattern Recognition*, 2010.

[15] G. Griffin, A. Holub, and P. Perona. Caltech-256 object category dataset. Technical Report 7694, California Institute of Technology, 2007.

[16] W. Hsu, L. Kennedy, , and S.-F. Chang. Reranking methods for visual search. *IEEE Multimedia*, 14:14–22, 2007.

[17] F. Jing and S. Baluja. Visualrank: Applying pagerank to large-scale image search. *IEEE Transactions on Pattern Analysis and Machine Intelligence*, 30:1877–1890, 2008.

[18] F. Jing, M. Li, H.-J. Zhang, and B. Zhang. A unified framework for image retrieval using keyword and visual features. *IEEE Transactions on Image Processing*, 2005.

[19] F. Jing, C. Wang, Y. Yao, K. Deng, L. Zhang, and W.-Y. Ma. Igroup: Web image search results clustering. In *Proceedings of the ACM Conference on Multimedia*, pages 377–384, 2006.

[20] L. S. Kennedy, S. F. Chang, and I. V. Kozintsev. To search or to label? predicting the performance of search-based automatic image classifiers. In *Proceedings of the ACM International Workshop on Multimedia Information Retrieval*, 2006.

[21] G. Li, M. Wang, Y. T. Zheng, Z.-J. Zha, H. Li, and T.-S. Chua. Shottagger: Tag location for internet videos. In *Proceedings of the ACM International Conference on Multimedia Retrieval*, 2011.

[22] X. Li, C. G. Snoek, and M. Worring. Learning social tag relevance by neighbor voting. *Pattern Recognition Letters*, 11(7), 2009.

[23] X. Li, C. G. Snoek, and M. Worring. Unsupervised multi-feature tag relevance learning for social image retrieval. In *Proceedings of the International Conference on Image and Video Retrieval*, 2010.

[24] D. Liu, X. C. Hua, M. Wang, and H. Zhang. Image retagging. In *Proceedings of the ACM Conference on Multimedia*, 2010.

[25] D. Liu, X.-S. Hua, L. Yang, M. Wang, and H.-J. Zhang. Tag ranking. In *Proceedings of the International Conference on World Wide Web*, 2009.

[26] D. Liu, X.-S. Hua, and H.-J. Zhang. Content-based tag processing for internet social images. *Multimedia Tools and Application*, 51:723–738, 2010.

[27] D. Liu, S. Yan, Y. Rui, and H. J. Zhang. Unified tag analysis with multi-edge graph. In *Proceedings of the ACM Conference on Multimedia*, 2010.

[28] X. Liu, B. Cheng, S. Yan, J. Tang, T. C. Chua, and H. Jin. Label to region by bi-layer sparsify priors. In *Proceedings of the ACM Conference on Multimedia*, 2009.

[29] X. Liu, S. Yan, J. Luo, J. Tang, Z. Huang, and H. Jin. Nonparametric label-to-region by search. In *Proceedings of the IEEE International Conference on Computer Vision and Pattern Recognition*, 2010.

[30] Y. Liu, T. Mei, and X.-S. Hua. Crowdreranking: Exploring multiple search engines for visual search reranking. In *Proceedings of the ACM SIGIR Conference*, 2009.

[31] T. Mei, Z.-J. Zha, Y. Liu, M. Wang, and et al. Msra at trecvid 2008: High-level feature extraction and automatic search. In *Proceedings of NIST TRECVID workshop*, 2008.

[32] S. J. Pan and Q. Yang. A survey on transfer learning. *IEEE Transactions on Knowledge and Data Engineering*, 22(10), 2010.

[33] G.-J. Qi, C. C. Aggarwal, and T. Huang. Towards semantic knowledge propagation from text corpus to web images. In *Proceedings of the International Conference on World Wide Web*, 2011.

[34] M. Rege, M. Dong, and J. Hua. Graph theoretical framework for simultaneously integrating visual and textual features for efficient

web image clustering. In *Proceedings of the International Conference on World Wide Web*, 2008.

[35] F. Schroff, A. Criminisi, and A. Zisserman. Harvesting images databases from the web. In *Proceedings of the International Conference on Computer Vision*, 2007.

[36] D. A. Shamma, R. Shaw, P. L. Shafton, and Y. Liu. Watch what i watch: using community activeity to understand content. In *Proceedings of the ACM Workshop on Multimedia Information Retrieval*, 2007.

[37] X. Shi, Q. Liu, W. Fan, P. S. Yu, and R. Zhu. Transfer learning on heterogenous feature spaces via spectral tranformation. In *Proceedings of the International Conference on Data Mining*, 2010.

[38] B. Sigurbjörnsson and R. V. Zwol. Flickr tag recommendation based on collective knowledge. In *Proceedings of International Conference on World Wide Web*, 2008.

[39] J. Smith and S.-F. Chang. Visually searching the web for content. *IEEE Multimedia*, 4:12–20, 1995.

[40] R. Srihari. Automatic indexing and content-based retrieval of captioned images. *IEEE Computer*, 28:49–56, 1995.

[41] A. Sun and S. S. Bhowmick. Quantifying tag representativeness of visual content of social images. In *Proceedings of the ACM Conference on Multimedia*, 2010.

[42] X. Tian, L. Yang, J. Wang, Y. Yang, X. Wu, and X.-S. Hua. Bayesian video search reranking. In *Proceedings of the ACM Conference on Multimedia*, 2008.

[43] A. Ulges, C. Schulze, D. Keysers, and T. M. Breuel. Identifying relevant frames in weakly labeled videos for training concept detectors. In *Proceedings of the International Conference on Image and Video Retrieval*, 2008.

[44] G. Wang and D. A. Forsyth. Object image retrieval by exploiting online knowledge resources. In *Proceedings of the IEEE Conference on Computer Vision and Pattern Recognition*, 2008.

[45] J. Wang, Y.-G. Jiang, and S.-F. Chang. Label diagnosis through self tuning for web image search. In *Proceedings of the IEEE Conference on Computer Vision and Pattern Recognition*.

[46] M. Wang, X. S. Hua, R. Hong, J. Tang, G. J. Qi, and Y. Song. Unified video annotation via multi-graph learning. *IEEE Transactions on Circuits and Systems for Video Technology*, 19(5), 2009.

[47] M. Wang, X. S. Hua, J. Tang, and R. Hong. Beyond distance measurement: Constructing neighborhood similarity for video annotation. *IEEE Transactions on Multimedia*, 11(3), 2009.

[48] M. Wang, B. Ni, X.-S. Hua, and T.-S. Chua. Assistive multimedia tagging: A survey of multimedia tagging with human-computer joint exploration. *ACM Computing Survey*, 2011.

[49] X.-J. Wang, W.-Y. Ma, G.-R. Xue, and X. Li. Multi-model similarity propagation and its application for web image retrieval. In *Proceedings of the ACM Conference on Multimedia*, pages 944–951, 2004.

[50] X.-J. Wang, W.-Y. Ma, L. Zhang, and X. Li. Iteratively clustering web images based on link and attribute reinforcements. In *Proceedings of the ACM Conference on Multimedia*, 2005.

[51] L. Wu, X.-S. Hua, N. Yu, W.-Y. Ma, and S. Li. Flickr distance. In *Proceedings of the ACM Conference on Multimedia*, 2008.

[52] H. Xu, J. Wang, X.-S. Hua, and S. Li. Tag refinement by regularized LDA. In *Proceedings of the ACM Conference on Multimedia*, 2009.

[53] R. Yan and A. G. Hauptmann. Co-retrieval: A boosted reranking approach for video retrieval. In *Proceedings of the ACM Conference on Image and Video Retrieval*, 2004.

[54] R. Yan, A. G. Hauptmann, and R. Jin. Multimedia search with pseudo-relevance feedback. In *Proceedings of the ACM Conference on Image and Video Retrieval*, 2003.

[55] K. Yang, X.-S. Hua, M. Wang, and H. C. Zhang. Tagging tags. In *Proceedings of the ACM Conference on Multimedia*, 2010.

[56] Q. Yang, Y. Chen, G.-R. Xue, W. Dai, and Y. Yu. Heterogeneous transfer learning from image clustering via the social web. In *Proceedings of the Joint Conference of the Annual Meeting of the ACL*, 2009.

[57] Y.-H. Yang, P. Wu, C. W. Lee, K. H. Lin, W. Hsu, and H. H. Chen. Contextseer: Context search and recommendation at query time for shared consumer photos. In *Proceedings of the ACM Conference on Multimedia*, 2008.

[58] Z.-J. Zha, X.-S. Hua, T. Mei, J. Wang, G.-J. Qi, and Z. Wang. Joint multi-label multi-instance learning for image classification. In *Proceedings of the IEEE Conference on Computer Vision and Pattern Recognition*, 2008.

[59] Z.-J. Zha, T. Mei, J. Wang, X.-S. Hua, and Z. Wang. Graph-based semi-supervised learning with multiple labels. *Journal of Visual Communication and Image Representation*, 2009.

[60] Z.-J. Zha, M. Wang, Y.-T. Zheng, Y. Yang, R. Hong, and T.-S. Chua. Interactive video indexing with statistical active learning. *IEEE Transactions on Multimedia*, 2011.

[61] Z.-J. Zha, L. Yang, T. Mei, M. Wang, and Z. Wang. Viusal query suggestion. In *Proceedings of the ACM Conference on Multimedia*, 2009.

[62] R. Zhang, Z. M. Zhang, M. Li, W.-Y. Ma, and H.-J. Zhang. A probabilistic semantic model for image annotation and multi-modal image retrieval. In *Proceedings of the International Conference on Computer Vision*, pages 846–851, 2005.

[63] R. Zhao and W. I. Grosky. Narrowing the semantic gap - improved text-based web document retireval using visual fetures. *IEEE Transactions on Multimedia*, 4, 2002.

[64] G. Zhu, S. Yan, and Y. Ma. Image tag refinement towards low-rank, content-tag prior and error sparsity. In *Proceedings of the ACM Conference on Multimedia*, 2010.

[65] Y. Zhu, Y. Chen, Z. Lu, S. J. Pan, G.-R. Xue, Y. Yu, and Q. Yang. Heterogeneous transfer learning for image classification. In *Proceedings of the AAAI Conference on Artificial Intelligence*, 2011.

Chapter 12

TEXT ANALYTICS IN SOCIAL MEDIA

Xia Hu
Computer Science and Engineering
Arizona State University
xiahu@asu.edu

Huan Liu
Computer Science and Engineering
Arizona State University
huanliu@asu.edu

Abstract   The rapid growth of online social media in the form of collaboratively-created content presents new opportunities and challenges to both producers and consumers of information. With the large amount of data produced by various social media services, text analytics provides an effective way to meet usres' diverse information needs. In this chapter, we first introduce the background of traditional text analytics and the distinct aspects of textual data in social media. We next discuss the research progress of applying text analytics in social media from different perspectives, and show how to improve existing approaches to text representation in social media, using real-world examples.

Keywords: Text Analytics, Social Media, Text Representation, Time Sensitivity, Short Text, Event Detection, Collaborative Question Answering, Social Tagging, Semantic Knowledge

1. Introduction

Social media such as blogs, microblogs, discussion forums and multimedia sharing sites are increasingly used for users to communicate breaking news, participate in events, and connect to each other anytime, from anywhere. The social media sites play a very important role in current web applications, which accounts for 50% of top 10 sites according to statistics from Alexa[1], as shown in Table 12.1. Besides that, the Twitter messages are even archived in the US Library of Congress[2]. These social media provides rich information of human interaction and collective behavior, thus attracting much attention from disciplines including sociology, business, psychology, politics, computer science, economics, and other cultural aspects of societies.

Table 12.1. Internet Traffic Report by Alexa on March 3rd, 2011

Rank	Website	Rank	Website
1	Google	6	**Blogger**
2	**Facebook**	7	Baidu
3	**Youtube**	8	**Wikipedia**
4	Yahoo!	9	**Twitter**
5	Windows Live	10	QQ.com

We present a definition of Social Media from a social media source, Wikipedia[3], as follows:

> "Social media are media for social interaction, using highly accessible and scalable communication techniques. It is the use of web-based and mobile technologies to turn communication into interactive dialogue."

Moturu [43] defines social media as:

> "Social Media is the use of electronic and Internet tools for the purpose of sharing and discussing information and experiences with other human beings in more efficient ways."

Traditional media such as newspaper, television and radio follow a unidirectional delivery paradigm, from business to consumer. The information is produced from media sources or advertisers and transmitted to media consumers. Different from this traditional way, web 2.0 technologies are more like consumer to consumer services. They allow users to interact and collaborate with each other in a social media dialogue of user-generated content in a virtual community. We categorize the most popular social media web sites into groups, shown in Table 12.2.

---

[1] www.alexa.com
[2] http://blogs.loc.gov/loc/2010/04/how-tweet-it-is-library-acquires-entire-twitter-archive/
[3] http://en.wikipedia.org/wiki/Social_media/

Table 12.2. Types of Social Media

Category	Representative Sites
Wiki	Wikipedia, Scholarpedia
Blogging	Blogger, LiveJournal, WordPress
Social News	Digg, Mixx, Slashdot
Micro Blogging	Twitter, Google Buzz
Opinion & Reviews	ePinions, Yelp
Question Answering	Yahoo! Answers, Baidu Zhidao
Media Sharing	Flickr ,Youtube
Social Bookmarking	Delicious, CiteULike
Social Networking	Facebook, LinkedIn, MySpace

From the table 12.2, social media web sites contain various types of services and thus create different formats of data, including text, image, video etc. For example, the media sharing sites Flickr and Youtube allow to observe what "ordinary" users do when given the ability to more readily incorporate images and video in their everyday activity [55]. We are seeing people engaged in the creation and sharing of their personal photography. As a result, a large amount of image and video data is archived in the sites. Besides, in blogging sites, the users post frequently and create a huge number of textual / text-based data; in social bookmarking sites, users share with each other tags and URLs.

Among the various formats of data exchanged in social media, text plays a important role. The information in most social media sites (the ones with bold font in Table 12.2) are stored in text format. For example, microblogging services allow users to post small amounts of text for communicating breaking news, information sharing, and participating in events. This emerging media has become a powerful communication channel, as evidenced by many recent events like "Egyptian Revolution" and the "Tohoku earthquake and tsunami".

On the other hand, there are also a lot of useful textual data containing in the sites (the ones without bold font in Table 12.2) which are concentrating on other domains. For instance, researchers proposed to utilize tag information in multimedia sharing sites to perform video retrieval [63] and community detection [59]. Under these scenarios, how to mine useful information from textual data presents great opportunities to social media research and applications.

Text Analytics (also as know as Text Mining) refers to the discovery of knowledge that can be found in text archives [49]. This field has received much attention due to its wide application as a multi-purpose tool, borrowing techniques from Natural Language Processing (NLP),

Data Mining (DM), Machine Learning (ML), Information Retrieval (IR) etc.

Text Analytics is defined in Wikipedia as follows:

> "Text Analytics describes a set of linguistic, statistical, and machine learning techniques that model and structure the information content of textual sources for business intelligence, exploratory data analysis, research, or investigation."

Text analytics techniques can help efficiently deal with textual data in social media for research and business purposes. The rest of this chapter is organized as follows: Section 2 introduces specialty for text analytics in social media by analyzing the features of textual data. Section 3 presents proposed approaches for several representative research issues. Section 4 introduces one example to illustrate in detail the process of text analytics methods to solve real world problems. Section 5 concludes the chapter with some possible directions of future work.

2. Distinct Aspects of Text in Social Media

Textual data in social media gives us insights into social networks and groups that were not previously possible in both scale and extent. Unfortunately, textual data in social media presents many new challenges due to its distinct characteristics. In this section, we first review traditional processes of text analytics and then discuss the distinctive features of text in social media, including *Time Sensitivity, Short Length, Unstructured Phrases, Abundant Information*.

2.1 A General Framework for Text Analytics

In this subsection, we briefly introduce the general framework of text analytics to process a text corpus. A traditional text analytics framework consists of three consecutive phases: Text Preprocessing, Text Representation and Knowledge Discovery, shown in Figure 12.1. We use an example to illustrate these methods in each step.

Given a text corpus which contains three microblogging messages, as shown below:

"watching the King's Speech"
"I like the King's Speech"
"they decide to watch a movie"

Text Preprocessing: Text preprocessing aims to make the input documents more consistent to facilitate text representation, which is

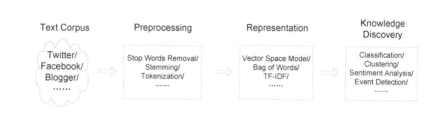

Figure 12.1. A Traditional Framework for Text Analytics

necessary for most text analytics tasks. Traditional text preprocessing methods include *stop word removal* and *stemming*. Stop word removal eliminates words using a stop word list[4], in which the words are considered more general and meaningless; Stemming [46] reduces inflected (or sometimes derived) words to their stem, base or root form. For example, "watch", "watching", "watched" are represented as "watch", so the words with variant forms can be regarded as same feature. The output of text preprocessing for the three microblogging messages are:

"watch King' Speech"

"King' Speech"

"decid watch movi"

Preprocessing methods depend on specific application. In many applications, such as Opinion Mining or NLP, they need to analyze the message from a syntactical point of view, which requires that the method retains the original sentence structure. Without this information, it is difficult to distinguish "Which university did the president graduate from?" and "Which president is a graduate of Harvard University?", which have overlapping vocabularies. In this case, we need to avoid removing the syntax-containing words.

Text Representation: The most common way to model documents is to transform them into sparse numeric vectors and then deal with them with linear algebraic operations. This representation is called "Bag Of Words" (BOW) or "Vector Space Model" (VSM). In these basic text representation models, the linguistic structure within the text is ignored and thus leads to "structural curse".

In BOW model, a word is represented as a separate variable having numeric weight of varying importance. The most popular weighting

---

[4]http://www.lextek.com/manuals/onix/

schema is Term Frequency / Inverse Document Frequency (TF-IDF):

$$tfidf(w) = tf * log\frac{N}{df(w)}, \quad (12.1)$$

where:
- tf(w) is term frequency (the number of word occurrences in a document)
- df(w) is document frequency (the number of documents containing the word)
- N is number of documents in the corpus
- tfidf(w) is the relative weight of the feature in the vector

Using BOW to model the three messages with a TF-IDF weight, the corpus can be represented as a words * documents matrix. Each row represents a word (5 distinct words in total) and each column represents a message, as shown below:

$$\begin{bmatrix} watch \\ King' \\ Speech \\ decid \\ movi \end{bmatrix} = \begin{bmatrix} 0.4055 & 0 & 0.4055 \\ 0.4055 & 0.4055 & 0 \\ 0.4055 & 0.4055 & 0 \\ 0 & 0 & 1.0986 \\ 0 & 0 & 1.0986 \end{bmatrix} \quad (12.2)$$

Knowledge Discovery: When we successfully transform the text corpus into numeric vectors, we can apply the existing machine learning or data mining methods like classification or clustering. For example, in machine learning, similarity is an important measure for many tasks. A widely used similarity measure between two messages V_1 and V_2 is cosine similarity, which can be computed as:

$$similarity(V_1, V_2) = cos(\theta) = \frac{V_1 * V_2}{||V_1|| ||V2||}, \quad (12.3)$$

By conducting text preprocessing, text representation and knowledge discovery methods, we can mine latent, useful information from the input text corpus, like similarity between two messages in our example. However, this presents challenges for traditional text analytics methods when applied directly to textual data in social media due to its distinct features. Now we analyze the new features of textual data in social media from four different perspectives: Time Sensitivity, Short Length, Unstructured Phrases, and Abundant Information.

2.2 Time Sensitivity

An important and common feature of many social media services is their real-time nature. Particularly, bloggers typically update their blogs every several days, while microblogging and social networking users may post news and information several times daily. Users may want to communicate instantly with friends about "What are you doing?" (Twitter) or "What is on your mind" (Facebook). When submitting a query to Twitter, the returned results are only several minutes old.

Besides communicating and sharing minds with each other, users post comments on recent events, such as new products, movies, sports, games, political campaigns, etc. The large number of real-time updates contain abundant information, which provides a lot of opportunities for detection and monitoring of an event. With these data, we are able to infer a user's interest in an event [37], and track information provenance from the user's communications [9]. For example, Sakaki et al. [47] investigate the real-time interaction of events such as earthquakes, and they propose an algorithm to monitor tweets and to detect a target event.

With the rapid evolution of content and communication styles in social media, text is changing too. Different from traditional textual data, the text in social media is not independent and identically distributed (i.i.d.) data anymore. A comment or post may reflect the user's interest, and a user is connected and influenced by his friends. People will not be interested in a movie after several months, while they may be interested in another movie released several years ago because of the recommendation from his friends; reviews of a product may change significantly after some issues, like the comments on Toyota vehicles after the break problem. All these problems originate from the time sensitivity of textual data in social media.

2.3 Short Length

Certain social media web sites restrict the length of user-created content such as microblogging messages, product reviews, QA passages and image captions, etc. Twitter allows users to post news quickly and the length of each tweet is limited to 140 characters. Similarly, Picasa comments are limited to 512 characters, and personal status messages on Windows Live Messenger are restricted to 128 characters. As we can see, data with a short length is ubiquitous on the web at present. As a result, these short messages have played increasing important roles in applications of social media. Successful processing short texts is essential to text analytics methods.

Short messages, as the most important data format, make people more efficient with their participate in social media applications. However, this brings new challenges to traditional fundamental research topics in text analytics, such as text clustering, text classification, information extraction and sentiment analysis. Unlike standard text with lots of words and their resulting statistics, short messages consist of few phrases or sentences. They cannot provide sufficient context information for effective similarity measure [45], the basis of many text processing methods [27].

To tackle the data sparseness problem, several traditional text analytics methods have been proposed, which can be generally categorized into two groups. The first is the basic representation of texts called surface representation [32, 36], which exploits phrases in the original text from different aspects to preserve the contextual information. However, NLP techniques such as parsing are not employed, as it is time consuming to apply such techniques to analyze the structure of standard text in detail. As a result, the methods fail to perform a deep understanding of the original text. Another limitation of such methods is that they did not use external knowledge, which has been found to be useful in dealing with the semantic gap in text representation [18]. For example, tag "Japan Earthquake" does not contain any words or phrases related to "Nuclear Crisis" while we learn that these two events are related from recent news. Because they have no common words or phrases, it is very difficult for BOW-based models and methods to build semantic connections between each other. One intuitive approach is to enrich the contexts of basic text segments by exploiting external resources, and such methods have been found to be effective in narrowing the *semantic gap* in different tasks [20, 54].

2.4 Unstructured Phrases

An important difference between the text in social media and traditional media is the variance in the quality of the content. First, the variance of quality originates from people's attitudes when posting a microblogging message or answering a question in a forum. Some users are experts for the topic and post information very carefully, while others do not post as high of quality. The main challenge posed by content in social media sites is the fact that the distribution of quality has high variance: from very high-quality items to low-quality, sometimes abusive content. This makes the tasks of filtering and ranking in such systems more complex than in other domains [5].

Second, when composing a message, users may use or coin new abbreviations or acronyms that seldom appear in conventional text documents. For example, messages like "How r u?", "Good 9t" are not really words, but they are intuitive and popular in social media. They provide users convenience in communicating with each other, however it is very difficult to accurately identify the semantic meaning of these messages. Besides the unstructured expressions, the text is sometimes "noisy" for a specific topic. For instance, one QA passage in Yahoo! Answers "I like sony" should be noisy data to a post that is talking about iPad 2 release. It is difficult to classify the passage into corresponding classes without considering its context information.

2.5 Abundant Information

Social media in general exhibit a rich variety of information sources. In addition to the content itself, there is a wide array of non-content information available. For example, Twitter allows users to utilize the "#" symbol, called hashtag, to mark keywords or topics in a Tweet (tag information); an image is usually associated with multiple labels which are characterized by different regions in the image [66]; users are able to build connection with others (link information) in Facebook and other social network sites; Wikipedia provides an efficient way for users to redirect to the ambiguity concept page or higher level concept page (semantic hierarchy information).

All these external information presents opportunities for traditional tasks. Previous text analytics sources always appear as <user, content> structure, while the text analytics in social media is able to derive data from various aspects, which include user, content, link, tag, time stamp etc. Recently, many research work utilizes link information in microblogging services to detect the popular event [37], distinguish the microblogging message is credible news or just rumor [42]. Also, with the user metadata (e.g. tags) mined from blogosphere and bookmarking sites, Wang et al. [59] take advantage of networking information between users and tags to discover overlapping communities. These successful applications motivated us to exploit more opportunities behind such abundant additional information available in social media.

3. Applying Text Analytics to Social Media

It presents great challenges to apply traditional methods to process textual data in social media. Recently, a number of methods have been proposed to handle the textual data with new features. In this section, we introduce a variety of applying text analytics to social media.

3.1 Event Detection

Event Detection aims to monitor a data source and detect the occurrence of an event that is captured within that source [40]. These data sources include images, video, audio, spatio-temporal data, text documents and relational data. Among them, event detection and evolution tracking of news articles [60], digital books [22] receives much attention. The volume of textual data in social media is increasing exponentially, thus providing us many opportunities for event detection and tracking.

In some sense, social text streams are sensors of the real world [67]. As the real-time nature of textual data in social media, a lot of work has been done to extract real world events from social text streams. One interesting application is to monitor real-time events. For example, when an earthquake or tsunami occurs, one convenient way to communicate updated news with others is to post messages related to the event via microblogging. Therefore, it provides possibility for us to promptly detect the occurrence of earthquake or tsunami, simply by mining the corresponding microblogging messages. Based on the above observation, Sakaki et al. [47] investigate the real-time interaction of events on Twitter. They consider each user as a sensor to monitor tweets posted recently and to detect earthquake or rainbow. To detect a target event, the work flow is as follows. First, a classifier is trained by using keywords, message length, and corresponding context as features to classify tweets into positive or negative cases. Second, they build a probabilistic spatio-temporal model for the target event to identify location of the event. As an application, the authors constructed an earthquake-reporting system in Japan, where has numerous earthquakes every year as well as a large number of active microblogging users.

One important direction of event detection in social media is to improve traditional news detection. A large number of news stories are generated from various news channels day after day. Among them, only a relatively few receive attention from users, which are recognized as "breaking news". Traditionally, editors of newspapers and websites decide which stories can be ranked higher and assigned in an important place, like the front page. In a similar way, web-based news aggregated services, such as Google News[5], give users access to broad perspectives on the important news stories being reported by grouping articles into related news events. Deciding automatically on which top stories to show is a challenging problem [39]. A poll conducted by Technorati found that 30% of bloggers consider themselves to be blogging about news-related topics [41].

Motivated by this observation, researchers proposed to utilize blogosphere to facilitate news detection and evaluation. Lee et al. present novel approaches to identify important news story headlines from the

---

[5]http://news.google.com/

blogosphere for a given day [34]. The proposed system consists of two components based on the language model framework, the query likelihood and the news headline prior. For the query likelihood, the authors propose several approaches to estimate the query language model and the news headline language model. They also suggest several criteria to evaluate the importance or newsworthiness of the news headline for a given day.

Tracking the diffusion and evolution of a popular event in social media is another interesting direction in this field. Different from i.i.d. textual data in traditional media, user generated content in social media is a mixture of a text stream and a network structure. Lin et al. take into account the burstiness of user interest, information diffusion in the network structure and the evolution of textual topics to model the popularity of events over time [37]. They tackle the problem of popular event tracking in online communities by studying the interplay between textual content and social networks.

Besides detecting events from pure textual data, some methods have been proposed to mine text information in social media to facilitate event detection. Chen et al. is to detect events from photos on Flickr by analyzing the tag of the photos [13]. In the proposed framework, the authors first analyze temporal and locational distributions of tag usage. Second, they identify tags related with events, and further distinguish if the tags are relevant to aperiodic events or periodic events. Afterwards, tags are clustered into their corresponding clusters. Each cluster represents an event, and consists of tags with similar temporal and locational distribution patterns as well as with similar associated photos. Finally, for each tag cluster, photos corresponding to the represented event are extracted.

3.2 Collaborative Question Answering

Collaborative question answering services begin to emerge with the blooming of social media. They bring together a network of self-declared "experts" to answer questions posted by other people. A large volume of questions are asked and answered every day on social Question and Answering (QA) web sites such as Yahoo! Answers. Collaborative question answering portals are a popular destination for users looking for advice with a particular situation, for gathering opinions, for sharing technical knowledge, for entertainment, for community interaction, and for satisfying one's curiosity about a countless number of things.

Over time, a tremendous amount of historical QA pairs have built up their databases, and this transformation gives users an alternative place to look for information, as opposed to a web search. Instead of looking through a list of potentially relevant documents from the Web or posting a new question in a forum, users may directly search for relevant historical questions or answers from QA archives. As a result, the corresponding best solutions could be explicitly extracted and returned.

This problem could be considered from two sides. On one hand, the most relevant questions semantically related to the query are returned, so that users can find similar questions and their corresponding answers. Wang et al. [57] propose a graph based approach to perform question retrieval by segmenting multi-sentence questions. The authors first attempt to detect question sentences using a classifier built from both lexical and syntactic features, and use similarity and co-reference chain based methods to measure the closeness score between the question and context sentences. On the other hand, systems provide corresponding quality QA pairs from answer's point of view. Adamic et al. [1] evaluate the quality of answers for specific question by analyzing Yahoo! Answer's knowledge sharing activity. First, forum categories are clustered according to the content characteristics and patterns of interaction among users. The interactions in different categories reveal different characteristics. Some categories are more like expertise sharing forums, while others incorporate discussion, everyday advice, and support. Similarly, some users focus narrowly on specific topics, while others participate across categories. Second, the authors utilize this feature to map related categories and characterize the entropy of the users' interests. Both user attributes and answer characteristics are combined to predict, within a given category, whether a particular answer will be chosen as the best answer by the asker.

In order to improve QA archives management, there are a number of works done by evaluating the quality of QA pairs. Harper et al. [25] tried to determine which questions and answers have archival value by analyzing the differences between conversational questions and informational questions. Informational questions refer to the questions with the intent of obtaining information the asker could learn from. An example is "Is drinking Coke good for health?". Conversational questions refer to the questions with the intent of stimulating discussion. In these questions, the users may aim at getting opinions or self-expression. An example is "Do you like drinking Coke?". The authors present evidence that conversational questions typically have much lower potential archival value than informational questions. Further, they used machine learning techniques to automatically classify questions as conversational or informational from perspectives of the process about categorical, linguistic, and social differences between different question types. Agichtein et al. [5]

introduced a general classification framework for combining the evidence from different sources, that can be tuned automatically for quality prediction of QA pairs. In particular, they exploit features of QA pairs that are intuitively correlated with quality, including intrinsic content quality, interactions between content creators and users, as well as the content usage statistics. Then a classifier is trained to appropriately select and weight the features for each specific type of item, task, and quality definition.

3.3 Social Tagging

Social tagging is a method for Internet users to organize, store, manage and search for tags / bookmarks (also as known as social bookmarking) of resources online. Unlike file sharing, the resources themselves aren't shared, merely the tags that describes them or bookmarks that reference them[6]. The rise of social tagging services presents a potential great deal of data for mining useful information on the web. The users of tagging services have created a large volume of tagging data which has attracted recent attention from the research community. From oceans of tags, it is difficult for a user to quickly locate the relevant resources he wants via browsing the tags. Typically, the tagging services provide keyword-based search which returns resources annotated by the given tags. However, the results returned by the search module are inadequate for users to discover interesting resources due to the short and unstructured nature of tags. First, it is very difficult to design an effective tag ranking algorithm due to the short length and sparseness of tags. Second, current systems are designed for keywords based search, which failed to capture the semantic relationship between two semantically related tags. For example, when a user searches for a recent event, such as "Egyptian Revolution", the systems will return results that are tagged as "Egyptian" or "Revolution". Among them, resources tagged with "Mubarak's resignation" and "Protest" which are highly related to "Egyptian Revolution" will be ignored. This "semantic gap" results in many valuable and interesting results overlooked and buried in disorganized resources.

Research work in social tagging services can be typically divided into two categories: one aims to improve the quality of tag recommendation and the other studies how to utilize social tagging resources to facilitate other applications. First, Sigurbjornsson and Van [48] investigate how to assist users during the tagging phase in multimedia sharing sites (Flickr).

---
[6]http://en.wikipedia.org/wiki/Social_bookmarking/

They present and evaluate tag recommendation strategies to support the user in the photo annotation task by recommending a set of tags that can be added to the photo. Yin et al. [61] address the problem of tag prediction by proposing a probabilistic model for personalized tag prediction. On the other hand, social tagging resources are exploited to improve other web applications, including web object classification [62], document recommendation [23], web search quality [26] etc.

3.4 Bridging the Semantic Gap

As we discussed in Section 2, the textual data in social media is short and unstructured. When processing this kind of data, traditional bag of words (BOW) approach is inherently limited, as it can only use pieces of information that are explicitly mentioned in the documents [18]. Consider one famous movie "The Dark Knight". By mining the original posts related to this movie, it is inadequate to build the semantic relationship with other relevant concepts due to the *semantic gap*. For example, "The Dark Knight" and "Batman" are different names of one movie, but they cannot be linked as the same concept without additional information from external knowledge. Specifically, this approach has no access to the wealth of world knowledge possessed by humans, and is easily puzzled by facts and terms not mentioned in the data set. Recently, researchers have proposed semantic knowledge bases to bridge the widely extant semantic gap in short text representation.

The aggregation of information in groups is often better than what could have been made by any single member of the group [52]. Wikipedia is a free, web-based, collaborative, multilingual encyclopedia project. Its 18 million articles have been written collaboratively by volunteers around the world, and almost all of its articles can be edited by anyone with access to the site[7]. Unlike other standard ontologies, such as WordNet or Mesh, Wikipedia is not a structured thesaurus edited by experts, but it was contributed collaboratively by users on the web. It is comprehensive, up to date and well-formed [29]. In Wikipedia, each article concentrates on one specific topic. The title of each article is a succinct phrase that resembles an ontology term. Equivalent concepts are grouped together by redirected links. Meanwhile, Wikipedia contains a hierarchical categorization system, in which each article belongs to at least one category. All these features are making Wikipedia a potential ontology for enhancing text representation.

---

[7]http://en.wikipedia.org/wiki/Wikipedia/

Some methods were proposed to tackle the problems of data sparseness and the semantic gap in short texts clustering and classification by exploiting semantic knowledge. Somnath et al. [8] proposed a method to enrich short text representation with additional features from Wikipedia. The method used titles of Wikipedia articles as additional external features, and it showed improvement in the accuracy of short texts clustering. Phan et al. [45] presented a framework for building classifiers that deal with short texts from the Web and achieved qualitative enhancement. The underlying idea of the framework is to collect large-scale data and then build a classifier on both labeled data and external semantics for each classification task. In addition, researchers [56, 18, 58] analyzed the documents and found related ontological concepts within WordNet and Wikipedia, in turn producing a set of features that augment standard BOW. Towards improving the management of Google snippets, existing methods focus either on classifying the web texts into smaller categories [28] or assigning labels for each category [10] with the help of Wikipedia.

3.5 Exploiting the Power of Abundant Information

Abundant information associated with textual information is ubiquitous in social media. On Twitter, for example, two microblogging messages can be linked together via their authors' follower, followee, retweet or reply relationship; two microblogging messages can be classified into the same or similar category when they share the same hashtag or contain same hyperlink; semantic similarity between two microblogging messages can be measured based on their posting time (time stamp), posting place (geotag), author's personal information (profile), etc. Similar phenomena can be observed in Facebook, LinkedIn, Wikipedia and other social media sites. Different from i.i.d. documents in traditional media, if one can utilize these abundant information available in social media, performance of many text analytics methods may be significantly improved.

To utilize the abundant information appearing along with text content in social media, recent methods have been proposed to integrate this into text analytics tasks, including classification, clustering etc.. Among these methods, a combination of link and text content for mining purposes is becoming popular. A major difference between these two kinds of methods is that traditional methods measure the similarity between documents purely based on attribute similarity (e.g. cosine similarity between two attribute vectors); while the methods for text in social media measures document similarity based on connectivity (e.g. the number of possible paths between authors of the documents) and structural similarity (e.g. the number of shared neighbors) [68], besides the attribute similarity of text content. Links clearly contain high-quality semantical clues that are lost in purely text-based methods, but exploiting link information is not easy. The major difficulty is the similarity measurement between each pair of objects, due to the characteristics of differing social networks:

- *Multi-dimensional social networks.* The connections between users in social media are often multi-dimensional [53]. Users can connect to each other for different reasons, e.g., alumni, colleagues, living in the same community, sharing similar interests, etc. Different types of links have different semantic meanings associated with their respective latent social dimensions.

- *Network representation.* Traditional text analytics methods utilize local features or attributes to represent documents. However, there is no natural feature representation for all types of network data [31]. When we use an adjacency matrix to represent a network, the matrix will be very sparse, highly dimensional and its equal weights cannot reflect tie strength well. Moreover, obtaining labels of objects in social network, which is very important for supervised learning methods, appears to be very expensive.

- *Dynamic networks.* Different from constant news collections or a documents corpus, social media is evolving continuously, with new users joining the network, extant users connecting with each other or becoming dormant. It is imperative to update the acquired community structure. As a result, how to efficiently integrate the updated network information is very important for many applications.

Many methods have been proposed to tackle the challenges and make use of link information sources. To our knowledge, the first topic classification system that simultaneously utilizes textual and link features was discussed in [11]. The authors aim to propose a statistical model and a relaxation labeling technique to build a classifier by exploiting link information form neighbors of the documents. Similarly Furnkranz [17] found that it is possible to classify documents more reliably with information originating from pages that point to the document than with features that are derived from the document itself. Later, Chakrabarti et al. [6] proposed a graph-based text classification method by learning from their neighbors. The difference between these two kinds of techniques is that the latter one considered more factors in social networks, including the network evolution (dynamic network), pruning of edges from the neighborhood graph, and weighing the influence of edges and edges themselves by content similarity measures. Recently, Aggarwal and Li [3] presented an efficient and scalable method to tackle the problem of node classification in dynamic information networks with both text content and links. To facilitate an effective classification process, different from previous models, they use a random walk approach in conjunction with the content of the network. This design makes the model more robust to variations in content and linkage structure. Aside from classification, link information has been also successfully integrated into the applications of clustering [68] and topic modeling [50]. It shows that the use of both link and text information achieved more effective results than a method based purely on either of the two [4].

In addition to integrating network information into text analytics tasks, researchers further exploit abundant information. In [51], a heterogeneous information network is defined as an information network composed of multiple types of object. The authors explored clustering of multi-typed heterogeneous networks with a star network schema, although clustering on homogeneous networks has been well studied over decades. Links across multi-typed objects are utilized to generate high-quality net-clusters. The general idea of the proposed framework is to avoid measuring the pairwise similarity between objects, which is hard in heterogeneous networks.Instead, it maps each target object into a low dimensional space defined by current clustering results. Then every target object in these clusters will be readjusted based on the new measure. The clustering results will be improved in each iteration until convergence.

3.6 Related Efforts

Aside from the topics discussed in the previous sections, even more attempts have been explored in mining social media resources. In Social Network Analysis, researchers utilize various information such as the posts, links, tags, etc., to identify influential users in the blogsphere [2] and microblogsphere [7], to understand user behavior in microblogsphere by analyzing the user intentions associated at a community level [30, 33]. In Sentiment Analysis, Conner et al. investigate several surveys on consumer confidence and political opinion, connect measures of public opinion measured from polls with sentiment measured from text [14]. Gerani et al. use a general opinion lexicon and propose using proximity information in order to capture opinion term relatedness to improve opinion retrieval in the blogsphere [21]. In Knowledge Management, Lerman and Hogg [35] use a model of social dynamics to predict the popularity of news. Incorporating aspects of web site design, the model improves on predictions based on simply extrapolating from early votes. Lu et al. exploit contextual information about the authors' identities and social networks for improving review quality prediction [38]. This model improves previous work, which addressed the problem by treating a review as a stand-alone document, extracting features from the review text, and learning a function based on these features for predicting the review quality.

4. An Illustrative Example

In this section, we present one real world application to further illustrate how to utilize text analytics to solve problems in social media applications. We now introduce an effective way to improve the short text representation quality by integrating semantic knowledge resources.

As we discussed in Section 2, textual data in social media has the problems of data sparseness and semantic gap. One effective way to solve these problems is to integrate semantic knowledge, which has been found to be useful in dealing with the semantic gap [18]. For example, the first search result returned by Google using "Friday" as the query does not contain any words or phrases related to "Rebecca Black", while we learn that the singer creates overnight sensations by sharing the song via YouTube. Because they have no words or phrases overlapping, this result can not be successfully build connection with Rebecca related content. Thus, one intuitive idea is to enrich the contexts of basic text representation by exploiting semantic resources.

Now, we follow the basic idea proposed in [28] to illustrate three steps of feature generation in detail: Seed Phrase Extraction from the original text corpus, Semantic Features Generation based on *seed phrases* and Feature Space Construction.

4.1 Seed Phrase Extraction

Given a text corpus, features can be derived by employing different techniques in NLP. The only requirement is that the extracted features could be informative to cover the key subtopics described in the short texts. Here we use shallow parsing [24] to divide sentences into a series of words that together compose a grammatical unit. To ensure the extracted features are able to cover main topics, we use these phrases generated by shallow parser, with the combination of sentences in the original text, to extract the *seed phrases*. However, there are redundancies between these two kinds of features. If we employ all these features as *seed phrases*, they would produce some duplicate information between each other. Therefore, to make the tradeoff between informativeness and effectiveness, we propose to measure the semantic similarity between sentence level features and phrase level features to eliminate information redundancy.

Several methods have been proposed to calculate the semantic similarity between associations [12] using web search. However, along with the increasing scale of the web, the page counts provided by some commercial search engines are not so reliable [15]. Thus instead of simply using the search engine page counts, we propose a phrase-phrase semantic similarity measuring algorithm using a co-occurrence double check in Wikipedia to reduce the semantic duplicates. For Wikipedia, we download the XML corpus [16], remove xml tags and create a Solr [8] index of all XML articles.

Let T denote a *sentence level* feature, $T = \{t_1, t_2, \ldots, t_n\}$, where t_i denotes the *phrase level* feature contained in T. The *sentence level* feature is too sparse to calculate its frequency directly. Therefore, we calculate the semantic similarity between t_i and $\{t_1, t_2, \ldots, t_n\}$ as $InfoScore(t_i)$ instead. We select the *phrase level* feature which has the largest similarity with other features in T and remove it as the redundant feature.

Given two phrases t_i and t_j, we use t_i and t_j separately as a query to retrieve top C Wikipedia pages from the built index. The total occurrences of t_i in the top C Wikipedia pages retrieved by query t_j is denoted as $f(t_i|t_j)$; and we define $f(t_j|t_i)$ in a similar manner. The total occurrences of t_i in the top C Wikipedia pages retrieved by query t_i is denoted as $f(t_i)$, and similarly for $f(t_j)$. The variants of three popular co-occurrence measures [15] are defined as below:

$$WikiDice(t_i, t_j) = \begin{cases} 0 & \text{if } f(t_i \mid t_j) = 0 \text{ or } f(t_j \mid t_i) = 0 \\ \frac{f(t_i|t_j)+f(t_j|t_i)}{f(t_i)+f(t_j)} & \text{otherwise} \end{cases}, \qquad (12.4)$$

where WikiDice is a variant of the Dice coefficient.

$$WikiJaccard(t_i, t_j) = \frac{min(f(t_i \mid t_j), f(t_j \mid t_i))}{f(t_i) + f(t_j) - max(f(t_i \mid t_j), f(t_j \mid t_i))}, \qquad (12.5)$$

where WikiJaccard is a variant of the Jaccard coefficient.

---

[8] http://lucene.apache.org/solr/

$$WikiOverlap(t_i, t_j) = \frac{min(f(t_i \mid t_j), f(t_j \mid t_i))}{min(f(t_i), f(t_j))}, \quad (12.6)$$

where WikiOverlap is a variant of the Overlap(Simpson) coefficient.

For ease of comparison, all the $\frac{n^2}{2}$ WikiDice similarity scores are normalized into values in $[0, 1]$ range using the linear normalization formula defined below:

$$WD_{ij} = \frac{WikiDice_{ij} - min(WikiDice_k)}{max(WikiDice_k) - min(WikiDice_k)}, \quad (12.7)$$

where k is from 1 to $\frac{n^2}{2}$. We again define WJ_{ij} and WO_{ij} in a similar manner. A linear combination is then used to incorporate the three similarity measures into an overall semantic similarity between two phrases t_i and t_j, as follows:

$$WikiSem(t_i, t_j) = (1 - \alpha - \beta)WD_{ij} + \alpha WJ_{ij} + \beta WO_{ij}, \quad (12.8)$$

where α and β weight the importance of the three similarity measures. Text clustering is an unsupervised method where we do not have any labeled data to tune the parameters. We thus empirically set α and β to equal weight.

For each *sentence level* feature, we rank the information score defined in Equation 12.9 for its child node features at *phrase level*.

$$InfoScore(t_i) = \sum_{j=1, j \neq i}^{n} WikiSem(t_i, t_j). \quad (12.9)$$

Finally, we remove the *phrase level* feature t^*, which delegates the most information duplicates to the *sentence level* feature T, defined as:

$$t^* = \arg \max_{t_i \in \{t_1, t_2, \ldots, t_n\}} InfoScore(t_i). \quad (12.10)$$

4.2 Semantic Feature Generation

After extracting the *seed phrases* from the first step, we obtain an informative and effective basic representation of the input text corpus. In this step, we discuss an algorithm to generate semantic features based on the *seed phrases* using Wikipedia as background knowledge.

4.2.1 Background Knowledge Base. Wikipedia, as background knowledge, has a wider knowledge coverage than other semantic knowledge bases and is regularly updated to reflect recent events. Under this scenario, we take Wikipedia as the semantic knowledge source to generate semantic concepts.

Gabrilovich and Markovitch [19], as well as Hu et al. [27] preprocessed the Wikipedia corpus to collect semantic concepts. Preprocessing

---
Algorithm 1: GenerateFeatures(S)
---
input : a set S of *seed phrases*
output: *semantic features* SF

$SF \leftarrow null$
for *seed phrase* $s \in S$ **do**
　if $s \in$ Sentence level **then**
　　s.Query $\leftarrow$ SolrSyntax(s, OR)
　else
　　s.Query $\leftarrow$ SolrSyntax(s, AND)
　WikiPages $\leftarrow$ Retrive(s.Query)
　$SF \leftarrow SF +$ Analyze(WikiPages)
return SF

Figure 12.2. Semantic feature generation scheme

Wikipedia is one way to build the concepts space. However, it ignores the valuable contextual information of Wikipedia plain texts and always encounters problems when mapping the original text to appropriate concepts. Therefore, in this study we introduce another way to process the Wikipedia corpus, it is to preserve the original pages of Wikipedia with the built-in Solr index.

4.2.2 Feature Generator. The *semantic feature* generation algorithm is illustrated in Figure 12.2. Given a *seed phrase*, we retrieve corresponding Wikipedia pages with the help of the Solr search engine. Then we extract semantic concepts from the retrieved Wikipedia pages.

In order to retrieve the appropriate pages from the large Wikipedia corpus, we derive queries based on *seed phrase* arising from *sentence level* or *phrase level* separately. As the key information of *seed phrases* from *phrase level* is more focused, we build the "AND" query which requires the retrieved pages to contain every term in the phrase. On the other hand, the *seed phrases* from *sentence level* are informative but sparse, we thus build "OR" query[9] which means there is no guarantee that the retrieved Wikipedia pages will contain every term in the phrase. We use these two kinds of queries to retrieve the top ω articles from the Wikipedia corpora. Similar to [8], we extract titles and bold terms (links) from the retrieved Wikipedia pages to serve as part

---
[9]For more details about "AND" and "OR" query syntax, please refer to http://wiki.apache.org/solr /SolrQuerySyntax/

of the *semantic features*. To discover the intrinsic concepts hidden in the plain texts, we adopt an effective key phrase extraction algorithm — Lingo [44], which uses algebraic transformations of the term-document matrix and implements frequent phrase extraction using suffix arrays. The key phrases extracted from the Wikipedia pages are added to the *semantic feature* space. By utilizing this method, we may obtain extrinsic concepts "Friday" for the phrase "Rebecca Black" and the intrinsic concepts like "Song", "Singer" and "Youtube" by mining the related pages. Therefore, we can build semantic relationships between the concepts of "Friday" and "Rebecca Black".

4.3 Feature Space Construction

As the construction of Wikipedia follows the non-binding guidelines and the data quality is only under social control by the community [65], it often introduces noise to the corpus. Meanwhile, a single text may generate a huge number of features. These overzealous *semantic features* bring adverse impact on the effectiveness and dilute the influence of valuable original information. Therefore, we conduct feature filtering to refine the unstructured or meaningless features and apply feature selection to avoid aggravating the "curse of dimensionality".

Feature Filtering: We formulate empirical rules to refine the unstructured features obtained from Wikipedia pages, some typical rules are as follows:

- Remove features generated from too general *seed phrase* that returns a large number (more than 10,000) of articles from the index corpus.

- Transform features used for Wikipedia management or administration, e.g. "List of hotels"→"hotels", "List of twins"→"twins".

- Apply phrase sense stemming using Porter stemmer [46], e.g. "fictional books"→"fiction book".

- Remove features related to chronology, e.g. "year", "decade" and "centuries".

Feature Selection: We need to select *semantic features* to construct feature space for various tasks. The number of *semantic features* we need to collect is determined by the specific task. In this chapter, we utilize a simple way to select the most frequent features.

First, the *tf-idf* weights of all generated features are calculated. One *seed phrase* $s_i (0 < i \leq m)$ may generate k *semantic features*, denoted by $\{f_{i1}, f_{i2}, \ldots, f_{ik}\}$. In order to explore the diversity of the *semantic features*, we select one feature for each *seed phrase*. Thus m features are collected as follows:

$$f_i^* = \arg \max_{f_{ij} \in \{f_{i1}, f_{i2}, \ldots, f_{ik}\}} tf\_idf(f_{ij}). \tag{12.11}$$

Second, the top n features are extracted from the remaining *semantic features* based on their frequency. These frequently appearing features, together with the features from the first step, are used to construct the $m + n$ *semantic features*.

Now we prepare the feature space for clustering, classification or other text analytics methods. From the discussion above, key idea of the framework is to introduce semantic knowledge base (Wikipedia) to build semantic connection between two short documents. This section provides a clear mind about how to apply text analytics methods in social media resources.

5. Conclusion and Future Work

Textual data in social media carries abundant information. User-generated content provides diverse and unique information in forms of comments, posts and tags. The useful information hidden in the text resources of social media provides opportunities for researchers of different disciplines to mine patterns and information of interest that might not be obvious. In this chapter, we discuss about the distinct aspects of textual data in social media and their challenges, and elaborate current work of utilizing text analytics methods to solve problems in social media.

This chapter has only discussed some essential issues. There are a number of interesting directions for further exploration.

- How to better make use of the real-time nature in social media? A real-time search system which can find, summarize and track updated breaking news or events in social communities will be very challenging but useful.

- How to handle textual data with short length in social media? As we discussed, short text plays a very important role in social media. On one hand, these textual data contains less information than standard documents; on the other hand, it provides possibility for us to use traditional syntax-based NLP models to perform fine-level textual analysis, which were very time consuming for standard text.

- How to exploit cross media data to facilitate social behavior analysis? Cross media data here refers data of different formats or data from different social media resources [64]. The variance types of data in social media, including text, image, link or even multilingual data, have latent relationships and interactions between each other. Also, an efficient and effective way to integrate these kinds of data will be very useful to address the data sparseness problem.

- How to process web scale data available in social media? The large volume and the compact but noisy presentation of textual data in social media hinders the accessibility of information for users to conveniently search, navigate and locate the specific messages one might be interested in. Finding an efficient way to handle these large scale data types is very challenging.

Acknowledgments

This work is, in part, supported by the grants NSF (#0812551), ONR (N000141010091) and AFOSR (FA95500810132). The authors would like to acknowledge all of the researchers in Arizona State University's Data Mining and Machine Learning Laboratory. The views expressed in this chapter are solely attributed to the authors and do not represent the opinions or policies of any of the funding agencies.

References

[1] L. Adamic, J. Zhang, E. Bakshy, and M. Ackerman. Knowledge sharing and yahoo answers: everyone knows something. In *Proceeding of the 17th international conference on World Wide Web*, pages 665–674. ACM, 2008.

[2] N. Agarwal, H. Liu, L. Tang, and P. S. Yu. Identifying the influential bloggers in a community. In *Proceedings of the international conference on Web search and web data mining*, WSDM '08, pages 207–218, New York, NY, USA, 2008. ACM.

[3] C. C. Aggarwal and N. Li. On node classification in dynamic content-based networks. In *The Eleventh SIAM International Conference on Data Mining*, pages 355–366, 2011.

[4] C. C. Aggarwal and H. Wang. Text mining in social networks. *Social Network Data Analytics*, pages 353–378, 2011.

[5] E. Agichtein, C. Castillo, D. Donato, A. Gionis, and G. Mishne. Finding high-quality content in social media. In *Proceedings of the international conference on Web search and web data mining*, WSDM '08, pages 183–194, New York, NY, USA, 2008. ACM.

[6] R. Angelova and G. Weikum. Graph-based text classification: learn from your neighbors. In *Proceedings of the 29th annual international ACM SIGIR conference on Research and development in information retrieval*, pages 485–492. ACM, 2006.

[7] E. Bakshy, J. Hofman, W. Mason, and D. Watts. Identifying influencers on twitter. In *Proceedings of the fourth ACM International Conference on Web Search and Data Mining*, 2011.

[8] S. Banerjee, K. Ramanathan, and A. Gupta. Clustering short texts using wikipedia. In *Proceedings of the 30th annual international ACM SIGIR conference on Research and development in information retrieval*, pages 787–788. ACM, 2007.

[9] G. Barbier and H. Liu. Information Provenance in Social Media. *Social Computing, Behavioral-Cultural Modeling and Prediction*, pages 276–283, 2011.

[10] D. Carmel, H. Roitman, and N. Zwerdling. Enhancing cluster labeling using wikipedia. In *Proceedings of the 32nd international ACM SIGIR conference on Research and development in information retrieval*, pages 139–146. ACM, 2009.

[11] S. Chakrabarti, B. Dom, and P. Indyk. Enhanced hypertext categorization using hyperlinks. In *ACM SIGMOD Record*, volume 27, pages 307–318. ACM, 1998.

[12] H.-H. Chen, M.-S. Lin, and Y.-C. Wei. Novel association measures using web search with double checking. In *Proceedings of the 21st International Conference on Computational Linguistics and the 44th annual meeting of the Association for Computational Linguistics*, pages 1009–1016. Association for Computational Linguistics, 2006.

[13] L. Chen and A. Roy. Event detection from Flickr data through wavelet-based spatial analysis. In *Proceeding of the 18th ACM conference on Information and knowledge management*, pages 523–532. ACM, 2009.

[14] B. Connor, R. Balasubramanyan, B. R. Routledge, and N. A. Smith. From tweets to polls: Linking text sentiment to public opinion time series. In *Proceedings of the International AAAI Conference on Weblogs and Social Media*, pages 122–129, 2010.

[15] B. Danushka, M. Yutaka, and I. Mitsuru. Measuring semantic similarity between words using web search engines. In *Proceedings of the 16th international conference on World Wide Web*, WWW '07, pages 757–766, 2007.

[16] L. Denoyer and P. Gallinari. The wikipedia xml corpus. *SIGIR Forum*, 40(1):64–69, 2006.

[17] J. F"urnkranz. Exploiting structural information for text classification on the www. *Advances in Intelligent Data Analysis*, pages 487–497, 1999.

[18] E. Gabrilovich and S. Markovitch. Feature generation for text categorization using world knowledge. In *International joint conference on artificial intelligence*, volume 19, page 1048, 2005.

[19] E. Gabrilovich and S. Markovitch. Overcoming the brittleness bottleneck using wikipedia: Enhancing text categorization with encyclopedic knowledge. In *Proceedings of the National Conference on Artificial Intelligence*, volume 21, page 1301, 2006.

[20] E. Gabrilovich and S. Markovitch. Computing semantic relatedness using wikipedia-based explicit semantic analysis. In *Proceedings of the 20th International Joint Conference on Artificial Intelligence*, pages 6–12, 2007.

[21] S. Gerani, M. J. Carman, and F. Crestani. Proximity-based opinion retrieval. In *Proceeding of the 33rd international ACM SIGIR conference on Research and development in information retrieval*, SIGIR '10, pages 403–410, New York, NY, USA, 2010. ACM.

[22] M. Gray, B. Team, J. Pickett, D. Hoiberg, D. Clancy, P. Norvig, J. Orwant, and S. Pinker. Quantitative Analysis of Culture Using Millions of Digitized Books. *science*, 1199644(176):331, 2011.

[23] Z. Guan, C. Wang, J. Bu, C. Chen, K. Yang, D. Cai, and X. He. Document recommendation in social tagging services. In *Proceedings of the 19th international conference on World wide web*, WWW '10, pages 391–400, New York, NY, USA, 2010. ACM.

[24] J. Hammerton, M. Osborne, S. Armstrong, and W. Daelemans. Introduction to special issue on machine learning approaches to shallow parsing. *Machine Learning Research*, 2:551–558, 2002.

[25] F. M. Harper, D. Moy, and J. A. Konstan. Facts or friends?: distinguishing informational and conversational questions in social qa sites. In *Proceedings of the 27th international conference on Human factors in computing systems*, CHI '09, pages 759–768, New York, NY, USA, 2009. ACM.

[26] P. Heymann, G. Koutrika, and H. Garcia-Molina. Can social bookmarking improve web search? In *Proceedings of the international conference on Web search and web data mining*, pages 195–206. ACM, 2008.

[27] J. Hu, L. Fang, Y. Cao, H. Zeng, H. Li, Q. Yang, and Z. Chen. Enhancing text clustering by leveraging Wikipedia semantics. In *Proceedings of the 31st annual international ACM SIGIR conference on Research and development in information retrieval*, pages 179–186. ACM, 2008.

[28] X. Hu, N. Sun, C. Zhang, and T.-S. Chua. Exploiting internal and external semantics for the clustering of short texts using world knowledge. In *Proceeding of the 18th ACM conference on Information and knowledge management*, pages 919–928. ACM, 2009.

[29] X. Hu, X. Zhang, C. Lu, E. K. Park, and X. Zhou. Exploiting wikipedia as external knowledge for document clustering. In *Proceedings of the 15th ACM SIGKDD international conference on Knowledge discovery and data mining*, pages 389–396. ACM, 2009.

[30] A. Java, X. Song, T. Finin, and B. Tseng. Why we twitter: understanding microblogging usage and communities. In *Proceedings of the 9th WebKDD and 1st SNA-KDD 2007 workshop on Web mining and social network analysis*, pages 56–65. ACM, 2007.

[31] M. Ji, Y. Sun, M. Danilevsky, J. Han, and J. Gao. Graph regularized transductive classification on heterogeneous information networks. *Machine Learning and Knowledge Discovery in Databases*, pages 570–586, 2010.

[32] G. Kumaran and J. Allan. Text classification and named entities for new event detection. In *Proceedings of the 27th annual international ACM SIGIR conference on Research and development in information retrieval*, pages 297–304. ACM, 2004.

[33] H. Kwak, C. Lee, H. Park, and S. Moon. What is twitter, a social network or a news media? In *Proceedings of the 19th international conference on World wide web*, WWW '10, pages 591–600, New York, NY, USA, 2010. ACM.

[34] Y. Lee, H.-y. Jung, W. Song, and J.-H. Lee. Mining the blogosphere for top news stories identification. In *Proceeding of the 33rd international ACM SIGIR conference on Research and development in information retrieval*, SIGIR '10, pages 395–402, New York, NY, USA, 2010. ACM.

[35] K. Lerman and T. Hogg. Using a model of social dynamics to predict popularity of news. In *Proceedings of the 19th international conference on World wide web*, WWW '10, pages 621–630, New York, NY, USA, 2010. ACM.

[36] D. Lewis and W. Croft. Term clustering of syntactic phrases. In *Proceedings of the 13th annual international ACM SIGIR confer-*

ence on Research and development in information retrieval, pages 385–404. ACM, 1989.

[37] C. Lin, B. Zhao, Q. Mei, and J. Han. Pet: a statistical model for popular events tracking in social communities. In *Proceedings of the 16th ACM SIGKDD international conference on Knowledge discovery and data mining*, pages 929–938. ACM, 2010.

[38] Y. Lu, P. Tsaparas, A. Ntoulas, and L. Polanyi. Exploiting social context for review quality prediction. In *Proceedings of the 19th international conference on World wide web*, WWW '10, pages 691–700, New York, NY, USA, 2010. ACM.

[39] C. Macdonald, I. Ounis, and I. Soboroff. Overview of the trec-2009 blog track. *Proceedings of TREC 2009*, 2010.

[40] D. Margineantu, W. Wong, and D. Dash. Machine learning algorithms for event detection. *Machine Learning*, 79(3):257–259, 2010.

[41] J. McLean. State of the Blogosphere, introduction, 2009.

[42] M. Mendoza, B. Poblete, and C. Castillo. Twitter Under Crisis: Can we trust what we RT? In *1st Workshop on Social Media Analytics (SOMA'10)*, 2010.

[43] S. Moturu. *Quantifying the Trustworthiness of User-Generated Social Media Content*. PhD thesis, Arizona State University, 2009.

[44] S. Osinski, J. Stefanowski, and D. Weiss. Lingo: Search results clustering algorithm based on singular value decomposition. In *Proceedings of the IIS: IIPWM'04 Conference*, page 359, 2004.

[45] X.-H. Phan, L.-M. Nguyen, and S. Horiguchi. Learning to classify short and sparse text & web with hidden topics from large-scale data collections. In *Proceeding of the 17th international conference on World Wide Web*, pages 91–100. ACM, 2008.

[46] M. F. Porter. An algorithm for suffix stripping. *Program*, 14(3):130–137, 1980.

[47] T. Sakaki, M. Okazaki, and Y. Matsuo. Earthquake shakes twitter users: real-time event detection by social sensors. In *Proceedings of the 19th international conference on World wide web*, pages 851–860. ACM, 2010.

[48] B. Sigurbjornsson and R. Van Zwol. Flickr tag recommendation based on collective knowledge. In *Proceeding of the 17th international conference on World Wide Web*, pages 327–336. ACM, 2008.

[49] A. Stavrianou, P. Andritsos, and N. Nicoloyannis. Overview and semantic issues of text mining. *ACM SIGMOD Record*, 36(3):23–34, 2007.

[50] Y. Sun, J. Han, J. Gao, and Y. Yu. itopicmodel: Information network-integrated topic modeling. In *Data Mining, 2009. ICDM'09. Ninth IEEE International Conference on*, pages 493–502. IEEE, 2009.

[51] Y. Sun, Y. Yu, and J. Han. Ranking-based clustering of heterogeneous information networks with star network schema. In *Proceedings of the 15th ACM SIGKDD international conference on Knowledge discovery and data mining*, pages 797–806. ACM, 2009.

[52] J. Surowiecki. *The wisdom of crowds: Why the many are smarter than the few and how collective wisdom shapes business, economies, societies, and nations*. Random House of Canada, 2004.

[53] L. Tang and H. Liu. Relational learning via latent social dimensions. In *Proceedings of the 15th ACM SIGKDD international conference on Knowledge discovery and data mining*, pages 817–826. ACM, 2009.

[54] L. Urena-Lopez, M. Buenaga, and J. Gomez. Integrating linguistic resources in TC through WSD. *Computers and the Humanities*, 35(2):215–230, 2001.

[55] N. Van House. Flickr and public image-sharing: distant closeness and photo exhibition. In *CHI'07 extended abstracts on Human factors in computing systems*, pages 2717–2722. ACM, 2007.

[56] J. Wang, Y. Zhou, L. Li, B. Hu, and X. Hu. Improving short text clustering performance with keyword expansion. In *The Sixth International Symposium on Neural Networks (ISNN 2009)*, pages 291–298. Springer, 2009.

[57] K. Wang, Z. Ming, X. Hu, and T. Chua. Segmentation of multi-sentence questions: towards effective question retrieval in cQA services. In *Proceeding of the 33rd international ACM SIGIR conference on Research and development in information retrieval*, pages 387–394. ACM, 2010.

[58] P. Wang and C. Domeniconi. Building semantic kernels for text classification using Wikipedia. In *Proceeding of the 14th ACM SIGKDD international conference on Knowledge discovery and data mining*, pages 713–721. ACM, 2008.

[59] X. Wang, L. Tang, H. Gao, and H. Liu. Discovering overlapping groups in social media. In *the 10th IEEE International Conference on Data Mining series (ICDM2010)*, Sydney, Australia, December 14 - 17 2010.

[60] X. Wang, C. Zhai, X. Hu, and R. Sproat. Mining correlated bursty topic patterns from coordinated text streams. In *Proceedings of the*

13th ACM SIGKDD international conference on Knowledge discovery and data mining, pages 784–793. ACM, 2007.

[61] D. Yin, Z. Xue, L. Hong, and B. D. Davison. A probabilistic model for personalized tag prediction. In *Proceedings of the 16th ACM SIGKDD international conference on Knowledge discovery and data mining*, KDD '10, pages 959–968, New York, NY, USA, 2010. ACM.

[62] Z. Yin, R. Li, Q. Mei, and J. Han. Exploring social tagging graph for web object classification. In *Proceedings of the 15th ACM SIGKDD international conference on Knowledge discovery and data mining*, KDD '09, pages 957–966, New York, NY, USA, 2009. ACM.

[63] J. Yuan, Z. Zha, Z. Zhao, X. Zhou, and T. Chua. Utilizing related samples to learn complex queries in interactive concept-based video search. In *Proceedings of the ACM International Conference on Image and Video Retrieval*, pages 66–73. ACM, 2010.

[64] R. Zafarani and H. Liu. Connecting Corresponding Identities across Communities. In *Proceedings of the 3rd International Conference on Weblogs and Social Media (ICWSM09)*, 2009.

[65] T. Zesch, C. Muller, and I. Gurevych. Extracting lexical semantic knowledge from wikipedia and wiktionary. In *Proceedings of the Conference on Language Resources and Evaluation (LREC)*, pages 1646–1652. Citeseer, 2008.

[66] Z. Zha, X. Hua, T. Mei, J. Wang, G. Qi, and Z. Wang. Joint multi-label multi-instance learning for image classification. In *Computer Vision and Pattern Recognition, 2008. CVPR 2008. IEEE Conference on*, pages 1–8. IEEE, 2008.

[67] Q. Zhao, P. Mitra, and B. Chen. Temporal and information flow based event detection from social text streams. In *Proceedings of the 22nd national conference on Artificial intelligence - Volume 2*, pages 1501–1506. AAAI Press, 2007.

[68] Y. Zhou, H. Cheng, and J. Yu. Graph clustering based on structural/attribute similarities. *Proceedings of the VLDB Endowment*, 2(1):718–729, 2009.

Chapter 13

A SURVEY OF OPINION MINING AND SENTIMENT ANALYSIS

Bing Liu
University of Illinois at Chicago
Chicago, IL
liub@cs.uic.edu

Lei Zhang
University of Illinois at Chicago
Chicago, IL
lzhang32@gmail.com

Abstract Sentiment analysis or opinion mining is the computational study of people's opinions, appraisals, attitudes, and emotions toward entities, individuals, issues, events, topics and their attributes. The task is technically challenging and practically very useful. For example, businesses always want to find public or consumer opinions about their products and services. Potential customers also want to know the opinions of existing users before they use a service or purchase a product.

With the explosive growth of *social media* (i.e., reviews, forum discussions, blogs and social networks) on the Web, individuals and organizations are increasingly using public opinions in these media for their decision making. However, finding and monitoring opinion sites on the Web and distilling the information contained in them remains a formidable task because of the proliferation of diverse sites. Each site typically contains a huge volume of opinionated text that is not always easily deciphered in long forum postings and blogs. The average human reader will have difficulty identifying relevant sites and accurately summarizing the information and opinions contained in them. Moreover, it is also known that human analysis of text information is subject to considerable biases, e.g., people often pay greater attention to opinions that are consistent with their own preferences. People also have difficulty, owing to their mental and physical limitations, producing consistent results when the amount of information to be processed is large. Automated opinion mining and summarization systems are thus needed, as subjective biases and mental limitations can be overcome with an objective sentiment analysis system.

In the past decade, a considerable amount of research has been done in academia [58,76]. There are also numerous commercial companies that provide opinion mining services. In this chapter, we first define the opinion mining problem. From the definition, we will see the key technical issues that need to be addressed. We then describe various key mining tasks that have been studied in the research literature and their representative techniques. After that, we discuss the issue of detecting opinion spam or fake reviews. Finally, we also introduce the research topic of assessing the utility or quality of online reviews.

Keywords: opinion mining, sentiment analysis

1. The Problem of Opinion Mining

In this first section, we define the opinion mining problem, which enables us to see a structure from the intimidating unstructured text and to provide a unified framework for the current research. The abstraction consists of two parts: opinion definition and opinion summarization [31].

1.1 Opinion Definition

We use the following review segment on iPhone to introduce the problem (an id number is associated with each sentence for easy reference):

"(1) I bought an iPhone a few days ago. (2) It was such a nice phone. (3) The touch screen was really cool. (4) The voice quality was clear too. (5) However, my mother was mad with me as I did not tell her before I bought it. (6) She also thought the phone was too expensive, and wanted me to return it to the shop ..."

The question is: what we want to mine or extract from this review? The first thing that we notice is that there are several opinions in this review. Sentences (2), (3) and (4) express some positive opinions, while sentences (5) and (6) express negative opinions or emotions. Then we also notice that the opinions all have some targets. The target of the opinion in sentence (2) is the iPhone as a whole, and the targets of the opinions in sentences (3) and (4) are "touch screen" and "voice quality" of the iPhone respectively. The target of the opinion in sentence (6) is the price of the iPhone, but the target of the opinion/emotion in sentence (5) is "me", not iPhone. Finally, we may also notice the holders of opinions.

The holder of the opinions in sentences (2), (3) and (4) is the author of the review ("I"), but in sentences (5) and (6) it is "my mother". With this example in mind, we now formally define the opinion mining problem. We start with the *opinion target*.

In general, opinions can be expressed about anything, e.g., a product, a service, an individual, an organization, an event, or a topic, by any person or organization. We use the *entity* to denote the target object that has been evaluated. Formally, we have the following:

DEFINITION 13.1 (ENTITY) *An entity e is a product, service, person, event, organization, or topic. It is associated with a pair, e : (T, W), where T is a hierarchy of components (or parts), sub-components, and so on, and W is a set of attributes of e. Each component or sub-component also has its own set of attributes.*

An example of an entity is as follows:

EXAMPLE 13.2 *A particular brand of cellular phone is an entity, e.g.,* iPhone. *It has a set of components, e.g.,* battery *and* screen, *and also a set of attributes, e.g.,* voice quality, size, *and* weight. *The battery component also has its own set of attributes, e.g.,* battery life, *and* battery size.

Based on this definition, an entity is represented as a tree or hierarchy. The root of the tree is the name of the entity. Each non-root node is a component or sub-component of the entity. Each link is a part-of relation. Each node is associated with a set of attributes. An opinion can be expressed on any node and any attribute of the node.

In practice, it is often useful to simplify this definition due to two reasons: First, natural language processing is a difficult task. To effectively study the text at an arbitrary level of detail as described in the definition is very hard. Second, for an ordinary user, it is too complex to use a hierarchical representation. Thus, we simplify and flatten the tree to two levels and use the term *aspects* to denote both components and attributes. In the simplified tree, the root level node is still the entity itself, while the second level nodes are the different aspects of the entity.

For product reviews and blogs, opinion holders are usually the authors of the postings. Opinion holders are more important in news articles as they often explicitly state the person or organization that holds an opinion [5, 13, 49]. Opinion holders are also called *opinion sources* [107].

There are two main types of opinions: *regular opinions* and *comparative opinions*. Regular opinions are often referred to simply as opinions in the research literature. A comparative opinion expresses a relation of similarities or differences between two or more entities, and/or a preference of the opinion holder based on some of the shared aspects of the entities [36, 37]. A comparative opinion is usually expressed using the *comparative* or *superlative* form of an adjective or adverb, although not always. The discussion below focuses only on regular opinions. Comparative opinions will be discussed in Sect. 6. For simplicity, the terms *regular opinion* and *opinion* are used interchangeably below.

An opinion (or regular opinion) is simply a positive or negative sentiment, attitude, emotion or appraisal about an entity or an aspect of the entity from an opinion holder. Positive, negative and neutral are called *opinion orientations* (also called *sentiment orientations, semantic orientations,* or *polarities*). We are now ready to define an opinion [58].

DEFINITION 13.3 (OPINION) *An opinion (or regular opinion) is a quintuple, $(e_i, a_{ij}, oo_{ijkl}, h_k, t_l)$, where e_i is the name of an entity, a_{ij} is an aspect of e_i, oo_{ijkl} is the orientation of the opinion about aspect a_{ij} of entity e_i, h_k is the opinion holder, and t_l is the time when the opinion is expressed by h_k. The opinion orientation oo_{ijkl} can be positive, negative or neutral, or be expressed with different strength/intensity levels. When an opinion is on the entity itself as a whole, we use the special aspect GENERAL to denote it.*

These five components are essential. Without any of them, it can be problematic in general. For example, if one says "*The picture quality is great*", and we do not know whose picture quality, the opinion is of little use. However, we do not mean that every piece of information is needed in every application. For example, knowing each opinion holder is not necessary if we want to summarize opinions from a large number of people. Similarly, we do not claim that nothing else can be added to the quintuple. For example, in some applications the user may want to know the sex and age of each opinion holder.

An important contribution of this definition is that it provides a basis for transforming unstructured text to structured data. The quintuple gives us the essential information for a rich set of qualitative and quantitative analysis of opinions. Specifically, the quintuple can be regarded as a schema for a database table. With a large set of opinion records mined from text, database management systems tools can be applied to slice and dice the opinions for all kinds of analyses.

Objective of opinion mining: Given a collection of opinionated documents D, discover all opinion quintuples $(e_i, a_{ij}, oo_{ijkl}, h_k, t_l)$ in D.

To achieve this objective, one needs to perform the following tasks:

- **Task 1** (entity extraction and grouping): Extract all entity expressions in D, and group synonymous entity expressions into entity clusters. Each entity expression cluster indicates a unique entity e_i.

- **Task 2** (aspect extraction and grouping): Extract all aspect expressions of the entities, and group aspect expressions into clusters. Each aspect expression cluster of entity e_i indicates a unique aspect a_{ij}.

- **Task 3** (opinion holder and time extraction): Extract these pieces of information from the text or unstructured data.

- **Task 4** (aspect sentiment classification): Determine whether each opinion on an aspect is positive, negative or neutral.

- **Task 5** (opinion quintuple generation): Produce all opinion quintuples $(e_i, a_{ij}, oo_{ijkl}, h_k, t_l)$ expressed in D based on the results of the above tasks.

We use an example blog to illustrate these tasks (a sentence id is associated with each sentence):

EXAMPLE 13.4 (BLOG POSTING) **Posted by: bigXyz on Nov-4-2010:** *(1) I bought a Motorola phone and my girlfriend bought a Nokia phone yesterday. (2) We called each other when we got home. (3) The voice of my Moto phone was unclear, but the camera was good. (4) My girlfriend was quite happy with her phone, and its sound quality. (5) I want a phone with good voice quality. (6) So I probably will not keep it.*

Task 1 should extract the entity expressions, "Motorola", "Nokia" and "Moto", and group "Motorola" and "Moto" together as they represent the same entity. Task 2 should extract aspect expressions "camera", "voice", and "sound", and group "voice" and "sound" together as they are synonyms representing the same aspect. Task 3 should find the holder of the opinions in sentence (3) to be bigXyz (the blog author), and the holder of the opinions in sentence (4) to be bigXyz's girlfriend. It should also find the time when the blog was posted, which is Nov-4-2010. Task 4 should find that sentence (3) gives a negative opinion to the voice quality of the Motorola phone, but a positive opinion to its camera. Sentence (4) gives positive opinions to the Nokia phone as a whole and also its sound quality. Sentence (5) seemingly expresses a positive opinion, but it does not. To generate opinion quintuples for sentence (4), we also need to know what "her phone" is and what "its" refers to. All these are challenging problems. Task 5 should finally generate the following four opinion quintuples:

(Motorola, voice_quality, negative, bigXyz, Nov-4-2010)
(Motorola, camera, positive, bigXyz, Nov-4-2010)

(Nokia, GENERAL, positive, bigXyz's_girlfriend, Nov-4-2010)
(Nokia, voice_quality, positive, bigXyz's_girlfriend, Nov-4-2010)

Before going further, let us discuss two other important concepts related to opinion mining and sentiment analysis, i.e., *subjectivity* and *emotion*.

DEFINITION 13.5 (SENTENCE SUBJECTIVITY) *An objective sentence presents some factual information about the world, while a subjective sentence expresses some personal feelings, views or beliefs.*

For instance, in the above example, sentences (1) and (2) are objective sentences, while all other sentences are subjective sentences. Subjective expressions come in many forms, e.g., opinions, allegations, desires, beliefs, suspicions, and speculations [87, 103]. Thus, a subjective sentence may not contain an opinion. For example, sentence (5) in Example 4 is subjective but it does not express a positive or negative opinion about anything. Similarly, not every objective sentence contains no opinion. For example, "*the earphone broke in two days*", is an objective sentence but it implies a negative opinion. There is some confusion among researchers to equate subjectivity with opinion. As we can see, the concepts of subjective sentences and opinion sentences are not the same, although they have a large intersection. The task of determining whether a sentence is subjective or objective is called subjectivity classification [105], which we will discuss in Sect. 3.

DEFINITION 13.6 (EMOTION) *Emotions are our subjective feelings and thoughts.*

According to [80], people have 6 primary emotions, i.e., love, joy, surprise, anger, sadness, and fear, which can be sub-divided into many secondary and tertiary emotions. Each emotion can also have different intensities. The strengths of opinions are related to the intensities of certain emotions, e.g., joy, anger, and fear, as these sentences show: (1) "I am very angry with this shop," (2) "I am so happy with my iPhone," and (3) "with the current economic condition, I fear that I will lose my job." However, the concepts of emotions and opinions are not equivalent. Many opinion sentences express no emotion (e.g., "the voice of this phone is clear"), which are called *rational evaluation sentences*, and many emotion sentences give no opinion, e.g., "I am so surprised to see you."

1.2 Aspect-Based Opinion Summary

Most opinion mining applications need to study opinions from a large number of opinion holders. One opinion from a single person is usually not sufficient for action. This indicates that some form of summary of opinions is desirable. Opinion quintuples defined above provide an excellent source of information for generating both qualitative and quantitative summaries. A common form of summary is based on aspects, and is called *aspect-based opinion summary* (or *feature-based opinion summary*) [31, 60]. Below, we use an example to illustrate this form of summary, which is widely used in industry.

EXAMPLE 13.7 *Assume we summarize all the reviews of a particular cellular phone, cellular phone 1. The summary looks like that in Fig. 13.1, which was proposed in [31] and is called a structured summary. In the figure,* GENERAL *represents the phone itself (the entity). 125 reviews expressed positive opinions about the phone and 7 expressed negative opinions. Voice quality and size are two product aspects. 120 reviews expressed positive opinions about the voice quality, and only 8 reviews expressed negative opinions. The <individual review sentences> link points to the specific sentences and/or the whole reviews that give the positive or negative opinions. With such a summary, the user can easily see how existing customers feel about the phone. If he/she is interested in a particular aspect, he/she can drill down by following the <individual review sentences> link to see why existing customers like it and/or dislike it.*

Cellular phone 1:
  Aspect: GENERAL
    Positive: 125   <individual review sentences>
    Negative: 7   <individual review sentences>
  Aspect: Voice quality
    Positive: 120   <individual review sentences>
    Negative: 8   <individual review sentences>
    ...

Figure 13.1. An aspect-based opinion summary

The aspect-based summary in Fig. 13.1 can be visualized using a bar chart and opinions on multiple products can also be compared in a visualization (see [60]).

Researchers have also studied opinion summarization in the tradition fashion, e.g., producing a short *text summary* [4, 11, 51, 89, 91]. Such a summary gives the reader a quick overview of what people think about a product or service. A weakness of such a text-based summary is that it is not quantitative but only qualitative, which is usually not suitable for analytical purposes. For example, a traditional text summary may say *"Most people do not like this product"*. However, a quantitative summary may say that 60% of the people do not like this product and 40% of them like it. In most opinion mining applications, the quantitative side is crucial just like in the traditional survey research. In survey research, aspect-based summaries displayed as bar charts or pie charts are commonly used because they give the user a concise, quantitative and visual view. Recently, researchers also tried to produce text summaries similar to that in Fig. 13.1 but in a more readable form [73, 81, 96].

2. Document Sentiment Classification

We are now ready to discuss some main research topics of opinion mining. This section focuses on *sentiment classification*, which has been studied extensively in the literature (see a survey in [76]). It classifies an opinion document (e.g., a product review) as expressing a positive or negative opinion or sentiment. The task is also commonly known as the *document-level sentiment classification* because it considers the whole document as the basic information unit.

DEFINITION 13.8 (DOCUMENT LEVEL SENTIMENT) *Given an opinionated document d evaluating an entity e, determine the opinion orientation oo on e, i.e., determine oo on aspect GENERAL in the quintuple $(e, GENERAL, oo, h, t)$. e, h, and t are assumed known or irrelevant.*

An important assumption about sentiment classification is as follows:
Assumption: Sentiment classification assumes that the opinion document d (e.g., a product review) expresses opinions on a single entity e and the opinions are from a single opinion holder h.

This assumption holds for customer reviews of products and services because each such review usually focuses on a single product and is written by a single reviewer. However, it may not hold for a forum and blog posting because in such a posting the author may express opinions on multiple products, and compare them using comparative sentences.

Most existing techniques for document-level sentiment classification are based on supervised learning, although there are also some unsupervised methods. We give an introduction to them below.

2.1 Classification based on Supervised Learning

Sentiment classification obviously can be formulated as a supervised learning problem with three classes, positive, negative and neutral. Training and testing data used in the existing research are mostly product reviews, which is not surprising due to the above assumption. Since each review already has a reviewer-assigned rating (e.g., 1 to 5 stars), training and testing data are readily available. For example, a review with 4 or 5 stars is considered a positive review, a review with 1 or 2 stars is considered a negative review and a review with 3 stars is considered a neutral review.

Any existing supervised learning methods can be applied to sentiment classification, e.g., naive Bayesian classification, and support vector machines (SVM). Pang et al. [78] took this approach to classify movie reviews into two classes, positive and negative. It was shown that using unigrams (a bag of individual words) as features in classification performed well with either naive Bayesian or SVM.

Subsequent research used many more features and techniques in learning [76]. As most machine learning applications, the main task of sentiment classification is to engineer an effective set of features. Some of the current features are listed below.

- *Terms and their frequency:* These features are individual words or word n-grams and their frequency counts. In some cases, word positions may also be considered. The TF-IDF weighting scheme from information retrieval may be applied too. These features have been shown quite effective in sentiment classification.

- *Part of speech:* It was found in many researches that adjectives are important indicators of opinions. Thus, adjectives have been treated as special features.

- *Opinion words and phrases:* Opinion words are words that are commonly used to express positive or negative sentiments. For example, *beautiful, wonderful, good, and amazing* are positive opinion words, and *bad, poor, and terrible* are negative opinion words. Although many opinion words are adjectives and adverbs, nouns (e.g., *rubbish, junk, and crap*) and verbs (e.g., *hate* and *like*) can also indicate opinions. Apart from individual words, there are also opinion phrases and idioms, e.g., *cost someone an arm and a leg*. Opinion words and phrases are instrumental to sentiment analysis for obvious reasons.

- *Negations:* Clearly, negation words are important because their appearances often change the opinion orientation. For example, the sentence "*I don't like this camera*" is negative. However, negation words must be handled with care because not all occurrences of such words mean negation. For example, "*not*" in "*not only ... but also*" does not change the orientation direction (see opinion shifters in Sect. 5.1).

- *Syntactic dependency:* Word dependency based features generated from parsing or dependency trees are also tried by several researchers.

Instead of using a standard machine learning method, researchers have also proposed several custom techniques specifically for sentiment classification, e.g., the score function in [15] based on words in positive and negative reviews. In [74], feature weighting schemes are used to enhance classification accuracy.

Manually labeling training data can be time-consuming and label-intensive. To reduce the labeling effort, *opinion words* can be utilized in the training procedure. In [95], Tan et al. used opinion words to label a portion of informative examples and then learn a new supervised classifier based on labeled ones. A similar approach is also used in [86]. In addition, opinion words can be utilized to increase the sentiment classification accuracy. In [68], Melville et al. proposed a framework to incorporate lexical knowledge in supervised learning to enhance accuracy.

Apart from classification of positive or negative sentiments, research has also been done on predicting the rating scores (e.g., 1-5 stars) of reviews [77]. In this case, the problem is formulated as regression since the rating scores are ordinal. Another interesting research direction is transfer learning or domain adaptation as it has been shown that sentiment classification is highly sensitive to the domain from which the training data is extracted. A classifier trained using opinionated documents from one domain often performs poorly when it is applied or tested on opinionated documents from another domain. The reason is that words and even language constructs used in different domains for expressing opinions can be quite different. To make matters worse, the same word in one domain may mean positive, but in another domain may mean negative. Thus, domain adaptation is needed. Existing research has used labeled data from one domain and unlabeled data from the target domain and general opinion words as features for adaptation [2, 7, 75, 112].

2.2 Classification based on Unsupervised Learning

It is not hard to imagine that opinion words and phrases are the dominating indicators for sentiment classification. Thus, using unsupervised learning based on such words and phrases would be quite natural. For example, the method in [93] uses known opinion words for classification, while [100] defines some phrases which are likely to be opinionated. Below, we give a description of the algorithm in [100], which consists of three steps:

Step 1: It extracts phrases containing adjectives or adverbs as adjectives and adverbs are good indicators of opinions. However, although an isolated adjective may indicate opinion, there may be insufficient context to determine its opinion orientation (called semantic orientation in [100]). For example, the adjective *"unpredictable"* may have a negative orientation in an automotive review, in such a phrase as *"unpredictable steering"*, but it could have a positive orientation in a movie review, in a phrase such as *"unpredictable plot"*. Therefore, the algorithm extracts two consecutive words, where one member of the pair is an adjective or adverb, and the other is a context word.

Two consecutive words are extracted if their POS tags conform to any of the patterns in Table 13.1. For example, the pattern in line 2 means that two consecutive words are extracted if the first word is an adverb and the second word is an adjective, but the third word cannot be a noun. NNP and NNPS are avoided so that the names of entities in the review cannot influence the classification.

EXAMPLE 13.9 *In the sentence "This camera produces beautiful pictures", "beautiful pictures" will be extracted as it satisfies the first pattern.*

Step 2: It estimates the semantic orientation of the extracted phrases using the pointwise mutual information (PMI) measure given in Equation 13.1:

$$PMI(term_1, term_2) = \log_2 \left(\frac{Pr(term_1 \wedge term_2)}{Pr(term_1) \cdot Pr(term_2)} \right) \quad (13.1)$$

Here, $Pr(term_1 \wedge term_2)$ is the co-occurrence probability of $term_1$ and $term_2$, and $Pr(term_1) \cdot Pr(term_2)$ gives the probability that the two terms co-occur if they are statistically independent. The ratio between $Pr(term_1 \wedge term_2)$ and $Pr(term_1) \cdot Pr(term_2)$ is thus a measure of the degree of statistical dependence between them. The log of this ratio is the amount of information that we acquire about the presence of one of the words when we observe the other. The semantic/opinion orientation (SO) of a phrase is computed based on its association with the positive reference word *"excellent"* and its association with the negative reference word *"poor"*:

$$SO(phrase) = PMI(phrase, \text{``excellent''}) - PMI(phrase, \text{``poor''}) \quad (13.2)$$

	First Word	Second Word	Third Word (Not Extracted)
1	JJ	NN or NNS	anything
2	RB, RBR, or RBS	JJ	not NN nor NNS
3	JJ	JJ	not NN nor NNS
4	NN or NNS	JJ	not NN nor NNS
5	RB, RBR, or RBS	VB, VBD, VBN, or VBG	anything

Table 13.1. Patterns of tags for extracting two-word phrases

The probabilities are calculated by issuing queries to a search engine and collecting the number of hits. For each search query, a search engine usually gives the number of relevant documents to the query, which is the number of hits. Thus, by searching the two terms together and separately, we can estimate the probabilities in Equation 13.1. The author of [100] used the AltaVista search engine because it has a NEAR operator, which constrains the search to documents that contain the words within ten words of one another in either order. Let hits(query) be the number of hits returned. Equation 13.2 can be rewritten as follows:

$$SO(phrase) = \log_2 \left(\frac{hits(\text{phrase NEAR ``excellent''}) hits(\text{``poor''})}{hits(\text{phrase NEAR ``poor''}) hits(\text{``excellent''})} \right) \quad (13.3)$$

To avoid division by 0, 0.01 is added to the hits.

Step 3: Given a review, the algorithm computes the average SO of all phrases in the review, and classifies the review as recommended if the average SO is positive, not recommended otherwise.

Final classification accuracies on reviews from various domains range from 84% for automobile reviews to 66% for movie reviews.

To summarize, we can see that the main advantage of document level sentiment classification is that it provides a prevailing opinion on an entity, topic or event. The main shortcomings are that it does not give details on what people liked and/or disliked and it is not easily applicable to non-reviews, e.g., forum and blog postings, because many such postings evaluate multiple entities and compare them.

3. Sentence Subjectivity and Sentiment Classification

Naturally the same document-level sentiment classification techniques can also be applied to individual sentences. The task of classifying a sentence as subjective or objective is often called *subjectivity classification* in the existing literature [30, 87, 88, 106, 109, 110, 113]. The resulting subjective sentences are also classified as expressing positive or negative opinions, which is called *sentence-level sentiment classification*.

DEFINITION 13.10 *Given a sentence s, two sub-tasks are performed:*

1. Subjectivity classification: *Determine whether s is a subjective sentence or an objective sentence,*

2. Sentence-level sentiment classification: *If s is subjective, determine whether it expresses a positive, negative or neutral opinion.*

Notice that the quintuple (e, a, oo, h, t) is not used in defining the problem here because sentence-level classification is often an intermediate step. In most applications, one needs to know what entities or aspects of the entities are the targets of opinions. Knowing that some sentences have positive or negative opinions but not about what, is of limited use. However, the two sub-tasks are still useful because (1) it filters out those sentences which contain no opinions, and (2) after we know what entities and aspects of the entities are talked about in a sentence, this step can help us determine whether the opinions about the entities and their aspects are positive or negative.

Most existing researches study both problems, although some of them focus only on one. Both problems are classification problems. Thus, traditional supervised learning methods are again applicable. For example, one of the early works reported in [104] performed subjectivity classification using the naive Bayesian classifier. Subsequent researches also used other learning algorithms.

Much of the research on sentence-level sentiment classification makes the following assumption:

Assumption: The sentence expresses a single opinion from a single opinion holder.

This assumption is only appropriate for simple sentences with a single opinion, e.g., "*The picture quality of this camera is amazing.*" However, for compound and complex sentences, a single sentence may express more than one opinion. For example, the sentence, "*The picture quality of this camera is amazing and so is the battery life, but the viewfinder is too small for such a great camera*", expresses both positive and negative opinions (it has mixed opinions). For "*picture quality*" and "*battery life*", the sentence is positive, but for "*viewfinder*", it is negative. It is also positive for the camera as a whole (i.e., the GENERAL aspect).

Many papers have been published on subjectivity classification and sentence-level sentiment classification. In [113], for subjectivity classification, it applied supervised learning. For sentiment classification of each subjective sentence, it used a similar method to that in Sect. 2.2 but with many more seed words, and the score function was log-likelihood ratio. The same problem was also studied in [30] considering gradable adjectives, and in [23] using semi-supervised learning. In [48-50], researchers also built models to identify some specific types of opinions.

As we mentioned earlier, sentence-level classification is not suitable for compound and complex sentences. It was pointed out in [109] that not only a single sentence may contain multiple opinions, but also both subjective and factual clauses. It is useful to pinpoint such clauses. It is also important to identify the strength of opinions. A study of automatic sentiment classification was presented to classify clauses of every sentence by the *strength* of the opinions being expressed in individual clauses, down to four levels deep (*neutral, low, medium,* and *high*). The strength of *neutral* indicates the absence of opinion or subjectivity. Strength classification thus subsumes the task of classifying a sentence as subjective versus objective. In [108], the problem was studied further using supervised learning by considering contextual sentiment influencers such as negation (e.g., *not* and *never*) and contrary (e.g., *but* and *however*). A list of influencers can be found in [82]. However, in many cases, identifying only clauses are insufficient because the opinions can be embedded in phrases, e.g., "*Apple is doing very well in this terrible economy.*" In this sentence, the opinion on "*Apple*" is clearly positive but on "*economy*" it is negative.

Besides analyzing opinion sentences in reviews, research has been done in threaded discussions, which includes forum discussions, emails, and newsgroups. In threaded discussions, people not only express their opinions on a topic but also interact with each other. However, the discussions could be highly emotional and heated with many emotional statements between participants. In [115], Zhai et al. proposed a method to identify those *evaluative sentences* from forum discussions, which only express people's opinions on entities or topics and their different aspects. In [28], Hassan et al. proposed an approach to find sentences with attitudes of participants toward one another. That is, it predicts whether a sentence contains an attitude toward a text recipient or not.

Finally, we should bear in mind that not all subjective sentences have opinions and those that do form only a subset of opinionated sentences. Many objective sentences can imply opinions too. Thus, to mine opinions from text one needs to mine them from both subjective and objective sentences.

4. Opinion Lexicon Expansion

In the preceding sections, we mentioned that opinion words are employed in many sentiment classification tasks. We now discuss how such words are generated. In the research literature, opinion words are also known as opinion-bearing words or sentiment words. Positive opinion words are used to express some desired states while negative opinion words are used to express some undesired states. Examples of positive opinion words are: *beautiful, wonderful, good,* and *amazing*. Examples of negative opinion words are *bad, poor,* and *terrible*. Apart from individual words, there are also opinion phrases and idioms, e.g., *cost someone an arm and a leg*. Collectively, they are called the *opinion lexicon*. They are instrumental for opinion mining for obvious reasons.

In order to compile or collect the opinion word list, three main approaches have been investigated: manual approach, dictionary-based approach, and corpus-based approach. The manual approach is very time-consuming and thus it is not usually used alone, but combined with automated approaches as the final check because automated methods make mistakes. Below, we discuss the two automated approaches.

4.1 Dictionary based approach

One of the simple techniques in this approach is based on bootstrapping using a small set of seed opinion words and an online dictionary, e.g., WordNet [69] or thesaurus[71]. The strategy is to first collect a small set of opinion words manually with known orientations, and then to grow this set by searching in the WordNet or thesaurus for their synonyms and antonyms. The newly found words are added to the seed list. The next iteration starts. The iterative process stops when no more new words are found. This approach is used in [31, 49]. After the process completes, manual inspection can be carried out to remove and/or correct errors. Researchers have also used additional information (e.g., glosses) in WordNet and additional techniques (e.g., machine learning) to generate better lists [1, 19, 20, 45]. Several opinion word lists have been produced [17, 21, 31, 90, 104].

The dictionary based approach and the opinion words collected from it have a major shortcoming. The approach is unable to find opinion words with domain and context specific orientations, which is quite common. For example, for a speaker phone, if it is quiet, it is usually negative. However, for a car, if it is quiet, it is positive. The corpus-based approach can help deal with this problem.

4.2 Corpus-based approach and sentiment consistency

The methods in the corpus-based approach rely on syntactic or co-occurrence patterns and also a seed list of opinion words to find other opinion words in a large corpus. One of the key ideas is the one proposed by Hazivassiloglou and McKeown [29]. The technique starts with a list of seed opinion adjectives, and uses them and a set of linguistic constraints or conventions on connectives to identify additional adjective opinion words and their orientations. One of the constraints is about the conjunction AND, which says that conjoined adjectives usually have the same orientation. For example, in the sentence, *"This car is beautiful and spacious,"* if *"beautiful"* is known to be positive, it can be inferred that *"spacious"* is also positive. This is so because people usually express the same opinion on both sides of a conjunction. The following sentence is rather unnatural, "This car is beautiful and difficult to drive". If it is changed to "This car is beautiful but difficult to drive", it becomes acceptable. Rules or constraints are also designed for other connectives, OR, BUT, EITHER-OR, and NEITHER-NOR. This idea is called *sentiment consistency*. Of course, in practice it is not always consistent. Learning is applied to a large corpus to determine if two conjoined adjectives are of the same or different orientations. Same and different-orientation links between adjectives form a graph. Finally, clustering is performed on the graph to produce two sets of words: positive and negative. In [46], Kanayama and Nasukawa expanded this approach by introducing the idea of intra-sentential (within a sentence) and inter-sentential (between neighboring sentences) sentiment consistency (called coherency in [46]). The intra-sentential consistency is similar to that in [29]. Inter-sentential consistency applies the idea to neighboring sentences. That is, the same opinion orientation (positive or negative) is usually expressed in a few consecutive sentences. Opinion changes are indicated by adversative expressions such as *but* and *however*. Some criteria to determine whether to add a word to the positive or negative lexicon are also proposed. This study was based on Japanese text. In Sect. 5.4, a related but also quite different method will be described. Other related work includes [43, 44].

In [17], Ding et al. explored the idea of intra-sentential and inter-sentential sentiment consistency further. Instead of finding domain dependent opinion words, they showed that the same word could indicate different orientations in different contexts even in the same domain. This fact was also clearly depicted by the basic rules of opinions in Sect. 5.2. For example, in the digital camera domain, the word *"long"* expresses opposite opinions in the two sentences: *"The battery life is long"* (positive) and *"The time taken to focus is long"* (negative). Thus, finding domain dependent opinion words is insufficient. They then proposed to consider both possible opinion words and aspects together, and use the pair (aspect, opinion_word) as the opinion context, e.g., the pair (*"battery life"*, *"long"*). Their method thus determines opinion words and their orientations together with the aspects that they modify. The above rules about connectives are still applied. The work in [24] adopted the same context definition but used it for analyzing comparative sentences. In [63], Lu et al. proposed an optimization framework to learn aspect-dependent sentiments in opinion context based on integer linear programming [14]. In fact, the method in [94, 100] can also be considered as a method for finding context specific opinions, but it does not use the sentiment consistency idea. Its opinion context is based on syntactic POS patterns rather than aspects and opinion words that modify them. All these context definitions, however, are still insufficient as the basic rules of opinions discussed in Sect. 5.2 show, i.e., many contexts can be more complex, e.g., consuming a large amount of resources. In [9], the problem of extracting opinion expressions with any number of words was studied. The Conditional Random Fields (CRF) method [52] was used as a sequence learning technique for extraction. In [84, 85], a double-propagation method was proposed to extraction both opinion words and aspects together. We describe it in Sect. 5.4.

Using the corpus-based approach alone to identify all opinion words, however, is not as effective as the dictionary-based approach because it is hard to prepare a huge corpus to cover all English words. However, as we mentioned above, this approach has a major advantage that the dictionary-based approach does not have. It can help find domain and context specific opinion words and their orientations using a domain corpus. Finally, we should realize that populating an opinion lexicon (domain dependent or not) is different from determining whether a word or phrase is actually expressing an opinion and what its orientation is in a particular sentence. Just because a word or phrase is listed in an opinion lexicon does not mean that it actually is expressing an opinion in a sentence. For example, in the sentence, *"I am looking for a good health insurance"*, *"good"* does not express either a positive or negative opinion on any particular insurance. The same is true for opinion orientation. We should also remember that opinion words and phrases are not the only expressions that bear opinions. There are many others as we will see in Sect. 5.2.

5. Aspect-Based Sentiment Analysis

Although classifying opinionated texts at the document level or at the sentence level is useful in many cases, it does not provide the necessary detail needed for many other applications. A positive opinionated document about a particular entity does not mean that the author has positive opinions on all aspects of the entity. Likewise, a negative opinionated document does not mean that the author dislikes everything. In a typical opinionated document, the author writes both positive and negative aspects of the entity, although the general sentiment on the entity may be positive or negative. Document and sentence sentiment classification does not provide such information. To obtain these details, we need to go to the aspect level. That is, we need the full model of Sect. 1.1, i.e., aspect-based opinion mining. Instead of treating opinion mining simply as a classification of sentiments, aspect-based sentiment analysis introduces a suite of problems which require deeper natural language processing capabilities, and also produce a richer set of results.

Recall that, at the aspect level, the mining objective is to discover every quintuple $(e_i, a_{ij}, oo_{ijkl}, h_k, t_l)$ in a given document d. To achieve the objective, five tasks need to be performed. This section mainly focuses on the following two core tasks and they have also been studied more extensively by researchers (in Sect. 7, we will briefly discuss some other tasks):

1. **Aspect extraction:** Extract aspects that have been evaluated. For example, in the sentence, "*The picture quality of this camera is amazing,*" the aspect is "*picture quality*" of the entity represented by "*this camera*". Note that "*this camera*" does not indicate the GENERAL aspect because the evaluation is not about the camera as a whole, but about its picture quality. However, the sentence "*I love this camera*" evaluates the camera as a whole, i.e., the GENERAL aspect of the entity represented by "*this camera*". Bear in mind whenever we talk about an aspect, we must know which entity it belongs to. In our discussion below, we often omit the entity just for simplicity of presentation.

2. **Aspect sentiment classification:** Determine whether the opinions on different aspects are positive, negative or neutral. In the first example above, the opinion on the "*picture quality*" aspect is positive, and in the second example, the opinion on the GENERAL aspect is also positive.

5.1 Aspect Sentiment Classification

We study the second task first, determining the orientation of opinions expressed on each aspect in a sentence. Clearly, the sentence-level and clause-level sentiment classification methods discussed in Sect. 3 are useful here. That is, they can be applied to each sentence or clause which contains some aspects. The aspects in it will take the opinion orientation of the sentence or clause. However, these methods have difficulty dealing with mixed opinions in a sentence and opinions that need phrase level analysis, e.g., *"Apple is doing very well in this terrible economy."* Clause-level analysis also needs techniques to identify clauses which itself is a challenging task, especially with informal text of blogs and forum discussions, which is full of grammatical errors. Here, we describe a *lexicon-based approach* to solving the problem [17, 31], which tries to avoid these problems and has been shown to perform quite well. The extension of this method to handling comparative sentences is discussed in Sect. 6. In the discussion below, we assume that entities and their aspects are known. Their extraction will be discussed in Sect. 5.3, 5.4, and 7.

The lexicon-based approach basically uses an opinion lexicon, i.e., a list of *opinion words* and *phrases*, and a set of rules to determine the orientations of opinions in a sentence [17, 31]. It also considers opinion shifters and but-clauses. The approach works as follows:

1 **Mark opinion words and phrases:** Given a sentence that contains one or more aspects, this step marks all opinion words and phrases in the sentence. Each positive word is assigned the opinion score of +1, each negative word is assigned the opinion score of -1.

2 **Handle opinion shifters:** Opinion shifters (also called valence shifters [82]) are words and phrases that can shift or change opinion orientations. Negation words like not, never, none, nobody, nowhere, neither and cannot are the most common type. Additionally, sarcasm changes orientation too, e.g., *"What a great car, it failed to start the first day."* Although it is easy to recognize such shifters manually, spotting them and handling them correctly in actual sentences by an automated system is by no means easy. Furthermore, not every appearance of an opinion shifter changes the opinion orientation, e.g., *"not only ... but also"*. Such cases need to be dealt with carefully.

3 **Handle but-clauses:** In English, *but* means contrary. A sentence containing *but* is handled by applying the following rule: the opinion orientation before *but* and after *but* are opposite to each

other if the opinion on one side cannot be determined. As in the case of negation, not every *but* means contrary, e.g., "*not only ... but also*". Such non-but phrases containing "*but*" also need to be considered separately. Finally, we should note that contrary words and phrases do not always indicate an opinion change, e.g., "*Car-x is great, but Car-y is better*". Such cases need to be identified and dealt with separately.

4. **Aggregating opinions:** This step applies an opinion aggregation function to the resulting opinion scores to determine the final orientation of the opinion on each aspect in the sentence. Let the sentence be s, which contains a set of aspects $\{a_1 \ldots a_m\}$ and a set of opinion words or phrases $\{ow_1 \ldots ow_n\}$ with their opinion scores obtained from steps 1, 2 and 3. The opinion orientation for each aspect a_i in s is determined by the following opinion aggregation function:

$$score(a_i, s) = \sum_{ow_j \in s} \frac{ow_j \cdot oo}{dist(ow_j, a_i)} \quad (13.4)$$

where ow_j is an opinion word/phrase in s, $dist(ow_j, a_i)$ is the distance between aspect a_i and opinion word ow_j in s. $ow_j.oo$ is the opinion score of ow_i. The multiplicative inverse is used to give lower weights to opinion words that are far away from aspect a_i. If the final score is positive, then the opinion on aspect a_i in s is positive. If the final score is negative, then the opinion on the aspect is negative. It is neutral otherwise.

This simple algorithm can perform quite well in many cases, but it is not sufficient in others. One main shortcoming is that opinion words and phrases do not cover all types of expressions that convey or imply opinions. There are in fact many other possible opinion bearing expressions. Most of them are also harder to deal with. Below, we list some of them, which we call the basic rules of opinions [58, 59].

5.2 Basic Rules of Opinions

An opinion rule expresses a concept that implies a positive or negative opinion [58, 59]. In actual sentences, the concept can be expressed in many different ways in natural language. We present these rules using a formalism that is similar to the BNF form. The top level rules are as follows:

1.   POSITIVE   ::= P
2. | PO
3. | orientation shifter N
4. | orientation shifter NE
5.   NEGATIVE   ::= N
6. | NE
7. | orientation shifter P
8. | orientation shifter PO

The non-terminals P and PO represent two types of *positive opinion expressions*. The non-terminal N and NE represent two types of *negative opinion expressions*. 'opinion shifter N' and 'opinion shifter NE' represent the negation of N and NE respectively, and 'opinion shifter P' and 'opinion shifter PO' represent the negation of P and PO respectively. We can see that these are not expressed in the actual BNF form but a pseudo-language stating some concepts. The reason is that we are unable to specify them precisely because for example, in an actual sentence, the opinion shifter may be in any form and can appear before or after N, NE, P, or PO. POSITIVE and NEGATIVE are the final orientations used to determine the opinions on the aspects in a sentence.

We now define N, NE, P and PO, which contain no opinion shifters. These opinion expressions are grouped into 6 conceptual categories based on their characteristics:

1 *Opinion word or phrase:* This is the most commonly used category, in which opinion words or phrases alone can imply positive or negative opinions on aspects, e.g., "*great*" in "*The picture quality is great*". These words or phrases are reduced to P and N.

9.   P   ::= a positive opinion word or phrase
10. N   ::= an negative opinion word or phrase

Again, the details of the right-hand-sides are not specified (which also apply to all the subsequent rules). It is assumed that a set of positive and negative opinion words/phrases exist for an application.

2 *Desirable or undesirable fact:* In this case, it is a factual statement, and the description uses no opinion words, but in the context of the entity, the description implies a positive or negative opinion. For example, the sentence "*After my wife and I slept on it for two*

weeks, I noticed a mountain in the middle of the mattress" indicates a negative opinion about the mattress. However, the word "mountain" itself does not carry any opinion. Thus, we have the following two rules:

  11. P ::= desirable fact
  12. N ::= undesirable fact

3 *High, low, increased and decreased quantity of a positive or negative potential item:* For some aspects, a small value/quantity of them is negative, and a large value/quantity of them is positive, e.g., *"The battery life is short"* and *"The battery life is long."* We call such aspects *positive potential items (PPI)*. Here *"battery life"* is a positive potential item. For some other aspects, a small value/quantity of them is positive, and a large value/quantity of them is negative, e.g., *"This phone costs a lot"* and *"Sony reduced the price of the camera."* We call such aspects *negative potential items (NPI)*. *"cost"* and *"price"* are negative potential items. Both positive and negative potential items themselves express no opinions, i.e., *"battery life"* and *"cost"*, but when they are modified by quantity adjectives or quantity change words or phrases, positive or negative opinions are implied.

  13. PO ::= no, low, less or decreased quantity of NPI
  14. | large, larger, or increased quantity of PPI
  15. NE ::= no, low, less, or decreased quantity of PPI
  16. | large, larger, or increased quantity of NPI
  17. NPI ::= a negative potential item
  18. PPI ::= a positive potential item

4 *Decreased and increased quantity of an opinionated item (N and P):* This set of rules is similar to the negation rules 3, 4, 7, and 8 above. Decreasing or increasing the quantity associated with an opinionated item (often nouns and noun phrases) can change the orientation of the opinion. For example, in the sentence *"This drug reduced my pain significantly"*, *"pain"* is a negative opinion word, and the reduction of *"pain"* indicates a desirable effect of the drug. Hence, decreased pain implies a positive opinion on the drug. The concept of decreasing also extends to removal and disappearance, e.g., *"My pain has disappeared after taking the drug."*

19.	PO	::=	less or decreased N
20.			| more or increased P
21.	NE	::=	less or decreased P
22.			| more or increased N

Rules 20 and 22 may not be needed as there is no change of opinion orientation, but they can change the opinion intensity. The key difference between this set of rules and the rules in the previous category is that no opinion words or phrases are involved in the previous category.

5 *Deviation from the norm or some desired value range:* In some application domains, the value of an aspect may have a desired range or norm. If it is above or below the normal range, it is negative, e.g., "*This drug causes low (or high) blood pressure*" and "*This drug causes my blood pressure to reach 200.*" Notice that no opinion word appeared in these sentences.

23.	PO	::=	within the desired value range
24.	NE	::=	above or below the desired value range

6 *Producing and consuming resources and wastes:* If an entity produces a lot of resources, it is positive. If it consumes a lot of resources, it is negative. For example, water is a resource. The sentence, "*This washer uses a lot of water*" gives a negative opinion about the washer. Likewise, if an entity produces a lot of wastes, it is negative. If it consumes a lot of wastes, it is positive.

25.	PO	::=	produce a large quantity of or more resource
26.			| produce no, little or less waste
27.			| consume no, little or less resource
28.			| consume a large quantity of or more waste
29.	NE	::=	produce no, little or less resource
30.			| produce some or more waste
31.			| consume a large quantity of or more resource
32.			| consume no, little or less waste

We should note that these rules are not the only rules that govern expressions of positive and negative opinions. With further research, additional new rules may be discovered.

5.3 Aspect Extraction

Existing research on *aspect extraction* (more precisely, *aspect expression extraction*) is mainly carried out in online reviews. We thus focus on reviews here. We describe some unsupervised methods for finding aspect expressions that are nouns and noun phrases. The first method is due to [31]. The method consists of two steps:

1. Find frequent nouns and noun phrases. Nouns and noun phrases (or groups) are identified by a POS tagger. Their occurrence frequencies are counted, and only the frequent ones are kept. A frequency threshold can be decided experimentally. The reason for using this approach is that when people comment on different aspects of a product, the vocabulary that they use usually converges. Thus, those nouns that are frequently talked about are usually genuine and important aspects. Irrelevant contents in reviews are often diverse, i.e., they are quite different in different reviews. Hence, those infrequent nouns are likely to be non-aspects or less important aspects.

2. Find infrequent aspects by exploiting the relationships between aspects and opinion words. The above step can miss many genuine aspect expressions which are infrequent. This step tries to find some of them. The idea is as follows: The same opinion word can be used to describe or modify different aspects. Opinion words that modify frequent aspects can also modify infrequent aspects, and thus can be used to extract infrequent aspects. For example, *"picture"* has been found to be a frequent aspect, and we have the sentence,
   "The pictures are absolutely amazing."
   If we know that *"amazing"* is an opinion word, then *"software"* can also be extracted as an aspect from the following sentence,
   "The software is amazing."
   because the two sentences follow the same dependency pattern and *"software"* in the sentence is also a noun. This idea of using the modifying relationship of opinion words and aspects to extract aspects was later generalized to using dependency relations [120], which was further developed into the double-propagation method for simultaneously extracting both opinion words and aspects [85]. The double-propagation method will be described in Sect. 5.4.

The precision of step 1 of the above algorithm was improved in [83]. Their algorithm tries to remove those noun phrases that may not be product aspects/features. It evaluates each noun phrase by computing a pointwise mutual information (PMI) score between the phrase and some *meronymy discriminators* associated with the product class, e.g., a scanner class. The meronymy discriminators for the scanner class are, "of scanner", "scanner has", "scanner comes with", etc., which are used to find components or parts of scanners by searching the Web.

$$PMI(a,d) = \frac{hits(a \wedge d)}{hits(a) \cdot hits(d)} \qquad (13.5)$$

where a is a candidate aspect identified in step 1 and d is a discriminator. Web search is used to find the number of hits of individual terms and also their co-occurrences. The idea of this approach is clear. If the PMI value of a candidate aspect is too low, it may not be a component of the product because a and d do not co-occur frequently.

Other related works on aspect extraction use existing knowledge, supervised learning, semi-supervised learning, topic modeling and clustering. For example, many information extraction techniques can also be applied, e.g., Conditional Random Fields (CRF) [33, 52], and Hidden Markov Models (HMM) [22, 34, 35], and sequential rule mining [60]. Wu et al. [111] used dependency tree kernels. Su et al. [92] proposed a clustering method with mutual reinforcement to identify implicit aspects.

Topic modeling methods have also been attempted as an unsupervised and knowledge-lean approach. Titov and McDonald [99] showed that global topic models such as LDA (Latent Dirichlet allocation [6]) might not be suitable for detecting rateable aspects. They proposed multi-grain topic models to discover local rateable aspects. Here each discovered aspect is a unigram language model, i.e., a multinomial distribution over words. Such a representation is thus not as easy to interpret as aspects extracted by previous methods, but its advantage is that different words expressing the same or related aspects (more precisely aspect expressions) can usually be automatically grouped together under the same aspect. However, Titov and McDonald [99] did not separate aspects and opinion words in the discovery. Lin and He [57] proposed a joint topic-sentiment model also by extending LDA, where aspect words and opinion words were still not explicitly separated. To separate aspects and opinion words using topic models, Mei et al. [67] proposed to use a positive sentiment model and a negative sentiment model in additional to aspect models. Brody and Elhadad [10] proposed to first identify aspects using topic models and then identify aspect-specific opinion words by considering adjectives only. Zhao et al. [119] proposed a MaxEnt-LDA hybrid model to jointly discover both aspect words and aspect-specific opinion words, which can leverage syntactic features to help separate aspects and opinion words. Topic modeling based approaches were also used by Liu et al. [62] and Lu et al. [65].

Another line of work is to associate aspects with opinion/sentiment ratings. It aims to predict ratings based on learned aspects or jointly model aspects and ratings. Titov and McDonald [98] proposed a statistical model that is able to discover aspects from text and to extract textual evidence from reviews supporting each aspect rating. Lu et al. [66] defined a problem of rated aspect summarization. They proposed to use the structured probabilistic latent semantic analysis method to learn aspects from a bag of phrases, and a local/global method to predict aspect ratings. Wang et al. [102] proposed to infer both aspect ratings and aspect weights at the level of individual reviews based on learned latent aspects. Jo and Oh [41] proposed an *Aspect* and *Sentiment Unification Model* (ASUM) to model sentiments toward different aspects.

5.4 Simultaneous Opinion Lexicon Expansion and Aspect Extraction

In [84, 85], a method was proposed to extract both opinion words and aspects simultaneously by exploiting some syntactic relations of opinion words and aspects. The method needs only an initial set of opinion word seeds as the input and no seed aspects are required. It is based on the observation that opinions almost always have targets. Hence there are natural relations connecting opinion words and targets in a sentence due to the fact that opinion words are used to modify targets. Furthermore, it was found that opinion words have relations among themselves and so do targets among themselves too. The opinion targets are usually aspects. Thus, opinion words can be recognized by identified aspects, and aspects can be identified by known opinion words. The extracted opinion words and aspects are utilized to identify new opinion words and new aspects, which are used again to extract more opinion words and aspects. This propagation or bootstrapping process ends when no more opinion words or aspects can be found. As the process involves propagation through both opinion words and aspects, the method is called double propagation. Extraction rules are designed based on different relations between opinion words and aspects, and also opinion words and aspects themselves. Specifically, four subtasks are performed:

1 extracting aspects using opinion words;

2 extracting aspects using the extracted aspects;

3 extracting opinion words using the extracted aspects;

4 extracting opinion words using both the given and the extracted opinion words.

Dependency grammar [97] was adopted to describe the relations. The algorithm uses only a simple type of dependencies called *direct dependencies* to model the relations. A direct dependency indicates that one word depends on the other word without any additional words in their dependency path or they both depend on a third word directly. Some constraints are also imposed. Opinion words are considered to be adjectives and aspects nouns or noun phrases.

For example, in an opinion sentence "*Canon G3 produces great pictures*", the adjective "*great*" is parsed as directly depending on the noun "*pictures*" through relation *mod*. If we know "*great*" is an opinion word and are given the rule 'a noun on which an opinion word directly depends through *mod* is taken as an aspect', we can extract "*pictures*" as an aspect. Similarly, if we know "*pictures*" is an aspect, we can extract "*great*" as an opinion word using a similar rule.

6. Mining Comparative Opinions

Directly or indirectly expressing positive or negative opinions about an entity and its aspects is only one form of evaluation. Comparing the entity with some other similar entities is another. Comparisons are related to but are also quite different from regular opinions. They not only have different semantic meanings, but also different syntactic forms. For example, a typical regular opinion sentence is "*The picture quality of this camera is great*", and a typical comparative sentence is "*The picture quality of Camera-x is better than that of Camera-y.*" This section first defines the problem, and then presents some existing methods to solve it [15, 18, 24, 37].

In general, a comparative sentence expresses a relation based on similarities or differences of more than one entity. The comparison is usually conveyed using the comparative or superlative form of an adjective or adverb. A comparative sentence typically states that one entity has more or less of a certain attribute than another entity. A superlative sentence typically states that one entity has the most or least of a certain attribute among a set of similar entities. In general, a comparison can be between two or more entities, groups of entities, and one entity and the rest of the entities. It can also be between an entity and its previous versions.

Two types of comparatives: In English, comparatives are usually formed by adding the suffix *-er* and superlatives are formed by adding the suffix *-est* to their base *adjectives* and *adverbs*. For example, in *"The battery life of Camera-x is longer than that of Camera-y"*, *"longer"* is the comparative form of the adjective *"long"*. In *"The battery life of this camera is the longest"*, *"longest"* is the superlative form of the adjective *"long"*. We call this type of comparatives and superlatives as *Type 1* comparatives and superlatives. Note that for simplicity, we often use comparative to mean both comparative and superlative if superlative is not explicitly stated.

Adjectives and adverbs with two syllables or more and not ending in *y* do not form comparatives or superlatives by adding *-er* or *-est*. Instead, *more, most, less* and *least* are used before such words, e.g., more beautiful. We call this type of comparatives and superlatives as *Type 2* comparatives and superlatives. Both Type 1 and Type 2 are called regular comparatives and superlatives. In English, there are also irregular comparatives and superlatives, i.e., *more, most, less, least, better, best, worse, worst, further/farther* and *furthest/farthest*, which do not follow the above rules. However, they behave similarly to Type 1 comparatives and are thus grouped under Type 1.

Apart from these standard comparatives and superlatives, many other words or phrases can also be used to express comparisons, e.g., *prefer* and *superior*. For example, the sentence, *"Camera-x's quality is superior to Camera-y"*, says that *"Camera-x is better or preferred."* In [36], Jindal and Liu identified a list of such words. Since these words behave similarly to Type 1 comparatives, they are also grouped under Type 1.

Types of comparative relations: Comparative relations or comparisons can be grouped into four main types. The first three types are called the *gradable comparisons* and the last one the *non-gradable comparisons*.

1. *Non-equal gradable comparisons:* Relations of the type *greater or less than* that express an ordering of some entities with regard to some of their shared aspects, e.g., *"The Intel chip is faster than that of AMD"*. This type also includes user preferences, e.g., *"I prefer Intel to AMD"*.

2. *Equative comparisons:* Relations of the type *equal to* that state two or more entities are equal with regard to some of their shared aspects, e.g., *"The performance of Car-x is about the same as that of Car-y."*

3. *Superlative comparisons:* Relations of the type *greater or less than all others* that rank one entity over all others, e.g., *"The Intel chip is the fastest"*.

4 *Non-gradable comparisons:* Relations that compare aspects of two or more entities, but do not grade them. There are three main sub-types:

- Entity A is similar to or different from entity B with regard to some of their shared aspects, e.g., *"Coke tastes differently from Pepsi."*
- Entity A has aspect a_1, and entity B has aspect a_2 (a_1 and a_2 are usually substitutable), e.g., *"Desktop PCs use external speakers but laptops use internal speakers."*
- Entity A has aspect a, but entity B does not have, e.g., *"Phone-x has an earphone, but Phone-y does not have."*

Comparative words used in non-equal gradable comparisons can be further categorized into two groups according to whether they express increased or decreased quantities, which are useful in opinion analysis.

• *Increasing comparatives:* Such a comparative expresses an increased quantity, e.g., *more* and *longer*.
• *Decreasing comparatives:* Such a comparative expresses a decreased quantity, e.g., *less* and *fewer*.

Objective of mining comparative opinions: Given a collection of opinionated documents D, discover in D all comparative opinion sextuples of the form (E_1, E_2, A, PE, h, t), where E_1 and E_2 are the entity sets being compared based on their shared aspects A (entities in E_1 appear before entities in E_2 in the sentence), $PE(\in \{E1, E2\})$ is the preferred entity set of the opinion holder h, and t is the time when the comparative opinion is expressed.

EXAMPLE 13.11 *Consider the comparative sentence "Canon's optics is better than those of Sony and Nikon." written by John in 2010. The extracted comparative opinion is:*
({Canon}, {Sony, Nikon}, {optics}, preferred: {Canon}, John, 2010)
The entity set E_1 is {Canon}, the entity set E_2 is {Sony, Nikon}, their shared aspect set A being compared is {optics}, the preferred entity set is {Canon}, the opinion holder h is John and the time t when this comparative opinion was written is 2010.

To mine comparative opinions, the tasks of extracting entities, aspects, opinion holders and times are the same as those for mining regular opinions. In [37], a method based on label sequential rules (LSR) is proposed to extract entities and aspects that are compared. A similar approach is described in [54] for extracting the compared entities. Clearly, the approaches discussed in previous sections are applicable as well, and so are many other information extraction methods. See [37, 24, 18] for some existing methods for performing sentiment analysis of comparative sentences, i.e., identifying comparative sentences and identifying the preferred entity set.

7. Some Other Problems

Besides the problems discussed in previous sections, there are many other challenges in opinion mining. This section gives an introduction to some of them. As we will see, most of these problems are related to their general problems that have been studied before but the opinion text provides more clues for their solutions and also has additional requirements.

Entity, opinion holder, and time extraction: In some applications, it is useful to identify and extract entities, opinion holders, and the times when opinions are given. These extraction tasks are collectively called Named Entity Recognition (NER). They have been studied extensively in the literature.

In the case of social media on the Web, the opinion holders are often the authors of the discussion postings, bloggers, or reviewers, whose login ids are known although their true identities in the real world may be unknown. The date and time when an opinion is submitted are also known and displayed on the page, so their extraction is easy [59].

For entity name extraction, there is a difference from NER. In a typical opinion mining application, the user wants to find opinions on some competing entities, e.g., competing products or brands. However, he/she often can only provide a few names because there are so many different brands and models. Furthermore, Web users also write names of the same product brands in many ways. For example, "*Motorola*" may be written as "*Moto*" or "*Mot*", and "*Samsung*" may be written as "*Sammy*". Product model names have even more variations. It is thus important for a system to automatically discover them from a relevant corpus. The key requirement is that the discovered entities must be of the same type as entities provided by the user (e.g., phone brands and models). In [55], this problem was modeled as a *set expansion problem* [25, 79], which expands a set of given seed entities (e.g., product names). Formally, the problem is stated as follows: Given a set Q of seed entities of a particular class C, and a set D of candidate entities, we wish to determine which of the entities in D belong to C. That is, we "grow" the class C based on the set of seed examples Q. Although this is a classification problem, in practice, the problem is often solved as a ranking problem, i.e., to rank the entities in D based on their likelihoods of belonging to C. It was shown that learning from positive and unlabeled examples provides a more effective method than the traditional distributional similarity methods [53, 79] and the machine learning technique *Bayesian Sets* [25] which was designed specifically for set expansion.

Objective expressions implying sentiments: Much of the research on sentiment analysis focuses on subjective sentences, which are regarded as opinion bearing. However, many objective sentences can bear opinions as well. For example, in a mattress review, the sentence *"Within a month, a valley formed in the middle of the mattress"* is not a subjective sentence, but an objective sentence. However, it implies a negative opinion about the mattress. Specifically, *"valley"* in this context indicates the quality of the mattress (a product aspect) and implies a negative opinion. Objective words (or sentences) that imply opinions are very difficult to recognize because their recognition typically requires the commonsense or world knowledge of the application domain. In [116], a method was proposed to deal with the problem of product aspects which are nouns and imply opinions using a large corpus. Our experimental results show some promising results. However, the accuracy is still low, and much further research is still needed.

Grouping aspect expressions indicating the same aspects: It is common that people use different words or phrases (which are called aspect expressions in Sect. 1) to describe the same aspect. For example, *photo* and *picture* refer to the same aspect in digital camera reviews. Identifying and grouping aspect expressions indicating the same aspect are essential for applications. Although WordNet [69] and other thesaurus dictionaries help to some extent, they are far from sufficient due to the fact that many synonyms are domain dependent. For example, *picture* and *movie* are synonyms in movie reviews, but they are not synonyms in digital camera reviews as *picture* is more related to *photo* while *movie* refers to *video*. It is also important to note that although most aspect expressions of an aspect are domain synonyms, they are not always synonyms. For example, *"expensive"* and *"cheap"* can both indicate the aspect price but they are not synonyms of price.

Carenini et al [12] proposed the first method to solve this problem in the context of opinion mining. Their method is based on several similarity metrics defined using string similarity, synonyms and distances measured using WordNet. It requires a taxonomy of aspects to be given for a particular domain. The algorithm merges each discovered aspect expression to an aspect node in the taxonomy. Experiments based on digital camera and DVD reviews showed promising results.

In [114], Zhai et al. proposed a semi-supervised learning method to group aspect expressions into the user specified aspect groups. Each group represents a specific aspect. To reflect the user needs, he/she first manually labels a small number of seeds for each group. The system then assigns the rest of the discovered aspect expressions to suitable groups using semi-supervised learning based on labeled seeds and unlabeled examples. The method used the Expectation-Maximization (EM) algorithm. Two pieces of prior knowledge were used to provide a better initialization for EM, i.e., (1) aspect expressions sharing some common words are likely to belong to the same group, and (2) aspect expressions that are synonyms in a dictionary are likely to belong to the same group.

Mapping implicit aspect expressions to aspects: There are many types of implicit aspect expressions. Adjectives are perhaps the most common type. Many adjectives modify or describe some specific attributes or properties of entities. For example, the adjective "heavy" usually describes the aspect weight of an entity. "Beautiful" is normally used to describe (positively) the aspect look or appearance of an entity. By no means, however, does this say that these adjectives only describe such aspects. Their exact meanings can be domain dependent. For example, *"heavy"* in the sentence *"the traffic is heavy"* does not describe the weight of the traffic. One way to map implicit aspect expressions to aspects is to manually compile a list of such mappings during training data annotation, which can then be used in the same domain in the future. However, we should note that some implicit aspect expressions are very difficult to extract and to map, e.g., *"fit in pockets"* in the sentence *"This phone will not easily fit in pockets"*.

Coreference resolution: This problem has been extensively studied in the NLP community. However, the sentiment analysis context has additional needs. In [16], the problem of entity and aspect coreference resolution was proposed. It determines which mentions of entities and/or aspects refer to the same entities. The key interesting points were the design and testing of two opinion-related features for machine learning. The first feature is based on opinion analysis of regular sentences and comparative sentences, and the idea of sentiment consistency. For example, we have the sentences, *"The Sony camera is better than the Canon camera. It is cheap too."* It is clear that *"It"* means *"Sony"* because in the first sentence, the opinion about *"Sony"* is positive (comparative positive), but it is negative (comparative negative) about *"Canon"*, and the second sentence is positive. Thus, we can conclude that "*It*" refers to "*Sony*" because people usually express sentiments in a consistent way. It is unlikely that "*It*" refers to "*Canon*". As we can see, to obtain this feature, the system needs to have the ability to determine positive and negative opinions expressed in regular and comparative sentences.

The second feature considers what entities and aspects are modified by what opinion words. Consider these sentences, "*The picture quality of the Canon camera is very good. It is not expensive either.*" The question is what "*It*" refers to, "*Canon camera*" or "*picture quality*". Clearly, we know that "*It*" refers to "*Canon camera*" because "*picture quality*" cannot be expensive. To obtain this feature, the system needs to identify what opinion words are usually associated with what entities or aspects, which means that the system needs to discover such relationships from the corpus. These two features can boost the coreference resolution accuracy.

Cross lingual opinion mining: This research involves opinion mining for a language corpus based on the corpora from other languages. It is needed in following scenarios. Firstly, there are many English sentiment corpora on the Web nowadays, but for other languages (e.g. Chinese), the annotated sentiment corpora are limited [101]. And it is not a trivial task to label them manually. Utilizing English corpora for opinion mining in Chinese can relieve the labeling burden. Secondly, there are many situations where opinion mining results need to be multilanguage-comparable. For example, global companies need to analyze customer feedback for their products and services from many countries in different languages [47]. Thus, cross-lingual opinion mining is necessary. The basic idea of the current research is to utilize available language corpora to train sentiment classifiers for the target language data. Machine translation is typically used [3, 8, 27, 47, 101].

8. Opinion Spam Detection

It has become a common practice for people to find and to read opinions on the Web for many purposes. For example, if one wants to buy a product, one typically goes to a merchant or review site (e.g., amazon.com) to read some reviews of existing users of the product. If one sees many positive reviews of the product, one is very likely to buy the product. However, if one sees many negative reviews, he/she will most likely choose another product. Positive opinions can result in significant financial gains and/or fames for organizations and individuals. This, unfortunately, gives good incentives for opinion spam, which refers to human activities (e.g., write spam reviews) that try to deliberately mislead readers or automated opinion mining systems by giving undeserving positive opinions to some target entities in order to promote the entities and/or by giving unjust or false negative opinions to some other entities in order to damage their reputation. Such opinions are also called fake opinions, bogus opinions, or fake reviews. The problem of detecting fake or spam opinions was introduced by Jindal and Liu in [38, 39].

Individual Spammers and Group Spammers: A spammer may act individually (e.g., the author of a book) or as a member of a group (e.g., a group of employees of a company).

Individual spammers: In this case, a spammer, who does not work with anyone else, writes spam reviews. The spammer may register at a review site as a single user, or as many fake users using different user-ids. He/she can also register at multiple review sites and write spam reviews.

Group spammers: A group of spammers works together to promote a target entity and/or to damage the reputation of another. They may also register at multiple sites and spam on these sites. Group spam can be very damaging because they may take control of the sentiment on a product and completely mislead potential customers.

8.1 Spam Detection Based on Supervised Learning

In general, spam detection can be formulated as a classification problem with two classes, spam and non-spam. However, manually labeling the training data for learning is very hard, if not impossible. The problem is that identifying spam reviews by simply reading the reviews is extremely difficult because a spammer can carefully craft a spam review that is just like any innocent review.

Since manually labeling training data is hard, other ways have to be explored in order to find training examples for detecting possible fake reviews. In [38], it exploited duplicate reviews. In their study of 5.8 million reviews, 2.14 million reviewers and 6.7 million products from amazon.com, a large number of duplicate and near-duplicate reviews were found. Certain types of duplicate and near-duplicate reviews were regarded as spam reviews, and the rest of the reviews as non-spam reviews.

In [38, 39], three sets of features were identified for learning:

1. *Review centric features:* These are features about the content of reviews. Example features include actual words in a review, the number of times that brand names are mentioned, the percentage of opinion words, the review length, and the number of helpful feedbacks.

2. *Reviewer centric features:* These are features about each reviewer. Example features include the average rating given by the reviewer, the standard deviation in rating, the ratio of the number of reviews that the reviewer wrote which were the first reviews of the products to the total number of reviews that he/she wrote, and the ratio of the number of cases in which he/she was the only reviewer.

3. *Product centric features:* These are features about each product. Example features include the price of the product, the sales rank of the product (amazon.com assigns sales rank to 'now selling products' according to their sales volumes), the average review rating of the product, and the standard deviation in ratings of the reviews for the product.

Logistic regression was used for model building. Experimental results showed some interesting results.

8.2 Spam Detection Based on Abnormal Behaviors

Due to the difficulty of manually labeling training data, treating opinion spam detection as a supervised learning problem is problematic because many non-duplicated reviews can be spam too. Here, we describe two techniques that try to identify atypical behaviors of reviewers for detecting spammers. For example, if a reviewer wrote all negative reviews for a brand but other reviewers were all positive about the brand, then this reviewer is naturally a spam suspect.

The first technique [56] identifies several unusual reviewer behavior models based on different review patterns that suggest spamming. Each model assigns a numeric spamming behavior score to a reviewer by measuring the extent to which the reviewer practices spamming behavior of the type. All the scores are then combined to produce a final spam score for each reviewer.

The second technique [40] identifies unusual reviewer behavior patterns via unexpected rule discovery. This approach formulates the problem as finding unexpected class association rules [59] from data. Four types of unexpected rules are found based on four unexpectedness definitions. Below, an example behavior is given for each type of unexpectedness definition. Their detailed definitions for these types of unexpectedness are involved [40]. Below, we briefly introduce them by giving an example behavior for each unexpectedness.

- **Confidence Unexpectedness:** Using this measure, we can find reviewers who give all high ratings to products of a brand, but most other reviewers are generally negative about the brand.

- **Support Unexpectedness:** Using this measure, we can find reviewers who write multiple reviews for a single product, while other reviewers only write one review.

- **Attribute Distribution Unexpectedness:** Using this measure, we can find that most positive reviews for a brand of products are from only one reviewer although there are a large number of reviewers who have reviewed the products of the brand.

- **Attribute Unexpectedness:** Using this measure, we can find reviewers who write only positive reviews to one brand, and only negative reviews to another brand.

Experimental results of both papers [40, 56] using amazon.com reviews showed that many spammers can be detected based on their behaviors.

8.3 Group Spam Detection

A group spam detection algorithm was reported in [72]. It finds groups of spammers who work together to promote or demote some products. The method works in two steps:

1. **Frequent pattern mining:** First, it extracts the review data to produce a set of transactions. Each transaction represents a unique product and consists of all the reviewers (their ids) who have reviewed that product. Using all the transactions, it performs frequent pattern mining. The patterns thus give us a set of candidate groups who might have spammed together. The reason for using frequent pattern mining is as follows: If a group of reviewers who only worked together once to promote or to demote a single product, it can be hard to detect based on their collective or group behavior. However, these fake reviewers (especially those who get paid to write) cannot be just writing one review for a single product because they would not make enough money that way. Instead, they work on many products, i.e., write many reviews about many products, which unfortunately also give them away. Frequent pattern mining can be used to find them working together on multiple products.

2 **Rank groups based on a set of group spam indicators:** The groups discovered in step 1 may not all be spammer groups. Many of the reviewers are grouped together in pattern mining simply due to chance. Then, this step first uses a set of indicators to catch different types of unusual group behaviors. These indicators including writing reviews together in a short time window, writing reviews right after the product launch, group content similarity, group rating deviation, etc (see [72] for details). It then ranks the discovered groups from step 1 based on their indicator values using SVM rank (also called Ranking SVM) [42].

9. Utility of Reviews

A related problem that has also been studied in the past few years is the determination of the usefulness, helpfulness or utility of each review [26, 50, 61, 118, 64, 70, 117]. This is a meaningful task as it is desirable to rank reviews based on utilities or qualities when showing reviews to the user, with the most useful reviews first. In fact, many review aggregation sites have been practicing this for years. They obtain the helpfulness or utility score of each review by asking readers to provide helpfulness feedbacks to each review. For example, in amazon.com, the reader can indicate whether he/she finds a review helpful by responding to the question *"Was the review helpful to you?"* just below each review. The feedback results from all those responded are then aggregated and displayed right before each review, e.g., *"15 of 16 people found the following review helpful"*. Although most review sites already provide the service, automatically determining the quality of a review is still useful because many reviews have few or no feedbacks. This is especially true for new reviews.

Determining the utility of reviews is usually formulated as a regression problem. The learned model assigns a utility value to each review, which can be used in review ranking. In this area of research, the ground truth data used for both training and testing are usually the user-helpfulness feedback given to each review, which as we discussed above is provided for each review at many review sites. So unlike fake review detection, the training and testing data here is not an issue.

Researchers have used many types of features for model building. Example features include review length, review rating (the number of stars), counts of some specific POS tags, opinion words, tf-idf weighting scores, wh-words, product attribute mentions, comparison with product specifications, comparison with editorial reviews, and many more. Subjectivity classification was also applied in [26]. In [61], Liu et al.

formulated the problem slightly differently. They made it a binary classification problem. Instead of using the original helpfulness feedback as the target or dependent variable, they performed manual annotation based on whether the review evaluates many product aspects or not.

Finally, we should note that review utility regression/classification and review spam detections are different concepts. Not-helpful or low quality reviews are not necessarily fake reviews or spam, and helpful reviews may not be non-spam. A user often determines whether a review is helpful or not based on whether the review expresses opinions on many aspects of the product. A spammer can satisfy this requirement by carefully crafting a review that is just like a normal helpful review. Using the number of helpful feedbacks to define review quality is also problematic because user feedbacks can be spammed too. Feedback spam is a sub-problem of click fraud in search advertising, where a person or robot clicks on some online advertisements to give the impression of real customer clicks. Here, a robot or a human spammer can also click on helpfulness feedback button to increase the helpfulness of a review. Another important point is that a low quality review is still a valid review and should not be discarded, but a spam review is untruthful and/or malicious and should be removed once detected.

10. Conclusions

This chapter introduced and surveyed the field of sentiment analysis and opinion mining. Due to many challenging research problems and a wide variety of practical applications, it has been a very active research area in recent years. In fact, it has spread from computer science to management science. This chapter first presented an abstract model of sentiment analysis, which formulated the problem and provided a common framework to unify different research directions. It then discussed the most widely studied topic of sentiment and subjectivity classification, which determines whether a document or sentence is opinionated, and if so whether it carries a positive or negative opinion. We then described aspect-based sentiment analysis which exploits the full power of the abstract model. After that we briefly introduced the problem of analyzing comparative sentences. Last but not least, we discussed opinion spam, which is increasingly becoming an important issue as more and more people are relying on opinions on the Web for decision making. Several initial algorithms were described. Finally, we conclude the chapter by saying that all the sentiment analysis tasks are very challenging. Our understanding and knowledge of the problem and its solution are still limited. The main reason is that it is a natural language processing task, and natural language processing has no easy problems. However, many significant progresses have been made. This is evident from the large number of start-up companies that offer sentiment analysis or opinion mining services. There is a real and huge need in the industry for such services because every company wants to know how consumers perceive their products and services and those of their competitors. These practical needs and the technical challenges will keep the field vibrant and lively for years to come.

References

[1] Andreevskaia, A. and S. Bergler. Mining WordNet for fuzzy sentiment: Sentiment tag extraction from WordNet glosses. In *Proceedings of Conference of the European Chapter of the Association for Computational Linguistics (EACL-06)*, 2006.

[2] Aue, A. and M. Gamon. Customizing sentiment classifiers to new domains: a case study. In *Proceedings of Recent Advances in Natural Language Processing (RANLP-2005)*, 2005.

[3] Banea, C., R. Mihalcea and J. Wiebe. Multilingual subjectivity: are more languages better?. In *Proceedings of International Conference on Computational Linguistics (COLING-2010)*, 2010.

[4] Beineke, P., T. Hastie, C. Manning, and S. Vaithyanathan. An exploration of sentiment summarization. In *Proceedings of AAAI Spring Symposium on Exploring Attitude and Affect in Text: Theories and Applications*, 2003.

[5] Bethard, S., H. Yu, A. Thornton, V. Hatzivassiloglou, and D. Jurafsky. Automatic extraction of opinion propositions and their holders. In *Proceedings of the AAAI Spring Symposium on Exploring Attitude and Affect in Text*, 2004.

[6] Blei, D., A. Ng, and M. Jordan. Latent dirichlet allocation. The *Journal of Machine Learning Research*, 2003, 3: p. 993-1022.

[7] Blitzer, J., M. Dredze, and F. Pereira. Biographies, bollywood, boomboxes and blenders: Domain adaptation for sentiment classification. In *Proceedings of Annual Meeting of the Association for Computational Linguistics (ACL-2007)*, 2007.

[8] Boyd-Graber, J. and P. Resnik. Holistic sentiment analysis across languages: multilingual supervised latent dirichlet allocation. In *Proceedings of the Human Language Technology Conference and the Conference on Empirical Methods in Natural Language Processing (HLT/EMNLP-2010)*, 2010.

[9] Breck, E., Y. Choi, and C. Cardie. Identifying expressions of opinion in context. In *Proceedings of the International Joint Conference on Artificial Intelligence (IJCAI-2007)*, 2007.

[10] Brody, S. and S. Elhadad. An unsupervised aspect-sentiment model for online reviews. In *Proceedings of The 2010 Annual Conference of the North American Chapter of the ACL*, 2010.

[11] Carenini, G., R. Ng, and A. Pauls. Multi-document summarization of evaluative text. In *Proceedings of the European Chapter of the Association for Computational Linguistics (EACL-2006)*, 2006.

[12] Carenini, G., R. Ng, and E. Zwart. Extracting knowledge from evaluative text. In *Proceedings of Third Intl. Conf. on Knowledge Capture (K-CAP-05)*, 2005.

[13] Choi, Y., C. Cardie, E. Riloff, and S. Patwardhan. Identifying sources of opinions with conditional random fields and extraction patterns. In *Proceedings of the Human Language Technology Conference and the Conference on Empirical Methods in Natural Language Processing (HLT/EMNLP-2005)*, 2005.

[14] Choi, Y. and C. Claire. Adapting a polarity lexicon using integer linear programming for domain-specific sentiment classification. In *Proceedings of the Conference on Empirical Methods in Natural Language Processing (EMNLP-2009)*, 2009.

[15] Dave, K., S. Lawrence, and D. Pennock. Mining the peanut gallery: Opinion extraction and semantic classification of product reviews. In *Proceedings of International Conference on World Wide Web (WWW-2003)*, 2003.

[16] Ding, X. and B. Liu. Resolving object and attribute coreference in opinion mining. In *Proceedings of International Conference on Computational Linguistics (COLING-2010)*, 2010.

[17] Ding, X., B. Liu, and P. Yu. A holistic lexicon-based approach to opinion mining. In *Proceedings of the Conference on Web Search and Web Data Mining (WSDM-2008)*, 2008.

[18] Ding, X., B. Liu, and L. Zhang. Entity discovery and assignment for opinion mining applications. In *Proceedings of ACM SIGKDD International Conference on Knowledge Discovery and Data Mining (KDD-2009)*, 2009.

[19] Esuli, A. and F. Sebastiani. Determining term subjectivity and term orientation for opinion mining. In *Proceedings of Conf. of the European Chapter of the Association for Computational Linguistics (EACL-2006)*, 2006.

[20] Esuli, A. and F. Sebastiani. Determining the semantic orientation of terms through gloss classification. In *Proceedings of ACM International Conference on Information and Knowledge Management (CIKM-2005)*, 2005.

[21] Esuli, A. and F. Sebastiani. SentiWordNet: a publicly available lexical resource for opinion mining. In *Proceedings of Language Resources and Evaluation (LREC-2006)*, 2006.

[22] Freitag, D. and A. McCallum. Information extraction with HMM structures learned by stochastic optimization. In *Proceedings of National Conf. on Artificial Intelligence (AAAI-2000)*, 2000.

[23] Gamon, M., A. Aue, S. Corston-Oliver, and E. Ringger. Pulse: Mining customer opinions from free text. *Advances in Intelligent Data Analysis VI*, 2005: p. 121-132.

[24] Ganapathibhotla, M. and B. Liu. Mining opinions in comparative sentences. In *Proceedings of International Conference on Computational Linguistics (COLING-2008)*, 2008.

[25] Ghahramani, Z. and K. Heller. Bayesian sets. *Advances in Neural Information Processing Systems (NIPS-2005)*, 2005.

[26] Ghose, A. and P. Ipeirotis. Designing novel review ranking systems: predicting the usefulness and impact of reviews. In *Proceedings of the International Conference on Electronic Commerce*, 2007.

[27] Guo, H., H. Zhu., X. Zhang., and Z. Su. OpinionIt: a text mining system for cross-lingual opinion analysis. In *Proceedings of ACM International Conference on Information and knowledge management (CIKM-2010)*, 2010.

[28] Hassan, A., V. Qazvinian., D. Radev. What's with the attitude? identifying sentences with attitude in online discussion. In *Proceedings of the Conference on Empirical Methods in Natural Language Processing (EMNLP-2010)*, 2010.

[29] Hatzivassiloglou, V. and K. McKeown. Predicting the semantic orientation of adjectives. In *Proceedings of Annual Meeting of the Association for Computational Linguistics (ACL-1997)*, 1997.

[30] Hatzivassiloglou, V. and J. Wiebe. Effects of adjective orientation and gradability on sentence subjectivity. In *Proceedings of International Conference on Computational Linguistics (COLING-2000)*, 2000.

[31] Hu, M. and B. Liu. Mining and summarizing customer reviews. In *Proceedings of ACM SIGKDD International Conference on Knowledge Discovery and Data Mining (KDD-2004)*, 2004.

[32] Huang, X. and W. B. Croft. A unified relevance model for opinion retrieval. In *Proceedings of ACM International Conference on Information and knowledge management (CIKM-2009)*, 2009.

[33] Jakob, N. and I. Gurevych. Extracting opinion targets in a single- and cross-domain setting with conditional random fields. In *Proceedings of Conference on Empirical Methods in Natural Language Processing (EMNLP-2010)*, 2010.

[34] Jin, W. and H. Ho. A novel lexicalized HMM-based learning framework for web opinion mining. In *Proceedings of International Conference on Machine Learning (ICML-2009)*, 2009.

[35] Jin, W. and H. Ho. OpinionMiner: a novel machine learning system for web opinion mining and extraction. In *Proceedings of ACM SIGKDD International Conference on Knowledge Discovery and Data Mining (KDD-2009)*, 2009.

[36] Jindal, N. and B. Liu. Identifying comparative sentences in text documents. In *Proceedings of ACM SIGIR Conf. on Research and Development in Information Retrieval (SIGIR-2006)*, 2006.

[37] Jindal, N. and B. Liu. Mining comparative sentences and relations. In *Proceedings of National Conf. on Artificial Intelligence (AAAI-2006)*, 2006.

[38] Jindal, N. and B. Liu. Opinion spam and analysis. In *Proceedings of the Conference on Web Search and Web Data Mining (WSDM-2008)*, 2008.

[39] Jindal, N. and B. Liu. Review spam detection. In *Proceedings of International Conference on World Wide Web (WWW-2007)*, 2007.

[40] Jindal, N., B. Liu, and E. Lim. Finding unusual review patterns using unexpected rules. In *Proceedings of ACM International Conference on Information and Knowledge Management (CIKM-2010)*, 2010.

[41] Jo, Y. and A. Oh. Aspect and sentiment unification model for online review analysis. In *Proceedings of the Conference on Web Search and Web Data Mining (WSDM-2011)*, 2011.

[42] Joachims, T. Optimizing search engines using clickthrough data. In *Proceedings of the ACM Conference on Knowledge Discovery and Data Mining (KDD-2002)*, 2002.

[43] Kaji, N. and M. Kitsuregawa. Automatic construction of polarity-tagged corpus from HTML documents. In *Proceedings of COLING/ACL 2006 Main Conference Poster Sessions (COLING-ACL-2006)*, 2006.

[44] Kaji, N. and M. Kitsuregawa. Building lexicon for sentiment analysis from massive collection of HTML documents. In *Proceedings of the Joint Conference on Empirical Methods in Natural Language Processing and Computational Natural Language Learning (EMNLP-2007)*, 2007.

[45] Kamps, J., M. Marx, R. Mokken, and M. De Rijke. Using WordNet to measure semantic orientation of adjectives. In *Proc. of LREC-2004*, 2004.

[46] Kanayama, H. and T. Nasukawa. Fully automatic lexicon expansion for domain-oriented sentiment analysis. In *Proceedings of Conference on Empirical Methods in Natural Language Processing (EMNLP-2006)*, 2006.

[47] Kim, J., J. Li., and J. Lee. Evaluating multilanguage-comparability of subjective analysis system. In *Proceedings of Annual Meeting of the Association for Computational Linguistics (ACL-2010)*, 2010.

[48] Kim, S. and E. Hovy. Crystal: analyzing predictive opinions on the web. In Proceedings of the Joint Conference on Empirical Methods in Natural Language Processing and Computational Natural Language Learning (EMNLP/CoNLL-2007), 2007.

[49] Kim, S. and E. Hovy. Determining the sentiment of opinions. In *Proceedings of Interntional Conference on Computational Linguistics (COLING-2004)*, 2004.

[50] Kim, S., P. Pantel, T. Chklovski, and M. Pennacchiotti. Automatically assessing review helpfulness. In *Proceedings of the Conference on Empirical Methods in Natural Language Processing (EMNLP-2006)*, 2006.

[51] Ku, L., Y. Liang, and H. Chen. Opinion extraction, summarization and tracking in news and blog corpora. In *Proceedings of AAAI-CAAW-2006*, 2006.

[52] Lafferty, J., A. McCallum, and F. Pereira. Conditional random fields: probabilistic models for segmenting and labeling sequence data. In *Proceedings of International Conference on Machine Learning (ICML-2001)*, 2001.

[53] Lee, L. Measures of distributional similarity. In *Proceedings of Annual Meeting of the Association for Computational Linguistics (ACL-1999)*, 1999.

[54] Li, S., C. Lin, Y. Song, and Z. Li. Comparable entity mining from comparative questions. In *Proceedings of Annual Meeting of the Association for Computational Linguistics (ACL-2010)*, 2010.

[55] Li, X., L. Zhang, B. Liu, and S. Ng. Distributional similarity vs. PU learning for entity set expansion. In *Proceedings of Annual Meeting of the Association for Computational Linguistics (ACL-2010)*, 2010.

[56] Lim, E., V. Nguyen, N. Jindal, B. Liu, and H. Lauw. Detecting product review spammers using rating behaviors. In *Proceedings of ACM International Conference on Information and Knowledge Management (CIKM-2010)*, 2010.

[57] Lin, C. and Y. He. Joint sentiment/topic model for sentiment analysis. In *Proceedings of ACM International Conference on Information and Knowledge Management (CIKM-2009)*, 2009.

[58] Liu, B. Sentiment analysis and subjectivity. In Handbook of Natural Language Processing, Second Edition, N. Indurkhya and F.J. Damerau, Editors. 2010.

[59] Liu, B. Web Data Mining: Exploring Hyperlinks, Contents, and Usage Data. Second Edition, Springer, 2011.

[60] Liu, B., M. Hu, and J. Cheng. Opinion observer: analyzing and comparing opinions on the web. In *Proceedings of International Conference on World Wide Web (WWW-2005)*, 2005.

[61] Liu, J., Y. Cao, C. Lin, Y. Huang, and M. Zhou. Low-quality product review detection in opinion summarization. In *Proceedings of the Joint Conference on Empirical Methods in Natural Language Processing and Computational Natural Language Learning (EMNLP-CoNLL-2007)*, 2007.

[62] Liu, Y., X. Huang, A. An, and X. Yu. ARSA: a sentiment-aware model for predicting sales performance using blogs. In *Proceedings of ACM SIGIR Conf. on Research and Development in Information Retrieval (SIGIR-2007)*, 2007.

[63] Lu, Y., M. Castellanos, U. Dayal, and C. Zhai. Automatic construction of a context-aware sentiment lexicon: an optimization approach. In *Proceedings of International Conference on World Wide Web (WWW-2011)*, 2011.

[64] Lu, Y., P. Tsaparas., A. Ntoulas., and L. Polanyi. Exploiting social context for review quality prediction. In *Proceedings of International Conference on World Wide Web (WWW-2010)*, 2010.

[65] Lu, Y. and C. Zhai. Opinion integration through semi-supervised topic modeling. In *Proceedings of International Conference on World Wide Web (WWW-2008)*, 2008.

[66] Lu, Y., C. Zhai, and N. Sundaresan. Rated aspect summarization of short comments. In *Proceedings of International Conference on World Wide Web (WWW-2009)*, 2009.

[67] Mei, Q., X. Ling, M. Wondra, H. Su, and C. Zhai. Topic sentiment mixture: modeling facets and opinions in weblogs. In *Proceedings of International Conference on World Wide Web (WWW-2007)*, 2007.

[68] Melville, P., W. Gryc., R. D. Lawrence. Sentiment analysis of blogs by combining lexical knowledge with text classification. In *Proceedings of ACM SIGKDD International Conference on Knowledge Discovery and Data Mining (KDD-2009)*, 2009.

[69] Miller, G., R. Beckwith, C. Fellbaum, D. Gross, and K. Miller. WordNet: an on-line lexical database. *Oxford Univ. Press.*, 1990.

[70] Mishne, G. and N. Glance. Predicting movie sales from blogger sentiment. In *Proceedings of AAAI Symposium on Computational Approaches to Analysing Weblogs (AAAI-CAAW-2006)*, 2006.

[71] Mohammad, S., C. Dunne., and B. Dorr. Generating high-coverage semantic orientation lexicons from overly marked words and a thesaurus. In *Proceedings of the Conference on Empirical Methods in Natural Language Processing (EMNLP-2009)*, 2009.

[72] Mukherjee, A., B. Liu, J. Wang, N. Glance, and N. Jindal. Detecting group review spam. In *Proceedings of International Conference on World Wide Web (WWW-2011)*, 2011.

[73] Nishikawa, H., T. Hasegawa, Y. Matsuo, and G. Kikui. Optimizing informativeness and readability for sentiment summarization. In *Proceedings of Annual Meeting of the Association for Computational Linguistics (ACL-2010)*, 2010.

[74] Paltoglou, G. and M. Thelwall. A study of information retrieval weighting schemes for sentiment analysis. In *Proceedings of Annual Meeting of the Association for Computational Linguistics (ACL-2010)*, 2010.

[75] Pan, S., X. Ni, J. Sun, Q. Yang, and Z. Chen. Cross-domain sentiment classification via spectral feature alignment. In *Proceedings of International Conference on World Wide Web (WWW-2010)*, 2010.

[76] Pang, B. and L. Lee. Opinion mining and sentiment analysis. *Foundations and Trends in Information Retrieval*, 2(1-2): p. 1-135, 2008.

[77] Pang, B. and L. Lee. Seeing stars: Exploiting class relationships for sentiment categorization with respect to rating scales. In *Proceedings of Meeting of the Association for Computational Linguistics (ACL-2005)*, 2005.

[78] Pang, B., L. Lee, and S. Vaithyanathan. Thumbs up?: sentiment classification using machine learning techniques. In *Proceedings of Conference on Empirical Methods in Natural Language Processing (EMNLP-2002)*, 2002.

[79] Pantel, P., E. Crestan, A. Borkovsky, A. Popescu, and V. Vyas. Web-scale distributional similarity and entity set expansion. In *Proceedings of Conference on Empirical Methods in Natural Language Processing (EMNLP-2009)*, 2009.

[80] Parrott, W. Emotions in social psychology: Essential readings. *Psychology Press*, 2001.

[81] Paul, M., C. Zhai, and R. Girju. Summarizing contrastive viewpoints in opinionated text. In *Proceedings of Conference on Empirical Methods in Natural Language Processing (EMNLP-2010)*, 2010.

[82] Polanyi, L. and A. Zaenen. Contextual valence shifters. In *Proceedings of the AAAI Spring Symposium on Exploring Attitude and Affect in Text*, 2004.

[83] Popescu, A. and O. Etzioni. Extracting product features and opinions from reviews. In *Proceedings of Conference on Empirical Methods in Natural Language Processing (EMNLP-2005)*, 2005.

[84] Qiu, G., B. Liu, J. Bu, and C. Chen. Expanding domain sentiment lexicon through double propagation. In *Proceedings of International Joint Conference on Artificial Intelligence (IJCAI-2009)*, 2009.

[85] Qiu, G., B. Liu, J. Bu, and C. Chen. Opinion word expansion and target extraction through double propagation. *Computational Linguistics*, 2011.

[86] Qiu, L., W. Zhang., C. Hu., and K. Zhao. SELC: A self-supervised model for sentiment classification. In *Proceedings of ACM International Conference on Information and knowledge management (CIKM-2009)*, 2009.

[87] Riloff, E., S. Patwardhan, and J. Wiebe. Feature subsumption for opinion analysis. In *Proceedings of the Conference on Empirical Methods in Natural Language Processing (EMNLP-2006)*, 2006.

[88] Riloff, E. and J. Wiebe. Learning extraction patterns for subjective expressions. In *Proceedings of Conference on Empirical Methods in Natural Language Processing (EMNLP-2003)*, 2003.

[89] Seki, Y., K. Eguchi, N. Kando, and M. Aono. Opinion-focused summarization and its analysis at DUC 2006. In *Proceedings of the Document Understanding Conference (DUC)*, 2006.

[90] Stone, P. The general inquirer: a computer approach to content analysis. *Journal of Regional Science*, 8(1), 1968.

[91] Stoyanov, V. and C. Cardie. Partially supervised coreference resolution for opinion summarization through structured rule learning. In *Proceedings of Conference on Empirical Methods in Natural Language Processing (EMNLP-2006)*, 2006.

[92] Su, Q., X. Xu, H. Guo, Z. Guo, X. Wu, X. Zhang, B. Swen, and Z. Su. Hidden sentiment association in chinese web opinion mining. In *Proceedings of International Conference on World Wide Web (WWW-2008)*, 2008.

[93] Taboada, M., J, Brooke, M. Tofiloski, K. Voll, and M. Stede, Lexicon-based methods for sentiment analysis. *Computational Intelligence*, 2010.

[94] Takamura, H., T. Inui, and M. Okumura. Extracting semantic orientations of phrases from dictionary. In *Proceedings of the Joint Human Language Technology/North American Chapter of the ACL Conference (HLT-NAACL-2007)*, 2007.

[95] Tan, S., Y. Wang., and X. Cheng. Combining learn-based and lexicon-based techniques for sentiment detection without using labeled examples. In *Proceedings of ACM SIGIR Conference on Research and Development in Information Retrieval (SIGIR-2008)*, 2008.

[96] Tata, S. and B. Di Eugenio. Generating fine-grained reviews of songs from album reviews. In *Proceedings of Annual Meeting of the Association for Computational Linguistics (ACL-2010)*, 2010.

[97] Tesniere, L. Elements de syntaxe structurale: Pref. de Jean Fourquet. 1959: C. Klincksieck.

[98] Titov, I. and R. McDonald. A joint model of text and aspect ratings for sentiment summarization. In *Proceedings of Annual Meeting of the Association for Computational Linguistics (ACL-2008)*, 2008.

[99] Titov, I. and R. McDonald. Modeling online reviews with multigrain topic models. In *Proceedings of International Conference on World Wide Web (WWW-2008)*, 2008.

[100] Turney, P. Thumbs up or thumbs down?: semantic orientation applied to unsupervised classification of reviews. In *Proceedings of Annual Meeting of the Association for Computational Linguistics (ACL-2002)*, 2002.

[101] Wan, X. Co-training for cross-lingual sentiment classification. In *Proceedings of Annual Meeting of the Association for Computational Linguistics (ACL-2009)*, 2009.

[102] Wang, H., Y. Lu, and C. Zhai. Latent aspect rating analysis on review text data: a rating regression approach. In *Proceedings of ACM SIGKDD International Conference on Knowledge Discovery and Data Mining (KDD-2010)*, 2010.

[103] Wiebe, J. Learning subjective adjectives from corpora. In *Proceedings of National Conf. on Artificial Intelligence (AAAI-2000)*, 2000.

[104] Wiebe, J., R. Bruce, and T. O'Hara. Development and use of a gold-standard data set for subjectivity classifications. In *Proceedings of the Association for Computational Linguistics (ACL-1999)*, 1999.

[105] Wiebe, J. and E. Riloff. Creating subjective and objective sentence classifiers from unannotated texts. *Computational Linguistics and Intelligent Text Processing*, p. 486-497, 2005.

[106] Wiebe, J., T. Wilson, R. Bruce, M. Bell, and M. Martin. Learning subjective language. *Computational Linguistics*, 30(3): p. 277-308, 2004.

[107] Wiebe, J., T. Wilson, and C. Cardie. Annotating expressions of opinions and emotions in language. *Language Resources and Evaluation*, 39(2): p. 165-210, 2005.

[108] Wilson, T., J. Wiebe, and P. Hoffmann. Recognizing contextual polarity in phrase-level sentiment analysis. In *Proceedings of the Human Language Technology Conference and the Conference on Empirical Methods in Natural Language Processing (HLT/EMNLP-2005)*, 2005.

[109] Wilson, T., J. Wiebe, and R. Hwa. Just how mad are you? finding strong and weak opinion clauses. In *Proceedings of National Conference on Artificial Intelligence (AAAI-2004)*, 2004.

[110] Wilson, T., J. Wiebe, and R. Hwa. Recognizing strong and weak opinion clauses. *Computational Intelligence*, 22(2): p. 73-99, 2006.

[111] Wu, Y., Q. Zhang, X. Huang, and L. Wu. Phrase dependency parsing for opinion mining. In *Proceedings of Conference on Empirical Methods in Natural Language Processing (EMNLP-2009)*, 2009.

[112] Yang, H., L. Si, and J. Callan. Knowledge transfer and opinion detection in the TREC2006 blog track. In *Proceedings of TREC*, 2006.

[113] Yu, H. and V. Hatzivassiloglou. Towards answering opinion questions: separating facts from opinions and identifying the polarity of opinion sentences. In *Proceedings of Conference on Empirical Methods in Natural Language Processing (EMNLP-2003)*, 2003.

[114] Zhai, Z., B. Liu, H. Xu, and P. Jia. Grouping product features using semi-supervised learning with soft-constraints. In *Proceedings of International Conference on Computational Linguistics (COLING-2010)*, 2010.

[115] Zhai, Z., B. Liu., L. Zhang., H. Xu., and P. Jia. Identifying evaluative sentences in online discussions. In *Proceedings of National Conf. on Artificial Intelligence (AAAI-2011)*, 2011.

[116] Zhang, L. and B. Liu. Identifying noun product features that imply opinions. In *Proceedings of Annual Meeting of the Association for Computational Linguistics (ACL-2011)*, 2011.

[117] Zhang, M. and X. Ye. A generation model to unify topic relevance and lexicon-based sentiment for opinion retrieval. In *Proceedings of ACM SIGIR Conference on Research and Development in Information Retrieval (SIGIR-2008)*, 2008.

[118] Zhang, Z. and B. Varadarajan. Utility scoring of product reviews. In *Proceedings of ACM International Conference on Information and Knowledge Management (CIKM-2006)*, 2006.

[119] Zhao, W., J. Jiang, H. Yan, and X. Li. Jointly modeling aspects and opinions with a MaxEnt-LDA hybrid In *Proceedings of Conference on Empirical Methods in Natural Language Processing (EMNLP-2010)* , 2010.

[120] Zhuang, L., F. Jing, and X. Zhu. Movie review mining and summarization. In *Proceedings of ACM International Conference on Information and Knowledge Management (CIKM-2006)*, 2006.

Chapter 14

BIOMEDICAL TEXT MINING: A SURVEY OF RECENT PROGRESS

Matthew S. Simpson
Lister Hill National Center for Biomedical Communications
United States National Library of Medicine, National Institutes of Health
simpsonmatt@mail.nih.gov

Dina Demner-Fushman
Lister Hill National Center for Biomedical Communications
United States National Library of Medicine, National Institutes of Health
ddemner@mail.nih.gov

Abstract   The biomedical community makes extensive use of text mining technology. In the past several years, enormous progress has been made in developing tools and methods, and the community has been witness to some exciting developments. Although the state of the community is regularly reviewed, the sheer volume of work related to biomedical text mining and the rapid pace in which progress continues to be made make this a worthwhile, if not necessary, endeavor. This chapter provides a brief overview of the current state of text mining in the biomedical domain. Emphasis is placed on the resources and tools available to biomedical researchers and practitioners, as well as the major text mining tasks of interest to the community. These tasks include the recognition of explicit facts from biomedical literature, the discovery of previously unknown or implicit facts, document summarization, and question answering. For each topic, its basic challenges and methods are outlined and recent and influential work is reviewed.

Keywords: Biomedical information extraction, named entity recognition, relations, events, summarization, question answering, literature-based discovery

1. Introduction

The state of biomedical text mining is reviewed relatively regularly. The recent surveys [238, 237], special journal issues [85, 29], and books [12] in this area indicate that general-purpose text and data mining tools are not well-suited for the biomedical domain because it is highly specialized.

Despite the restricted nature of the domain, biomedical text mining is of interest not only to researchers but to the general public as well (perhaps unbeknownst to them). The recent biomedical advances that have prevented or altered the course of many diseases are undoubtedly valued by all. Progress in biomedicine is attributable to advances in the understanding of disease mechanisms and the societal and commercial value of researching these mechanisms as well as the approaches for the prevention and cure of diseases.

Biomedical text mining holds the promise of, and in some cases delivers a reduction in cost and an acceleration of discovery, providing timely access to needed facts and explicit and implicit associations among facts.

Due to the specific goals of biomedical text mining, biologists and clinicians are better positioned to define useful text mining tasks. Cohen and Hunter [33] note that the most fruitful approaches to biomedical text mining will combine the efforts and leverage the abilities of both biologists and computational linguists. Biologists and clinicians will leverage their ability to focus on specific tasks and experience in using the unparalleled publicly available domain-specific knowledge sources whereas text mining specialists will provide system components and design and evaluate methods.

The sheer size of the so-called bibliome (the entirety of the texts relevant to biology and medicine) dictates a stepwise approach to biomedical text mining. The goal of the first step is to reduce the set of text documents to be mined. This reduction is most commonly achieved using domain-specific information retrieval approaches, as described in *Information Retrieval: A Health and Biomedical Perspective* [65]. Alternatively, document sets can be selected using clustering and classification [98, 177, 22]. As discussed later in this chapter, the meaning and grammar of biomedical texts are so intertwined that all surveys dedicate a section to natural language preprocessing and grammatical analysis. However, this chapter presents these methods (e.g., tokenization, part-of-speech tagging, parsing, etc.) as needed to describe the reviewed text mining approaches.

This survey of recent advances in biomedical text mining begins with a discussion of the resources available for mining the biomedical literature. It then proceeds to describe the basic tasks of named entity recognition and relation and event extraction. The more complex tasks of summarization, question answering, and literature based discovery are described thereafter. The chapter concludes with a discussion of open tasks and potentially high-impact avenues for further development of the domain.

2. Resources for Biomedical Text Mining

The primary resource for biomedical text mining is obviously text, and this section introduces some widely-used text collections in the biomedical domain. Although text mining does not require the use of specialized or annotated corpora, manually annotated collections are often more useful than the original texts alone. For example, the original conception of literature-based discovery [189] was facilitated by the use of Medical Subject Headings (MeSH®), which are controlled vocabulary terms added to bibliographic citations during the process of MEDLINE® indexing. With the growth of publicly available annotated collections, the biomedical language processing community has begun focusing on common interchangeable annotation formats, guidelines, and standards, which this section also discusses. After describing these resources, the section concludes with a description of equally important lexical and knowledge-based repositories, widely-used biomedical text mining tools and frameworks, and registries that provide overviews and links to text collections and other resources.

2.1 Corpora

Whether text mining is viewed in the strict sense of discovery or in the broader sense that includes all text processing and retrieval steps leading towards discovery, MEDLINE was the first—and remains the primary—resource in biomedical text mining. The MEDLINE database contains bibliographic references to journal articles in the life sciences with a concentration on biomedicine, and it is maintained by the U.S. National Library of Medicine® (NLM®). The 2011 MEDLINE contains over 18 million references published from 1946 to the present in over 5,500 journals worldwide.

Abstracts of biomedical literature can be obtained in a variety of different ways. For text mining purposes, MEDLINE/PubMed® records can be downloaded using the Entrez Programming Utilities [131]. Alternatively, subsets of MEDLINE citations can be obtained from the archives of community-wide evaluations that use MEDLINE, as well as individual research groups that share their annotations. Such collections include the historic OHSUMED [200] set containing all MEDLINE citations in 270 medical journals published over a five-year period (1987–1991) and a more recent set of TREC Genomics Track data [201] that contains ten years of MEDLINE citations (1994–2003). Stand-off annotations supporting information retrieval relevance, document classification, and question answering are available for portions of these collections. Whereas TREC collections provide access to MEDLINE spans over a given time period, other collections are task-oriented. For example, the GENIA corpus [90] contains 1,999 MEDLINE abstracts retrieved using the MeSH terms "human," "blood cells," and "transcription factors." The GENIA corpus is currently the most thoroughly annotated collection of MEDLINE abstracts. It is annotated for part-of-speech, syntax, coreference, biomedical concepts and events, cellular localization, disease-gene associations, and pathways. In addition, the GENIA corpus is one of the three constituents of the BioScope corpus [217], which provides GENIA MEDLINE abstracts, five full-text articles, and a collection of radiology reports annotated with negation and modality cues as well as scope. Other topically-annotated collections of MEDLINE abstracts include the earlier BioCreAtIve collections [69, 97] and the PennBioIE corpus [105, 106]. The PennBioIE corpus contains 1100 abstracts for cytochrome P-450 enzymes and 1157 oncology abstracts with annotations for paragraphs, sentences, tokens, parts-of-speech, syntax, and biomedical entities. Finally, the Collaborative Annotation of a Large Biomedical Corpus (CALBC) initiative [26] has proposed the creation of a "silver standard" corpus that contains MEDLINE abstracts that have been automatically annotated with biomedical entities by the initiative participants. This corpus has just recently become publicly available.

Being informative and undoubtedly useful for text mining, MEDLINE abstracts do not contain all the information presented in full-text articles. Some information (e.g., the exact settings of an experiment or the discussion of the results) is almost exclusively contained in the body of an article. The promise of a qualitative increase in the amount of useful information brought about several full-text collections. For example, the TREC Genomics Track dataset contains about 160,000 full-text articles from about 49 genomics-related journals, which were obtained in HTML format from the Highwire Press [66] electronic distribution of the journals. Another collection of full-text articles annotated with relevance to patients' case descriptions was developed in the ImageCLEF evaluations [127, 84]. The Colorado Richly Annotated Full Text Corpus [38] adds to the growing body of semantically and syntactically annotated full text collections (including the full-text portion of the BioScope collection mentioned above). Finally, the largest publicly available source of original, full-text articles is the Open Access subset of PubMed Central [154].

With the growing interest in clinical text mining and biosurveillance, several public collections of clinical text have recently become available. These collections include reports in the Multiparameter Intelligent Monitoring in Intensive Care (MIMIC II) database [171], the Pittsburgh collection of clinical reports [211], and the annotated i2b2 collections [214, 213, 215, 216]. Several recent studies used the Web (i.e, Twitter and health-related blogs and community sites) as a corpus, but it is not clear if the collections created for these studies are publicly available or not.

2.2 Annotation

The annotation of biomedical text adds information to a document collection that can later be exploited for text mining purposes. In general, document annotation in the biomedical domain follows the principles set forth in open-domain natural language processing (NLP) by adding annotations at multiple levels of linguistic analysis. The various aspects involve grammatical (including morphological and syntactical), semantic, and pragmatic annotations [103]. Grammar and meaning are so intertwined that most annotation efforts combine the two. For example, corpus creators might decide to annotate named entities of interest only in noun phrases. As an alternative, Wilbur et al. [222] focus on annotating the "information-bearing fragments within scientific text" without specifying any grammar restrictions. The authors define the following five annotation axes: Focus, Polarity, Certainty, Evidence, and Direction. These classes are primarily used at the sentence level, and sentences may be broken as needed if a change in one of the annotations aspects is detected. However, even meaning-centric annotations cannot be completely grammar-free. For example, one of the clues for annotating fragments as *Evidence* is a past tense verb indicating an observation or finding. The guidelines published by the authors [178] are a good starting point for developing other text-mining annotation guidelines in the biomedical domain.

There are three approaches to the annotation of biomedical text. These methods include (1) a complete manual annotation that is based on annotators' knowledge; (2) an assisted annotation, in which the output of an annotation tool is manually corrected; and (3) an ontology-based annotation—either manual or assisted—in which only terms and relations present in an existing knowledge source are annotated. Each of these approaches has its strengths and weaknesses. For example, an assisted annotation is usually more consistent, but it may be biased. Similarly, an ontology-based annotation will likely be biased towards known facts. Having more than one annotator for each text document and having various annotator groups can compensate for such biases [15].

In addition to generic information extraction tools that can be used to assist in annotation (described below), several text mining tools have been developed to specifically support the annotation process. Examples of widely-used tools for annotating biomedical text include Knowtator [140] and eHOST (Extensible Human Oracle Suite of Tools) [48], the later of which is increasingly used for the annotation of clinical text. In order for such tools to be useful, they must be easy to use, support various annotation types, and allow collaborative annotation, among other factors [115, 47].

2.3 Knowledge Sources

The biomedical domain offers a rich set of knowledge sources supporting text mining applications. The Unified Medical Language System® (UMLS®) [111], a compendium of controlled vocabularies that is maintained by NLM, is the most comprehensive resource, unifying over 100 dictionaries, terminologies, and ontologies in its Metathesaurus. It also provides a semantic network that represents relations between Metathesaurus entries, a lexicon that contains lexicographic information about biomedical terms and common English words, and a set of lexical tools. Overall, NLM provides over 200 knowledge sources and tools that can be used for text mining [210]. Other sets of ontologies are maintained by collaborative effort in the OBO Foundry [143] and the National Center for Biomedical Ontology (NCBO) [129]. The NCBO ontologies are accessed and shared through BioPortal [130]. Other major centers that maintain specialized resources for biomedical text mining include the British National Centre for Text Mining [132] and the European Bioinformatics Institute [52].

In addition to these broad-coverage resources, the biomedical domain offers in-depth knowledge sources focused on specific subdomains of biomedicine. For example, the Pharmacogenomics Knowledge Base [152] is a collection of scientific publications annotated with primary genotype and phenotype data, gene variants, and gene-drug-disease relationships. The annotations are downloadable for individual research purposes. Another specialized source, the Neuroscience Information Framework [134], includes an ontology covering brain anatomy, cells, organisms, diseases, techniques, and other areas of neuroscience.

The best knowledge source for a given text mining task is determined by the nature of the problem at hand. For example, mining the scientific literature for relations between genes, diseases, and drugs first requires recognizing instances of these entities. To aid in this task, a researcher might rely on knowledge of the terms' corresponding semantic types in the UMLS or instead may chose to use individual knowledge sources, such as the Gene Ontology [16], SNOMED Clinical Terms® [188], or the FDA Approved Drug Products with Therapeutic Equivalence Evaluations (Orange Book) [209]. Approaches to the various text mining tasks in the biomedical domain make extensive use of the resources described in this section and sometimes derive meta-resources for a specific task. For example, Rinaldi et al. [163] define several entity types needed for mining the literature for protein interactions (protein/gene names, chemical compounds, cell lines, etc.) and then automatically aggregate terms extracted from curated resources such as the UMLS, Affymetrix identifiers for micro array probes, organism databases, and others into a list of 2,347,734 terms.

2.4 Supporting Tools

The variety and purpose of the tools supporting biomedical text mining echoes that of the knowledge sources described above. The following discussion of text mining tools omits applications described in recent surveys and instead focuses on the basic, widely used tools for identifying named entities and relations and the platforms that allow building text mining pipelines.

The most widely used tool for named entity recognition that is based upon the UMLS is MetaMap [14]. MetaMap is a highly configurable application that identifies UMLS Metathesaurus concepts in free text. Because MetaMap provides a wide range of configuration options and relies on the entire UMLS Metathesaurus, it is not easy to determine the best configuration for a given task. However, exploring the options using the interactive MetaMap website may aid with such choices. MetaMap, which was provided as service until recently, is now open source and available for download. Two statistical tools widely used for biological named entity recognition are ABNER [176] and BANNER [101]. Both ABNER and BANNER are based on conditional random fields and rely on a wide array of features. Unlike ABNER, BANNER avoids semantic features, but it uses syntactic features. Both systems exploit such domain-specific language characteristics as capitalization, word shapes, prefixes, suffixes, and Greek letters.

Tools for relation extraction are not yet as readily accessible as entity recognition tools. Kabiljo et al. [83] compared available tools for identifying biomedical relations (AkanePPI, Whatizit, and OpenDMAP) to a simple, regular expression-based approach and found that the simple approach performed surprisingly well. The authors conclude that high recall (around 90%) is achievable for extracting gene-protein relations when the available tools are combined.

A recent trend in tool development and use is the assemblage of pipelines based on open-source frameworks, such as the Generalized Architecture for Text Engineering (GATE) [39] and the Unstructured Information Management Architecture (UIMA) [54]. The most mature system for clinical text processing (ranging from identifying patients' problems to events) is MedLEE [58]. Descriptions of other systems and clinical text mining tasks can be found in a recent review [41].

This section has presented only a snapshot of open-domain biomedical text mining resources. By its nature, the information contained herein will become dated sooner than the other material presented in this chapter. To compensate for the rapid progress of research related to biomedical text mining, many researchers maintain websites with links to useful resources (e.g., BioNLP [19]). Realizing that this task is too time consuming for individual researchers, the U.S. Department of Veterans Affairs and NLM provide a registry of biomedical text mining tools, known as ORBIT, which is maintained by the research community [144].

3. Information Extraction

A goal of many biomedical text mining tasks is the identification of explicitly stated facts. *Information extraction* refers to the process by which structured facts are automatically derived from unstructured or semi-structured text. In the biomedical domain, unstructured text commonly includes scientific articles appearing in the biomedical literature as well as clinical narratives found in electronic health records or other clinical information systems. Although the information extracted from these sources can be the target of information retrieval systems, information extraction is often performed as an initial processing step for other biomedical text ming applications (Sections 4-6).

Biomedical information extraction technology has undergone rapid development in recent years, spurred in part by community-wide evaluations that have been focussed specifically on text mining within the biomedical domain. Some examples of recent evaluation forums include BioCreAtivE [69, 97], BioNLP [89, 88], i2b2 [214, 213, 215, 216], JNLPBA [91], and LLL [133] shared tasks. The strong interest in community-wide evaluation efforts such as these is reflective of the growing volume of unstructured biomedical text available electronically in databases such as MEDLINE or in clinical information systems.

Three major subtasks of information extraction are particularly relevant for processing biomedical text. First, named entity recognition is a task that seeks to identify and classify biomedical entities into predefined categories such as the names of proteins, genes, or diseases. Often, extracted entities are normalized to canonical, unambiguous representations with the aid of ontological resources and further classified into semantic categories. The second subtask of information extraction relevant to the biomedical domain is relation extraction, which aims to detect binary relationships among named entities. Examples include gene-disease relationships, protein-protein interactions, and medical problem-treatment relationships. Finally, the third major subtask, event extraction, seeks to identify highly complex relations among extracted entities. Events relevant to the biomedical domain include, for example, gene expression and regulation and protein binding.

Although each of these subtasks are distinct in the type of information they aim to extract, they achieve their goals by employing similar methods, which include machine learning, statistical analysis and other techniques of natural language processing. Challenges and approaches to the subtasks of biomedical information extraction are discussed below.

3.1 Named Entity Recognition

Biomedical Named Entity Recognition (NER) refers to the task of automatically identifying occurrences of biological or medical terms in unstructured text. Common entities of interest include gene and protein names, medical problems and treatments, drug names and their dosages, and other semantically well-defined data classifiable within the biomedical domain [104]. Although commonly discussed as a single task, NER is typically a three-step process that involves determining an entity's substring boundaries within the text, assigning the entity to a predefined class or category, and selecting the preferred name or unique identifier of the concept that the entity names. This last subtask, entity normalization, is sometimes addressed as a separate problem from NER, but it is briefly discussed here in the context of describing the many issues that make NER a challenging task in the biomedical domain.

NER is particularly challenging for biomedical text due to a variety of reasons. The most basic obstacle results from the dynamic nature of scientific discovery. In the biomedical domain, there exists a vast amount of semantically relevant entities that is constantly and rapidly increasing as new scientific discoveries are made [226]. This ever-growing list of relevant terms is problematic for NER systems that rely only on a dictionary of known terms or other curated resources to identify named entities since these resources can never be complete as long as scientific progress continues.

Another challenge to biomedical NER is synonymy. In biomedical literature, the same concept may be expressed using different words. For example, "heart attack" and "myocardial infarction" refer to the same medical problem so an NER system should recognize these terms as instances of the same concept, despite being expressed differently. When many synonyms for a particular concept are in use, it becomes difficult to integrate knowledge from multiple sources without a comprehensive synonymy resource such as the UMLS Metathesaurus or Gene Ontology. However, given the rapidly increasing number of biomedical entities, these resources are unlikely to be complete at any given moment, resulting in some synonymy relationships that may not be captured.

Finally, the abundant use of acronyms and abbreviations in biomedical literature make it difficult to automatically identify the concepts to which these terms refer. Often, successful acronym and abbreviation resolution depends greatly on the context in which the terms appear since the same term can refer to different concepts. For example, the abbreviation *RA* can refer to "right atrium," "rheumatoid arthritis," "refractory anemia," "renal artery," or one of several other concepts [148]. To address the challenges associated with acronyms, abbreviations, and synonymy, NER systems typically perform some form of entity normalization.

Entity normalization is a subtask of NER and refers to the process of mapping entity occurrences to their canonical, preferred names. Although a challenging task itself, entity normalization can help resolve issues resulting from synonymous terms and ambiguous acronyms and abbreviations by associating these entities with unique, unambiguous representations. Often, since there may not be community-wide agreement on the preferred name for a given entity, the goal of entity normalization is to map an entity instance to the unique identifier of a concept in a terminology resource. In general, entity normalization requires the existence of such terminology resources, though they may be incomplete. Since normalization is such a crucial component of many NER systems, it is often an implied processing step after identifying entity boundaries and assigning them to a category. However, the entity normalization subtask may be evaluated independently of these subtasks, as was the case in recent BioCreAtivE shared task evaluations [67, 126].

For NER systems that analyze large amounts of biomedical text, it is important to consider the quality that can be expected of the methods being utilized. Typically, the performance of NER systems is measured in terms of precision, recall, and F-score. However, a variety of issues make these measurements difficult to reliably obtain and compare.

One issue is the availability of large, high-quality annotated corpora to serve as the ground truth on which to base NER system evaluations. The ground truth corpora must be large enough to allow the extrapolation of experimental results to large text collections, such as the entirety of MEDLINE, and the annotations should exhibit high inter-annotator agreement and reflect expert-level judgement. However, while the size of a ground truth data set is crucial, annotation errors do not necessarily pose an insurmountable problem to system evaluation, especially if the data set is sufficiently large. For example, Uzuner et al. [216] demonstrated that errors in the ground truth for a recent i2b2 shared task evaluation could affect the relative performance of competing NER systems by 0.05% at most.

Another issue to consider when evaluating NER systems is how to define the boundaries of a correctly identified entity. A strict evaluation requires both the left and right boundaries of an extracted entity to exactly match those of the ground truth annotations while a loose evaluation requires only that the extracted entity boundaries overlap those of the annotations [104]. Olsson et al. [142] showed that the choice of a strict or loose evaluation affects the relative performance of NER systems and suggested several scoring criteria for different application needs.

Recent community-wide evaluations have demonstrated that NER systems are typically capable of achieving favorable results. For example, the best performing systems achieved F-scores of 0.83 and 0.87 for the first [226] and second [187] BioCreAtIve gene mention recognition tasks, 0.85 for the i2b2 concept extraction task [216], and 0.73 for the JNLPBA bio-entity recognition task [91]. Although NER systems may be tailored for a particular information extraction task, their primary methods can broadly be grouped as following one of several basic approaches, which are discussed below.

Dictionary-based methods, one of the most basic biomedical NER approaches, utilize comprehensive lists of biomedical terms in order to identify entity occurrences in text. Such systems determine whether a word or group of words selected from the text exactly matches a term from some biomedical resource. When used as stand-alone methods, dictionary-based approaches generally exhibit reasonably high precision, but they suffer from poor recall due to the existence of spelling mistakes and morphological variants [207]. However, low precision is also possible due to homonymy [68]. For example, many gene names and abbreviations (e.g. "an," "by," and "can") share lexical representations with common English words [99]. For these reasons, some form of inexact string matching is commonly utilized to improve the precision and recall of dictionary-based approaches. Some methods improve performance by first generating spelling variants for the terms in a biomedical resource, and then by appending these additional terms to the underlying word lists [205, 204]. The methods are then able perform exact matching using the augmented resource. Other methods utilize algorithms such as BLAST® [10, 11] to perform approximate string matching instead of exact matching [100]. Despite these improvements, dictionary-based methods are most often used in conjunction with more advanced NER approaches.

Another approach to NER is to define rules that describe the composition patterns of named biomedical entities and their context. Examples of rule-based approaches include the EMPathIE and PASTA systems [78, 61], which use context free grammars that recognize enzyme interactions and protein structures. Other systems utilize pattern-based rules that exploit the orthographic and lexical characteristics of targeted entity classes in order to recognize protein [59] and chemical [128] names. These simpler methods may be improved by additionally considering contextual information [70] and the results of syntactic parsing for determining entity boundaries [57]. However, while rule-based approaches typically achieve better performance than dictionary-based approaches, manual generation of the required rules is a time-consuming process, and, since the rules are usually very specific in order to achieve high precision, they are difficult to extend to other entity classes.

It is increasingly common for NER approaches to rely on statistical methods instead of, or in combination with, dictionary- and rule-based approaches. Unlike the previously described approaches, statistical methods typically rely on some form of machine learning algorithm to identify biomedical entities. While supervised machine learning approaches must be trained with observations taken from large annotated corpora, recent work has investigated the automatic generation of training data for the NER task through the use of bootstrapping and other semi-supervised statistical techniques [218, 125, 212]. Common statistical methods used for NER can be grouped as either classification- or sequence-based approaches.

Classification-based approaches transform the NER task into a classification problem, which can either be applied to individual words or groups of words. Common classifiers used for biomedical NER include

Naïve Bayes [139] and Support Vector Machine (SVM) [86, 118, 196, 224] classifiers. Although it is possible to classify multi-word phrases, a popular approach follows the BIO tagging scheme [157], where individual tokens are classified as being at the beginning (B) of an entity, inside (I) the boundaries of an entity, or outside (O) the boundaries of an entity. However, despite its success, this tagging scheme can be problematic if entity boundaries overlap, and several authors have addressed the problem of recognizing nested biomedical entities [62, 8]. The performance of classification-based approaches is highly dependent on the choice of features used for training, and many authors have explored various feature combinations. For example, Kazama et al. [86] and Mitsumori et al. [118], consider morpho-syntactic properties of named entities, Takeuchi and Collier [196] consider orthographic and head-noun features, and Yamamoto et al. [224] explore a variety of features encompassing boundary, morpho-lexical, and syntactic properties as well as a dictionary-based feature that indicates whether a word appears in a biomedical resource. Given the sensitivity of classification-based approaches to the choice of features, automatic feature selection is an important consideration. Hakenberg et al. [63] perform a systematic evaluation of common features and discuss their influence on the predictive quality of classification-based NER systems.

Unlike classification-based approaches, sequence-based NER systems consider complete sequences of words instead of only individual words or phrases. They are trained on tagged corpora and aim to predict the mostly likely tags for a given sequence of observations. A common statistical framework used for biomedical NER is the Hidden Markov Model (HMM) [36, 179, 124, 93]. Methods based on the Maximum Entropy Markov Model are also common [55, 37]. However, Conditional Random Fields (CRF) [141, 175] are often demonstrated to be superior statistical frameworks for biomedical NER. For example, CRFs were utilized by the best performing system on the i2b2 medical concept extraction task [216] and by highly ranked systems on the BioCreAtIve gene mention recognition tasks [226, 187] and the JNLPBA bio-entity recognition task [91]. Like other statistical methods, sequence-based approaches can be trained on a variety of features including orthographic features [36, 124], prefix and suffix information [179], and part-of-speech tag sets augmented to include tags for entity classes [93].

Many approaches do not just utilize a single method for performing biomedical NER and instead rely on multiple techniques and various resources. These hybrid approaches are often quite successful at combining dictionary- or rule-based approaches with statistical methods. As evidence of the advantages of hybrid approaches, Abacha et al. [2] compared the performance of common rule-based and statistical approaches to medical entity recognition and concluded that hybrid approaches utilizing machine learning and domain knowledge perform best. There are numerous hybrid biomedical NER systems. For example, Sasaki et al. [173] use a dictionary-based approach to identify known protein names in parallel with part-of-speech tagging. They then use a CRF-based approach to reduce the number of false positives and false negatives in the resulting tagged sequence. Other methods create meta-learners from multiple statistical methods. For example, Zhou et al. [236] utilize a meta-learner composed of two HMMs trained on different corpora whose outputs are combined with one SVM to recognize protein and gene names. Similarly, Mika and Rost [117] compose a meta-learner to recognize protein names from three SVMs trained on different copora and feature sets whose outputs are then combined with a fourth SVM. Finally, Cai and Cheng [25] present an approach to biomedical NER that utilizes three different classifiers to improve the generalization ability of the system.

A more thorough analysis of NER approaches in the biomedical domain can be found in the several literature surveys dedicated to the subject [99, 104].

3.2 Relation Extraction

Most information extraction tasks in the biomedical domain go beyond simply identifying named entities and, in addition, involve determining relationships among those entities. In their simplest form, associations among biomedical entities are binary, involving only the pair-wise relations between two entities. However, biomedical relationships can involve more than just two entities, and these complex associations are discussed later with the event extraction task. The goal of the relation extraction task, therefore, is to identify occurrences of particular types of relationships between pairs of given entities. Although common entity classes (e.g., genes or drugs) are generally quite specific, the types of identified relationships may be broad, including any type of biomedical association, or they may be specific, for example, by characterizing only gene regulatory associations.

A variety of biomedical relations have been the subject of information extraction tasks in the literature. In the current genomic era, much of this work has focussed on automatically extracting interactions between genes and proteins. In particular, because of its critical role in understanding biological processes, Protein-Protein Interaction (PPI) has been one of the most widely researched topics in biomedical information extraction. Other associations of interest include interactions between proteins and point mutations [102], proteins and their binding sites [28], genes and diseases [31], and genes and phenotypic context [113]. In the clinical domain, relationships between patients' presented medical problems and the tests or treatments they may undergo [216] is an increasingly important type of relation, especially considering the growing prominence of electronic health record systems.

Biomedical relation extraction faces many of the same challenges as NER, including the creation of high quality annotated corpora for training and evaluating relation extraction systems. Compared with the annotation of named entities, the annotation of relations is considerably more complicated since relations are generally expressed as discontinuous spans of text and the types of relations considered are usually application-specific [13]. Additionally, since there is often little consensus regarding how to best annotate given types of relations, the resulting resources are largely incompatible, and, as a result, the quality of the methods utilizing these resources is difficult to evaluate. For example, Pyysalo et al. [155] performed a comparative analysis of five PPI corpora and found that the performance of state-of-the-art PPI extraction systems, measured in terms of F-score, varied on average by 19 percentage points and by as much as 30 percentage points on the evaluated corpora. Participation in community-wide evaluations that are dedicated to the relation extraction task is indispensable for obtaining annotated corpora.

Relation extraction tasks have been a component of several recent evaluation forums, and these tasks include the LLL genic interaction challenge [133], the BioCreAtIve PPI extraction task [96], and the i2b2 relation extraction task [216]. The purpose of the LLL challenge was to extract protein and gene relationships from abstracts contained in MEDLINE, and the best-performing system achieved an F-score of 0.54 identifying these associations. The BioCreAtIve task consisted of four subtasks related to PPI extraction. These challenges included the classification of PubMed abstracts as to whether they were relevant for PPI annotation, the identification of binary protein-protein interactions from full-text articles, the extraction of protein interaction methods, and the retrieval of textual evidence describing the interactions. The best-performing system achieved a precision of 0.37 at recall 0.33 for extracting binary PPI relations. Finally, the aim of the i2b2 relation extraction challenge was to identify medical problem-treatment, problem-test, and problem-problem relationships in clinical notes. Participants were tasked, for example, with determining whether two co-occurring problem and treatment concepts were related, and if so, whether the patient's treatment improved, worsened, or caused the medical problem. The best-performing system on the i2b2 relation extraction challenged achieved an F-score of 0.74. Like the forums dedicated to evaluating the NER task, community-wide evaluations such as these have been instrumental in the development and evolution of relation extraction approaches.

Relation extraction approaches have shown an evolution from simple systems that rely solely on co-occurrence statistics to complex systems utilizing syntactic analysis and dependency parsing. Some recent approaches to the relation extraction task are described below. An accounting of additional methods can be found in other biomedical text mining surveys that cover the relation extraction task [13, 32, 238].

The simplest method of identifying relations between biomedical entities is to collect instances where the entities co-occur. If the entities are repeatedly mentioned together, then there is a greater chance that they may be related in some way, although the type and direction of this relation typically cannot be determined by co-occurrence statistics alone. For example, Chen et al. [30] apply co-occurrence statistics to compute the degree of association between diseases and drugs extracted from clinical records and biomedical literature. Co-occurrence approaches commonly exhibit high recall and low precision.

Rule-based approaches describe the linguistic patterns exhibited by particular relations. Unlike the systems based on term co-occurrences, rule-based approaches typically demonstrate high precision and low recall. The rules used for relation extraction can be manually defined by domain experts [172], or they can be derived from annotated copora by machine learning algorithms [64].

Classification-based approaches are also commonly used to identify relations, particularly those involving medical entities. Roberts et al. [168] describe a supervised machine learning system, trained on shallow features extracted from oncology reports, that detects various clinical relationships in patient narratives. Similarly, Rink et al. [167] describe a system that discovers relations between medical problems, treatments, and tests mentioned in electronic medical records. The system relies on supervised machine learning and lexical, syntactic, and semantic context features. Bundschus et al. [23] utilize CRFs to identify and classify relations between diseases and treatments extracted from PubMed abstracts and relations between genes and diseases in the human GeneRIF database. Finally, Abach and Zweigenbaum [1] describe a hybrid approach that utilizes patterns developed by domain experts as well as SVM classification to extract relations that occur between diseases and treatments in medical texts.

An important advance in the evolution of relation extraction methods has been the consideration of syntactic structures. In particular, dependency parsing is capable of producing informative syntactic descriptions of biomedical text, in the form of dependency trees or graphs, which encode grammatical relations between phrases or words. Fundel et al. [60] produce dependency trees from MEDLINE abstracts. Their system then applies three relation extraction rules to the syntactic structures in order to identify gene and protein associations. Similarly, Rinaldi et al. [164] combine syntactic patterns obtained from dependency tree structures in order to support querying the biomedical literature for interactions between genes and proteins. Miyao et al. [121] perform deep parsing to annotate predicate-argument structures in MEDLINE abstracts. Their system then relies on the structural matching of the semantic annotations to identify and retrieve relational concepts. In other work, Miyao et. al [122] evaluate various parsers and their output representations on their ability to improve accuracy when used as a component of a PPI extraction system.

With the growing availability of large corpora containing relational annotations, many approaches utilize machine learning algorithms to extract useful information from syntactic structures rather than applying manually derived patterns. In the context of kernel-based machine learning, several authors have proposed kernels capable of measuring the similarity between syntactic parse trees or graphs. Airola et al. [7] describe an all-paths graph kernel for computing the similarity between dependency graphs. The kernel function is then used in training a least squares support vector machine to identify protein-protein interactions. Kim et al. [92] suggest four genic relation extraction kernels defined on the shortest syntactic dependency path between two named entities. Finally, Miwa et al. [120] describe a framework for combining the outputs of multiple kernels and syntactic parsers to extract protein-protein interactions.

Syntactic analysis is often complemented by semantic role labeling, a natural language processing technique that identifies the semantic roles of words or phrases in sentences and expresses them as predicate-argument structures. Tsai et al. [202] construct a role labeling system that uses a maximum entropy machine learning model to extract biomedical relations from a prepared portion of the GENIA corpus. As discussed below, the annotation of semantic roles for named biomedical entities has enabled the extraction of a variety of complex entity associations.

3.3 Event Extraction

Recently, there has been a shift in biomedical information extraction from recognizing binary relations to the more ambitious task of identifying complex, nested event structures. Events are typically characterized by verbs or nominalized verbs. For example, in the sentence "*glnAP2* may be activated by *NifA*," the verb *activated* specifies the event, and *glnAP2* and *NifA* are the event's arguments. Unlike the case of simple binary relations, both concept labels and semantic roles are assigned to an event and its arguments. In this example, the verb *activated* indicates a positive regulation type event, which expects a protein (*NifA*) to act as the event's cause and a gene (*glnAP2*) to act as the event's theme [13].

Another important distinction between the extraction of binary relations and complex events is that events can be nested, with one event functioning as a participant of another event. For example, in the sentence "RFLAT-1 activates RANTES gene expression" two events are present [13]. One event is indicated by the nominalized verb *expression* whose theme is *RANTES*, a gene, and the other event is indicated by the verb *activates* whose cause is *RFLAT-1*, a protein, and whose theme is the gene expression event itself. Thus, event representations, unlike binary relations, are capable of capturing many different types of associations with an arbitrary number of entities and events related by a variety of semantic roles.

Due to the complexity of biomedical events, effective event extraction typically requires a thorough analysis of sentence structure. Event extraction is particularly aided by the use of semantic processing and deep parsing techniques, which are capable of analyzing both the syntactic and semantic structure of biomedical text. Dependency parsing is an especially useful technique for capturing semantics such as predicate-argument relationships, which have been shown to be an effective representation for event extraction [219]. Despite the complexity of the task, event extraction has broad applicability in the biomedical domain, and it is increasingly being used for the annotation of biomedical pathways, Gene Ontology annotation, and the enrichment of biological databases.

The growing interest in event extraction has largely been driven by the introduction, mostly in the domain of systems biology, of corpora containing the annotations necessary for the training and evaluation of statistical event extraction methods. The BioInfer corpus [156] was the first publicly available corpus in the biomedical domain to incorporate event annotations. Other annotated event corpora include the GENIA Event Corpus [92] and the Gene Regulation Event Corpus [198]. Notably, the GENIA corpus remains one of the most widely used resources in biomedical text mining, and the data for the BioNLP shared tasks on event extraction [89, 88] were prepared based on this resource.

The BioNLP '09 shared task [89] was the first-of-its-kind community-wide evaluation of event extraction methods. The primary challenge was to extract event types related to protein biology from MEDLINE abstracts. Targeted event types included, among others, gene expression, transcription, localization, binging, and regulation. The binding event type was more complex than the others since it required the detection of an arbitrary number of arguments, and the regulation event types were notable for allowing other events to act as their cause or theme. The best-performing system obtained an F-score of 0.52 on the primary event extraction task. The BioNLP '11 shared task [88] repeated the evaluation from the previous meeting, but also included additional tasks targeting event types in other subdomains of biology. On the subtask comparable with that of that of the first meeting, the best-performing system achieved an F-score of 0.57, which demonstrated a significant improvement in the community. Successful systems at the BioNLP shared task meetings relied on a variety of techniques including machine learning, Markov logic networks, and dependency parsing. Several approaches to biomedical event extraction are described below.

Most event extraction systems follow a pipelined approach that divides the task into a sequence of three stages. Fist, the systems predict a candidate set of event trigger words. Trigger words are often the verbs or nominalized verbs that indicate a particular event type, such as "phosphorylation," "activates," or "inhibits." Then, the systems seek to determine whether any recognized named entities or trigger words are instantiations of event arguments. The final stage in the process is a semantic post-processing step that attaches arguments to event triggers following constraints on the type and number arguments allowable for a given event type.

This basic architecture is a common approach to the event extraction task. Björne et al. [21] describe the best-performing system on the BioNLP '09 event extraction task. Their method trains separate multi-class SVMs for detecting event triggers and arguments using an extensive set of features, especially those derived from dependency parse graphs. Their system then uses a rule-based approach for attaching arguments to their corresponding events. This approach has been combined with BANNER to perform event extraction on an unlabeled subset of citations from PubMed [20]. Miwa et al. [119] describe an event extraction approach similar to that of Björne et al., but instead of relying on a rule-based approach to attach event participants to trigger words, they obtain an improvement by utilizing a classifier and additional features for this step. Buyko et al. [24] describe a system that relies on a dictionary-based approach to identify event triggers and an ensemble of feature- and kernel-based classifiers trained using "trimmed" dependency graphs to identify event participants. Kilicoglu and Bergler [87] also use a dictionary-based approach to identify event riggers, but they develop rules based on syntactic dependency paths to detect event participants. Finally, Cohen et al. [34] describe a pattern-based approach to event extraction that utilizes the OpenDMAP system [79] to define entity and event types as well as the constraints on event arguments.

Recently, joint prediction approaches have been proposed that seek to overcome the problem of cascading errors, which some of the above approaches allow. For example, by separating the event trigger and argument detection tasks, a system may not correctly extract an event if it fails to detect a trigger word in the first stage of the process. Poon and Vanderwende [153] propose a method based on Markov logic networks that jointly predicts events and arguments. For each word, the system predicts whether it is an event trigger word, and for each syntactic dependency edge, the system predicts whether it is an argument path leading to an event theme or cause. Additionally, Riedel and McCallum [161] propose a family of three joint prediction models based on Markov logic that are less computationally complex than previous work [160] and lead to better event extraction results.

4. Summarization

Information extraction techniques are often utilized as a first step in other biomedical text mining tasks. One such task is the automatic summarization of biomedical documents. *Automatic summarization* refers to the process by which the salient aspects of one or more documents is identified and presented succinctly and coherently. Due to the enormous growth of unstructured information in the form of scientific articles and electronic health records, a means for clinicians and researches to quickly and reliably assimilate knowledge from a multitude of biomedical sources is desirable. Automatic summarization is one approach to determine and make accessible the important information contained in an increasingly large and diverse volume of biomedical text.

In the biomedical domain, document summaries are commonly application-oriented, and can serve a variety of purposes. Summaries may be either a generic assimilation of facts or they may be targeted [3]. Generic summaries consider all the information contained in a document or set of documents while targeted summaries aim to satisfy a specific information need, which is usually presented to a system in the form of a query. For example, a targeted summary of the biomedical literature might seek to determine the best treatments for a given disease [56], whereas a generic summary might aim to extract from articles key sentences related to results or conclusions [169]. Additionally, a summary is considered indicative if its purpose is to inform a reader of the contents of a document or set of documents, or it is informative if its purpose is to supplant those contents in terms of information coverage [3].

Depending on their purpose, several different types of document summaries can be produced. Single-document summaries seek to summarize the contents of individual sources, whereas multi-document summaries consider the information contained in a collection of sources [3]. Often, document clustering is utilized when generating multi-document summaries in order to produce a topical account of a particular group of documents. Summaries may also be extractive or abstractive [3]. Extractive summaries are created by identifying the salient textual components of documents (e.g., their important sentences or paragraphs) and then presenting this information as the summary. The representative textual components are determined by statistical methods that rank them according to relevance or by graph-based methods that organize them according to their similarity. Alternatively, abstractive summaries are created by structuring document information in a way that can be processed by a natural language generation system to produce the summary. Salient information is typically generated through prior knowledge of the documents' structure or by utilizing ontological resources to produce semantic representations of the documents.

Considering both the various types of summaries that may be generated and their intended applications, the evaluation of summarization techniques within the biomedical domain is a challenging issue. This difficulty is due, in part, to the subjective aspect of determining whether a summary is of "good" quality or not. Existing evaluation criteria consider the intrinsic aspects of a summary, such as its coherence, conciseness, grammaticality, and readability. Other extrinsic evaluation criteria measure, for example, whether a reader is able to comprehend the content of a summary [3]. However, manual evaluations of summaries are time-consuming and expensive to perform. A popular automatic summary evaluation methodology is ROUGE [107]. ROUGE is an acronym for Recall-Oriented Understudy for Gisting Evaluation, and it determines the quality of an automatically generated summary by computing statistics based on n-gram co-occurrences and common subsequences between it and ideal human-produced summaries. ROUGE has been shown to correlate well with human evaluations of single-document summaries.

A related method is based on the Jensen-Shannon divergence of distributions between an automatically generated summary and reference summaries and is more effective for the multiple document summarization task [108].

Recent biomedical text summarization techniques have been shown to be effective tools for assimilating information from a diverse collection of sources. While most approaches in the biomedical domain aim to produce targeted or topic-specific summaries, the types of generated summaries are generally more diverse and include both single- and multi-document summaries as well as extractive and abstractive summaries. However, given the rapidly expanding volume of published biomedical literature, multi-document summaries are increasingly viewed as important. Examples of recent text summarization approaches and their applications are described below.

One of the most basic approaches to biomedical text summarization involves the classification of individual sentences into a given set of categories. These categories may be specific to the biomedical domain, but they are often representative of the general rhetorical categories commonly encountered in scientific literature. Agarwal and Yu [4] train a Naïve Bayes classifier to classify sentences in full-text biomedical articles as being related to the introduction, methods, results, and discussion rhetorical categories. Their system achieves an overall annotation agreement of 0.76 kappa with human annotators. Ruch et al. [169] describe a similar approach that classifies sentences in MEDLINE abstracts as being related to an article's purpose, methods, results, or conclusions. Finally, Demner-Fushman and Lin [45] produce extractive summaries for clinical information needs by extracting sentences from MEDLINE abstracts relating to the outcomes of a clinical study.

While some of the above approaches apply generic summarization methods to biomedical articles, most applications are targeted and seek a concise description of a specific type of information. Since the understanding of gene regulation and expression is crucial in current biomedical research, a variety of targeted methods have been proposed to generate multi-document gene summaries. Ling et al. [112] propose a method for generating abstractive multi-document gene summaries from biomedical literature. Their two-stage approach to gene summarization first retrieves articles that mention a particular gene, and it then identifies text within those articles that pertains to several gene-related semantic categories, which include expression, sequence, and phenotypic information. Similarly, Yang et al. [225] describe an extractive approach to gene summarization that first clusters genes into functional groups based on their mentions in MEDLINE abstracts. Their system then presents summaries for each functional group by ranking and extracting sentences from the abstracts.

A challenge facing many automatic summarization techniques is the accurate semantic interpretation of the text. To address this issue, several summarization methods utilize domain knowledge in order to produce ontology-based document summaries. Reeve et al. [158] describes a single-document abstractive approach that utilizes MetaMap to map text to concepts in the UMLS Metathesaurus. Their approach then discovers strong thematic chains of UMLS semantic types and extracts the corresponding sentences. Yoo et al. [229] describe an approach to multi-document summarization that first clusters articles into topical groups and then produces summaries for each cluster. Their system uses a graph-based method for both document clustering and summarization that is enriched with concepts from the MeSH ontology. Morales et al. [123] describe a similar graph-based approach to single-document summarization that represents documents using UMLS concepts. Finally, Fiszman et al. [56] utilize SemRep [165] to produce multi-document summaries of MEDLINE citations according to disease-treatment relationships relevant to user-specified topics. Their approach has become an integral component of Semantic MEDLINE [166].

In addition to the text found in biomedical articles, the figures they contain also convey essential information. However biomedical images are seldom self-evident, and much of the information required for their comprehension is found elsewhere in an article. Figure captions, article titles and abstracts, and snippets of text from within the bodies of articles all contribute to image understanding [230]. Given that figures are a crucial source of information in the biomedical literature, many methods seek to incorporate image-related text into document summaries. However, since the number of such approaches is so large, and their methods are diverse, a full accounting of the use of image-related text in bioinformatics warrants a separate review.

A few representative examples of figure summarization and the use figure captions for producing document summaries include the following. Similar to their approach for full-text summarization, Agarwal and Yu [5] produce figure summaries consisting of one sentence each from an article's introduction, methods, results, and discussion rhetorical categories. Yu and Lee [232] produce figure summaries by extracting sentences from article abstracts that are similar to figure captions, and Simpson et al. [181] utilize image-related text to produce full-text summaries in support of case-based article retrieval.

Several user-oriented systems have been developed for supporting biomedical document summarization. PERSIVAL [116, 49] is a clinical system that seeks to provide access to medical literature and consumer health information. For clinicians, the system produces targeted, multi-document summaries containing sentences, extracted from full-text biomedical articles, that relate to experimental results. For users of the system that are patients, PERSIVAL provides indicative summaries of information that is commonly repeated across a set of consumer health documents. Anne O'Tate [184] is another user-oriented system capable of producing summaries of biomedical literature. Anne O'Tate is a web-based tool that provides navigable, extractive multi-document summaries of article citations retrieved by PubMed. The tool presents import words and authors mentioned in the results and can cluster the retrieved citations by topic.

5. Question Answering

Another biomedical text mining task that builds upon information extraction techniques is question answering. Unlike traditional information retrieval, where a set of potentially relevant documents is returned for a given query, *question answering* refers to the process of providing direct and precise answers to natural language questions. Like automatic summarization, question answering is a task directed towards aiding researchers and health care professionals in managing the continuous growth of information in the biomedical domain. Since question answering requires the use of complex natural language processing techniques in order to produce accurate responses, question answering systems are often regarded as the next generation of search engines.

The basic processing steps required of a question answering system are well-understood. The input to such a system is natural language text. A question processing stage uses linguistic analysis and question classification techniques to determine the type of question being posed to the system and the type of response it should generate. It then constructs a query from the input text to be fed into a document processing stage. In the document processing stage, the system inputs the query into a search engine, which retrieves a set of documents, and from these documents, extracts relevant passages or snippets of text as potential answers. An answer processing stage ranks the candidate answers according to the degree to which they match the expected answer type that was determined in the question processing stage. The output of a question answering system is the top-ranked answer.

Several characteristics of this process distinguish question answering in the biomedical domain from general, open-domain question answering systems. First, biomedical question answering is both challenged and advantaged by a prominent use of domain-specific terminology. Although terminological variations and synonymy make text mining difficult in general for the biomedical domain (Section 3), question answering systems may benefit from the specificity and limited scope of potential questions that a domain-specific terminology provides. Second, the multitude of domain-specific corpora and the tools and methods required for exploiting the semantic information they contain (Section 2) allow for deep question processing. Lastly, agreement on domain-specific structures in which to organize questions—especially clinical questions—allows for answer processing strategies that can be tailored to specific question types.

Due to the unique characteristics of biomedicine as an application domain for question answering, recently proposed systems have increasingly sought to incorporate deep semantic knowledge throughout their processing stages in order to produce more precise responses. The remainder of the discussion in this section surveys biomedical question answering techniques, and organizes the methods according to the recent review by Athenikos and Han [17], in which the authors classify biomedical "semantic knowledge-based" systems into semantics-based, inference-based, and logic-based approaches. Semantics-based approaches produce answers to biomedical questions by exploiting the semantic metadata encoded in structured knowledge resources and ontologies; inference-based approaches derive responses by exploiting extracted semantic relationships, and logic-based approaches utilize explicit logical forms and theorem proving techniques to produce answers. The approaches can further be divided into those that support medical question answering and those that support biological question answering.

5.1 Medical Question Answering

A dominant theme of work related to medical (or clinical) question answering is the use of the evidence-based medicine framework. Evidence-based medicine [170] seeks to apply the best information garnered from scientific inquiry to clinical decision making. For determining the best available evidence supporting an answer to a given clinical question, the evidence-based paradigm suggests questions be structured according to the PICO [159] format. PICO is an acronym for Patient/Problem, Intervention, Comparison, and Outcome. Clinical questions containing elements that pertain to each of these semantic roles are considered well-formed. In addition to the structure of clinical questions, taxonomies of questions in the evidence-based framework have also been proposed. Ely et al. [50] describe a generic taxonomy for clinical questions that distinguishes among questions that are potentially answerable and those that are not. The authors claim that questions involving a search for evidence are among the answerable ones.

The first step towards answering a clinical question is processing the question so as to determine the type of answer to produce. Several authors in the medical domain have investigated question classification as a means of analyzing and filtering clinical questions. Huang et al. [77] describe a manual classification of primary care clinical questions as a means to evaluate the effectives of the PICO framework. The authors conclude that PICO is a useful organizing structure for clinical questions, but they suggest it is less suitable for questions that do not involve therapy elements. Additionally, Yu et al. [234, 235, 231] investigate various machine learning approaches for question filtering that automatically determine whether a clinical question is answerable according to the evidence taxonomy proposed by Ely et al., which was described above.

Most approaches to medical question answering in some way make use of domain-specific semantic knowledge for information extraction and retrieval. Jacqumart et al. [40, 81] describe a semantics-based approach for the development of a French-language medical question answering system. Their approach is notable for the use of pattern-based semantic models of medical questions and the use of UMLS concepts, semantic types and relations for identifying named entities and extracting answers. Niu et al. [135, 136, 138, 137] propose a PICO-based question answering approach within the EPoCare system. Their methods locate potential answers by identifying, in both the question and answer texts, semantic roles that correspond to the four elements of the PICO framework. The semantic roles identified in the question are then compared with those identified in candidate answers to select a response. Similarly, Demner-Fushman et al [45, 44, 42, 109, 43] propose an approach to clinical question answering based on the semantic unification of a query PICO frame with those of candidate answers. Making extensive us of MetaMap and SemRep, the authors describe semantic knowledge extractors for identifying PICO elements in medical texts, a semantic matcher for scoring and ranking MEDLINE citations according to a query PICO frame, and an answer generator for extracting answers from the scored citations. Weiming et al. [221] describe a question answering approach that represents questions and documents using UMLS concepts, semantic types, and semantic relations. Their approach is notable for incorporating a semantic clustering phase into the answer processing stage so as to organize potential answers according to their hierarchical relationships in the UMLS Metathesaurus. Finally AskHERMES [27] is an online clinical question answering system capable of processing long and complex questions. The system uses machine learning techniques with a variety of lexical, syntactic, and UMLS-derived features to classify questions and topically group and rank candidate answers. A preliminary version of AskHERMES, known as the MedQA [233] system, was a non-semantic-knowledge-based approach capable of answering definitional questions.

Few approaches to question answering in the medical domain are inference- or logic-based. Terol et al. [197] describe an approach based on comparing the formal logic forms derived from a natural language question with those of candidate answers. Their technique utilizes a pattern-based method for question classification, and it identifies medical entities in both questions and answers based on UMLS concepts and semantic types.

5.2 Biological Question Answering

Whereas evidence-based medicine provides a means to structure clinical questions and answers, work in the biological domain has yet to adopt such a prominent framework. However, systems targeting the biological domain still follow the general architecture of questioned answering systems outlined previously. A review of recent work related to biological question answering is presented below.

Like their use for medical question answering, semantics-based approaches are also commonly employed for answering questions in the biological domain. Takahashi et al. [195] describe an approach that utilizes the UMLS Metathesaurus and other biological dictionaries and thesauri for analyzing questions and generating queries. Their system then uses semantic information of terms selected from the retrieved documents to assimilate and rank candidate answers. Lin et al. [110] propose a system for answering questions about biomolecular events, including interactions between genes and proteins (Section 3). Their approach involves the use of semantic role labeling for extracting predicate-argument structures and the use of semantic features for ranking candidate answers. The system provides answer responses in the form of biomedical named entities. Finally, the BioSquash [180] system is a targeted, multi-document, semantic graph-based summarization system oriented towards answering biological questions.

Like the use inference- and logic-based methods for medical question answering, few approaches in the biological domain make use of these techniques. Kontus et al. [94, 95] describe the AROMA inference-based system for biological question answering. AROMA extracts rhetorical and causal relationships from multiple biological texts, combines the extracted text with manually entered domain knowledge, and encodes this information as Prolog facts. The system generates answers to questions by applying inference rules over the encoded facts. Rinaldi et al. [162] describe a logic-based approach to question answering in the genomics domain. Using deep linguistic and terminological information, the system derives logical forms for text taken from documents in the GENIA corpus and a subset of full-text documents indexed in MEDLINE. Natural language questions are processed with the same mechanism, and the system derives an answer using a theorem proving process.

6. Literature-Based Discovery

While the extraction of explicit relations and events among biomedical entities can be used to produce rich document summaries and enable complex question answering systems, an exciting use of these methods aims to uncover relationships that are not present in the text, but that can be inferred from other information. *Literature-based discovery* refers to the task of utilizing scientific literature to uncover "hidden," previously unknown or neglected relationships between existing knowledge. The goal of discovering these implicit relationships is to identify relations worthy of further scientific investigation or to find evidence supporting suspected relations.

As a technique useful for biomedical text mining, literature-based discovery was pioneered by the work of Swanson in the 1980s. Swanson suggested that novel information could be uncovered by systematically reviewing "complementary but disjoint" bodies of literature [192]. In what has become the prototypical example of literature-based discovery, Swanson linked fish oil, a substance widely-understood to have potential cardiovascular benefits, with Raynaud's syndrome, a vasospastic disorder causing the narrowing of blood vessels [189]. The discovery suggests fish oil supplements may help to control the symptoms of Raynaud's syndrome. To further demonstrate the feasibility of his ideas, Swanson later found evidence for relationships between migraine and magnesium [190], somatomedin C and arginine [191], and viruses and their potential use as biological weapons [194].

The basic premise of Swanson's approach is that there exists two scientific communities that do not communicate. A portion of the knowledge in one community may be related to or complement knowledge in the other one, but this relationship is unknown to either community. For example, suppose a scientific community has researched the relationship between a medical finding or characteristic B and a disease C. Further suppose that a separate community has studied the affects of substance A on characteristic B. The use of literature-based discovery techniques may suggest an *A-C* relationship, indicating in this example that substance *A* may potentially treat disease *C*.

Weeber et al. [220] distinguish between two modes of discovery. A "closed discovery," or hypothesis testing study, begins with known *A*- and *C*-terms. Thus, the discovery concerns finding novel *B*-terms that may explain the observed *A-C* association or hypothesis. On the other hand, an "open discovery," or hypothesis generation study, begins with known *A-B* associations in one domain and seeks to discover *B-C* relations in another domain, thereby suggesting or generating a potential *A-C* association.

Since the pioneering work of Swanson, literature-based discovery techniques have seen widespread use. Existing approaches can be grouped by the way in which they identify potentially novel relationships. There are those that depend exclusively on the co-occurrence of terms or concepts, those that make use of semantic information to inform the processing of co-occurring terms, and those that construct interaction networks of individual relations whose paths can reveal hidden associations. Some recent methods following these general approaches are reviewed below. Unlike other text mining tasks, measuring the performance of literature-based discovery tools is not straightforward, and a discussion of system evaluation follows as well.

Co-occurrence-based methods are among the simplest, although less precise, approaches to literature-based discovery. Like the most basic approaches to the relation extraction task (Section 3), these methods seek to identify terms that frequently occur together. However, whereas approaches to relation extraction identify first-order term co-occurrences, approaches to literature-based discovery explore second-order co-occurrences—the shared co-occurrences of two given biomedical entities [238].

Most of the earliest approaches to literature-based discovery and many modern approaches rely on entity co-occurrence statistics. The Arrowsmith [193, 185, 182, 183, 199, 186] two-node search tool implements Swanson's original approach to find biologically meaningful links between two sets of articles in PubMed using title words and phrases. Recent work related to this project has developed a method to estimate and rank the relevance of associations. BITOLA [74, 73, 72, 75] is a similar literature-based discovery system, but instead of identifying relations using title words, it represents documents using their MeSH terms and recognized gene symbols. Additionally, BITOLA uses association rules [6] as a measure of concept relatedness instead of word frequencies. LitLinker [227] also utilizes MeSH terms; however, it uses a statistical approach based on the background distribution of term probabilities to identify correlated concepts. Jelier et al. [82] describe a system that identifies functional associations between genes and other biomedical concepts. Their approach measures the strength of association of co-occurring concepts using a log likelihood ratio. RaJoLink [151] provides semi-automated suggestions for links between two sets of articles based on rare terms identified in the literature. FACTA+ [203] uses an information theoretic score to rank indirectly associated concepts. It identifies explicit associations among biomedical entities using methods inherited from an earlier version of the system [206]. Finally, unlike other literature-based discovery methods that rely on associations explicit in scientific literature, Benton et al. [18] use a corpus of posts to Internet breast cancer message boards to discover adverse drug effects.

Because systems relying solely on co-occurrence statistics tend to produce a large number of spurious relations, recent approaches increasingly rely on semantic information to identify hidden relations or augment the processing of co-occurring entities. Hristovski et al. [71] describe an improvement to BITOLA that uses the semantic predications produced by SemRep and BioMedLEE [114] to enable users to eliminate uninteresting or incorrect relations. A similar approach is used in the EpiphaNet system [35], an interactive visualisation tool for exploring associations between concepts found in MEDLINE. EpiphaNet makes extensive use of MetaMap and SemRep for identifying explicit relations. Other systems, including Weeber et al.'s DAD-system [220], filter candidate relations based on the UMLS semantic type of identified B-terms. Recall that for hypothesis generation, B-terms are used to uncover hidden A-C relations from explicit A-B and B-C associations. Hu et al. [76] describe a literature-based discovery method that uses association rules as a measure of concept relatedness but also filters potential relations using UMLS semantic types.

Another approach to discovering hidden relationships among biomedical entities involves the construction of interaction networks whose paths can reveal indirect associations. Seki and Mostafa [174] build an inference network [208] to predict implicit gene-disease associations. Genes and diseases are connected within the graph by intermediary nodes representing gene functions and phenotypes. Similarly, Özgür et al. [145, 146] build a gene-interaction network by collecting an initial set of known disease-related genes from biomedical texts using dependency parsing and SVMs. They then use network centrality metrics to predict gene-disease associations. Finally, Palakal et al. [149] describe BioMap, a directed graph that is constructed from explicit relationships between biomedical entities identified within text. Users are able to query the graph to uncover implicit associations among the entities.

Due to the nature of uncovering novel information, there is no ground truth available for evaluating literature-based discovery systems, and comparing the relative performance of alternative approaches is difficult. A common method for evaluating an automatic discovery technique is to use the system to replicate known discoveries, such as Swanson's linking of Raynaud's syndrome with fish oil or migraine with magnesium [220]. However, Yetisgen-Yildiz and Pratt [228] suggest this approach is uninformative of the overall performance of a system. They describe an alternative methodology that divides the abstracts in MEDLINE into two sets: those that were published before a given cut-off date, and those that were published after this date. Literature-based discovery methods are then applied to the older set of abstracts as hypotheses generating systems and to the newer set as hypotheses testing systems, using the generated associations from the older set as input. The performance of a system can then be quantified using standard information retrieval evaluation methods.

7. Conclusion

The past several years have seen some exciting developments in biomedical text mining. Progress was made in (1) defining and attempting more challenging tasks, such as event extraction and clinical text mining; (2) increasing the public availability of and community investment in resources, such as the MIMIC II database and the ORBIT registry; and (3) development and use of common frameworks, such as UIMA.

It is interesting to compare the development of the field to the desirable directions outlined by the leading researchers in 2008 [9]. At that time, the researchers were asked about the importance of text mining for biology, the utility of the text mining systems, and future directions.

The first suggested avenue for future research was *fusing literature and biological databases through text mining*. Understandably, this requires engaging the publishers of scientific literature and realizing potentially additional efforts by the publications' authors. To that end, Elsevier is piloting a tool, Reflect-Network [147], developed in partnership with the European Molecular Biology Laboratory and the Novo Nordisk Foundation Center for Protein Research. Reflect tags proteins and chemicals in documents and generates a graphical representation displaying interactions between entities and additional details about them.

The second proposed research direction was *interactivity and user interfaces*. This direction requires identifying more potential user groups and tasks. Progress was made in developing tools for database curation [223, 150]; however, more research is still needed in identifying user groups and tasks in parallel with tool development for known users.

The authors noted that success in the third direction, *tool scalability and integration into workflows*, depends on commonly accepted and used stable standards for the exchange and integration of information derived from text mining. Despite major initiatives towards seamless data exchange and interoperability (e.g., the i2b2 hive [80] or the eMERGE Network [51]) and pilot applications being included into workflows (e.g., NLM InfoBot [46]), this direction remains challenging. The efforts needed to make a system scalable and capable of handling real-time workflow interactions were recently demonstrated in the IBM DeepQA project [53].

The last direction, *development of text mining resources*, is an ongoing activity. Existing lexicons, standards, and ontologies are maintained— and new resources and community-wide evaluations emerge—following the progress in biology and medicine.

Acknowledgements

M.S.S. was supported in part by an appointment to the NLM Research Participation Program sponsored by the U.S. National Library of Medicine and administered by the Oak Ridge Institute for Science and Education.

References

[1] A. B. Abacha and P. Zweigenbaum. A hybrid approach for the extraction of semantic relations from MEDLINE abstracts. In A. Gelbukh, editor, *Computational Linguistics and Intelligent Text Processing*, volume 6609 of *Lecture Notes in Computer Science*, pages 139–150. Springer Berlin / Heidelberg, 2011.

[2] A. B. Abacha and P. Zweigenbaum. Medical entity recognition: A comparison of semantic and statistical methods. In *Proceedings of BioNLP 2011 Workshop*, pages 56–64, 2011.

[3] S. Afantenos, V. Karkaletsis, and P. Stamatopoulos. Summarization from medical documents: A survey. *Artificial Intelligence in Medicine*, 33(2):157–177, 2005.

[4] S. Agarwal and H. Yu. Automatically classifying sentences in fulltext biomedical articles into introduction, methods, results and discussion. *Bioinformatics*, 25(23):3174–3180, 2009.

[5] S. Agarwal and H. Yu. FigSum: Automatically generating structured text summaries for figures in biomedical literature. In *AMIA*

Annual Symposium Proceedings, pages 6–10, 2009.

[6] R. Agrawal, H. Mannila, R. Srikant, H. Toivonen, and A. I. Verkamo. Fast discovery of association rules. In U. M. Fayyad, G. Piatetsky-Shapiro, P. Smyth, and R. Uthurusamy, editors, *Advances in Knowledge Discovery and Data Mining*, pages 307–328. American Association for Artificial Intelligence, 1996.

[7] A. Airola, S. Pyysalo, J. Bjorne, T. Pahikkala, F. Ginter, and T. Salakoski. All-paths graph kernel for protein-protein interaction extraction with evaluation of cross-corpus learning. *BMC Bioinformatics*, 9(Suppl 11):S2, 2008.

[8] B. Alex, B. Haddow, and C. Grover. Recognising nested named entities in biomedical text. In *Proceedings of the Workshop on BioNLP 2007: Biological, Translational, and Clinical Language Processing*, pages 65–72, 2007.

[9] R. B. Altman, C. M. Bergman, J. Blake, C. Blaschke, A. Cohen, F. Gannon, L. Grivell, U. Hahn, W. Hersh, L. Hirschman, L. J. Jensen, M. Krallinger, B. Mons, S. I. O'Donoghue, M. C. Peitsch, D. Rebholz-Schuhmann, H. Shatkay, and A. Valencia. Text mining for biology - the way forward: opinions from leading scientists. *Genome Biology*, 9(Suppl 2):S7, 2008.

[10] S. F. Altschul, W. Gish, W. Miller, E. W. Myers, and D. J. Lipman. Basic local alignment search tool. *Journal of Molecular Biology*, 215(3):403–410, 1990.

[11] S. F. Altschul, T. L. Madden, A. A. Schäffer, J. Zhang, Z. Zhang, W. Miller, and D. J. Lipman. Gapped BLAST and PSI-BLAST: A new generation of protein database search programs. *Nucleic Acids Research*, 25(17):3389–3402, 1997.

[12] S. Ananiadou and J. Mcnaught. *Text Mining for Biology And Biomedicine*. Artech House, Inc., 2005.

[13] S. Ananiadou, S. Pyysalo, J. Tsujii, and D. B. Kell. Event extraction for systems biology by text mining the literature. *Trends in Biotechnology*, 28(7):381–390, 2010.

[14] A. R. Aronson and F.-M. Lang. An overview of MetaMap: historical perspective and recent advances. *Journal of the American Medical Informatics Association*, 17(3):229–236, 2010.

[15] R. Artstein and M. Poesio. Inter-coder agreement for computational linguistics. *Computational Linguistics*, 34(4):555–596, 2008.

[16] M. Ashburner, C. A. Ball, J. A. Blake, D. Botstein, H. Butler, J. M. Cheryy, A. P. Davis, K. Dolinski, S. S. Dwight, J. T. Eppig, M. A. Harris, D. P. Hill, L. Issel-Tarver, A. Kasarskis, S. Lewis,

J. C. Matese, J. E. Richardson, M. Ringwald, G. M. Rubin, and G. Sherlock. Gene ontology: Tool for the unification of biology. *Nature Genetics*, 25(1):25–29, 2000.

[17] S. J. Athenikos and H. Han. Biomedical question answering: A survey. *Computer Methods and Programs in Biomedicine*, 99(1):1–24, 2010.

[18] B. Benton, L. Ungar, S. Hill, S. Hennessy, J. Mao, A. Chung, C. E. Leonard, and J. H. Holmes. Identifying potential adverse effects using the web: A new approach to medical hypothesis generation. In Press, 2011.

[19] BioNLP. http://www.bionlp.org/.

[20] J. Björne, F. Ginter, S. Pyysalo, J. Tsujii, and T. Salakoski. Complex event extraction at PubMed scale. *Bioinformatics*, 26(12):i382–i390, 2010.

[21] J. Björne, J. Heimonen, F. Ginter, A. Airola, T. Pahikkala, and T. Salakoski. Extracting complex biological events with rich graph-based feature sets. In *Proceedings of the Workshop on Current Trends in Biomedical Natural Language Processing: Shared Task*, pages 10–18, 2009.

[22] K. W. Boyack, D. Newman, R. J. Duhon, R. Klavans, M. Patek, J. R. Biberstine, B. Schijvenaars, A. Skupin, N. Ma, and K. Borner. Clustering more than two million biomedical publications: Comparing the accuracies of nine text-based similarity approaches. *PLoS ONE*, 6(3):e18029, 2011.

[23] M. Bundschus, M. Dejori, M. Stetter, V. Tresp, and H.-P. Kriegel. Extraction of semantic biomedical relations from text using conditional random fields. *BMC Bioinformatics*, 9(1):207, 2008.

[24] E. Buyko, E. Faessler, J. Wermter, and U. Hahn. Event extraction from trimmed dependency graphs. In *Proceedings of the Workshop on Current Trends in Biomedical Natural Language Processing: Shared Task*, pages 19–27, 2009.

[25] Y. Cai and X. Cheng. Biomedical named entity recognition with tri-training learning. In *Proceedings of the 2009 2nd International Conference on Biomedical Engineering and Informatics*, pages 1–5, 2009.

[26] CALBC challenge. http://www.calbc.eu/.

[27] Y. Cao, F. Liu, P. Simpson, L. Antieau, A. Bennett, J. J. Cimino, J. Ely, and H. Yu. AskHERMES: An online question answering system for complex clinical questions. *Journal of Biomedical Informatics*, 44(2):277–288, 2011.

[28] D. T.-H. Chang, Y.-Z. Weng, J.-H. Lin, M.-J. Hwang, and Y.-J. Oyang. Protemot: Prediction of protein binding sites with automatically extracted geometrical templates. *Nucleic Acids Research*, 34(suppl 2):W303–W309, 2006.

[29] W. W. Chapman and K. B. Cohen. Current issues in biomedical text mining and natural language processing. *Journal of Biomedical Informatics*, 42(5):757–759, 2009.

[30] E. S. Chen, G. Hripcsak, H. Xu, M. Markatou, and C. Friedman. Automated acquisition of disease-drug knowledge from biomedical and clinical documents: An initial study. *Journal of the American Medical Informatics Association*, 15(1):87–98, 2008.

[31] H. W. Chun, Y. Tsuruoka, J. D. Kim, R. Shiba, N. Nagata, T. Hishiki, and J. Tsujii. Extraction of gene-disease relations from MEDLINE using domain dictionaries and machine learning. In *Pacific Symposium on Biocomputing*, pages 4–15, 2006.

[32] A. M. Cohen and W. R. Hersh. A survey of current work in biomedical text mining. *Briefings in Bioinformatics*, 6(1):57–71, 2005.

[33] K. B. Cohen and L. Hunter. Getting started in text mining. *PLoS Computational Biology*, 4(1):e20, 2008.

[34] K. B. Cohen, K. Verspoor, H. L. Johnson, C. Roeder, P. V. Ogren, W. A. Baumgartner, Jr., E. White, H. Tipney, and L. Hunter. High-precision biological event extraction with a concept recognizer. In *Proceedings of the Workshop on Current Trends in Biomedical Natural Language Processing: Shared Task*, pages 50–58, 2009.

[35] T. Cohen, G. K. Whitfield, R. W. Schvaneveldt, K. Mukund, and T. Rindflesch. EpiphaNet: An interactive tool to support biomedical discoveries. *Journal of Biomedical Discovery and Collaboration*, 5:21–49, 2010.

[36] N. Collier, C. Nobata, and J.-i. Tsujii. Extracting the names of genes and gene products with a hidden Markov model. In *Proceedings of the 18th Conference on Computational Linguistics - Volume 1*, pages 201–207, 2000.

[37] P. Corbett and A. Copestake. Cascaded classifiers for confidence-based chemical named entity recognition. *BMC Bioinformatics*, 9(Suppl 11):S4, 2008.

[38] CRAFT: The colorado richly annotated full text corpus. http://bionlp-corpora.sourceforge.net/CRAFT/index.shtml.

[39] H. Cunningham, D. Maynard, K. Bontcheva, V. Tablan, N. Aswani, I. Roberts, G. Gorrell, A. Funk, A. Roberts, D. Daml-

janovic, T. Heitz, M. A. Greenwood, H. Saggion, J. Petrak, Y. Li, and W. Peters. *Text Processing with GATE (Version 6)*. GATE, 2011.

[40] T. Delbecque, P. Jacquemart, and P. Zweigenbaum. Indexing UMLS semantic types for medical question-answering. In R. Engelbrecht, A. Geissbuhler, C. Lovis, and G. Mihalas, editors, *Connecting Medical Informatics and Bio-Informatics: Proceedings of MIE2005 - The XIXth International Congress of the European Federation for Medical Informatics*, pages 805–810. IOS Press, 2005.

[41] D. Demner-Fushman, W. W. Chapman, and C. J. McDonald. What can natural language processing do for clinical decision support? *Journal of Biomedical Informatics*, 42(5):760–772, 2009.

[42] D. Demner-Fushman, B. Few, S. E. Hauser, and G. Thoma. Automatically identifying health outcome information in MEDLINE records. *Journal of the American Medical Informatics Association*, 13(1):52–60, 2006.

[43] D. Demner-Fushman and J. Lin. Knowledge exraction for clinical question answering: Preliminary results. In *Proceedings of the AAAI 2005 Workshop on Question Ansering in Restricted Domains*, 2005.

[44] D. Demner-Fushman and J. Lin. Answer extraction, semantic clustering, and extractive summarization for clinical question answering. In *Proceedings of the 21st International Conference on Computational Linguistics and the 44th Annual Meeting of the Association for Computational Linguistics*, pages 841–848, 2006.

[45] D. Demner-Fushman and J. Lin. Answering clinical questions with knowledge-based and statistical techniques. *Computational Linguistics*, 33(1):63–103, 2007.

[46] D. Demner-Fushman, C. Seckman, C. Fisher, S. E. Hauser, J. Clayton, and G. R. . Thoma. A prototype system to support evidence-based practice. In *AMIA Annual Symposium Proceedings*, pages 151–155, 2008.

[47] S. Dipper, M. Götze, and M. Stede. Simple annotation tools for complex annotation tasks: An evaluation. In *Proceedings of the LREC Workshop on XML-Based Richly Annotated Corpora*, pages 54–62, 2004.

[48] eHOST: The extensible human oracle suite of tools. `http://code.google.com/p/ehost/`.

[49] N. Elhadad, M.-Y. Kan, J. L. Klavans, and K. R. McKeown. Customization in a unified framework for summarizing medical literature. *Artificial Intelligence in Medicine*, 33(2):179–198, 2005.

[50] J. W. Ely, J. A. Osheroff, M. H. Ebell, M. L. Chambliss, D. C. Vinson, J. J. Stevermer, and E. A. Pifer. Obstacles to answering doctors' questions about patient care with evidence: qualitative study. *British Medical Journal*, 324(7339):710, 2002.

[51] Electronic medical records and genomics. `https://www.mc.vanderbilt.edu/victr/dcc/projects/acc/index.php/Main_Page`.

[52] European bioinformatics institute. `http://www.ebi.ac.uk/`.

[53] D. Ferrucci, E. Brown, J. Chu-Carroll, J. Fan, D. Gondek, A. A. Kalyanpur, A. Lally, J. W. Murdock, E. Nyberg, J. Prager, N. Schlaefer, and C. Welty. Building Watson: An overview of the DeepQA project. *AI Magazine*, 31(3):59–79, 2010.

[54] D. Ferrucci and A. Lally. UIMA: An architectural approach to unstructured information processing in the corporate research environment. *Natural Language Engineering*, 10(3-4):327–348, 2004.

[55] J. Finkel, S. Dingare, H. Nguyen, M. Nissim, C. Manning, and G. Sinclair. Exploiting context for biomedical entity recognition: From syntax to the web. In *Proceedings of the International Joint Workshop on Natural Language Processing in Biomedicine and its Applications*, pages 88–91, 2004.

[56] M. Fiszman, D. Demner-Fushman, H. Kilicoglu, and T. C. Rindflesch. Automatic summarization of MEDLINE citations for evidence-based medical treatment: A topic-oriented evaluation. *Journal of Biomedical Informatics*, 42(5):801–813, 2009.

[57] K. Franzén, G. Eriksson, F. Olsson, L. Asker, P. Lidén, and J. Cöster. Protein names and how to find them. *International Journal of Medical Informatics*, 67(1-3):49–61, 2002.

[58] C. Friedman, G. Hripcsak, L. Shagina, and H. Liu. Arepresenting information in patient reports using natural language processing and the extensible markup language. *Journal of the American Medical Informatics Association*, 6:76–87, 1999.

[59] K. Fukuda, A. Tamura, T. Tsunoda, and T. Takagi. Toward information extraction: Identifying protein names from biological papers. In *Pacific Symposium on Biocomputing*, pages 707–718, 1998.

[60] K. Fundel, R. Küffner, and R. Zimmer. RelEx—relation extraction using dependency parse trees. *Bioinformatics*, 23(3):365–371, 2007.

[61] R. Gaizauskas, G. Demetriou, P. J. Artymiuk, and P. Willett. Protein structures and information extraction from biological texts: The PASTA system. *Bioinformatics*, 19(1):135–143, 2003.

[62] B. Gu. Recognizing nested named entities in GENIA corpus. In *Proceedings of the Workshop on Linking Natural Language Processing and Biology: Towards Deeper Biological Literature Analysis*, pages 112–113, 2006.

[63] J. Hakenberg, S. Bickel, C. Plake, U. Brefeld, H. Zahn, L. Faulstich, U. Leser, and T. Scheffer. Systematic feature evaluation for gene name recognition. *BMC Bioinformatics*, 6(Suppl 1):S9, 2005.

[64] J. Hakenberg, C. Plake, and U. Leser. LLL'05 challenge: Genic interaction extraction - identification of language patterns based on alignment and finite state automata. In *In Proceedings of the ICML 2005 Workshop on Learning Language in Logic*, pages 38–45, 2005.

[65] W. Hersh. *Information Retrieval: A Health and Biomedical Perspective*. Health Informatics. Springer, third edition, 2005.

[66] HighWire press. http://highwire.org/.

[67] L. Hirschman, M. Colosimo, A. Morgan, and A. Yeh. Overview of BioCreAtIvE task 1B: Normalized gene lists. *BMC Bioinformatics*, 6(Suppl 1):S11, 2005.

[68] L. Hirschman, A. A. Morgan, and A. S. Yeh. Rutabaga by any other name: Extracting biological names. *Journal of Biomedical Informatics*, 35(4):247–259, 2002.

[69] L. Hirschman, A. Yeh, C. Blaschke, and A. Valencia. Overview of BioCreAtIvE: Critical assessment of information extraction for biology. *BMC Bioinformatics*, 6(Suppl 1):S1, 2005.

[70] W.-J. Hou and H.-H. Chen. Enhancing performance of protein name recognizers using collocation. In *Proceedings of the ACL 2003 Workshop on Natural Language Processing in Biomedicine - Volume 13*, pages 25–32, 2003.

[71] D. Hristovski, C. Friedman, T. C. Rindflesch, and B. Peterlin. Exploiting semantic relations for literature-based discovery. In *AMIA Anual Symposium Proceedings*, pages 349–353, 2006.

[72] D. Hristovski, B. Peterlin, S. Džeroski, and J. Stare. Literature-based discovery support system and its application to disease gene identification. In S. Džeroski and L. Todorovski, editors, *Computational Discovery of Scientific Knowledge*, volume 4660 of *Lecture Notes in Computer Science*, pages 307–326. Springer Berlin / Heidelberg, 2007.

[73] D. Hristovski, B. Peterlin, J. A. Mitchell, and S. M. Humphrey. Improving literature-based discovery support by genetic knowledge integration. *Studies in Health Technogy and Informatics*, 95:68–73, 2003.

[74] D. Hristovski, B. Peterlin, J. A. Mitchell, and S. M. Humphrey. Using literature-based discovery to identify disease candidate genes. *International Journal of Medical Informatics*, 74(2-4):289–298, 2005.

[75] D. Hristovski, J. Stare, B. Peterlin, and S. Džeroski. Supporting discovery in medicine by association rule mining in MEDLINE and UMLS. In V. L. Patel, R. Rogers, and R. Haux, editors, *Proceedings of the 10th World Congress on Medical Informatics*, volume 84/2001 of *Studies in Health Technology and Informatics*, pages 1344–1348. IOS Press, 2001.

[76] X. Hu, X. Zhang, I. Yoo, X. Wang, and J. Feng. Mining hidden connections among biomedical concepts from disjoint biomedical literature sets through semantic-based association rule. *International Journal of Intelligent Systems*, 25(2):207–223, 2010.

[77] X. Huang, J. Lin, and D. Demner-Fushman. Evaluation of PICO as a knowledge representation for clinical questions. In *AMIA Annual Symposium Proceedings*, pages 359–363, 2006.

[78] K. Humphreys, G. Demetriou, and R. Gaizauskas. Two applications of information extraction to biological science yournal articles: Enzyme interactions and protein structures. In *Pacific Symposium on Biocomputing*, pages 502–513, 2000.

[79] L. Hunter, Z. Lu, J. Firby, W. Baumgartner, H. Johnson, P. Ogren, and K. B. Cohen. OpenDMAP: An open source, ontology-driven concept analysis engine, with applications to capturing knowledge regarding protein transport, protein interactions and cell-type-specific gene expression. *BMC Bioinformatics*, 9(1):78, 2008.

[80] Informatics for integrating biology and the bedside. https://www.i2b2.org/resrcs/hive.html.

[81] P. Jacqumart and P. Zweigenbaum. Towards a medical question-answering system: A feasibility study. *Studies in Health Technology and Informatics*, 95:463–468, 2003.

[82] R. Jelier, G. Jenster, L. Dorssers, B. Wouters, P. Hendriksen, B. Mons, R. Delwel, and J. Kors. Text-derived concept profiles support assessment of DNA microarray data for acute myeloid leukemia and for androgen receptor stimulation. *BMC Bioinformatics*, 8(1):14, 2007.

[83] R. Kabiljo, A. B. Clegg, and A. J. Shepherd. A realistic assessment of methods for extracting gene/protein interactions from free text. *BMC Bioinformatics*, 10:233, 2008.

[84] J. Kalpathy-Cramer, H. Müler, S. Bedrick, I. Eggel, A. de Herrera, and T. Tsikrika. The CLEF 2011 medical image retrieval and classification tasks. In *CLEF 2011 Working Notes*, 2011.

[85] H. Karsten and H. Suominen. Mining of clinical and biomedical text and data. *International Journal of Medical Informatics*, 78(12):786–787, 2009.

[86] J. Kazama, T. Makino, Y. Ohta, and J. Tsujii. Tuning support vector machines for biomedical named entity recognition. In *Proceedings of the ACL-02 Workshop on Natural Language Processing in the Biomedical Domain - Volume 3*, pages 1–8, 2002.

[87] H. Kilicoglu and S. Bergler. Syntactic dependency based heuristics for biological event extraction. In *Proceedings of the Workshop on Current Trends in Biomedical Natural Language Processing: Shared Task*, pages 119–127, 2009.

[88] J.-D. Kim, T. Ohta, N. Nguyen, S. Pyysalo, R. Bossy, and J. Tsujii. Overview of BioNLP shared task 2011. In *Proceedings of the BioNLP Shared Task 2011 Workshop*, pages 1–6, 2011.

[89] J.-D. Kim, T. Ohta, S. Pyysalo, Y. Kano, and J. Tsujii. Overview of BioNLP'09 shared task on event extraction. In *Proceedings of the Workshop on Current Trends in Biomedical Natural Language Processing: Shared Task*, pages 1–9, 2009.

[90] J.-D. Kim, T. Ohta, Y. Tateisi, and J. Tsujii. GENIA corpus—a semantically annotated corpus for bio-textmining. *Bioinformatics*, 19(Suppl 1):i180–i182, 2003.

[91] J.-D. Kim, T. Ohta, Y. Tsuruoka, Y. Tateisi, and N. Collier. Introduction to the bio-entity recognition task at JNLPBA. In *Proceedings of the International Joint Workshop on Natural Language Processing in Biomedicine and its Applications*, pages 70–75, 2004.

[92] S. Kim, J. Yoon, and J. Yang. Kernel approaches for genic interaction extraction. *Bioinformatics*, 24(1):118–126, 2008.

[93] S. Kinoshita, K. B. Cohen, P. Ogren, and L. Hunter. BioCreAtIvE task 1A: Entity identification with a stochastic tagger. *BMC Bioinformatics*, 6(Suppl 1):S4, 2005.

[94] J. Kontos, J. Lekakis, I. Malagardi, and J. Peros. Grammars for question answering systems based on intelligent text mining in biomedicine. In *Proceedings of the 7th Hellenic Europeoan Conference on Computer Mathematics and its Applications*, 2005.

[95] J. Kontos, I. Malagardi, and J. Peros. Question answering and rhetoric analysis of biomedical texts in the AROMA system. In *Proceedings of the 7th Hellenic Europeoan Conference on Computer Mathematics and its Applications*, 2005.

[96] M. Krallinger, F. Leitner, C. Rodriguez-Penagos, and A. Valencia. Overview of the protein-protein interaction annotation extraction task of BioCreAtIve II. *Genome Biology*, 9(Suppl 2):S4, 2008.

[97] M. Krallinger, A. Morgan, L. Smith, F. Leitner, L. Tanabe, J. Wilbur, L. Hirschman, and A. Valencia. Evaluation of text-mining systems for biology: Overview of the second BioCreAtIvE community challenge. *Genome Biology*, 9(Suppl 2):S1, 2008.

[98] M. Krallinger, A. Valencia, and L. Hirschman. Linking genes to literature: text mining, information extraction, and retrieval applications for biology. *Genome biology*, 9(Suppl 2):S8, 2008.

[99] M. Krauthammer and G. Nenadic. Term identification in the biomedical literature. *Journal of Biomedical Informatics*, 37(6):512–526, 2004.

[100] M. Krauthammer, A. Rzhetsky, P. Morozov, and C. Friedman. Using BLAST for identifying gene and protein names in journal articles. *Gene*, 259(1-2):245–252, 2000.

[101] R. Leaman and G. Gonzalez. BANNER: An executable survey of advances in biomedical named entity recognition. In *Pacific Symposium on Biocomputing*, pages 652–663, 2008.

[102] L. C. Lee, F. Horn, and F. E. Cohen. Automatic extraction of protein point mutations using a graph bigram association. *PLoS Computational Biology*, 3(2):e16, 2007.

[103] G. Leech. Adding linguistic annotation. In M. Wynne, editor, *Developing Linguistic Corpora: A Guide to Good Practice*, pages 17–29. Oxbow Books, 2005.

[104] U. Leser and J. Hakenberg. What makes a gene name? named entity recognition in the biomedical literature. *Briefings in Bioinformatics*, 6(4):357–369, 2005.

[105] M. Liberman, M. Mandel, and GlaxoSmithKline Pharmaceuticals R&D. PennBioIE CYP 1.0, 2008.

[106] M. Liberman, M. Mandel, and P. White. PennBioIE Oncology 1.0, 2008.

[107] C.-Y. Lin. ROUGE: A package for automatic evaluation of summaries. In *Proceedings of the Workshop on Text Summarization Branches Out*, 2004.

[108] C.-Y. Lin, G. Cao, J. Gao, and J.-Y. Nie. An information-theoretic approach to automatic evaluation of summaries. In *Proceedings of the Human Language Technology Conference of the North American Chapter of the Association of Computational Linguistics*, pages 463–470, 2006.

[109] J. Lin and D. Demner-Fushman. The role of knowledge in conceptual retrieval: A study in the domain of clinical medicine. In *Proceedings of the 29th Annual International ACM SIGIR Conference on Research and Development in Information Retrieval*, pages 99–106, 2006.

[110] R. T. K. Lin, J. Liang-Te Chiu, H.-J. Dai, M.-Y. Day, R. T.-H. Tsai, and W.-L. Hsu. Biological question answering with syntactic and semantic feature matching and an improved mean reciprocal ranking measurement. In *Proceedings of the 2008 IEEE International Conference on Information Reuse and Integration*, pages 184–189, 2008.

[111] D. A. Lindberg, B. L. Humphreys, and A. T. McCray. The unified medical language system. *Methods of Information in Medicine*, 32(4):281–291, 1993.

[112] X. Ling, J. Jiang, X. He, Q. Mei, C. Zhai, and B. Schatz. Generating gene summaries from biomedical literature: A study of semi-structured summarization. *Information Processing & Management*, 43(6):1777–1791, 2007.

[113] Y. Lussier, T. Borlawsky, D. Rappaport, Y. Liu, and C. Friedman. PheneGo: Assigning phenotypic context to gene ontology annotations with natural language processing. In *Pacific Symposium on Biocomputing*, pages 64–75, 2006.

[114] Y. Lussier, T. Borlawsky, D. Rappaport, Y. Liu, and C. Friedman. PhenoGo: Assigning phenotypic context to Gene Ontology annotations with natural language processing. In *Pacific Symposium on Biocomputing*, pages 64–75, 2006.

[115] D. Maynard. D1.2.2.1.3 benchmarking of annotation tools, 2007. http://knowledgeweb.semanticweb.org/semanticportal/deliverables/D1.2.2.1.3.pdf.

[116] K. R. McKeown, S.-F. Chang, J. Cimino, S. K. Feiner, C. Friedman, L. Gravano, V. Hatzivassiloglou, S. Johnson, D. A. Jordan, J. L. Klavans, A. Kushniruk, V. Patel, and S. Teufel. PERSIVAL, a system for personalized search and summarization over multimedia healthcare information. In *Proceedings of the 1st ACM/IEEE-CS Joint Conference on Digital Libraries*, pages 331–340, 2001.

[117] S. Mika and B. Rost. Protein names precisely peeled off free text. *Bioinformatics*, 20(suppl 1):i241–i247, 2004.

[118] T. Mitsumori, S. Fation, M. Murata, K. Doi, and H. Doi. Gene/protein name recognition based on support vector machine using dictionary as features. *BMC Bioinformatics*, 6(Suppl 1):S8, 2005.

[119] M. Miwa, R. Sætre, and J.-D. Kim. Event extraction with complex event classification using rich features. *Journal of Bioinformatics and Computational Biology*, 8(1):131–146, 2010.

[120] M. Miwa, R. Sætre, Y. Miyao, and J. Tsujii. Protein-protein interaction extraction by leveraging multiple kernels and parsers. *International Journal of Medical Informatics*, 78(12):e39–e46, 2009.

[121] Y. Miyao, T. Ohta, K. Masuda, Y. Tsuruoka, K. Yoshida, T. Ninomiya, and J. Tsujii. Semantic retrieval for the accurate identification of relational concepts in massive textbases. In *Proceedings of the 21st International Conference on Computational Linguistics and the 44th Annual Meeting of the Association for Computational Linguistics*, pages 1017–1024, 2006.

[122] Y. Miyao, K. Sagae, R. Sætre, T. Matsuzaki, and J. Tsujii. Evaluating contributions of natural language parsers to protein-protein interaction extraction. *Bioinformatics*, 25(3):394–400, 2009.

[123] L. P. Morales, A. D. Esteban, and P. Gervás. Concept-graph based biomedical automatic summarization using ontologies. In *Proceedings of the 3rd Textgraphs Workshop on Graph-Based Algorithms for Natural Language Processing*, pages 53–56, 2008.

[124] A. Morgan, L. Hirschman, A. Yeh, and M. Colosimo. Gene name extraction using FlyBase resources. In *Proceedings of the ACL 2003 Workshop on Natural Language Processing in Biomedicine - Volume 13*, pages 1–8, 2003.

[125] A. A. Morgan, L. Hirschman, M. Colosimo, A. S. Yeh, and J. B. Colombe. Gene name identification and normalization using a model organism database. *Journal of Biomedical Informatics*, 37(6):396–410, 2004.

[126] A. A. Morgan, Z. Lu, X. Want, A. M. Cohen, J. Fluck, P. Ruch, A. Divoli, K. Fundel, R. Leaman, J. Hakenberg, C. Sun, H.-h. Liu, R. Torres, M. Krauthammer, W. W. Lau, H. Liu, C.-N. Hsu, M. Scheumie, K. B. Cohen, and L. Hirschman. Overview of BioCreAtIvE II: Gene normalization. *Genome Biology*, 9(Suppl 2):S3, 2008.

[127] H. Müller, J. Kalpathy-Cramer, I. Eggel, S. Bedrick, C. E. Charles E. Kahn, Jr., and W. Hersh. Overview of the clef 2010 medical image retrieval track. In *Working Notes of CLEF 2010*, 2010.

[128] M. Narayanaswamy, K. E. Ravikumar, and K. Vijay-Shanker. A biological named entity recognizer. In *Pacific Symposium on Biocomputing*, pages 427–438, 2003.

[129] National center for biomedical ontology. http://www.bioontology.org/.

[130] NCBO BioPortal. http://bioportal.bioontology.org/.

[131] National Center for Biotechnology Information. *Entrez Programming Utilities Help*, 2010. http://www.ncbi.nlm.nih.gov/books/NBK25501/.

[132] National centre for text mining. http://www.nactem.ac.uk/.

[133] C. Nédellec. Learning language in logic - genic interaction extraction challenge. In *In Proceedings of the ICML 2005 Workshop on Learning Language in Logic*, pages 31–37, 2005.

[134] Neuroscience information framework. http://neuinfo.org/.

[135] Y. Niu and G. Hirst. Analysis and semantic classes in medical text for question answering. In *Proceedings of the ACL 2004 Workshop on Question Answering in Restricted Domains*, 2004.

[136] Y. Niu, G. Hirst, G. McArthur, and R.-G. P. Answering clinical questions with role identification. In *Proceedings of the ACL 2003 Workshop on Natural Language Processing in Biomedicine*, pages 73–80, 2003.

[137] Y. Niu, X. Zhu, and G. Hirst. Using outcome polarity in sentence extraction for medical question-answering. In *AMIA Anual Symposium Proceedings*, pages 599–603, 2006.

[138] Y. Niu, X. Zhu, J. Li, and G. Hirst. Analysis of polarity information in medical text. In *AMIA Anual Symposium Proceedings*, pages 570–574, 2005.

[139] C. Nobata, N. Collier, and J.-i. Tsujii. Automatic term identification and classification in biology texts. In *Proceedings of the Natural Language Pacific Rim Symposium*, pages 369–374, 1999.

[140] P. V. Ogren. Knowtator: A protégé plug-in for annotated corpus construction. In *Proceedings of the 2006 Conference of the North American Chapter of the Association for Computational Linguistics on Human Language Technology*, pages 273–275, 2006.

[141] D. Okanohara, Y. Miyao, Y. Tsuruoka, and J. Tsujii. Improving the scalability of semi-Markov conditional random fields for named

entity recognition. In *Proceedings of the 21st International Conference on Computational Linguistics and the 44th Annual Meeting of the Association for Computational Linguistics*, pages 465–472, 2006.

[142] F. Olsson, G. Eriksson, K. Franzén, L. Asker, and P. Lidén. Notions of correctness when evaluating protein name taggers. In *Proceedings of the 19th International Conference on Computational Linguistics - Volume 1*, pages 1–7, 2002.

[143] Open biological and biomedical ontologies. http://www.obofoundry.org/.

[144] ORBIT project. http://orbit.nlm.nih.gov/.

[145] A. Özgür, T. Vu, G. Erkan, and D. R. Radev. Identifying gene-disease associations using centrality on a literature mined gene-interaction network. *Bioinformatics*, 24(13):i277–i285, 2008.

[146] A. Özgür, Z. Xiang, D. R. Radev, and Y. He. Literature-based discovery of IFN-γ and vaccine-mediated gene interaction networks. *Journal of Biomedicine & Biotechnology*, page 426479, 2010.

[147] E. Pafilis, S. O'Donoghue, L. Jensen, H. Horn, M. Kuhn, N. Brown, and R. Schneider. Reflect - augmented browsing for the life scientist. *Nature Biotechnology*, 27:508–510, 2009.

[148] S. Pakhomov. Semi-supervised maximum entropy based approach to acronym and abbreviation normalization in medical texts. In *Proceedings of the 40th Annual Meeting on Association for Computational Linguistics*, pages 160–167, 2002.

[149] M. Palakal, J. Bright, T. Sebastian, and S. Hartanto. A comparative study of cells in inflammation, EAE and MS using biomedical literature data mining. *Journal of Biomedical Science*, 14(1):67–85, 2007.

[150] V. Petri, M. Shimoyama, G. Hayman, J. Smith, M. Tutaj, J. de Pons, M. Dwinell, D. Munzenmaier, S. Twigger, and H. Jacob. The rat genome database pathway portal. *Database*, 2011.

[151] I. Petrič, U. Tanja, B. Cestnik, and M. Macedoni-Lukšič. Literature mining method RaJoLink for uncovering relations between biomedical concepts. *Journal of Biomedical Informatics*, 42(2):219–227, 2009.

[152] Pharmacogenomics knowledge base. http://www.pharmgkb.org/.

[153] H. Poon and L. Vanderwende. Joint inference for knowledge extraction from biomedical literature. In *Human Language Technologies: The 2010 Annual Conference of the North American Chapter*

of the *Association for Computational Linguistics*, pages 813–821, 2010.

[154] PubMed central open access subset. http://www.ncbi.nlm.nih.gov/pmc/tools/openftlist/.

[155] S. Pyysalo, A. Airola, J. Heimonen, J. Bjorne, F. Ginter, and T. Salakoski. Comparative analysis of five protein-protein interaction corpora. *BMC Bioinformatics*, 9(Suppl 3):S6, 2008.

[156] S. Pyysalo, F. Ginter, J. Heimonen, J. Bjorne, J. Boberg, J. Jarvinen, and T. Salakoski. BioInfer: A corpus for information extraction in the biomedical domain. *BMC Bioinformatics*, 8(1):50, 2007.

[157] L. A. Ramshaw and M. P. Marcus. Text chunking using transformation-based learning. In *3rd ACL SIGDAT Workshop on Very Large Corpora*, pages 82–94, 1995.

[158] L. H. Reeve, H. Han, and A. D. Brooks. The use of domain-specific concepts in biomedical text summarization. *Information Processing & Management*, 43(6):1765–1776, 2007.

[159] W. S. Richardson, M. C. Wilson, J. Nishikawa, and R. S. Hayward. The well-built clinical question: A key to evidence-based decisions. *ACP Journal Club*, 123(3):A12–A13, 1995.

[160] S. Riedel, H.-W. Chun, T. Takagi, and J. Tsujii. A Markov logic approach to bio-molecular event extraction. In *Proceedings of the Workshop on Current Trends in Biomedical Natural Language Processing: Shared Task*, pages 41–49, 2009.

[161] S. Riedel and A. McCallum. Fast and robust joint models for biomedical event extraction. In *Proceedings of the 2011 Conference on Emperical Methods in Natural Language Processing*, pages 1–12, 2011.

[162] F. Rinaldi, J. Dowdall, G. Schneider, and A. Persidis. Answering questions in the genomics domain. In *Proceedings of the ACL 2004 Workshop on Question Answering in Restricted Domains*, 2005.

[163] F. Rinaldi, K. Kaljurand, and R. Saetre. Terminological resources for text mining over biomedical scientific literature. *Artificial Intelligence in Medicine*, 52(2):107–114, 2011.

[164] F. Rinaldi, G. Schneider, K. Kaljurand, M. Hess, C. Andronis, O. Konstandi, and A. Persidis. Mining of relations between proteins over biomedical scientific literature using a deep-linguistic approach. *Artificial Intelligence in Medicine*, 39(2):127–136, 2007.

[165] T. C. Rindflesch and M. Fiszman. The interaction of domain knowledge and linguistic structure in natural language processing:

Interpreting hypernymic propositions in biomedical text. *Journal of Biomedical Informatics*, 36(6):462–477, 2003.

[166] T. C. Rindflesch, H. Kilicoglu, M. Fiszman, G. Rosemblat, and D. Shin. Semantic MEDLINE: An advanced information management application for biomedicine. *Information Services & Use*, 31:15–21, 2011.

[167] B. Rink, S. Harabagiu, and K. Roberts. Automatic extraction of relations between medical concepts in clinical texts. *Journal of the American Medical Informatics Association*, 18(5):594–600, 2011.

[168] A. Roberts, R. Gaizauskas, and M. Hepple. Extracting clinical relationships from patient narratives. In *Proceedings of the Workshop on Current Trends in Biomedical Natural Language Processing*, pages 10–18, 2008.

[169] P. Ruch, C. Boyer, C. Chichester, I. Tbahriti, A. Geissbühler, P. Fabry, J. Gobeill, V. Pillet, D. Rebholz-Schuhmann, C. Lovis, and A.-L. Veuthey. Using argumentation to extract key sentences from biomedical abstracts. *International Journal of Medical Informatics*, 76(2-3):195–200, 2007.

[170] D. L. Sackett, W. M. C. Rosenberg, J. A. M. Gray, and R. B. Haynes. Evidence based medicine: What it is and what it isn't. *British Medical Journal*, 312(7023):71–72, 1996.

[171] M. Saeed, M. Villarroel, A. Reisner, G. Clifford, L. Lehman, G. Moody, T. Heldt, T. Kyaw, B. Moody, and R. Mark. Multiparameter intelligent monitoring in intensive care II (MIMIC-II): A public-access intensive care unit database. *Crit Care Med*, 39(5):952–960, 2011.

[172] J. Šarić, L. J. Jensen, R. Ouzounova, I. Rojas, and P. Bork. Extraction of regulatory gene/protein networks from MEDLINE. *Bioinformatics*, 22(6):645–650, 2006.

[173] Y. Sasaki, Y. Tsuruoka, J. McNaught, and S. Ananiadou. How to make the most of NE dictionaries in statistical NER. *BMC Bioinformatics*, 9(Suppl 11):S5, 2008.

[174] J. Seki, K. Mostafa. Discovering implicit associations between genes and hereditary diseases. In *Pacific Symposium on Biocomputing*, pages 316–327, 2007.

[175] B. Settles. Biomedical named entity recognition using conditional random fields and rich feature sets. In *Proceedings of the International Joint Workshop on Natural Language Processing in Biomedicine and its Applications*, pages 104–107, 2004.

[176] B. Settles. ABNER: an open source tool for automatically tagging genes, proteins and other entity names in text. *Bioinformatics*, 21(4):3191–3192, 2005.

[177] H. Shatkay, F. Pan, A. Rzhetsky, and W. Wilbur. Multidimensional classification of biomedical text: toward automated, practical provision of high-utility text to diverse users. *Bioinformatics*, 24(18):2086–2093, 2008.

[178] H. Shatkay, J. W. Wilbur, and A. Rzhetsky. Annotation guidelines, 2005. http://www.ncbi.nlm.nih.gov/CBBresearch/Wilbur/AnnotationGuidelines.pdf.

[179] D. Shen, J. Zhang, G. Zhou, J. Su, and C.-L. Tan. Effective adaptation of a hidden markov model-based named entity recognizer for biomedical domain. In *Proceedings of the ACL 2003 Workshop on Natural Language Processing in Biomedicine - Volume 13*, pages 49–56, 2003.

[180] Z. Shi, G. Melli, Y. Wang, Y. Liu, B. Gu, M. Kashani, A. Sarkar, and F. Popowich. Question answering summarization of multiple biomedical documents. In Z. Kobti and D. Wu, editors, *Advances in Artificial Intelligence*, volume 4509 of *Lecture Notes in Computer Science*, pages 284–295. Springer Berlin / Heidelberg, 2007.

[181] M. S. Simpson, D. Demner-Fushman, and G. R. Thoma. Evaluating the importance of image-related text for ad-hoc and case-based biomedical article retrieval. In *AMIA Annual Symposium Proceedings*, pages 752–756, 2010.

[182] N. Smalheiser. The Arrowsmith project: 2005 status report. In A. Hoffmann, H. Motoda, and T. Scheffer, editors, *Discovery Science*, volume 3735 of *Lecture Notes in Computer Science*, pages 26–43. Springer Berlin / Heidelberg, 2005.

[183] N. Smalheiser, V. Torvik, A. Bischoff-Grethe, L. Burhans, M. Gabriel, R. Homayouni, A. Kashef, M. Martone, G. Perkins, D. Price, A. Talk, and R. West. Collaborative development of the arrowsmith two node search interface designed for laboratory investigators. *Journal of Biomedical Discovery and Collaboration*, 1(1):8, 2006.

[184] N. Smalheiser, W. Zhou, and V. Torvik. Anne O'Tate: A tool to support user-driven summarization, drill-down and browsing of PubMed search results. *Journal of Biomedical Discovery and Collaboration*, 3(1):2, 2008.

[185] N. R. Smalheiser and D. R. Swanson. Using Arrowsmith: A computer-assisted approach to formulating and assessing scien-

tific hypotheses. *Computer Methods and Programs in Biomedicine*, 57(3):149–153, 1998.

[186] N. R. Smalheiser, V. I. Torvik, and W. Zhou. Arrowsmith two-node search interface: A tutorial on finding meaningful links between two disparate sets of articles in MEDLINE. *Computer Methods and Programs in Biomedicine*, 94(2):190–197, 2009.

[187] L. Smith, L. Tanabe, R. Johnson nee Ando, C.-J. Kuo, I.-F. Chung, C.-N. Hsu, Y.-S. Lin, R. Klinger, C. Friedrich, K. Ganchev, M. Torii, H. Liu, B. Haddow, C. Struble, R. Povinelli, A. Vlachos, W. Baumgartner, L. Hunter, B. Carpenter, R. Tzong-Han Tsai, H.-J. Dai, F. Liu, Y. Chen, C. Sun, S. Katrenko, P. Adriaans, C. Blaschke, R. Torres, M. Neves, P. Nakov, A. Divoli, M. Mana-Lopez, J. Mata, and W. Wilbur. Overview of BioCreAtIve II: Gene mention recognition. *Genome Biology*, 9(Suppl 2):S2, 2008.

[188] M. Q. Stearns, C. Price, K. A. Spackman, and A. Y. Wang. SNOWMED clinical terms: Overview of the development process and project status. In *Proceedings of the AMIA Symposium*, pages 662–666, 2001.

[189] D. R. Swanson. Fish oil, Raynaud's syndrome, and undiscovered public knowledge. *Perspectives in Biology and Medicine*, 30(1):7–18, 1986.

[190] D. R. Swanson. Migraine and magnesium: Eleven neglected connections. *Perspectives in Biology and Medicine*, 31(4):526–557, 1988.

[191] D. R. Swanson. Somatomedin C and arginine: Implicit connections between mutually isolated literatures. *Perspectives in Biology and Medicine*, 33(2):157–186, 1990.

[192] D. R. Swanson. Complementary structures in disjoint science literatures. In *Proceedings of the 14th Annual International ACM SIGIR Conference on Research and Development in Information Retrieval*, pages 280–289, 1991.

[193] D. R. Swanson and N. R. Smalheiser. An interactive system for finding complementary literatures: A stimulus to scientific discovery. *Artificial Intelligence*, 91(2):183–203, 1997.

[194] D. R. Swanson, N. R. Smalheiser, and A. Bookstein. Information discovery from complementary literatures: Categorizing viruses as potential weapons. *Journal of the American Society for Information Science and Technology*, 52(10):797–812, 2001.

[195] K. Takahashi, A. Koike, and T. Takagi. Question answering system in biomedical domain. In *Proceedings of the 15th International Conference on Genome Informatics*, pages 161–162, 2004.

[196] K. Takeuchi and N. Collier. Bio-medical entity extraction using support vector machines. *Artificial Intelligence in Medicine*, 33(2):125–137, 2005.

[197] R. M. Terol, P. Martínez-Barco, and M. Palomar. A knowledge based method for the medical question answering problem. *Computers in Biology and Medicine*, 37(10):1511–1521, 2007.

[198] P. Thompson, S. Iqbal, J. McNaught, and S. Ananiadou. Construction of an annotated corpus to support biomedical information extraction. *BMC Bioinformatics*, 10(1):349, 2009.

[199] V. I. Torvik and N. R. Smalheiser. A quantitative model for linking two disparate sets of articles in MEDLINE. *Bioinformatics*, 23(13):1658–1665, 2007.

[200] TREC-9 filtering track collections. http://trec.nist.gov/data/t9_filtering.html.

[201] TREC genomics track data. http://ir.ohsu.edu/genomics/data.html.

[202] R. Tsai, W.-C. Chou, Y.-S. Su, Y.-C. Lin, C.-L. Sung, H.-J. Dai, I. Yeh, W. Ku, T.-Y. Sung, and W.-L. Hsu. BIOSMILE: A semantic role labeling system for biomedical berbs using a maximum-entropy model with automatically generated template features. *BMC Bioinformatics*, 8(1):325, 2007.

[203] Y. Tsuruoka, M. Miwa, K. Hamamoto, J. Tsujii, and S. Ananiadou. Discovering and visualizing indirect associations between biomedical concepts. *Bioinformatics*, 27(13):i111–i119, 2011.

[204] Y. Tsuruoka and J. Tsujii. Boosting precision and recall of dictionary-based protein name recognition. In *Proceedings of the ACL 2003 Workshop on Natural Language Processing in Biomedicine - Volume 13*, pages 41–48, 2003.

[205] Y. Tsuruoka and J. Tsujii. Probabilistic term variant generator for biomedical terms. In *Proceedings of the 26th Annual International ACM SIGIR Conference on Research and Development in Informaion Retrieval*, pages 167–173, 2003.

[206] Y. Tsuruoka, J. Tsujii, and S. Ananiadou. FACTA: A text search engine for finding associated biomedical concepts. *Bioinformatics*, 24(21):2559–2560, 2008.

[207] O. Tuason, L. Chen, L. H., and C. Friedman. Biological nomenclatures: A source of lexical knowledge and ambiguity. In *Pacific Symposium on Biocomputing*, pages 238–249, 2004.

[208] H. Turtle and W. B. Croft. Evaluation of an inference network-based retrieval model. *ACM Transactions on Information Systems*, 9:187–222, 1991.

[209] Orange book: Approved drug products with therapeutic equivalence evaluations. http://www.accessdata.fda.gov/scripts/cder/ob/default.cfm.

[210] Databases, resources & APIs. http://wwwcf2.nlm.nih.gov/nlm_eresources/eresources/search_database.cfm.

[211] University of Pittsburgh NLP repository. http://www.dbmi.pitt.edu/nlpfront.

[212] Y. Usami, H.-C. Cho, N. Okazaki, and J. Tsujii. Automatic acquisition of huge training data for bio-medical named entity recognition. In *Proceedings of BioNLP 2011 Workshop*, pages 65–73, 2011.

[213] O. Uzuner. Recognizing obesity and comorbidities in sparse data. *Journal of the American Medical Informatics Association*, 16(5):561–570, 2009.

[214] O. Uzuner, I. Goldstein, Y. Luo, and I. Kohane. Identifyingn patient smoking status from medical discharge records. *Journal of the American Medical Informatics Association*, 15(1):14–24, 2008.

[215] O. Uzuner, I. Solti, and E. Cadag. Extracting medication information from clinical text. *Journal of the American Medical Informatics Association*, 17(5):514–518, 2010.

[216] O. Uzuner, B. R. South, S. Shen, and S. L. DuVall. 2010 i2b2/VA challenge on concepts, assertions, and relations in clinical text. *Journal of the American Medical Informatics Association*, 18(5):552–556, 2011.

[217] V. Vincze, G. Szarvas, R. Farkas, G. Mora, and J. Csirik. The BioScope corpus: Biomedical texts annotated for uncertainty, negation and their scopes. *BMC Bioinformatics*, 9(Suppl 11):S9, 2008.

[218] A. Vlachos and C. Gasperin. Bootstrapping and evaluating named entity recognition in the biomedical domain. In *Proceedings of the HLT-NAACL BioNLP Workshop on Linking Natural Language and Biology*, pages 138–145, 2006.

[219] T. Wattarujeekrit, P. Shah, and N. Collier. PASBio: Predicate-argument structures for event extraction in molecular biology. *BMC Bioinformatics*, 5(1):155, 2004.

[220] M. Weeber, H. Klein, L. T. W. de Jong-van den Berg, and R. Vos. Using concepts in literature-based discovery: Simulating Swanson's Raynaud-fish oil and migraine-magnesium discoveries. *Journal of the American Society for Information Science and Technology*, 52(7):548–557, 2001.

[221] W. Weiming, D. Hu, M. Feng, and L. Wenyin. Automatic clinical question answering based on UMLS relations. In *Third International Conference on Semantics, Knowledge and Grid*, pages 495–498, 2007.

[222] J. W. Wilbur, A. Rzhetsky, and H. Shatkay. New directions in biomedical text annotation: Definitions, guidelines and corpus construction. *BMC Bioinformatics*, 7:356, 2006.

[223] G. Williams, P. Davis, A. Rogers, T. Bieri, P. Ozersky, and J. Spieth. Methods and strategies for gene structure curation in wormbase. *Database*, 2011.

[224] K. Yamamoto, T. Kudo, A. Konagaya, and Y. Matsumoto. Protein name tagging for biomedical annotation in text. In *Proceedings of the ACL 2003 Workshop on Natural Language Processing in Biomedicine - Volume 13*, pages 65–72, 2003.

[225] J. Yang, A. M. Cohen, and W. Hersh. Automatic summarization of mouse gene information by clustering and sentence extraction from MEDLINE abstracts. In *AMIA Annual Symposium Proceedings*, pages 831–835, 2007.

[226] A. Yeh, A. Morgan, M. Colosimo, and L. Hirschman. BioCreAtIvE task 1A: Gene mention finding evaluation. *BMC Bioinformatics*, 6(Suppl 1):S2, 2005.

[227] M. Yetisgen-Yildiz and W. Pratt. Using statistical and knowledge-based approaches for literature-based discovery. *Journal of Biomedical Informatics*, 39(6):600–611, 2006.

[228] M. Yetisgen-Yildiz and W. Pratt. A new evaluation methodology for literature-based discovery systems. *Journal of Biomedical Informatics*, 42(4):633–643, 2009.

[229] I. Yoo, X. Hu, and I.-Y. Song. A coherent graph-based semantic clustering and summarization approach for biomedical literature and a new summarization evaluation method. *BMC Bioinformatics*, 8(Suppl 9):S4, 2007.

[230] H. Yu, S. Agarwal, M. Johnston, and A. Cohen. Are figure legends sufficient? Evaluating the contribution of associated text to biomedical figure comprehension. *Journal of Biomedical Discovery and Collaboration*, 4(1):1, 2009.

[231] H. Yu and Y.-G. Cao. Automatically extracting information needs from ad hoc clinical questions. In *AMIA Annual Symposium Proceedings*, pages 96–100, 2008.

[232] H. Yu and M. Lee. Accessing bioscience images from abstract sentences. *Bioinformatics*, 22(14):e547–e556, 2006.

[233] H. Yu, M. Lee, D. Kaufman, J. Ely, J. A. Osheroff, G. Hripcsak, and J. Cimino. Development, implementation, and a cognitive evaluation of a definitional question answering system for physicians. *Journal of Biomedical Informatics*, 40(3):236–251, 2007.

[234] H. Yu and C. Sable. Being Erlang Shen: Identifying answerable questions. In *Proceedings of the Nineteenth International Joint Conference on Artificial Intelligence on Knowledge and Reasonin for Answering Questions*, pages 6–14, 2005.

[235] H. Yu, C. Sable, and H. Zhu. Classifying medical questions based on an evidence taxonomy. In *Proceedings of the AAAI 2005 Workshop on Question Answering in Restricted Domains*, 2005.

[236] G. Zhou, D. Shen, J. Zhang, J. Su, and S. Tan. Recognition of protein/gene names from text using an ensemble of classifiers. *BMC Bioinformatics*, 6(Suppl 1):S7, 2005.

[237] P. Zweigenbaum and D. Demner-Fushman. Advanced literature-mining tools. In D. Edwards, J. Stajich, and D. Hansen, editors, *Bioinformatics: Tools and Applications*, pages 347–380. Springer, 2009.

[238] P. Zweigenbaum, D. Demner-Fushman, H. Yu, and K. B. Cohen. Frontiers of biomedical text mining: Current progress. *Briefings in Bioinformatics*, 8(5):358–375, 2007.

Index

χ^2-statistic, 170
k-Means Clustering, 93
k-Medoids Clustering, 92

Agglomerative Clustering, 90
Aspect Sentiment Classification, 433
Aspect-Based Opinion Summary, 420

Background Knowledge Base, 404
Bagging, 210
Baum=Welch Algorithm, 281
Bayesian Network Applications, 277
Bayesian Networks, 276
Bayesian Topic Models, 53
Bernoulli Multivariate Model, 183
Biological Question Answering, 491
Biomedical Information Extraction, 472
Biomedical Mining, 465
Biomedical Relation Extraction, 472
Biomedical Text Annotation, 469
Biomedical Text Mining Resources, 467
BioNLP, 472
Boosting, 210
Boosting in Transfer Learning, 233
BOW Toolkit, 78, 164
Browsing in Scatter/Gather, 97
Buckshot, 95
Bursty Features for Stream Clustering, 113, 304

C4.5 Classification, 178
Centroid Summarization, 50
Chaining, 91
Chinese Restaurant Process, 269
Classification Improvement with Unlabeled Data, 192
Classifying Text Streams, 312
CLASSIT, 92, 299
Cluster Refinement Hypothesis, 97
Clustering Text in Networks, 115
Clustering Text Streams, 110
Clustering with Frequent Phrases, 105
Co-clustering with Graph Partitioning, 104

Co-clustering Words and Documents, 103
Co-training, 191
COBWEB, 92, 299
Collaborative Question Answering, 395
Collective Classification of Emails, 208
Community Detection with Edge Content, 118
Community Detection with Node Content, 117
Comparative Opinions, 418, 441
Complete Linkage Clustering, 92
Composite Kernels, 28
Concept Decomposition for Clustering, 85
Conditional Random Fields, 20, 285
Constrained Probabilistic Models, 287
Context in Summarization, 56
Corpus-based Approach for Opinion Mining, 430
Correlated Bursty Topic Patterns, 311
Cost-Sensitive Classification, 212
CRF Applications, 286
Cross Domain Text Classification, 225
Cross Text and Visual Content Mining, 374
Cross-Domain Ensemble Learning, 232
Cross-Lingual Information Retrieval, 324
Cross-Lingual Mining, 324

Decision Trees, 176
Dependency Grammar, 441
Dictionary-based Approach for Opinion Mining, 429
Dimensionality Reduction, 130
Dirichlet Distribution, 133
Dirichlet Process, 270
Dirichlet Process Mixture Model, 271
Distance-based Clustering Algorithms, 89
Distance-based Partitioning Algorithms, 92
DNF Rules for Classification, 179
Document Sentiment Classification, 422
Dynamic Topic Modeling, 151

EM Algorithm, 266

EM for Classification, 192
EM for Semi-Supervised Clustering, 119
Email Network Classification, 208
Email Spam Filtering, 165, 186
Email Summarization, 59
EMPathIE, 476
Ensemble Learning, 209
Entropy-based Unsupervised Feature Selection, 83
Evaluation of Information Extraction Methods, 33
Event Detection in Social Media, 393
Event Extraction from Biomedical Data, 482
Evolution Analysis in Text Streams, 316
Evolution in Blog Streams, 317
Extractive Summarization, 44

Feature Compression with Clustering, 85
Feature Selection for Text Classification, 167
Feature Selection for Text Clustering, 81
Feature Transformation for Text Clustering, 81
Feature-based Transfer Learning, 235
Fisher's Linear Discriminant, 173
Fractionation, 95
Frequency-based Feature Selection, 81
Frequent Word Patterns for Clustering, 100

Gaussian Mixture Model, 263
Generalized Singular Value Decomposition, 175
Generative Process, 133
Gibbs Sampling, 145, 268
Gini Index, 168
Global Summary Selection, 65
Graph-based Semi-Supervised Clustering, 119
Graph-based Summarization, 61
Graphical Models, 275
Group Spam Detection in Opinions, 450
Group-Average Linkage Clustering, 91

Heterogeneous Label Space, 243
Heterogeneous Transfer Learning, 239
Hidden Markov Models, 278
Hidden Markov Models in Named Entity Recognition, 18
Hierarchical Classification, 186
Hierarchical Clustering, 90
HMM Applications, 281
HMRF for Semi-Supervised Clustering, 119
Hyperlinks for Translingual Mining, 351

Incremental LSI, 137

Indicator Representation for Summarization, 60
Information Extraction, 11
Information Gain, 169
Information-Theoretic Co-clustering, 105
Instance-based Transfer, 231
Inverse Document Frequency, 80
iTopicModel, 118

Joint Visual and Text Mining, 370

Kernel Methods, 26

Lagrangian Optimization in NMF, 87
LAIR2, 98
Latent Dirichlet Allocation, 108, 130, 142, 264
Latent Semantic Analysis, 52
Latent Semantic Indexing, 52, 84, 130, 133
LDA, 108, 130, 142, 264
LDA Model Training, 144
Lemur Toolkit, 78
Lexical Chains for Summarization, 50
Lexical Translation Model, 327
Linear Classifiers, 176, 193
Linear Discriminant Analysis, 173
Link and Text Analytics in Social Media, 400
Linked Data Classification, 203
Literature-based Discovery from Biomedical Text, 492
LLL Genic Extraction Challenge, 479
LLSF Method, 196
Local Feature Selection, 104
Locality Sensitive Hashing, 311
LSH, 311
LSI, 84, 130, 133
LSI and PCA Relationship, 85
LSI and PLSI Relationships, 139, 140
LSI Implementation Issues, 135

Machine Learning for Summarization, 60
Machine Translation in Translingual Mining, 325
Map Reduce in Text Mining, 289
Markov Logic Network, 286
Markov Random Fields, 282
Matching Parallel Pages, 339
Matrix Factorization, 86
Matrix Factorization for Linked Text Classification, 209
Maximal Margin Relevance for Summarization, 64
Maximum Entropy Markov Models, 19
Medical Question Answering, 489
MetaMap, 471

INDEX

Microblogging, 388
Mixture Modeling for Classification, 191
Mixture Modeling for Text Classification, 190
Mixture Models, 261
Mixture of Unigrams, 263
MRF Applications, 284
MRF for Linked Text Clustering, 116
Multimedia Text Mining, 362
Multinomial Distribution, 132, 188
Mutual Information for Clustering, 105
Mutual Information for Feature Selection, 169

Naive Bayes Classifier, 181
Naive Bayes for Linked Text Classification, 205
Named Entity Recognition, 11, 15, 473
Nearest Neighbor Classifier, 201
NetPLSA, 118
Neural Network Classifiers, 197, 315
Neural networks for Stream Classification, 315
News Filtering, 164
NMF, 86
Non-negative Matrix Factorization, 86

Objective-Centered Classification, 212
OCFS, 175
One-Class Stream Classification, 314
Online Spherical k-Means Algorithm, 110, 299
Open Information Extraction, 32
Opinion Lexicon Expansion, 429
Opinion Mining, 416
Opinion Orientations, 418
Opinion Spam Detection, 447
Opinion Spam Learning, 448
Opinion Utility, 451
Optimal Orthogonal Centroid Feature Selection, 175
ORBIT, 472
OSKM, 110, 299

Pachinko Machine, 187
Parallel Learning Algorithms, 288
PASTA, 476
PCA, 84
Perceptron Algorithm, 198
PERSIVAL, 487
Phrase-based Models for Translingual Mining, 329
Pitman-Yor Process, 274
PLSI, 108, 140, 264
Polarities, 418
Principal Component Analysis, 84
Probabilistic Document Clustering, 107

Probabilistic Latent Semantic Indexing, 108, 140, 264
Probabilistic Models with Constraints, 287
Probabilistic Techniques for Text Mining, 259
Projected Clustering, 103
Projections for Document Clustering, 98
Prototype Hierarchies for Supervision, 120
Proximity-based Classifiers, 200

Query-Focused Summarization, 58
Question Answering for Biomedical Data, 488
Question Answering in Social Media, 395

Random Walks for Linked Text Classification, 206
Regular Opinions, 417
Regularization of Linked Text, 207
Relation Discovery and Template Induction, 31
Relation Extraction, 11, 22
Relational Topic Models, 153
Relaxation Labeling for Linked Text Classification, 204
Review Utility, 451
RIPPER, 181
Rocchio Method, 202
Rule-based Approach in Named Entity Recognition, 16
Rule-based Classifiers, 178

Scatter/Gather, 94
Scientific Article Summarization, 58
Seed Phrase Extraction, 402
Semantic Feature Generation, 404, 405
Semantic Gap in Social Media, 398
Semi-Supervised Clustering, 118
Sentence Clustering for Summarization, 55
Sentence Subjectivity in Sentiment Classification, 426
Sentiment Analysis, 416
Sentiment Classification, 422
Sentiment Consistency, 430
Sentiment Orientations, 418
Sequence-based Kernels, 26
Side Information in Text Clustering, 116
Single Linkage Clustering, 91
Singular Value Decomposition, 84, 133
Sleeping Experts, 181, 315
Social Media, 386
Social Streams, 310, 394
Social Tagging, 397
Spam Filtering, 165, 186, 313

Statistical Learning Approach in Named Entity Recognition, 17
Statistical Machine Translation, 325
Stop Words, 81
Stream Mining, 297
Subspace Clustering, 103
Summarization of Biomedical Data, 484
Summary Sentences, 64
Supervised Clustering for Dimensionality Reduction, 172
Supervised LSI, 171
Support Vector Machines, 194, 225
Surrounding Text Mining in Multimedia, 364
SVD, 84, 133
SVM, 194, 225
Syntax-based Models for Translingual Mining, 333

Tag Information Enrichment, 369
Tag Mining in Multimedia, 366
Tag Refinement in Multimedia, 367
TAPER, 187
TDT, 307
Temporal Decay in Stream Clustering, 300
Term Context in LSI, 137
Term Contribution for Feature Selection, 83
Term Strength for Feature Selection, 82
Text Clustering, 77
Text Mining in Social Media, 385
Text Streams, 299
Text Summarization, 44
Theme Evolution Graph, 316
Theme Shifts, 317
Topic Detection and Tracking, 307
Topic Modeling Applications, 154

Topic Modeling in Networked Data, 152
Topic Modeling in Text Streams, 151, 307
Topic Models, 107, 130, 139
Topic Representation for Summarization, 46
Topic Signatures, 114
Topic Words for Summarization, 46
Topical Difference Factor Analysis, 174
Transfer Learning for Text Classification, 225
Transfer Learning in Text Mining, 224
Translingual Mining, 324
Translingual Relations from Monolingual Texts, 349
Tree-based Kernels, 27

Unified Medical Language System, 470
Universum for Supervised Clustering, 120
Unlabeled Data in Classification, 191
Unsupervised Feature Selection, 81
Unsupervised Information Extraction, 30
Unsupervised Sentiment Classification, 424

Variational Approximation for LDA, 145
Variational EM for LDA, 147
Visual Re-ranking, 371

Web Data Classification, 203
Web Structure for Crosslingual Mining, 337
Web Summarization, 57
WHIRL, 201
Word and Phrase-based Clustering, 99
Word Clusters, 99
Word-based Models for Translingual Mining, 327